T0310126

# Body Sensor Networking, Design and Algorithms

# Body Sensor Networking, Design and Algorithms

*Saeid Sanei*
Nottingham Trent University
Nottingham, UK

*Delaram Jarchi*
University of Essex
Colchester, UK

*Anthony G. Constantinides*
Imperial College London
London, UK

This edition first published 2020


*Registered Offices*
John Wiley & Sons, Inc., 111 River Street, Hoboken, NJ 07030, USA
John Wiley & Sons Ltd, The Atrium, Southern Gate, Chichester, West Sussex, PO19 8SQ, UK

*Editorial Office*
The Atrium, Southern Gate, Chichester, West Sussex, PO19 8SQ, UK

For details of our global editorial offices, customer services, and more information about Wiley products visit us at www.wiley.com.

Wiley also publishes its books in a variety of electronic formats and by print-on-demand. Some content that appears in standard print versions of this book may not be available in other formats.



*Library of Congress Cataloging-in-Publication Data*

Names: Sanei, Saeid, author. | Jarchi, Delaram, author. | Constantinides, Anthony G., 1983- author.
Title: Body sensor networking, design, and algorithms / Saeid Sanei, University of Surrey, Guildford, UK, Anthony G. Constantinides, Imperial College, London, UK, Delaram Jarchi, University of Essex, Colchester, UK.
Description: First edition. | Hoboken, NJ : John Wiley & Sons, Inc., 2020. | Includes bibliographical references and index.
Identifiers: LCCN 2019057875 (print) | LCCN 2019057876 (ebook) | ISBN 9781119390022 (hardback) | ISBN 9781119390046 (adobe pdf) | ISBN 9781119390015 (epub)
Subjects: LCSH: Body area networks (Electronics)
Classification: LCC TK5103.35 .S26 2020 (print) | LCC TK5103.35 (ebook) | DDC 681/.2–dc23
LC record available at https://lccn.loc.gov/2019057875
LC ebook record available at https://lccn.loc.gov/2019057876

Cover Design: Wiley
Cover Images: © imaginima/Getty Images, © gremlin/Getty Images

Set in 9.5/12.5pt STIXTwoText by SPi Global, Chennai, India
Printed and bound in Singapore by Markono Print Media Pte Ltd

10 9 8 7 6 5 4 3 2 1

# Contents

# Preface

The increasing multiplicity of data gathering from living systems both in volume and modalities provides a wealth of information. From a usefulness perspective it becomes imperative to understand the mechanisms of such systems where the information is kept and processed efficiently and robustly. These mechanisms are, by and large, highly complex and exquisite and require effective and robust sensors and sensor networks for data collection, and specific frameworks for communication and processing.

Sensors and sensor networks are perhaps the most dominating and fast-growing areas of research and development nowadays and cover a vast range of applications from automation to medication. New technologies make available new sensors on a continual basis, while more advanced mobile and stationary computing platforms become available to cater for better and faster recording, archiving, mining, processing, and recognition of the data of various modalities.

Over the past decade miniaturised wearable devices, particularly those embedded within mobile systems such as handphones, have attracted considerable interest. Such devices allow for the measurement of physical, behavioural, and physiological information from the human body mainly for monitoring patients, athletes, drivers, and many others.

The multiplicity of body (and area) sensors and their functionalities have rendered the area a highly significant research topic and, of course, raised more questions to be answered and problems to be solved. This area of research became even more fascinating when the captured data are to be transferred to other parties, such as hospitals, remotely and through the existing wireless communication media. As a result, wireless body sensor networking and the associated wearable technologies have marked another peak in living systems, in particular for human monitoring, and have fuelled further enthusiasm for solving many more problems.

Current research and applications in body sensor networking include monitoring the patients outside hospitals, assistive technology, hazard prevention for drivers and pilots, multiparty gaming, and enhancing the ability of the athletes. These applications have great impact on daily human life.

Sensor design is a pivotal problem to be tackled by the researchers in this domain. In-body, on-body, and off-body sensors are designed for different applications. The sensors have different sizes, technologies, power requirements, and communication modalities. The information captured by these sensors may have different qualities in terms of noise, interference, and strength. Suitable hardware and software platforms enable effective ways to access and communicate (mostly wireless) the sensor data.

Scientists need to make sense of the received data. They need to be able to combine (or fuse) the data from different sensors, to extract the required diagnostic information from the data, to diagnose any specific health-related problems, and to make the necessary decisions. This interactive framework of actions requires to a great extent research into signal processing and machine learning. Fortunately, many of the advanced signal processing algorithms – especially those designed for processing biomedical signals, artefact removal, or learning from the data – can be applied to the sensor data. In addition, both centralised and distributive signal processing techniques widely developed for other applications can be exploited and deployed to the aggregated information from the sensors.

Efficient communication of data across a network has also become a significant research area, owing to the complexity of communication channel. Human movement, change in the environment, noise, interference, data traffic, temperature, humidity, tolerable data rates by the sensors, number of network segments or hubs, and the on-board or remote processing systems significantly affect the complexity of the communication channel. As a result, standard physical, link (MAC), and network layers need to be modified to better cater for such applications. For this purpose a number of new routing algorithms have been introduced to optimise the flow of information for different networks in different scenarios.

In applications where the sensors need long-lasting power supplies, such as in-body sensors, the problem of energy harvesting becomes crucial. Hence, researchers have become sensitive to such potential limitations which need to be taken into consideration for future development. Energy sources from biochemical or even metabolic reactions, movement, and heat have already been utilised for some wireless sensor networks. These are becoming applicable to body sensor networks, too. Currently, body movement and heat appear to be the two major sources for energy harvesting with applications to body sensor networks.

Last, but not least, the security of human information is vital, and therefore effective measures towards protecting such data need to be taken. This becomes even more crucial when the data are to be transferred over the Internet or public cloud. Both, data security, through the encryption of the data, and network security, through effective authentication, have to be in place for all types of human-related information.

We decided to write this book when we came together in Beijing, China, for the 21st Conference on Digital Signal Processing, in 2016. At that time, a couple of editorials and reviews on wireless sensor networks had been produced. Thus, the need for a coherent monograph was very evident. To fulfil such a requirement, in this monograph we try to provide some food for thought on almost all the different areas of research related to body sensor networks.

The authors wish to acknowledge the help and dedication of Ahmadreza Hosseini-Yazdi, Ales Prochazka, Andrew Pierson, Funminiyi Olajide, and Samaneh Kouchaki for their help in the provision of materials, proofreading, and constructive comments and advice throughout the preparation of this book. We also thank Sara Sanei for her help in typing and organising the materials.

*Saeid Sanei, Anthony G. Constantinides and Delaram Jarchi*

## About the Companion Website

This book is accompanied by a companion website:

**www.wiley.com/go/sanei/algorithm-design**

The website includes:

- Software codes
- Videos
- Colour images
- A list of links to access data banks

Scan this QR code to visit the companion web site:

# 1 Introduction

## 1.1 History of Wearable Technology

Earlier in history, it would take hundreds of years between breakthroughs such as eyeglasses being developed in 1286 and the abacus ring being manufactured in 1600. Today, new wearable tech innovations happen on a monthly basis, if not weekly. In the last 10 years, we have had the Google Glass, Fitbit, Oculus Rift, and countless others.

The Nuremberg egg manufactured in 1510 by Peter Henlein was one of the early portable mechanical timekeeping devices (like a watch) which had a chain to hang over the neck. An air-conditioned top hat was a wearable designed by a Victorian in the nineteenth century. In 1890, a lighting company in New York used to send girls with wearable lights onto the performance stage and to light up houses during ceremonies. In the1960s, the wearers of roulette shoes, created by Edward Thorp and Claude Shannon, used to observe the rotations of the roulette ball, tap the shoe accordingly, and then receive a vibration telling them which number to bet on. In 1963, a small portable TV screen was worn as a glass. The aviator Alberto Santos-Dumont pioneered the use of the wristwatch in 1904 as it allowed him to have his hands free while flying. This also led people to start using wristwatches. Calculator watches came onto market in 1975 and the first low-cost Walkman stereo was offered by Sony in 1979. In the 1990s, interest in the Internet of Things (IoT) started to rise. In December 1994, Steve Mann, a Canadian researcher, developed the wearable wireless Webcam. Despite its bulk, it paved the way for future IoT technologies. This required advances in artificial intelligence, which started to flourish in the 2000s.

The Sony Walkman was a clear commercial success. The Walkman and subsequent Sony Discman helped the company become an entertainment powerhouse. Over 400 million Walkman portable music players have been sold with about 200 million of those being cassette players.

However, not all products launched with a fanfare are destined for success. The commercial potential of many wearable technologies introduced in recent years are not always predictable or even achieved.

Fitbit filed for a $100 million initial public offering, but it now has to compete against a plethora of other fitness trackers on the market. The Apple Watch was been launched amidst a great deal of publicity, but it comes with no guarantees for Apple – a company that needs a lot of new revenues on a product to move the needle. Finally, the creation of the Oculus Rift virtual reality (VR) headsets could finally bring VR to the masses. The company

*Body Sensor Networking, Design and Algorithms,* First Edition. Saeid Sanei, Delaram Jarchi and Anthony G. Constantinides.

Companion Website: www.wiley.com/go/sanei/algorithm-design

has already been bought by Facebook for over $2 billion. Garmin, as a global positioning system (GPS), and Samsung Galaxy Gear, as a smart watch, are other popular wearables.

What is clear is that, based on the history of wearable technology, devices that move the masses are far and between. The successes that do make it, however, can change the world and generate chart-topping returns. Meanwhile, people's needs change over time, and include entertainment, activity, sport, and now most importantly health. This brings wearables such as Quell to the market. When strapped on the body Quell predicts and detects the onset of chronic pain and stimulates nerves to block pain signals to the brain. Other wearables to measure blood alcohol content, athletic performance, blood sugar, heart rate, and many other bioindicators rapidly came to the market as the desire for health monitoring grew. This may become more demanding as the interest in personal medicine grows.

## 1.2 Introduction to BSN Technology

Wearable technology including sensors, sensor networks, and the associated devices has opened its space in a variety of applications. Long-term, noninvasive, and nonintrusive monitoring of the human body through collecting as much biometric data and state indicators as possible is the major goal of healthcare wearable technology developers. Patients suffering from diabetes need a simple noninvasive tool to monitor their blood sugar on an hourly basis. Those suffering from seizure require the necessary instrumentation to alarm them before any seizure onset to prevent them from fall injury. The stroke patients need their heart rate recorded constantly. These are only a small number of examples which show how crucial and necessary wearable healthcare systems can be.

At the Wearable Technology Conference in 2018, the winners of seven wearable device producers were introduced. These winners include the best ones in Lifestyle with the objective of 'play stress away'; Sports and Fitness for making a football performance device, healthcare for developing a smart eyewear with assistive artificial intelligence capabilities for the blind and visually impaired; Industrial for designing a unique smart and connected industry 4.0 safety shoe; Smart Clothing Challenge for the nonintrusive acquisition of heart signals that will enable pervasive health monitoring, emotional state assessment, drowsiness detection, and identity recognition; Smart Lamp, which allows you to move the light in any direction without moving the lamp; and Connected Living Challenge, for creating accessories linking braintech with fashion design. Headpieces and earrings use electroencephalography (EEG) technology, capturing and providing users with brain data, allowing them to be conscious of their mental state in real time, for example for reducing anxiety and depression or increasing focus or relaxation of the user [1]. This simple example together with the above examples clearly show the diversity in applications of wearable technology. The aim of this book is therefore to familiarise readers with sensors, connections, signal processing tools and algorithms, electronics, communication systems, and networking protocols as well as many applications of wearable devices for the monitoring of mental, metabolic, physical, and physiological states of the human body.

Disease prevention, patient monitoring, and disable and elderly homecare have become the major objectives for investment in social health and public wellbeing. According to the World Health Organization (WHO), an ageing population is becoming a significant problem and degenerative brain diseases, such as dementia and depression, are increasingly seen in people while a bad lifestyle is causing millions of people to suffer from obesity or chronic diseases. It is thus reasonable to expect that this circumstance will only contribute to an ongoing decline in the quality of services (QoSs) provided by an already overloaded healthcare system [2]. A remote low-cost monitoring strategy, therefore, would significantly promote social and clinical wellbeing. This can only be achieved if sufficient reliably recorded information from the human body is available. Such information may be metabolic, biological, physiological, behavioural, psychological, functional, or motion-related.

On the other hand, the development of mobile telephone systems since the early 1990s and its improvement till now together with the availability of large size archiving and wideband communication channels significantly increase the chance of achieving the above objectives without hospitalising the caretakers in hospitals and care units for a long time. This may be considered a revolution in human welfare. More effective and efficient data collection from the human body has therefore a tremendous impact and influence on healthcare and the technology involved. The state of a patient during rest, walking, working, and sleeping can be well recognised if all the biomarkers of the physiological, biological, and behavioural changes of human body can be measured and processed. This requirement sparks the need for deployment of a multisensor and multimodal data collection system on the body. A body sensor network (BSN) therefore is central to a complete solution for patient monitoring and healthcare. Several key applications benefit from the advanced integration of BSNs, often called body area networks (BANs), with the new mobile communication technology [3, 4].

The main applications of BSNs are expected to appear in the healthcare domain, especially for the continuous monitoring and logging of vital parameters of elderly people or patients suffering from degenerative diseases such as dementia or chronic diseases such as diabetes, asthma, and heart attacks. As an example, a BAN network on a patient can alert the hospital, even before they have a heart attack, through measuring changes in their vital signs, or placing it on a diabetic patient could auto-inject insulin through a pump as soon as their insulin level declines.

The IEEE 802.15 Task Group 6 (BAN) is developing a communication standard optimised for reliable low-power devices and operation on, in, or around the human body (but not limited to humans) to serve a variety of applications including medical, consumer electronics/personal entertainment, and security [5]. This was approved on 22 July 2011 and the first meeting of IEEE 802.15 wireless personal area network (WPAN) was held on 3 March 2017.

The BSN technology benefits from developments in various areas of sensors, automation, communications, and more closely the vast advances in wired and wireless sensor networks (WSNs) for short- and long-range communications and industrial control. For interconnecting multiple appliances, for example, some developed their own personal area network (PAN). One was by Massachusetts Institute of Technology (MIT) which was later expanded by Thomas G. Zimmerman to interconnect different body sensors and actuators to locate the human through the measures performed by electric field sensors. He introduced the PAN technology by exploiting the body as a conductor. Neil Gershenfeld, a physician at MIT, did the major work on near-field coupling of the field and human body tissue for localisation [6]. By fixing pairs of antennas on the body, for example around the elbow and hand, and applying an electric current through them, they showed that the system is capable of tracking the person. They learnt that as one moves a capacitance in their circuit is charged. So, they can locate the antennas in places where there is maximum change in the movement between them.

BSNs have their root within WSNs. Like many advanced technologies, the origin of WSNs can be seen in military and heavy industrial applications. The first wireless network which had some similarity with a modern WSN is the sound surveillance system (SOSUS), developed by the United States military in the 1950s to detect and track Soviet submarines. This network used submerged acoustic sensors – hydrophones – distributed in the Atlantic and Pacific oceans. This sensing technology is still in service, though for many different objectives, from monitoring undersea wildlife to volcanic activity [7]. Echoing the investments made in the 1960s and 1970s to develop the hardware for today's Internet, the United States Defense Advanced Research Projects Agency (DARPA) started the Distributed Sensor Network (DSN) programme in 1980 to formally explore the challenges in implementing distributed WSNs. With the birth of DSN and its progression into academia through partnering universities such as Carnegie Mellon University and the MIT Lincoln Laboratory, WSN technology soon found its place in academia and civilian scientific research.

Governments and universities eventually began using WSNs in applications such as air quality monitoring, forest fire detection, natural disaster prevention, weather stations, and structural monitoring. Then as engineering students made their way into the corporate world of the technology giants of the day, such as IBM and Bell Labs, they began promoting the use of WSNs in heavy industrial applications such as power distribution, wastewater treatment, and specialised factory automation.

Although BSNs' objective and technology have their own requirements, they owe their birth and early development, particularly with regards to data communication, to the WSN technologies, which enable fruitful use of permitted wireless communication features and frequency range.

BSNs are also called wireless body area networks (WBANs) as often the transmission is through wireless systems. In their current form, BSNs are wireless networks of wearable devices with recording and some processing capabilities [4, 7–9]. Such devices may be embedded inside the body, implants, surface-mounted on the body in fixed positions, or carried in one way or another [10]. From its start of development, there have been tremendous attempts in reducing the size and cost, and increasing the flexibility, of such devices–particularly those with direct contact with the human body [11, 12]. The development of BSN technology started in 1995 around the idea of using WPAN technologies to implement communications on, near, and around the human body. Later in early 2000, the term 'BAN' came to refer to the systems where communication is entirely within, on, and in the immediate proximity of a human body [13, 14]. A WBAN further expands WPAN wireless technologies as gateways to reach longer ranges. Through gateway devices, it is possible to connect wearable devices on the human body to the Internet. This allows medical professionals to access patient data online using the Internet independent of patient location [15].

BSNs have opened two important fronts in research and technology: one as a measuring tool in health and the other as an integral part of the public network. Such networks have tremendous applications in healthcare [16–18], sports, entertainment [19–21], industry, the military, and surveillance [22], assistive technology [23], and interactive and collaborative computer games [24] and other social public fields [25–27]. In parallel with introducing and supplying new sensors, embedding electronic circuits as well as mobile applications and gadgets (Google glass, wristband, armband, headband, watch, and mobile with more biological data recording capabilities), which can be conveniently mounted on human body, the research and development in BSN technology continue apace. The key BSN applications, stated above, benefit from the advanced integration of BANs and emerging wireless technologies. For example, in remote health/fitness monitoring, health and motion information are monitored in real-time and delivered to nearby diagnostic or storage devices, through which the data can be forwarded to off-site clinical unites for further inspection. In military and sports training the motion sensors can be worn on both hands and elbows for tracking the movement and accurate feature extraction of sports players' movements. In interactive gaming, body sensors enable players to simulate and perform actual body movements, such as boxing and shooting, that can be fed back to the gaming console, thereby enhancing their entertainment experiences. Or for personal information sharing any private or business information can be stored in body sensors for many daily life applications such as shopping, activity monitoring, and information exchange. Finally, in secure authentication both physiological and behavioural biometrics – such as facial patterns, fingerprints, and iris recognition – can be restored and shared with authorities all over the world. In such cases, potential problems, such as forgery and duplicability, have motivated investigations into more and new physical/behavioural characteristics of the human body, by means of other measurements, such as EEG, gait information, and multimodal biometric systems.

BSNs may also be considered a subset of WSN often used in various industrial applications to monitor a large connected system. In many cases, however, each group of sensors, such as those for an EEG, can be wired up to a central recording system, such as the EEG machine, which can then be processed together. For BSNs the sensors often sample the physiological and metabolic variables from human body. Using BSNs for health monitoring, the necessary warning or alarming states for risk prevention can be generated and the diagnostic data for long-term inspection by clinicians can be recorded and archived.

The main components of the BSN technology are sensors, data processing, data fusion, machine learning, and low- and long-rage communication systems. Groups of researchers in sensor design, microelectronics, integrated circuit fabrication, data processing, machine learning, short- and long- range communications, security, data science, and computer networking, as well as clinicians, have to work together to design an efficient and usable BSN.

The advances in sensor technology, data analytics for large datasets, distributed systems, new generation of communication systems, mobile technology, and cooperative networks have opened a vast research platform in BSN as an emerging technology and an essential tool for the future development of ubiquitous healthcare monitoring systems [28]. Researchers should (i) enable seamless data transfer through standards such as Bluetooth, ZigBee, or ultrawideband (UWB) Wi-Fi to promote information exchange and the efficiency of migration across networks and uninterrupted connectivity, (ii) the sensors used in an BSN should be of low complexity, small size lightweight, easy to use, reconfigurable, and compatible with the existing tools and software, (iii) the transmission should be secure and reliable, and (iv) the sensors should be convenient to use and ethically approved.

On the other hand, agile solutions for clinical problems require access to multimodal physiological, biological, and metabolic data as well as those related to body motion, behaviour, mode, etc., which may be captured by cameras. The fusion of multimodal information is itself a fascinating area of research within both computer science and engineering communities.

Looking at the BSN with respect to WSN, WSNs have more general applications. For example, they can be deployed to inaccessible environments, such as forests, sea vessels, swamps, or mountains. In such cases, many redundant or spare nodes may be placed in the environment, making more dense distribution of the sensors to avoid any negative impact of node failures. In BSNs, however, the nodes are located in clinically more informative zones around or even inside the human body. This makes the total number of nodes limited, and generally rarely more than a few dozen. Each node is mounted properly to ensure more robust and accurate results [29]. However, there are cases where the sensors are movable and deployed for short duration recordings. An example of such sensors is endoscopic capsules, also called esophagogastroduodenoscopy (EGD), for monitoring human intestine and internal abdomen tissues.

Also, in terms of functionality attributes, the nodes in WSNs often record data of the same modality (although, in recent applications, different modalities such as sound and video have been taken into account by WSNs), whereas, in BSNs, various sensors collect different physiological and biological data.

Some limitations in sensor design – such as their geometrical dimensions, weight, shape, appearance, and size – may be less important for the WSN nodes than those of BSNs. Different sensor types are used in a BSN for recording various data types from the human body [8]. For a WSN there may be large-size sensors which are very resistive to a rough and hostile environment. In BSNs the nodes are supported by more robust electronic circuits which are less sensitive to noise, such as well-tuned differential amplifiers, to enable the recording of very low amplitude signals such as scalp EEG or surface electromyography (EMG). The sensors are often small and delicate enough to be wearable, less intrusive, easily deployable within the human body, and in many cases biocompatible [30].

There are other considerations and limitations for BSNs, for example in many applications the human body is in motion and the BSN nodes move accordingly. Also, unlike for WSNs, where the nodes are powered by many sources such as the national grid, wind turbine, and solar cells, for conventional BSNs, the consumable energy should be optimised and batteries with limited power (though rechargeable) used [31, 32]. On the other hand, with regards to data transmission, the nodes in a WSN often transfer the data with similar rates as long as the data modality is the same. This is, however, not the case for a BSN, as various sensors sample and transfer the data at rates appropriate to the underlying physiological variables under examination.

Another concern about the data type in BSNs is that the human body is nonhomogeneous and each part is modelled as an entirely nonlinear system. Also, the physiological signals are inherently highly nonstationary, i.e. their statistical properties vary over time. Therefore, accurate analysis of such data is significantly more challenging than for other types of data, and many linear signal processing methods, therefore, are likely to fail to capture and analyse the true features of the data.

Additionally, BSNs are generally meant for monitoring human physiological, biological, and motion data, which are related to user's personal safety and privacy as well as other ethical issues. Therefore, some means of QoS, privacy protection, integrity, prosperity, and security in archiving and real-time data transmission must be considered [33, 34].

In terms of data communication through conventional wireless systems, WBANs support a variety of real-time health monitoring and consumer electronics applications. The latest standardization of WBANs is the IEEE 802.15.6 standard [35] which aims to provide an international standard for low-power, short-range, and extremely reliable wireless communication within the surrounding area of the human body, supporting a vast range of data rates for different applications. The security association in this standard includes four elliptic curve-based key agreement protocols that are used for generating a master key.

The Federal Communications Commission (FCC) has approved the allocation of 40 MHz of spectrum bandwidth for medical BAN low-power, wide-area radio links at the 2360–2400 MHz band. This allows off-loading WBAN communication from the already saturated standard Wi-Fi spectrum to a standard band [36].

Apart from 2390–2400 MHz band which is not subject to registration or coordination and may be used in all areas including residential, the 2360–2390 MHz frequency range is available on a secondary basis. The FCC will expand the existing Medical Device Radiocommunication (MedRadio) Service in Part 95 of its rules. WBAN devices using this band can operate on a 'licence-by-rule' basis, which eliminates the need to apply for individual transmitter licences. Usage of the 2360–2390 MHz frequencies is restricted to indoor operation at healthcare facilities and subject to registration and site approval by coordinators to protect aeronautical telemetry primary usage [37].

## 1.3 BSN Architecture

The general architecture of a BSN is shown in Figure 1.1. Sensor nodes which are placed around and possibly inside the body collect physiological data and perform preliminary processing. The data are then gathered by a sink node and transmitted to a local PC or

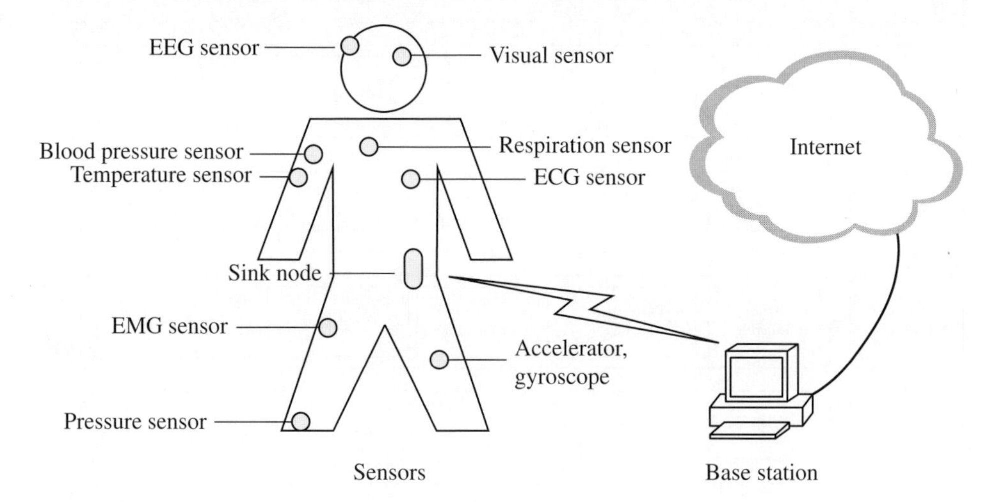


**Figure 1.1** Overall architecture of a BSN.

mobile system for personal or local use (such as alarming) or the base station of a public network to share with relevant bodies over the Internet. The recipient of the BSN data can be healthcare units, social welfare, emergency units within hospitals, or other experts in clinical, experimental, and sport departments. For any one of the above cases there are many design and technical challenges to tackle.

A more detail architecture, which are discussed in the corresponding chapter of this book, involves various levels and modes of communications between the body-mounted sensors and the corresponding clinical or social agencies. Such data transfer systems inherently include in-house or short-range media (approximately 2–3 m), often called intra-BAN communication, between personal and public network (inter-BAN communication), and those entirely within public wireless communication system (beyond-BAN communication).

Figure 1.2 summarises the main research areas of BSNs. Some research has been in progress on how to design wearable sensors [38], fault diagnosis of the BSN and how to mitigate the faults and avoid their impacts [39], energy consumption and energy harvesting [40], and sensor deployment [41]. Without doubt, tremendous research in signal processing – particularly on denoising [42], artefact removal, feature detection, data decomposition, estimation, feature extraction [43], and data compression [44] – has been carried out intensively for various applications. Such valuable experiences can be directly exploited and integrated within the design of BSNs.

Data fusion, as another BSN direction of research, has been under vast development as new techniques in multimodal data recording, analysis, and multiagent distributed systems and networks have been introduced.

Moreover, machine learning techniques have powered up BSN research by developing new techniques in clustering, classification [45], anomaly detection, and decision making as well as many other approaches in big data analytics, to suit the corresponding data.

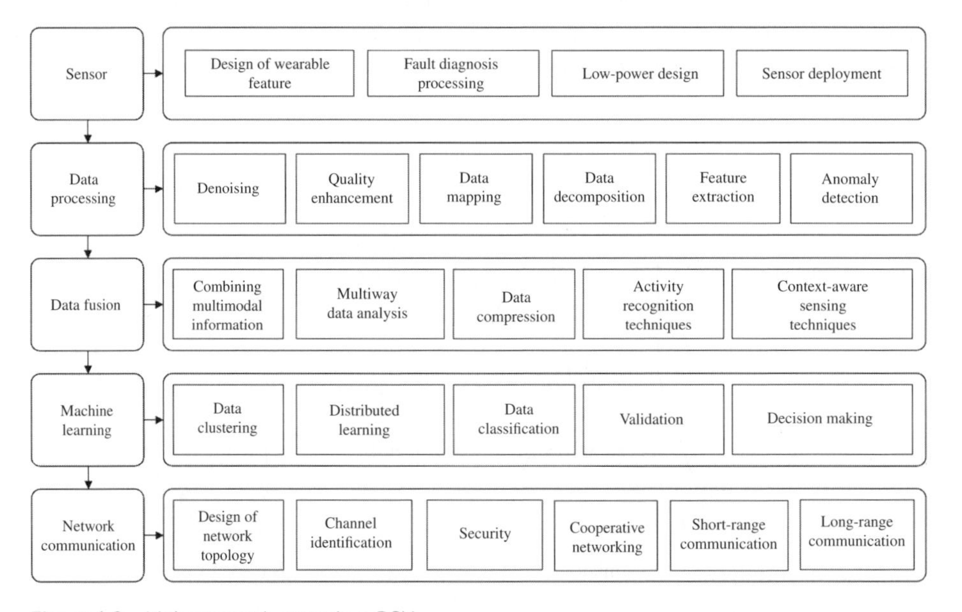


**Figure 1.2** Main research areas in a BSN.

BSNs have been looked at through different angles by a growing number of scholars in sensor technology, data processing, and communications. Some researchers have combined situational awareness and data fusion technologies to enable human activity recognition [46, 47]. Others have tried to understand the data by developing sophisticated signal processing algorithms to deal with multichannel biomedical data [48, 49]. Indeed, the design and provision of supercomputers, availability of large memory clusters, and accessibility of the cloud have been crucial to the expansion of sensor networks. Moreover, new data processing and machine learning methods based on tensor factorisation, cooperative learning, graph theory, kernel-based classification, deep learning neural networks, and distributed systems together with pervasive computing have revolutionised the assessment of the information collected from multisensor networks, particularly when the dataset is large. Network communication, on the other hand, involves network topology design [50], channel characterization [51], channel access control [52, 53], routing algorithm design [54], lightweight communication protocols design, energy harvesting in a network, and many other issues related to short- and long-range communications. These key technologies must be considered and further developed for building a complete BSN system.

To enable long-term data collection from the human body, biocompatible sensors and devices need to be designed. This field of research brings new areas of engineering researchers in biotechnology, biomaterials, bioelectronics, and biomechanics together to develop practical sensors.

The demand for a green environment pushes for the optimization of energy harvesting and an effective solution to energy consumption together with enhancing the QoS. For WSNs the plethora of battery technologies available today enables system designers to tailor their energy storage devices to the needs of their applications. The latest lithium battery technologies allow optimization for any operating lifetime or environment. For applications with small temperature variation and short lifetime, lithium manganese dioxide ($LiMnO_2$) batteries provide solid performance at cost-effective prices, while applications demanding large temperature ranges and multidecade lifetimes are satisfied with batteries based on lithium thionyl chloride ($LiSOCL_2$) chemistry [7].

While batteries represent the preferred low-cost energy storage technology, energy harvesting/scavenging devices are beginning to emerge as viable battery replacements in some applications. For example, power can be generated from temperature differences through thermoelectric and pyroelectric effects, kinetic motion of piezoelectric materials, photovoltaic cells that capture sunlight, or even the direct conversion of RF (radio frequency) energy through specialised antennas and rectification. Examples of energy scavenging/harvesting devices coming to market today include piezoelectric light, solar batteries, and doorbell switches. Although the above technology can solve BSN problems, further research is needed for designing biologically powered systems and biocompatible batteries which can last longer while being attached to body internal tissues.

Finally, secure connections and data security are vital, particularly when personal information is analysed or communicated. In parallel with increasing complexity in data hacking algorithms, there is great demand for producing more sophisticated data encryption and network security.

In dealing with BSNs, machine-centric and human-centric challenges confront researchers. The machine-centric problems, as mentioned before, include security [55], compatibility or interoperability; sensor design and sensor validity; data consistency, as the data residing on multiple mobile devices and wireless patient notes need to be collected and analysed in a seamless fashion; interference; and data management [56, 57].

Besides hardware-centric challenges, human-centric challenges include cost, constant monitoring, deployment constraints, and performance limitations [13, 58–61], which need to be taken care of in any BSN design. After all, the wearable system should be acceptable, convenient, and user friendly.

## 1.4 Layout of the Book

This monograph consists of 15 chapters and has been designed to cover all aspects of BSNs, starting with human body measurable or recordable biomarkers. Chapter 2 is dedicated to understanding these biomarkers, including physical, physiological, and biological measurable quantities. In Chapter 3, sensors, sensor classification, and the quantities measured by different sensors are described. In this chapter, the structures of the sensors for the applications listed in Chapter 2 are detailed. In Chapter 4, more popular and ambulatory sensor systems used in clinical departments and intensive care units are discussed and some examples of their recordings and analysis explained. This discussion continues in Chapter 5, where sleep, as a specific state of human body, is analysed. This chapter includes discussion of various sleep measurement modalities which are more popular and of interest today to researchers. Chapter 6 covers the area of noninvasive, intrusive, and nonintrusive measurement approaches. The objective of this chapter is to introduce the techniques and sensors for nonintrusive or contactless monitoring of major human vital signs, such as breathing, heart rate, and blood oxygen saturation level.

Next, Chapter 7, covers the important concept of gait analysis, recognition, and monitoring. The outcome of this study has a major application in assistive technology, rehabilitation, and assistive robotics. Chapter 8 brings together a wide range of techniques and research approaches in health monitoring. These address the important daily assisted living problems of disabled and older people, and patients with degenerative diseases, and the solutions being currently researched. In Chapter 9 numerous machine learning techniques used for both sensor networks and bioinformatics are explained. This chapter includes most popular, advanced, and very recent machine learning methods for clustering, classification, and feature learning. Support vector machines, reinforcement learning, and different deep neural networks are also included in this chapter. Some examples of machine learning for sensor networks conclude this chapter.

Signal processing techniques and their wide range of applications are covered in Chapter 10. This long chapter explains useful approaches in time, frequency, and multidimensional spaces. Multiresolution analysis, synchro-squeezing wavelet transform, and adaptive cooperative filtering are addressed. At the end of this chapter various signal processing platforms established recently are reviewed. Chapter 11 is devoted to communication systems for BSNs. Short-range communication methods, limitations, and barriers; problems with communication channels and their modelling; linking between body and public networks; and routing methods for BSNs are extensively explained in this chapter. The important topic of energy harvesting for sensor networks is covered in Chapter 12. Advanced techniques in harvesting kinetic, radiant, thermal, chemical, and biochemical energy are explored. At the end of this chapter, topology control and energy prediction, two hot research topics, are introduced.

Chapter 13 covers a wide range of studies of and practical approaches to solving information and network security problems. In addition to general security and privacy preserving techniques, some problems related to patient and clinical data and their importance are discussed in this chapter. QoS, as the major requirement for BSNs are emphasised in this chapter. In Chapter 14 various hardware and software platforms for developing sensor networks currently employed for BSN design are discussed and some practical examples reviewed. Finally, in Chapter 15, the book is summarised, the main topics highlighted, and some suggestions for future research in BSN proposed.

## References


**1** Lai, X., Liu, Q., Wei, X. et al. (2013). A survey of body sensor networks. *Sensors (Basel)* 13: 5406–5447.

**2** Younis, M, Akkaya, K., Eltoweissy, M., Wadaa, A. (2004) On handling QoS traffic in wireless sensor networks. *Proceedings of the 37th Annual Hawaii International Conference on System Sciences*, Big Island, HI (5–8 January 2004).

**3** Liolios, C., Doukas, C., Fourlas, G., and Maglogiannis, I. (2010). An overview of body sensor networks in enabling pervasive healthcare and assistive environments. In: *Proceedings of the 3rd International Conference on Pervasive Technologies Related to Assistive Environments*, Samos, Greece (23–25 June 2005). University of Texas at Arlington.

**4** Ullah, S., Higgins, H., Braem, B. et al. (2010). A comprehensive survey of wireless body area networks: on PHY, MAC, and network layers solutions. *Journal of Medical Systems* 36 (3): 1065–1094.

**5** IEEE 802.15 WPAN™ Task Group 6 (TG6) 2019 Body area networks. http://www.ieee802.org/15/pub/TG6.html (accessed 17 July 2019).

**6** Gershenfeld, N. (2012). How to make almost anything, the digital fabrication revolution. *Foreign Affairs* 91 (6): 43–57.

**7** Haver Samara, M., Fournet Michelle, E.H., Dziak Robert, P. et al. (2019). Comparing the Underwater Soundscapes of Four U.S. National Parks and Marine Sanctuaries. *Frontiers in Marine Science* 6 https://doi.org/10.3389/fmars.2019.00500.

**8** Chen, M., Gonzalez, S., Vasilakos, A. et al. (2010). Body area networks: a survey. *Mobile Networks and Applications* 16: 171–193.

**9** Movassaghi, S., Abolhasan, M., Lipman, J. et al. (2014). Wireless body area networks: a survey. *IEEE Communications Surveys and Tutorials* 16 (3): 1658–1686.

**10** Poslad, S. (2009). *Ubiquitous Computing Smart Devices, Smart Environments and Smart Interaction*. Wiley.

**11** Schmidt, R., Norgall, T., Mörsdorf, J. et al. (2002). Body area network BAN: a key infrastructure element for patient-centered medical applications. *Biomedizinische Technik* 47 (1): 365–368.

**12** O'Donovan, T., O'Donoghue, J., Sreenan, C., O'Reilly, P., Sammon, D. and O'Connor, K. (2009) A context aware wireless body area network (BAN), *Proceedings of the Pervasive Health Conference*, London (1–3 April 2009).

**13** Yuce, M.R. (2010). Implementation of wireless body area networks for healthcare systems. *Sensors and Actuators A: Physical* 162: 116–129.

**14** Jones, V.M., Bults, R.G.A., Konstantas, D., and Vierhout, P.A.M. (2001). *Body Area Networks for Healthcare*, 1–6. Faculty of Electrical Engineering, Mathematics & Computer Science Biomedical Signals and Systems.

**15** Yuce, M.R. and Khan, J.Y. (2011). *Wireless Body Area Networks: Technology, Implementation, and Applications*. Jenny Stanford Publishing.

**16** Yoo, J., Yan, L., Lee, S. et al. (2009). A wearable ECG acquisition system with compact planar-fashionable circuit board-based shirt. *IEEE Transactions on Information Technology in Biomedicine* 13: 897–902.

**17** Harrison, B.L., Consolvo, S., and Choudhury, T. (2009). Using multi-modal sensing for human activity modeling in the real world. In: *Handbook of Ambient Intelligence and Smart Environments*, 1e (eds. H. Nakashima, H. Aghajan and J. Carlos Augusto), 463–478. New York: Springer.

**18** Aziz, O., Lo, B., King, R. et al. (2006). Pervasive body sensor network: an approach to monitoring the post-operative surgical patient. In: *Proceedings of the International Workshop on Wearable and Implantable Body Sensor Networks*, 4–18. Cambridge, MA: IEEE.

**19** Conroy, L., Ó'Conaire, C., Coyle, S. et al. (2009). TennisSense: a multi-sensory approach to performance analysis in tennis. In: *Proceedings of the 27th International Society of Biomechanics in Sports Conference*, Limerick, Ireland (17–21 August 2009), 17–21. Biomechanics Research Unit, University of Limerick.

**20** Pansiot, J., Lo, B., and Yang, G.Z. (2010). Swimming stroke kinematic analysis with BSN. In: *Proceedings of the 2010 International Conference on Body Sensor Networks (BSN)*, Biopolis, Singapore (7–9 June 2010), 153–158. IEEE.

**21** Burchfield, R. and Venkatesan, S.A. (2010). Framework for golf training using low-cost inertial sensors. In: *Proceedings of the 2010 International Conference on Body Sensor Networks (BSN)*, Biopolis, Singapore (7–9 June 2010), 267–272. IEEE.

**22** Nesime, T., Mark, B., Reed, H. et al. (2004). Confidence-based data management for personal area sensor networks. In: *Proceedings of the First Workshop on Data Management for Sensor Networks (DMSN2004)*, Toronto, Canada (30 August 2004), 24–31. ACM.

**23** Shi, W. (2015). Recent advances of sensors for assistive technologies. *Journal of Computer and Communications* (3): 80–87.

**24** Jovanov, E., Milenkovic, A., Otto, C., and de Groen, P.C. (2005). A wireless body area network of intelligent motion sensors for computer assisted physical rehabilitation. *Journal of NeuroEngineering and Rehabilitation* 2: 1–10.

**25** Li, H.B., Takizawa, K., and Kohno, R. (2008). Trends and standardization of body area network (BAN) for medical healthcare. In: *Proceedings of the 2008 European Conference on Wireless Technology*, 1–4. Amsterdam: IEEE.

**26** Osmani, V., Balasubramaniam, S., and Botvich, D. (2007). Self-organising object networks using context zones for distributed activity recognition. In: *Proceedings of the*

*ICST 2nd International Conference on Body Area Network*, Florence, Italy (June 2007), 1–9. ICST.

**27** Wai, A.A.P., Fook, F.S., Jayachandran, M. et al. (2010). Implementation of context-aware distributed sensor network system for managing incontinence among patients with dementia. In: *Proceedings of the 2010 International Conference on Body Sensor Networks (BSN)*, Biopolis, Singapore (7–9 June 2010), 102–105. IEEE.

**28** Kong, J.-W. (2011) Research progress and application prospect of BSN. Shanghai Industries Intelligence Services. http://www.hyqb.sh.cn/publish/portal2/tab227/info4154.htm (accessed 27 June 2012).

**29** Latrè, B., Braem, B., Blondia, G. et al. (2011). A survey on wireless body area networks. *Wireless Networks* 11: 1–18.

**30** Garg, M.K., Kim, D.J., Turaga, D.S., and Prabhakaran, B. (2010). Multimodal analysis of body sensor network data streams for real-time healthcare. In: *Proceedings of the 11th ACM SIGMM International Conference on Multimedia Information Retrieval*, https://doi.org/10.1145/1743384.1743467. Philadelphia, PA (March 2010), 469–478. ACM.

**31** Cooney, M.J., Svoboda, V., Lau, C. et al. (2008). Enzyme catalysed biofuel cells. *Energy & Environmental Science* 1: 320–337.

**32** Yoo, J., Yan, L., Lee, S. et al. (2010). A 5.2 mW self-configured wearable body sensor network controller and a 12 μW wirelessly powered sensor for a continuous health monitoring system. *IEEE Journal of Solid-State Circuits* 45: 178–188.

**33** Bui, F.M. and Hatzinakos, D. (2011). Quality of service regulation in secure body area networks: system modeling and adaptation methods. *EURASIP Journal on Wireless Communications and Networking* 2011: 56–69.

**34** Kumar, P. and Lee, H.J. (2012). Security issues in healthcare applications using wireless medical sensor networks: a survey. *Sensors (Basel)* 12: 55–91.

**35** IEEE P802.15.6-2012, (2012) Standard for wireless body area networks. http://standards.ieee.org/findstds/standard/802.15.6-2012.html (accessed 25 November 2019)

**36** Mearian, L. 2012 'Body area networks' should free hospital bandwidth, untether patients. https://www.computerworld.com/article/2503882/-body-area-networks--should-free-hospital-bandwidth--untether-patients.html (accessed 25 November 2019).

**37** B. Butler, 2012 FCC dedicates spectrum enabling medical body area networks. https://www.fcc.gov/document/fcc-dedicates-spectrum-enabling-medical-body-area-networks (accessed 25 November 2019).

**38** Jafari, R. and Lotfian, R. (2011). A low power wake-up circuitry based on dynamic time warping for body sensor networks. In: *Proceedings of the 2011 International Conference on Body Sensor Network (BSN), University of Dallas at Texas*, 83–88. Dallas, TX: IEEE.

**39** Luo, X., Dong, M., and Huang, Y. (2006). On distributed fault-tolerant detection in wireless sensor networks. *IEEE Transactions on Computers* 55: 58–70.

**40** Kwong, J., Ramadass, Y.K., Verma, N., and Chandrakasan, A.P. (2009). A 65 nm sub-*Vt* microcontroller with integrated SRAM and switched capacitor DC-DC converter. *IEEE Journal of Solid-State Circuits* 44: 115–126.

**41** Atallah, L., Lo, B., King, R., and Yang, G.Z. (2010). Sensor placement for activity detection using wearable accelerometers. In: *Proceedings of the 2010 International Conference on Body Sensor Networks (BSN)*, Biopolis, Singapore (7–9 June 2010), 24–29. IEEE.

**42** Song, K.T., and Wang, Y.Q. (2005) Remote activity monitoring of the elderly using a two-axis accelerometer. *Proceedings of the CACS Automatic Control Conference*, Tainan, Taiwan (18–19 November 2005).

**43** Preece, S.J., Goulermas, J.Y., Kenney, L.P.J. et al. (2009). Activity identification using body-mounted sensors – a review of classification techniques. *Physiological Measurement* 30: 353–363.

**44** Akyildiz, I.F., Su, W., Sankarasubramaniam, Y., and Cayirci, E. (2002). Wireless sensor networks: a survey. *Computer Networks* 38: 393–422.

**45** Akin, A., Stephan, B., Mihai, M.P. et al. (2010). Activity recognition using inertial sensing for healthcare, wellbeing and sports applications: a survey. In: *Proceedings of the 23rd International Conference on Architecture of Computing Systems (ARCS)*, Hannover, Germany (22–25 February 2010), 1–10. IEEE.

**46** Korel, B.T. and Koo, S.G.M. (2007). Addressing context awareness techniques in body sensor networks. In: *Proceedings of the 21st International Conference on Advanced Information Networking and Applications Workshops*, Niagara Falls, ON (21–23 May 2007), 798–803. IEEE.

**47** Thiemjarus, S., Lo, B., and Yang, G.Z. (2006). A spatio-temporal architecture for context aware sensing. In: *Proceedings of the International Workshop on Wearable and Implantable Body Sensor Networks*, Cambridge, MA (3–5 April 2006). IEEE https://doi.org/10.1109/BSN.2006.5.

**48** Sanei, S. and Hassani, H. (2015). *Singular Spectrum Analysis of Biomedical Signals*. CRC Press.

**49** Sanei, S. (2013). *Adaptive Processing of Brain Signals*. Wiley.

**50** Natarajan, A., Motani, M., de Silva, B. et al. (2007). Investigating network architectures for body sensor networks. In: *Proceedings of the 1st ACM SIGMOBILE International Workshop on Systems and Networking Support For Healthcare and Assisted Living Environments (HealthNet 2007)*, San Juan, Puerto Rico (June 2007), 19–24. ACM.

**51** Tachtatzis, C., Graham, B., Tracey, D. et al. (2011). On-body to on-body channel characterization. In: *Proceedings of 2011 IEEE Sensors Conference*, Limerick, Ireland (28–31 October 2011), 908–911. IEEE.

**52** Ullah, S., Shen, B., Riazul, I.S. et al. (2010). A study of MAC protocols for WBANs. *Sensors (Basel)* 10: 128–145.

**53** Seo, S.H., Gopalan, S.A., Chun, S.M. et al. (2010). An energy-efficient configuration management for multi-hop wireless body area networks. In: *Proceedings of 3rd IEEE International Conference on Broadband Network and Multimedia Technology*, Beijing, China (26–28 October 2010), 1235–1239. IEEE.

**54** Benoît, L., Eli, D.P., Ingrid, M., and Piet, D. (2007). MOFBAN: a lightweight modular framework for body area networks. *Lecture Notes in Computer Science* 4808: 610–622.

**55** Toorani, M. (2015). On vulnerabilities of the security association in the IEEE 802.15.6 Standard. In: *Financial Cryptography and Data Security* (eds. M. Brenner, N. Christin, B. Johnson and K. Rohloff). Berlin: Springer.

**56** O'Donoghue, J., Herbert, J. and Fensli, R., (2006) Sensor validation within a pervasive medical environment, *Proceedings of the 5th IEEE Conference on Sensors*, Daegu, South Korea (22–25 October 2006), 972–975.

**57** O'Donoghue, J., Herbert, J., and Kennedy, R. (2006). Data consistency within a pervasive medical environment. In: *Proceedings of IEEE Sensors*. IEEE.

**58** Garcia, P., Virginia Pilloni, V., Franceschelli, F. et al. (2018). Deployment of Applications in Wireless Sensor Networks: A gossip-based lifetime maximization approach. *IEEE Transactions on Control Systems Technology* 24 (5): 1828–1836.

**59** O'Donoghue, J. and Herbert, J. (2012). Data management within mHealth environments: patient sensors, mobile devices, and databases. *Journal of Data and Information Quality* 4 (1): 5.

**60** Lai, D., Begg, R.K., and Palaniswami, M. (eds.) (2011). *Healthcare Sensor Networks: Challenges toward Practical Implementation*. CRC Press.

**61** O'Donovan, T., O'Donoghue, J., Sreenan, C., et al., (2019) A context aware wireless body area network (BAN). https://www.ucc.ie/en/media/research/misl/2009publications/pervasive09.pdf (accessed 25 November 2019).

# 2

# Physical, Physiological, Biological, and Behavioural States of the Human Body

## 2.1 Introduction

The identification and measurement of human body biomarkers is a major goal in clinical diagnosis and disease monitoring. Nevertheless, prior to any measurement, advances in medical science to a large extent help in the recognition of abnormalities by looking at the symptoms and peripheral information. As an example, a number of procedures and measurements are needed to find out if the tiny medial temporal discharges originating within the hippocampus indicate any impending seizure. These clinical operations may involve imaging of the head using MRI (magnetic resonance imaging), taking multichannel electroencephalography (EEG) or magnetoencephalography (MEG) from the scalp, observing a patient's behaviour and movement for a substantial period of time, implanting subdural electrodes within the patient's brain, and checking their biological and even psychological reactions.

This chapter elaborates on the most popular physical, physiological, biological, and behavioural symptoms; abnormalities; and diseases which mostly can be measured and quantified by means of multiple body sensors.

## 2.2 Physical State of the Human Body

In addition to the simple and obvious characteristics to describe a person, such as what they look like, their geometry, hair, and skin colour, there are additional attributes and perhaps more demanding factors in terms of their quantification such as those used in describing them and their actions. Among these factors are those often called biometrics. These include facial features, fingerprint, gait, voice, and other particular markers, such as skin spots.

Face, gait, and joint face-gait recognition have been well researched by groups of researchers around the world [1]. Gait includes static features such as height, stride length, and silhouette bounding box lengths plus some dynamic features such as the frequency or time-frequency domain parameters like frequency and phase of a walk. Gait as a biometric can be used at long distances, it is nonintrusive, noninvasive, and hard to disguise [2].

A useful demarcation between the types of gait analyses are described in [3]. The study of human gait was traditionally part of medical studies and was diagnostic in nature. It forms

*Body Sensor Networking, Design and Algorithms,* First Edition. Saeid Sanei, Delaram Jarchi and Anthony G. Constantinides.

Companion Website: www.wiley.com/go/sanei/algorithm-design

part of the field of biomechanics and kinesiology and is related to other medical fields, like podiatry. Gait metrics are derived from sensors attached to a human being and in some cases video sequences of gait are captured. In this respect, we may term this clinical gait analysis. This field of study has provided much of the terminology used in gait analysis as well as the initial experiments on recognition, for example various phases of a walk and various parts of the body used in walking.

Extensive studies on biomechanical and clinical aspects of hundreds of limbs, joints, and muscles working together indicate that we can derive a reliable description of a person, unique to their way of walking. Moreover, gait can not only reveal the presence of certain sicknesses or moods, but also distinguish between genders. The variability of gait for a person is fairly consistent and not easily changed, while allowing for differentiation with others.

Two different data recording modalities are normally used for gait analysis. One modality involves mounting or attaching proper sensors to the human body, while another uses frontal, lateral, or frontolateral video cameras to take the video of the walking subject and analyse it. The former type is intrusive and may affect the true gait motion. However, since the body movement is recorded more accurately, many applications in rehabilitative assessment effectively exploit that. In addition, using wired or wireless links between the sensors, the cooperation between the sensor signals can be a new platform in sensor networking research.

In one research attempt [4] the number of sensors used for gait analysis has been reduced to one, which can be mounted above the ear. It has been shown that stride length and walking speed can be accurately estimated. Thus, during the rehabilitation process, the pattern of walking demonstrates the rehabilitation progress of a subject with a prosthetic limb.

The use of a video camera requires skill in image processing in order to enable accurate extraction of movement features [5]. Furthermore, for video-based recognition of biometrics often more than one camera is needed to overcome occlusion problems. In practice, various combinations of biometric sensors may also be employed. Usually, two cameras are needed if frontoparallel gait is used. Many other biometrics can be extracted from the frontonormal plane. The problems of alignment and synchronisation are significant. If possible, single camera or monocular capture of video is preferred even if less data are recorded. To overcome this, Zhou and Bhanu use a profile view of a face with gait in order to use one camera at 3.3 m from the subject [6]. Of note is the work by Bazin that includes the ear and footfall as biometrics [7]. As another example, the frontonormal view allows one to use face and iris with gait for a robust recognition system, though some other problems, such as looming effect, make this modality a challenging and difficult case.

Research about human gait has been extended to rehabilitative assessment for various disorders such as stroke, cerebral palsy, paralysis, dementia, age, and Parkinson's.

As an application example, the ability to perform activities considered normal is important to someone who suffers from limb disability. This is particularly common in patients with stroke. The inability to have an independent life requires constant medical attention, resources, and often hiring a caregiver. Where it is possible to rehabilitate the limbs, a customised regimen of exercises needs to be tailored to the needs of the person, depending on the extent of the disability. At the same time, the treatment progress needs to be monitored in order to assess its effectiveness. Presently, these are labour-intensive tasks requiring trained therapists to record data, interpret them, and keep track of what are often repetitive exercises. Compounding this is the lack of clinical skills at home, which means only a limited transfer of the needs of care are permissible, and this may hamper the rehabilitation process [8]. One way to encapsulate the experience of healthcare practitioners is in the form of tests for limb function for tasks deemed essential in the activities of daily living (ADLs). There are many of such established tests which involve movements of a patient and their interactions with various objects [9].

The use of sensors in consumer devices such as mobile phones and gaming consoles allows for a better user experience as the processors in these devices deduce the intention of the user by their movements. Embedding sensors and processors into objects – often in daily use – is referred to as an instrumented object approach.

## 2.3 Physiological State of Human Body

Human physiology is a very complicated and vast area of science. It covers a large number of conditions and abnormalities in humans as well as animals. Wakefulness is taken to be a prime physiological state during which a person is conscious and aware of their environment. Sleep or coma are other states during which the person is not aware of their environment. Wakefulness may also be slightly different from being alert or vigilant as each state has its own characteristics and applications.

There are, however, states which describe abnormalities in the human body. Being deaf, dumb, lam, or blind are often considered abnormalities. Automatic techniques in instrumentation and data processing are able to distinguish most of these abnormalities from the human normal state. For example, a limp is a type of asymmetric abnormality of the gait and can be assessed or monitored using gait analysis techniques. There are many other state types of abnormalities or disabilities which require scrutinised medical examinations for their diagnosis and monitoring. Diverse examples include acathexia, an inability to retain bodily secretion; anhidrosis, the failure of the sweat glands; anaesthesia, the loss of bodily sensation with or without loss of consciousness; and elastosis, loss of elasticity in the skin of elderly people that results from degeneration of connective tissue.

There are some mechanical states of the body such as those related to joint flexion and extension. Gait analysis fully describes such states. Nevertheless, in a number of research works particularly related to brain–computer interfacing (BCI) applications for imaginary movement or related to amputated organs, electromyography (EMG) or EEG signals have been exploited.

Hypercarbia and hypocapnia (acapnia) represent abnormally high or low levels of carbon dioxide in the blood, respectively. It is not so difficult to measure and evaluate the carbon dioxide level in the blood using noninvasive sensors called oximeters, a relatively cheap monitoring modality widely available in hospitals and intensive care units (ICUs). The oximeters can also be used equivalently for measuring the blood oxygen level as well as the oxygen–carbon dioxide exchange during breathing caused by chocking, drowning, electric shock, or inhaling chemical gas.

Hyperthermia refers to high body temperature. The temperature can be measured using thermometers or nonintrusively using thermal imaging systems. Those who suffer from hyperthermia may have abnormalities in their heart rate, breathing, muscle activity, or brain rhythms, and the patterns of these signals can confirm the diagnosis.

Upon some mental or physical diseases such as Parkinson's or stroke, the patients may gradually lose their ability to walk, move their hands, or speak. In many such cases their muscles become weaker and they develop so-called myasthenia. In addition to measurement (using accelerometers) and characterisation of gait, surface or invasive EMG is one of the recording modalities often used to measure the muscle activity and its response to various stimuli.

Many heart diseases and abnormalities can be diagnosed by analysing the electrocardiogram (ECG) measuring heart muscle electrical activities, or stethoscope, which is used for auscultation and recording the heart sound.

Heart diseases refer to conditions that involve the heart, its vessels, muscles, valves, or internal electric pathways responsible for muscular contraction. Common heart diseases are coronary artery disease, heart failure, cardiomyopathy, heart valve disease, and arrhythmias.

In the case of heart attack, a coronary artery is blocked (usually by a blood clot), meaning an area of heart tissue loses its blood supply. This reduction of blood can quickly damage or kill the heart tissue, so quick treatments in an emergency department or catheterization suite are necessary to reduce the loss of heart tissue [10]. However, over 70 000 people die from heart disease in the UK each year, which accounts for approximately 25% of the total number of deaths.

Coronary artery disease occurs when a plaque, a sticky substance such as lipid or calcium compound, narrows or partially obstructs coronary arteries and can cause reduced blood flow. This consequently leads to chest pain (angina), a warning sign of potential heart problems such as a heart attack. The plaque may also trap small blood clots, causing full blockage of a coronary artery and resulting in a heart attack [11]. Figure 2.1 shows a schematic of the blocking of heart arteries with different severities.

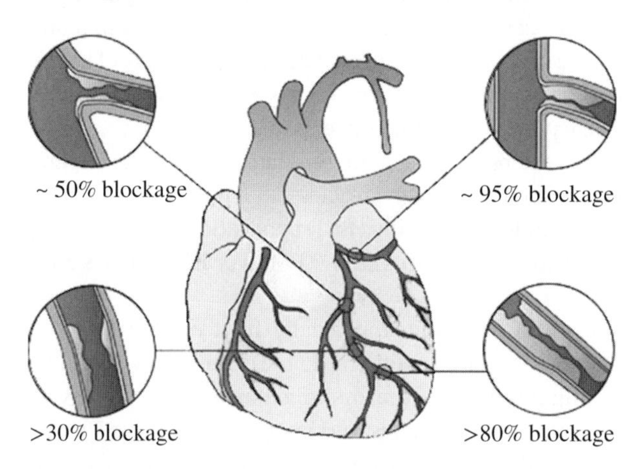


**Figure 2.1** Schematic of some possible blocking of heart arteries with different severities. (*See color plate section for color representation of this figure*)

A heart attack can cause other problems which may lead to more complication than the original problem of blocked arteries. For example, sudden cardiac death may occur when the ECGs become erratic. When the heart tissue that is responsible for regular electrical stimulus of heart muscle contractions is damaged, the heart stops effectively pumping the blood. Cardiomyopathy is a condition indicated by abnormal heart muscle. Abnormal muscles make it harder for the heart to pump blood to the rest of the body.

With cardiomyopathy, the abnormal muscles make it harder for the heart to pump blood to the rest of the body. There are three main types of cardiomyopathy: dilated, owing to stretched and thinned muscles, which lessens the heart's ability to pump blood; hypertrophic, meaning thickened heart muscle; or restrictive, which is a rare problem where the heart muscle does not stretch normally so the chambers do not sufficiently fill with blood.

Heart failure, or congestive heart failure, is when the heart's pumping action cannot meet the body's demand for blood. Heart failure has the same symptoms and signs as those seen with cardiomyopathy.

There are also many types of congenital heart defects. A congenital heart defect is a defect in the development of the heart that is usually first noticed at birth, though some abnormal cases are not diagnosed until adulthood. Some people with such defects don't need any treatment, but others may need medication or surgical repair.

In almost all heart diseases and abnormalities the ECG or heart sound patterns captured using a stethoscope deviate from the normal waveforms, and therefore researches have been undertaken to recognise or classify these patterns by applying signal processing and machine learning algorithms.

On the other hand, in auscultation of heart sound, for example to detect heart murmur, often the lung sound due to breathing is also heard. Examining the lung sound can assist clinicians in the detection of wheezing, lung infection, asthma, and many other pulmonary diseases.

In addition to the ECG (or EKG) and stethoscope, there are other anatomical or functional screening modalities to test and diagnose a heart disease. For example, a continual multichannel ECG system may be used to check the level of stress. This test measures the ability of a person's heart to respond to the body's demand for more blood during stress (exercise or work). Combination of ECG, heart rate, and blood pressure information may be examined as a person's exercise is gradually increased on a treadmill. The information helps to show how well the heart responds to the body's demands and may provide information to help diagnose and treat the defects. It can also be used to monitor the effects of heart treatment.

Many people have intermittent symptoms such as intermittent chest pain or occasional feelings of their heart beating faster or irregularly. In such cases, ECG changes may not be sufficiently useful for monitoring the heart over a long period or outside clinics. Instead, a device called a Holter monitor can be used and worn for a longer time. This device acts similarly to a normal ECG.

As another heart screening modality, an echocardiogram is a real-time moving picture of a functioning heart made by using ultrasound waves and generating meaningful images. It can show how well the heart chambers and heart valves operate (for example, effective or poor pumping action as the blood flows through the valves), before and after treatments, as well as other features.

Calcium build-up (plaque), blood clot, or lipid in coronary arteries, which are indeed life-threatening, can also be detected from three-dimensional (3D) computerised tomography (CT) scan of the heart.

Body muscles, on the other hand, translate thoughts to actions and therefore are considered as the most important movement related component of human body. EMG recordings check the health of both the muscles and the nerves controlling the muscles. The EMG signals can be used to detect the problems with muscles during rest or activity. Disorders or conditions that cause abnormal EMG patterns can be neurological, pathological, or biological. These disorders include [12]:

- alcoholic neuropathy: damage to nerves from drinking too much alcohol;
- amyotrophic lateral sclerosis (ALS): disease of the nerve cells in the brain and spinal cord that control muscle movement;
- axillary nerve dysfunction: damage of the nerve that controls shoulder movement and sensation;
- Becker muscular dystrophy: weakness of the legs and pelvis muscles;
- brachial plexopathy: problem affecting the set of nerves that leave the neck and enter the arm;
- carpal tunnel syndrome: problem affecting the median nerve in the wrist and hand;
- cubital tunnel syndrome: problem affecting the ulnar nerve in the elbow;
- cervical spondylosis: neck pain from wear on the disks and bones of the neck;
- common peroneal nerve dysfunction: damage of the peroneal nerve leading to loss of movement or sensation in the foot and leg;
- denervation: reduced nerve stimulation of a muscle;
- dermatomyositis: muscle disease that involves inflammation and a skin rash;
- distal median nerve dysfunction: problem affecting the median nerve in the arm;
- Duchenne muscular dystrophy: inherited disease that involves muscle weakness;
- facioscapulohumeral muscular dystrophy (Landouzy–Dejerine): disease of muscle weakness and loss of muscle tissue;
- familial periodic paralysis: disorder that causes muscle weakness and sometimes a lower than normal level of potassium in the blood;
- femoral nerve dysfunction: loss of movement or sensation in parts of the legs due to damage to the femoral nerve;
- Friedreich ataxia: inherited disease that affects areas in the brain and spinal cord that control coordination, muscle movement, and other functions;
- Guillain-Barré syndrome: autoimmune disorder of the nerves that leads to muscle weakness or paralysis;
- Lambert–Eaton myasthenic syndrome: autoimmune disorder of the nerves that causes muscle weakness;
- multiple mononeuropathy: a nervous system disorder that involves damage to at least two separate nerve areas;
- mononeuropathy: damage to a single nerve that results in loss of movement, sensation, or other function of that nerve;
- myopathy: muscle degeneration caused by a number of disorders, including muscular dystrophy;
- myasthenia gravis: autoimmune disorder of the nerves that causes weakness of the voluntary muscles;
- peripheral neuropathy: damage of nerves away from the brain and spinal cord;

- polymyositis: muscle weakness, swelling, tenderness, and tissue damage of the skeletal muscles;
- radial nerve dysfunction: damage of the radial nerve causing loss of movement or sensation in the back of the arm or hand;
- sciatic nerve dysfunction: injury to or pressure on the sciatic nerve that causes weakness, numbness, or tingling in the leg;
- sensorimotor polyneuropathy: condition that causes a decreased ability to move or feel because of nerve damage;
- Shy–Drager syndrome: nervous system disease that causes body-wide symptoms;
- thyrotoxic periodic paralysis: muscle weakness from high levels of thyroid hormone;
- tibial nerve dysfunction: damage of the tibial nerve causing loss of movement or sensation in the foot.

Surface EMGs are very noisy and often include the effects of heart pulsation, movement, and system noise. Therefore, computerised systems and algorithms for recognition of the abnormalities should be robust against noise. In some cases, however, EMG is taken invasively by inserting a needle electrode into the muscle.

A physical or physiological condition caused by a disease is often called the pathological state. Such a state is also evaluated by examining biochemistry and chemical metabolism of the human body often through sampling and laboratory analysis of human blood, exhale, tissue, urine, or faeces.

## 2.4 Biological State of Human Body

Biology is a general term which describes various organs of living organisms, including human, and their behaviour starting from biochemical reactions in tissue molecules and tissue cell behaviour in circulatory, respiratory, and nervous systems.

Current techniques in system biology focus on complex interactions within biological systems, using a holistic approach to biological research. One of the aims of systems biology is to model and discover emergent properties of cells, tissues, and organisms, collectively functioning as a system, whose theoretical description is only possible using techniques which fall under the remit of systems biology [13]. These typically involve metabolic or cell signalling networks [14] available in all live human organs.

Biological abnormalities refer to either genetic disorders, which may cause secondary problems, or infectious, bacterial, and immunodeficiency diseases. Tuberculosis, cholera, malaria, and AIDS as one of the deadliest one caused by the human immunodeficiency virus (HIV) are some examples of biological diseases.

Genetic disorders are caused by point mutation or gene damage due to insertion or deletion of a gene bond, deletion of a gene or genes, missing chromosomes, or trinucleotide repeat disorder. There are many types of such disorders, including colour blindness, cystic fibrosis, Down syndrome, haemophilia, neurofibromatosis, polycystic kidney disease, and spinal muscular atrophy.

Monitoring the human body for diseases often starts from noninvasive examinations and data recordings followed by taking and testing blood or urine samples. In the later stages, however, invasive tissue sampling by means of biopsy or various imaging modalities may become necessary. As an example, for monitoring a tumour in the body, the examination may be extended to biopsy operation and radiography to enable accurate diagnosis.

A class of biological diseases called autoimmune diseases are among the top mortality causes according to the American Autoimmune Related Diseases Association (AARDA). These diseases are the result of an abnormal immune response of the body against substances and tissues normally present in the body [15]. To diagnose autoimmune diseases, high resolution imaging of the affected organ through indirect immunofluorescence (IIF) is needed. IIF is an imaging technique that captures images of human epithelial type-2 (HEp-2) cells [16]. Using this imaging technique, antinuclear antibody (ANA), a type of autoantibody binding to the contents of the cell nucleus, is considered a hallmark of autoimmune diseases.

In the IIF test, antibodies are first stained in HEp-2 tissue and then bound to a fluorescent chemical compound. In the cells containing ANAs, the antibodies bound to the nucleus demonstrate different patterns that can be captured and seen via microscope imaging. Categorising the patterns in the HEp-2 cell images can be used to distinguish the phase and severity of autoimmune diseases [17]. The IIF imaging test consists of five different stages [18] where the first stage is autofocus image acquisition to reduce photobleaching effects [19]. The second stage is automatic segmentation of the cells using methods such as the similarity-based watershed and adaptive edge-based segmentation [20, 21]. This is followed by the mitotic cell segmentation using morphological and textural features and local binary patterns (LBPs) [22]. In the fourth stage, the intensity level images are classified into three classes of negative, intermediate, and positive intensities [23]. Finally, the cell staining patterns are classified into centromere (Ce), coarse-speckled (Cs), cytoplasmatic (Cy), fine-speckled (Fs), homogeneous (H), nucleolar (N), and Golgi (G), corresponding to different types of autoimmune diseases.

Computer-aided diagnosis (CAD) systems, developed by engineers and computer scientists for automatic classification of HEp-2 cells, have attracted much interest in the diagnosis of autoimmune diseases. The systems reduce the cost and time of the diagnosis process and provide repeatability of the test for different physicians.

## 2.5 Psychological and Behavioural State of the Human Body

Body functions under brain control, thus any factor affecting the human brain, consequently influence the behavioural and physical states of the body. The new technology, the tremendous research, and the conceptual advances in the behavioural, biological, and medical sciences can certainly aid recognition of bidirectional and multilevel relationships between behaviour and health. Psychological, neurological, and anatomical diagnoses often involve different screening and testing procedures. For example, epileptic seizure has different symptoms including whole body movement, heart rate variation, and most importantly changes in the EEG dynamics and waveforms. On the other hand, an anatomical problem, often caused by a brain tumour, should be diagnosed through medical imaging followed by pathological tests.

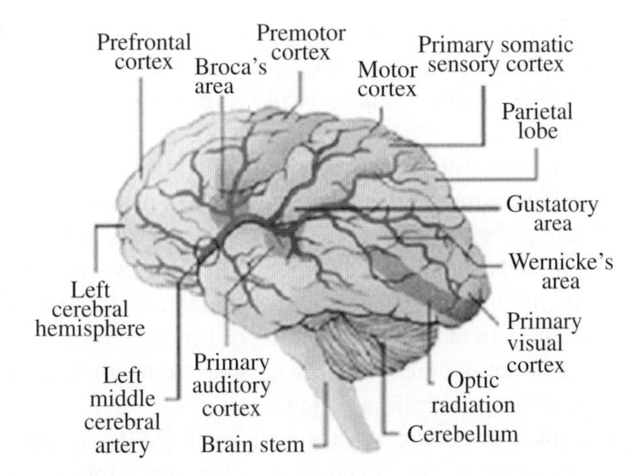


**Figure 2.2** Different brain sensory zones. (*See color plate section for color representation of this figure*)

Although the brain's action is unpredictable during wakefulness and is influenced by internal, such as emotions, and external, such as event and movement-related stimuli, during sleep the brain follows a number of well-defined states, generating predictable continuous or intermittent rhythms.

The abnormality in the brain may originate from different brain zones (Figure 2.2). Hence, the assessment of the human brain through various screening and imaging modalities during different behavioural states promotes our understanding of the links between human behaviour and basic neurological and neurochemical processes or specific neuroanatomic pathways.

The number of diseases and disorders in the brain is probably more than that of any other organ in the body. Most brain abnormalities manifest themselves in well-defined patterns in multichannel EEG recordings. However, not all these abnormalities have been studied through EEG analysis. The most important and popular brain abnormalities and diseases are listed alphabetically:

- Amnesia (amnestic syndrome): the loss of memories, such as facts, information, and experiences. Though it generally doesn't cause a loss of self-identity and those with amnesia are usually lucid and know who they are, they may have trouble learning new information and forming new memories. Amnesia can be caused by damage to areas of the brain that are vital for memory processing. Unlike a temporary episode of memory loss, amnesia can be permanent. There's no specific treatment for amnesia,
- Amyotrophic lateral sclerosis (ALS): a motor neuron disease and a rare group of neurological diseases that mainly involve the nerve cells (neurons) responsible for controlling voluntary muscle movement such as for chewing, walking, breathing, and talking. The disease is progressive and therefore the symptoms become worse over time. Currently, there is no cure for ALS and no effective treatment to halt or reverse its progression. ALS belongs to a wider group of disorders known as motor neuron diseases caused by the gradual deterioration (degeneration) and death of motor neurons. Motor neurons are nerve cells that extend from the brain to the spinal cord and to muscles throughout the body. These motor neurons initiate and provide vital communication links between the brain and the voluntary muscles. Early symptoms of ALS usually include muscle weakness

or stiffness. All muscles under voluntary control are gradually affected, and individuals lose their strength and the ability to speak, eat, move, and even breathe and mostly die from respiratory failure, usually within three to five years from when the symptoms first appear.

- Ataxia: neurological symptoms (rather than disorders) related to the movement and control of posture and balance, resulting in poor coordination. Ataxia can be due to many different causes. Cerebellar ataxia means unsteadiness due to pathology in the cerebellum, which is a leaf-like structure in the back part of the brain.
- Attention deficit hyperactivity disorder (ADHD): a brain disorder diagnosed by an ongoing pattern of severe inattention or hyperactivity impulsivity which affects the functioning or development of humans often from childhood and the impulsivity in action and behaviour continues into old age.
- Autism or autism spectrum disorder (ASD): a neurodevelopmental disorder characterised by deficits in communication, social interaction, and the presence of restricted, repetitive behaviours. Social communication deficits include impairments in aspects of joint attention and social reciprocity, as well as difficulties in the use of verbal and nonverbal communicative behaviours for social interaction.
- Bipolar disorder (used to be called manic depression): a mental abnormality that causes extreme mood swings including emotional highs (mania or hypomania) and lows (depression). When depressed, the subject may feel sad or hopeless and lose interest or pleasure in most activities. When the mood shifts to mania or hypomania (less extreme than mania) the patient may feel euphoric, full of energy, or unusually irritable. These mood swings can affect sleep, energy, activity, judgement, behaviour, and the ability to concentrate.
- Cancer: there are mainly two types of brain cancers called primary and secondary brain cancers. The former appears as benign or metastatic tumours such as glioblastoma and the latter is due to spreading cancers originated in other parts of the body into the brain. Headache, feeling sick, and seizure are the typical symptoms of brain cancer. Some examples of brain tumours are explained in the related part of this section.
- Central nervous system disease: a broad category of conditions in which the brain and the nerves in the spinal cord do not function normally, limiting health and the ability to function. This can be due to an inherited metabolic disorder, the result of damage from an infection, a degenerative condition, stroke, a brain tumour, or arise from unknown or multiple factors. Movement disorders such as Parkinson's disease, dystonia, and essential tremor are central nervous system conditions. What they have in common is the loss of sufficiently intact nervous system circuits that govern functions as varied as memory formation (in Alzheimer's) or voluntary motion (in movement disorders).
- Cerebral palsy: a condition that affects muscle control and movement. It is usually caused by an injury or infection to the brain before, during, or after birth. It may also be due to lack or shortage of oxygen in the fetus brain, genetic problem, or other abnormal brain development. Children diagnosed with cerebral palsy often have difficulties in controlling muscles and movements as they grow and develop.
- Cerebrovascular disease: the result of disease in the arteries and blood vessels in the brain which can cause blockage of food and oxygen supply to parts of the brain, leading to stroke.

- Creutzfeldt–Jakob disease (CJD) (mad cow disease): a fatal neurodegenerative disease often with a lifetime of less than one year. Early symptoms include dementia, change of personality, and hallucinations [24]. The symptoms of CJD are caused by the progressive death of the brain's nerve cells, which is associated with the build-up of abnormal prion protein molecules forming amyloids [25].
- Dementia (Alzheimer's, mild cognitive impairment): a worldwide problem which affects women more than men, and is the result of a number of brain diseases and abnormalities which lead to deterioration of memory, consciousness, and consequently physical disability and early death of the brain. It may also cause behavioural change, confusion and disorientation, delusion and hallucination, communication problems, problems in judging speed and distances, and even craving for particular foods. Alzheimer's is the most common dementia type but there are other dementias, including vascular dementia, dementia with Lewy bodies, and frontotemporal dementia. Research has shown that dementia causes loss of communication between brain cells and brain zones [17, 26].
- Depression: a common mental disorder that causes people to experience depressed mood, loss of interest or pleasure, feelings of guilt or low self-worth, disturbed sleep or appetite, low energy, and poor concentration. It is different from feeling down or sad and affects people of every age. A person experiencing depression often has intense emotions of anxiety, hopelessness, negativity, and helplessness, and the feelings stay with them for a long time [24].
- Encephalomyelitis: inflammation of the brain or spinal cord which can be the result of various diseases such as viral disease, mosquito bite, AIDS, or syndromes.
- Encephalopathy: various brain disorders, and also reflects structural and anatomical defects in the brain.
- Hydrocephalus: a condition in which there is an abnormal accumulation of cerebrospinal fluid in the brain [27]. This typically causes increased pressure inside the skull. Hydrocephalus can be due to birth defect or because of an injury in older age. In babies there may be a rapid increase in head size. Other symptoms may include vomiting, sleepiness, seizures, and downward pointing of the eyes. Older people may have headaches, double vision, poor balance, urinary incontinence, personality changes, or mental impairment.
- Huntington's disease (after George Huntington 1872 – also known as Huntington's chorea): an inherited brain disorder which causes death of brain cells [28]. It starts with occasional mood swings or mental abilities [28]. It is followed with lack of coordination and unsteady movement [29]. As the disease advances, uncoordinated, jerky body movements become more apparent [28]. The physical abilities gradually worsen until movement coordination becomes severe and the person is unable to walk [28, 29]. Mental abilities generally decline into dementia [30]. The specific symptoms vary somewhat between people [28]. The disease symptoms usually begin between 30 and 50 years of age but can start at any age [30, 31].
- Idiopathic intracranial hypertension: a neurological condition of unknown cause, with symptoms very similar to those of a brain tumour, defined by increased intracranial pressure around the brain without the presence of tumour or disease.
- Meningitis: an inflammation of the meninges (the protective membranes that cover the brain and spinal cord) caused by an infection. The inflammation can cause damage to the brain and spinal cord. Acute bacterial meningitis is rapidly developing inflammation of

the tissue layers and of the fluid-filled space between the meninges (subarachnoid space) when it is caused by bacteria.

- Migraine: affects approximately 15% of the population. It is a primary headache disorder characterised by recurrent headaches mostly severe [32]. Typically, the headaches affect one side of the head, are pulsating in nature, and last for 2 to 72 hours [32]. The symptoms can be nausea, vomiting, and sensitivity to light, sound, or smell [33]. The pain is generally made worse by physical activity [34]. Up to one-third of people have an aura: typically a short period of visual disturbance which announces the onset of headache in advance [34]. Occasionally, an aura can occur with little or no headache after that [35]. Migraines are believed to be due to a mixture of environmental and genetic factors [36].
- Multiple sclerosis (MS): a disease affecting or damaging the myelin which is the covering layer of nerve cells in the brain or spinal cord [37]. This damage disrupts communication between cells in the nervous system resulting in many signs and symptoms, including physical, mental, and sometimes psychiatric problems [38–40]. Double-vision, blindness of one eye, muscle weakness, trouble with sensation, and trouble with coordination are the specific symptoms [37]. MS is a degenerative disease and gradually goes to complexity. Brain MS is more severe. Although the actual cause is not clear, the underlying mechanism is thought to be either destruction of the immune system or failure of the myelin-producing cells.
- Paralysis: the inability to move a part of the body temporarily or permanently. In almost all cases, paralysis is due to nerve damage, not to an injury to the affected region. Often an injury in the middle or lower regions of the spinal cord is likely to disrupt function below the injury, including the ability to move the feet or feel sensations, even though the actual structures are completely healthy. In this situation, the brain is unable to relay a signal to an area of the body due to injuries to the brain. However, in some cases the brain is able to sense touch and other sensations in the body, but is unable to effectively relay a response due to injuries in the spinal cord.
- Parkinson's: a degenerative brain disorder. The cause is generally unknown but believed to be genetic or due to environmental factors [41]. Physiologically, the dopamine generators of the brain fail in generating sufficient dopamine as the result of death of cells in the substantia nigra in the midbrain region [42]. Diagnosis of typical cases is mainly based on symptoms such as tremor, rigidity, slowness of movement, difficulty in walking, dementia, depression, and anxiety.
- Pick's disease: a rare form of dementia and similar to Alzheimer's, except that it often affects only certain brain zones. It has unknown causes and people with Pick's disease have abnormal substances (called Pick bodies and Pick cells) inside nerve cells in the damaged areas of the brain.
- Seizure and epileptic seizure;: a temporary loss of control often, but not always, accompanied by convulsions, unconsciousness, or both. Most common types are epileptic seizures, or seizures, are caused by sudden abnormal electrical discharges in the brain. An epileptic seizure, also known as an epileptic fit, seizure, or fit, manifests itself in the form of a brief episode of signs or symptoms due to abnormal excessive or synchronous neuronal activity in the brain [43]. The outward effect can vary from uncontrolled jerking movement (tonic–clonic seizure) to something as subtle as a momentary loss of awareness (absence seizure). Diseases of the brain characterised by an enduring predisposition to

generate epileptic seizures are collectively called epilepsy [44]. On the other hand, for nonepileptic seizures the brain activity remains normal. Nonepileptic seizures have no identifiable physical cause, but they are believed to be physical reactions to psychological stress, change of emotions, in some cases due to tumour in the brain, or as symptoms of hypertension.

- Brain tumour: a growth of cells in the brain that multiplies in an abnormal, uncontrollable way. Brain tumours can be malignant, slow growing, or benign. The most common brain tumour is a glioma, which has different types including astrocytomas, glioblastomas, oligodendrogliomas, mixed gliomas, and ependymomas. Some grow slowly while others grow more quickly.
- Meningioma: a common brain tumour. It starts in the meninges, the tissue covering the brain and spinal cord, and is usually benign. A lymphoma is a cancer of the lymphatic system. It rarely starts in the brain. An acoustic neuroma is a benign tumour of the hearing nerve. Another benign brain tumour that starts in cells lining blood vessels is a haemangioblastomas.
- Tumours that start in the pituitary gland, which helps control hormones, are also benign. There are also tumours which start in the spinal cord and they are usually benign. Tumours of the pineal gland, such as germinomas and teratomas, are rare. They can be slow or fast growing. Medulloblastoma tumours are rare in adults but more common in children.

In the next chapters of this book we will see how single or multiple sensor systems can record and detect the clinically important features related to most of the above abnormalities.

Human development, often referred to as developmental psychology, explains the changes in human cognitive, emotional, and behavioural capabilities and functioning over the entire life.

On the other hand, the availability of monoclonal antibodies, routine production of genetically altered animals, and new understanding of the genetic code have contributed to the exploration of how genetics interacts with development and early experiences to influence both vulnerability to disease and resistance to age-related decline.

The combination of biology and society makes us what we are and what we do. The three main elements of biology contributing to human behaviour are: (i) self-preservation; (ii) the reason for self-preservation, reproduction; and (iii) a method to enhance self-preservation and reproduction [45].

As another important biological effect, biological rhythms are related to the changes in mood and consequently the human behaviour. These rhythms control much of the body's normal functions, including performance, sleep, and endocrine rhythms as well as behaviour. These functions are primarily regulated by the circadian clock, a cluster of nerves located in the hypothalamus in the brain. The circadian clock relies on environmental cues to regulate its function, primarily light cues from the day/night cycle. Any shift in these cues, such as by travel resulting in jet lag, can alter the sleep cycle and have a detrimental effect on normal circadian rhythms. In addition, season changes, which are accompanied by a decrease in the number of daylight hours, can negatively impact the function of the circadian clock, primarily the secretion of melatonin to induce sleep. If the alterations in biological rhythms are sufficiently strong, they may lead to mood disorders including mild depression and seasonal affective disorder [46].

There are two major categories of biological rhythms: endogenous and exogenous. Endogenous rhythms come from within the organism and are regulated by the organism itself, for example the body temperature cycle, brain rhythms, or heart rate. Exogenous rhythms are the result of external factors, such as a change in the seasons or transition from day to night. The environmental stimuli referred also to as zeitgebers, from the German for 'time givers', help to maintain these cycles. They include sunlight, noise, food, and even social interaction and help the biological clock maintain a 24-hour day.

There are many factors influencing biological rhythms. A cluster of approximately 10 000 nerve cells located on the suprachiasmatic nuclei (SCN) found on the hypothalamus in the brain. The circadian clock's primary function is to interpret external changes of light and darkness, as well as social contact, in order to establish diurnal rhythms. It is not uncommon for the circadian clock to be disrupted temporarily; events such as changes in work schedule from day to night, changing time zones, and to some extent old age can impact the consistency of circadian rhythms.

The circadian clock relies heavily on changes in light to determine day/night transitions. During the night, SCN emits melatonin hormone, which induces sleep. The process of wake to sleep itself has its own stages and each stage has its own duration [17].

Another major disruptive factor related to the circadian clock's interpretation of light is seasonal change. During the winter months, there are fewer daylight hours. As a result, the level of melatonin secretion increases along with the number of hours of darkness. The normal cycles may also be interrupted by changing one's daily habits, for example changing feeding time, following a gradual force-to-sleep or sleep depriving.

In addition to these major influences there are a variety of other environmental factors that may have an impact on biological rhythms. One of them is caffeine. A series of experiments on caffeine revealed differences in the effects of the drug depending on time of day. In the morning caffeine has been shown to hinder low impulsiveness, while the opposite is true in the evening [47]. This finding suggests that low impulsiveness and high impulsiveness differ in the phase of their diurnal rhythms, resulting in a difference in the effects of caffeine.

By establishing an understanding of various environmental factors that influence biological rhythms it is possible to draw connections between the significant time shifts and changes in nature and mood disorders.

Moreover, the influence of those factors may be quantified by developing a hybrid measurement system incorporating measures of brain activity, heart rate, and respiration as well as changes in the level of adrenalin in the blood over time.

## 2.6 Summary and Conclusions

Physical, biological, and mental biomarkers of the human body can well describe its state. Body movement, heart rate variability, and the brain responses to various internal and external stimuli can reveal the symptoms and causes of many abnormalities in the state of human body. To differentiate these abnormalities and disease indicators, however, a variety of tests and measurements by means of suitable sensors need to be undertaken. These indicators can be quantified using data-processing and intelligent systems for better and quicker diagnosis of the abnormalities in humans. Although sensors and sensory networks have facilitated recording and quantification of many of the states indicating variables, there is still a long way to go to cover all factors involved in the full recognition of human body states.

## References

1 Lee, T.K.M., Belkhatir, M., and Sanei, S. (2014). A comprehensive review of past and present vision-based techniques for gait recognition. *Multimedia Tools and Applications* 72 (3): 2833–2869.

2 Lee, T.K.M., Belkhatir, M., Lee, P.A., and Sanei, S. (2008). Nonlinear characterisation of fronto-normal gait for human recognition. In: *Advances in Multimedia Information Processing – PCM 2008, Lecture Notes in Computer Science* (eds. Y.-M.R. Huang et al.), 466–475. Berlin: Springer-Verlag.

3 Caldas, R., Mundt, M., Potthast, W., Buarque de Lima Neto, F., and Markert, B. (2017) A systematic review of gait analysis methods based on inertial sensors and adaptive algorithms. *Gait Posture* 57: 204–210.

4 Jarchi, D., Wong, C., Kwasnicki, R.M. et al. (2014). Gait parameter estimation from a miniaturised ear-worn sensor using singular spectrum analysis and longest common subsequence. *IEEE Transactions on Biomedical Engineering* 61 (4): 1261–1273.

5 Kumar, P., Mukherjee, S., Saimi, R. et al. (2019). Multimodal gait recognition with inertial sensor data and video using evolutionary algorithm. *IEEE Transactions on Fuzzy Systems* 27 (5): 956–965.

6 Zhou, X. and Bhanu, B. (2006). Feature fusion of face and gait for human recognition at a distance in video. *Proceedings of the 18th International Conference on Pattern Recognition* 4: 529–532.

7 Bazin, A. I. (2006) On probabilistic methods for object description and classification. PhD thesis, University of Southampton.

8 Andrews, K. and Steward, J. (1978). Stroke recovery: he can but does he? *Rheumatology* 16 (1): 43–48.

9 Lee, T. K. M., Gan, S.S.W., Sanei, S., and Kouchaki, S., (2013) Assessing rehabilitative reach and grasp movements with singular spectrum analysis. *Proceedings of the 21st European Signal Processing Conference (EUSIPCO)*, Marrakech, Morocco (9–13 September 2013).

10 Jenkins, K.J., Correa, A., Feinstein, J.A. et al. (2007). Noninherited risk factors and congenital cardiovascular defects: current knowledge: American Heart Association Council on cardiovascular disease in the young: endorsed by the American Academy of Pediatrics. *Circulation* 115 (23): 2995–3014.

11 Mendis, S. and Puska, P. (2011). *Global Atlas on Cardiovascular Disease Prevention and Control* (ed. World Health Organization), 3. World Health Organization in collaboration with the World Heart Federation and the World Stroke Organization.

12 Shelat, A. M., (2016) Electromyography, https://medlineplus.gov/ency/article/003929.htm (accessed 25 November 2019).

13 Longo, G. and Montévil, M. (2014). *Perspectives on Organisms*. Springer.
14 Bu, Z. and Callaway, D.J. (2011). Proteins MOVE! Protein dynamics and long-range allostery in cell signalling. *Advances in Protein Chemistry and Structural Biology* 83: 163–221.
15 Cotsapas, C. and Haer, D.A. (2013). Immune-mediated disease genetics: the shared basis of pathogenesis. *Trends in Immunology* 34: 22–26.
16 Meroni, P.L. and Schur, P.H. (2010). Ana screening: an old test with new recommendations. *Annals of the Rheumatic Diseases* 69: 1420–1422.
17 Sanei, S. (2013). *Adaptive Processing of Brain Signals*. Wiley.
18 Hiemann, R., Hilger, N., Sack, U., and Weigert, M. (2006). Objective quality evaluation of fluorescence images to optimize automatic image acquisition. *Cytometry Part A* 69: 182–184.
19 Soda, P., Rigon, A., Afeltra, A., and Iannello, G. (2006). Automatic acquisition of immunofluorescence images: algorithms and evaluation. In: *19th IEEE International Symposium on Computer-Based Medical Systems, CBMS*, 386–390. IEEE.
20 Huang, Y.-L., Chung, C.-W., Hsieh, T.-Y., and Jao, Y.-L. (2008). Outline detection for the HEp-2 cell in indirect immunofluorescence images using watershed segmentation. In: *2008 IEEE International Conference on Sensor Networks, Ubiquitous and Trustworthy Computing (SUTC'08)*, 423–427. IEEE.
21 Huang, Y.-L., Jao, Y.-L., Hsieh, T.-Y., and Chung, C.-W. (2008). Adaptive automatic segmentation of HEp-2 cells in indirect immunofluorescence images. In: *2008 IEEE International Conference on Sensor Networks, Ubiquitous and Trustworthy Computing (SUTC'08)*, 418–422. IEEE.
22 Foggia, P., Percannella, G., Soda, P., and Vento, M. (2010). Early experiences in mitotic cells recognition on HEp-2 slides. In: *IEEE 23rd International Symposium on Computer-Based Medical Systems (CBMS)*, 38–43. IEEE.
23 Soda, P. and Iannello, G. (2006). A multi-expert system to classify fluorescent intensity in antinuclear autoantibodies testing. In: *19th IEEE International Symposium on Computer-Based Medical Systems (CBMS 2006)*, 219–224. IEEE.
24 Murray, E.D., Buttner, N., and Price, B.H. (2012). Depression and psychosis in neurological practice. In: *Neurology in Clinical Practice*, 6e (eds. W.G. Bradley, R.B. Daroff, G.M. Fenichel and J. Jankovic). Butterworth Heinemann.
25 Sattar, H. (2011). *Fundamentals of Pathology*. Pathoma.
26 Escudero, J., Sanei, S., Jarchi, D. et al. (2011). Regional coherence evaluation in mild cognitive impairment and Alzheimer's disease based on adaptively extracted magnetoencephalogram rhythms. *Physiological Measurements* 32 (8): 1163–1180.
27 National Institute of Neurological Disorders and Stroke (NINDS). (2016). Hydrocephalus Fact Sheet. https://www.ninds.nih.gov/Disorders/Patient-Caregiver-Education/Fact-Sheets/Hydrocephalus-Fact-Sheet (accessed 25 November 2019).
28 Dayalu, P. and Albin, R.L. (2015). Huntington disease: pathogenesis and treatment. *Neurologic Clinics* 33 (1): 101–114.
29 Caron, N.S., Wright, G.E.B., and Hayden, M.R. (1998). Huntington Disease. In: *GeneReviews*® (eds. M.P. Adam, H.H. Ardinger, R.A. Pagon, et al.). Seattle: University of Washington.
30 Frank, S. (2014). Treatment of Huntington's disease. *Neurotherapeutics* 11 (1): 153–160.

31 National Institute of Neurological Disorders and Stroke (NINDS). (2019). Huntington's Disease Information Page. https://www.ninds.nih.gov/Disorders/All-Disorders/Huntingtons-Disease-Information-Page (accessed 25 November 2019).

32 World Health Organization (2020). Headache disorders. https://www.who.int/news-room/fact-sheets/detail/headache-disorders (accessed 1 January 2020).

33 Aminoff, M.J., Greenberg, D.A., and Simon, R.P. (2009). *Clinical Neurology*, 7e. New York: Lange Medical Books/McGraw-Hill.

34 Headache Classification Subcommittee of the International Headache Society (2004). The international classification of headache disorders: 2nd edition. *Cephalalgia* 24 (Suppl 1): 9–160.

35 Pryse-Phillips, W. (2003). *Companion to Clinical Neurology*, 2e, 587. Oxford: Oxford University Press.

36 Piane, M., Lulli, P., Farinelli, I. et al. (2007). Genetics of migraine and pharmacogenomics: some considerations. *Journal of Headache and Pain* 8 (6): 334–339.

37 National Institute of Neurological Disorders and Stroke (NINDS). 2019. NINDS Multiple Sclerosis Information Page. https://www.ninds.nih.gov/Disorders/All-Disorders/Multiple-Sclerosis-Information-Page, (accessed 25 November 2019).

38 Compston, A. and Coles, A. (2008). Multiple sclerosis. *Lancet* 372 (9648): 1502–1517.

39 Compston, A. and Coles, A. (2002). Multiple sclerosis. *Lancet* 359 (9313): 1221–1231.

40 Murray, E.D., Buttner, E.A., and Price, B.H. (2012). Depression and psychosis in neurological practice. In: *Bradley's Neurology in Clinical Practice*, 6e (eds. R. Daroff, G. Fenichel, J. Jankovic and J. Mazziotta). Philadelphia: Elsevier/Saunders.

41 Kalia, L.V. and Lang, A.E. (2015). Parkinson's disease. *Lancet* 386 (9996): 896–912.

42 National Institute of Neurological Disorders and Stroke (NINDS). (2016). Parkinson's Disease Information Page. https://www.ninds.nih.gov/Disorders/All-Disorders/Parkinsons-Disease-Information-Page (accessed 25 November 2019).

43 Fisher, R., van Emde Boas, W., Blume, W. et al. (2005). Epileptic seizures and epilepsy: definitions proposed by the International League Against Epilepsy (ILAE) and the International Bureau for Epilepsy (IBE). *Epilepsia* 46 (4): 470–472.

44 Fisher, R.S., Acevedo, C., Arzimanoglou, A. et al. (2014). ILAE official report: a practical clinical definition of epilepsy. *Epilepsia* 55 (4): 475–482.

45 Taflinger, R. F., (1996) Taking ADvantage: The biological basis of human behavior, http://public.wsu.edu/~taflinge/biology.html (accessed 25 November 2019).

46 Marino, P. C. (2005) Biological rhythms as a basis for mood disorders, Http://www.personalityresearch.org/papers/marino.html (accessed 25 November 2019).

47 Revelle, W., Humphreys, M.S., Simon, L., and Gilliland, K. (1980). The interactive effect of personality, time of day, and caffeine: a test of the arousal model. *Journal of Experimental Psychology: General* 109: 1–31.

# 3

# Physical, Physiological, and Biological Measurements

## 3.1 Introduction

Nowadays, many of the physical, physiological, biological, and behavioural states of the human body can be measured, evaluated, and described by means of wearable sensors. These sensors can monitor the state of the human body for longer than an expert's observation. Often, the fusion of data modalities collected using different sensors is used for diagnostic purposes.

Although the physical state of the human body can be observed in detail using video cameras or in some cases listened to using microphones, such modalities are subject to breach of privacy, costly to deploy, and are less fascinating for automated analysis body movement. Therefore, in this chapter we ignore these two modalities and investigate the cases where humans can wear sensors for a longer time to enable long-term monitoring. Here, the most popular methods for measuring very common human body states are explained and the advanced approaches described. The details as well as experimental considerations are described in later chapters.

## 3.2 Wearable Technology for Gait Monitoring

As described in Chapter 2, many physical or mental diseases or abnormalities directly or indirectly affect human gait. Stroke, Parkinson's, and leg amputation readily come to mind. Thus, gait analysis can be used to monitor both the cause and the symptoms of a wide range of such abnormalities. The state of gait can be measured using a number of sensing modalities (video, audio, footstep, acceleration, gravitational force, directionality, etc.). Among them, acceleration measurement is reasonably accurate, robust, cheap, and easy to do. It has been well established that in an unrestricted environment the most widely used method for effective gait analysis is performed using an accelerometer. This sensor is often combined with a gyro and magnetometer in a small and compatible inertial measurement unit (IMU).

The measuring instruments for quantitative gait analysis have been integrated into human recognition as well as clinical decision-making systems for assessing pathologies manifested by gait abnormalities. Recent advances in wearable sensors, especially inertial body sensors, have paved the path for a promising future gait analysis [1]. Possibly the

*Body Sensor Networking, Design and Algorithms,* First Edition. Saeid Sanei, Delaram Jarchi and Anthony G. Constantinides.

Companion Website: www.wiley.com/go/sanei/algorithm-design

most important advantage of using gait sensors compared to gait observation using video cameras is that they allow the subject to enjoy the free-living environment over a long period while being monitored.

Despite some recent techniques in human recognition through gait analysis using wearable sensors, e.g. [2–4], a large number of current studies have been dedicated to patient monitoring, such as those with ankle fracture [5], fall injuries [6], osteoarthrosis [7], ataxic [8], multiple sclerosis [9], hip arthroplasty [10], geriatric [11], post recovery [12], and Parkinson's [13–16]. In addition, gait and posture monitoring for athletes has become a significant area of research in sport science [17, 18] mainly for enhancing the capability of athletes and preventing their injuries.

### 3.2.1 Accelerometer and Its Application to Gait Monitoring

Traditional optical motion capture systems, by means of video cameras, have long been used for human gait analysis, and some clinics and motion laboratories have adopted them as a classic way of motion monitoring. These systems include a stationary system such as a treadmill surrounded by a number of video cameras. Obviously, the main limitation for using such sensors is that the subjects have to move and perform in a limited space within the laboratory so they can be captured by a number of cameras installed around that region. In addition, although such systems can perform highly accurate human motion analysis, they are a relatively expensive platform and require expert operation [19]. Finally, for patient monitoring and assistive technology, such systems are subject to privacy breach and therefore have limited application, particularly when the system is to be deployed for monitoring a wider community of patients, athletes, and so on.

In [20] a comprehensive review of accelerometers has been provided. Presently, by itself or as part of an IMU, accelerometers find applications in consumer devices such as digital cameras, health trackers, smartphones, cars, and interactive computer games. In less prominent applications, accelerometers are used in some industrial environments, such as for measuring vibrations between motors and their mountings and for measuring tilt. All these require accelerometers to provide higher sensitivity, robustness, and smaller size at a lower cost. As a result, accelerometers have proven very useful for *in vivo* movement analysis.

Sensor deployment to both environmental and human body communication networks has led to many important applications. For example, the method of placement and attachment of the sensor to human body for optimal sensing and analysis of the captured data has to be studied and taken into account in the design of the whole system.

Commonly used accelerometer types are piezoelectric, thermal, and capacitive, the last one is usually fabricated using micro-electromechanical sensor (MEMS) technology, which is also the lowest in cost because of so-called economies of scale.

A piezoelectric accelerometer exploits the piezoelectric effect of certain materials to measure dynamic changes in mechanical variables. Figure 3.1 shows the operation mechanism of such a device [21].

Unlike piezoelectric or capacitance type MEMS accelerometers, which use solid mass structures, the thermal method uses heated gas and thermocouples, and thus has no moving parts. It measures the temperature difference between the cooler and the warmer air [22].

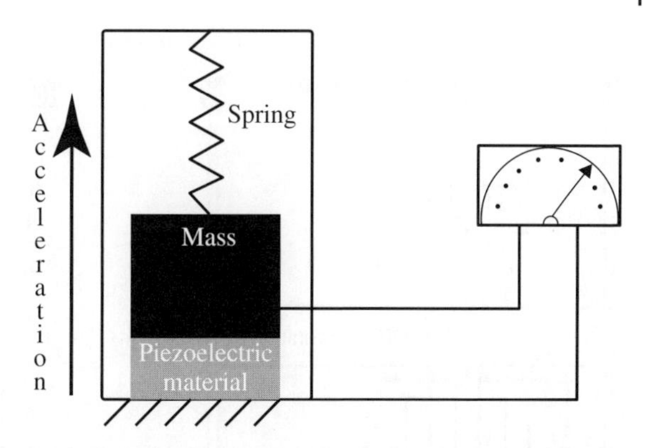


**Figure 3.1** The mechanism of a piezoelectric accelerometer.

Currently, most work into the simulation of accelerometers deals with verifying a design at the semiconductor fabrication level.

There are three major challenges in the design and use of accelerometers. The first is the effect of drift or change in the internal mechanical or electrical properties, which can manifest itself as a bias or an offset in readings. The second is noise from amplified microscopic mechanical motions, which needs to be reduced if not eliminated. The third is the effect of gravity, which is ever present. While this is strictly not a defect, we have to consider that the gravitational force vector is projected and superimposed along the axes of sensitivity of the accelerometer. Thus, the movements along these axes experience a confounding gravitational effect. Furthermore, with movements at frequencies beyond 10 kHz, the internal mechanical parts of an accelerometer move nonlinearly, resulting in dynamic errors.

#### 3.2.1.1 How Accelerometers Operate

A capacitive accelerometer measures acceleration by changes in the internal capacitance of the device. These devices are typically fabricated using MEMS technology on microstructures built into polysilicon. Some microstructures are fixed and some are movable, suspended from fixed points. By impressing a voltage between them, a capacitive effect arises from the electronic charge stored in these structures which is proportional to the area and the physical distances between these structures. External physical movements cause the distances between the microstructures to change, which in turn results in changes to the capacitance. These variations eventually cause changes to the voltage.

Acceleration has two main components: the first being the inertial acceleration that is corresponding to the changes in speed, resulting from rotation or translation, or both. The second component consists of gravitational acceleration arising from the microstructures being deflected in proportion to their static orientation to the gravitational field. Removing the confounding effect of this kind of acceleration [23] can be cumbersome.

In the simplified diagram presented in Figure 3.2 the parameter changes corresponding to acceleration happen between a proof mass to which some plates are attached. The proof mass is suspended from the body or frame of the accelerometer device and secured through anchor points. Another set of plates which are fixed to the frame and the changes in the distance between the two induce a change in the voltage between the plates. When the accelerometer frame moves, the inertia of the proof mass induces a reactionary force which is applied to the springs (often fabricated from polysilicon). These springs are affixed to the

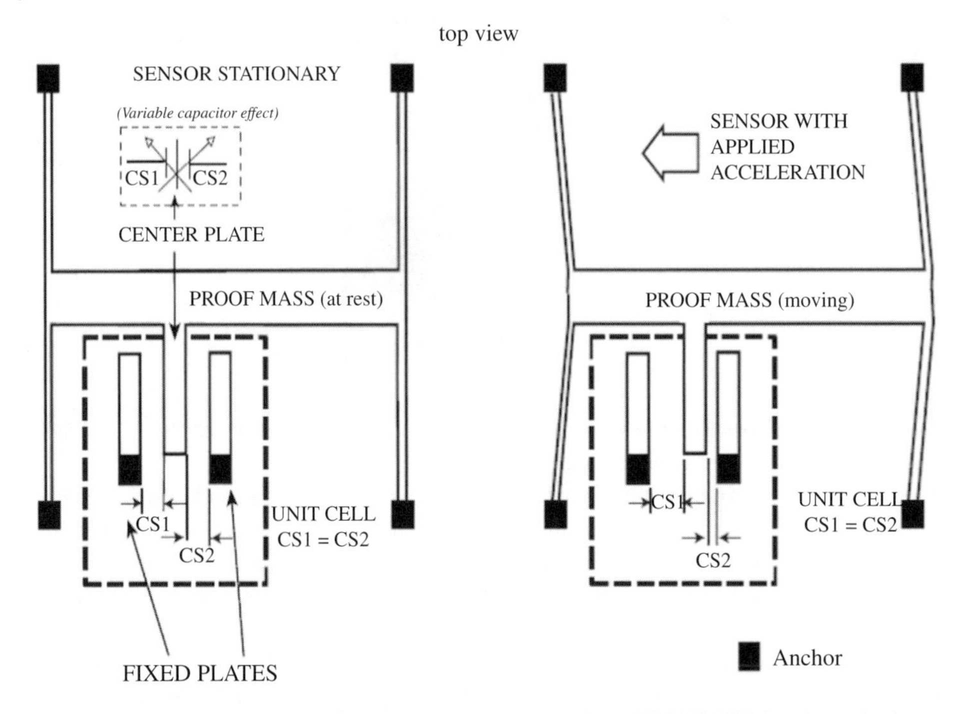


**Figure 3.2** Top view of a simplified accelerometer sensor from ADXL50 [49] datasheet. Anchor points move with the device frame, while inertia of the proof mass causes the distance between fixed plates to vary, changing the capacitance [20]. Source: Courtesy of Jarchi, D., Pope, J., Lee, T.K.M., Tamjidi, L., Mirzaei, A. and Sanei, S.

frame of the device at anchor points. The spring deformation is assumed to be linear so that the accelerometer obeys the familiar mass–spring–damper equation derived from Newton's and Hooke's Laws [24]:

$$M\frac{d^2\mathbf{E}}{d^2t} = M\frac{d^2\mathbf{D}}{d^2t} + b\frac{d\mathbf{D}}{dt} + k\mathbf{D} \tag{3.1}$$

where $M$ is the mass of proof mass, **E** and **D** are the distances travelled by the proof mass with respect to the earth and to its anchors respectively, $b$ is the damping factor, and $k$ the spring constant. The differentiation is with respect to time $t$. In addition, there are electrostatic forces between the fixed and moving plates to be considered. In order to measure the speed of movement, consider the relative motion between the stationary plates and the moving proof mass which changes the capacitance between them.

However, to measure the acceleration, the transfer of charge is amplified electronically by mixing with a high-frequency signal using a double sideband suppressed carrier modulation technique. Finally, the amplitude-modulated signal is synchronously demodulated and amplified.

There are several ways to evaluate the movement resulting from the changes in capacitance. Early methods used closed loop feedback by impressing a voltage on the capacitor plates so the proof mass stays at its original position [25]. Changes in this voltage thus reflect the change in position. More recent methods are open loop, where the capacitance is obtained using switched-capacitor techniques.

Despite the widespread use of accelerometers, there are limitations of current devices which prevent them from being used in even more applications. A prime example is that of deriving the speed and distance moved by performing mathematical integration of the measured acceleration. In practice, the measurements drift and the integration causes accumulation of these errors rendering the readings useless. To overcome this, it is necessary to periodically recalibrate the accelerometer readings. One popular method is zero-velocity update point (ZUPT). Other schemes depend on the instances where an external event indicates an instantaneous null in the movement pattern, for example in between footsteps [25].

Another source of accelerometer disturbance is the noise arising from mechanical sources due to the proof mass being subjected to Brownian motion. Nevertheless, electronic sources themselves generate considerable noise. The conversion of minute capacitance changes to usable voltages requires high electronic gain and, with that, an increase in noise. A widely used method benefits from switched-capacitor techniques, and the oscillation frequency can be superimposed to the main signal, resulting in an aliasing effect. This may be worsened when using chopper amplification with synchronous demodulation, often applied to moderate the effect of drift.

#### 3.2.1.2 Accelerometers in Practice

Accelerometers have gained even wider use by being part of consumer devices like smartphones and tablets. It is useful to differentiate between such installed units and those available as standalone units, where the accelerometer integrated circuit has settings for force sensitivity. The external components may filter part of the operating frequencies too. These units are small in size and consume little current but need to have proper packaging.

By way of contrast, in ready-to-use consumer devices, there is no control over how an accelerometer is mounted in its physical environment. This affects its operating conditions in terms of temperature, humidity, and electronic interference. In addition, inevitable smoothing of the measurements by means of digital filtering and processing may introduce further latencies. Nevertheless, in many situations, in order to achieve device independence [26], the accelerometer is treated as a 'black box', where the digital readout is assumed to be a measure of the acceleration. Then, the imperfections are treated statistically and data calibration would need to be employed for demanding applications. In spite of this, the use of accelerometers in consumer devices has proven to be adequate for many clinical purposes, mainly due to their compatibility with many computing systems and microcontrollers and also their flexibility in terms of price, size, and application as seen in various publications, such as [27]. In such applications the measurements are compared with gold standard equipment like motion capture cameras and accelerometers built in instruments designed for clinical use.

The use of accelerometry for gait analysis has been significantly increased due to their ease of use, portability, compatibility, and the capability to be integrated into low-powered wireless embedded platforms. However, in addition to accelerometers, there are other popular existing wearable technologies for gait analysis which are briefly described here.

### 3.2.2 Gyroscope and IMU

The gyroscope is arguably the next most commonly used motion detector after the accelerometer. It can be attached to the feet to measure the angular velocity of the foot for detecting different gait phases [28]. The MEMS technology used in accelerometers is used to drive down the cost of gyroscopes, so they are often featured alongside accelerometers in many devices and in IMUs. In contrast to accelerometers, which can directly measure linear acceleration, gyroscopes measure the angular movement about a given axis.

Accelerometers by themselves can measure angular rotation but they cannot give as good a result as gyroscopes, as shown in [29]. Thus, the gyroscope can be used to correct the accelerometer readings, or have its output fused with those of accelerometer when deployed together, such as in an IMU.

The miniaturisation technologies for fibre optic and MEMS progresses quickly. Except for gyroscopes working on optical principles, what the other types have in common is a mass that is constantly moving within the device in order to measure the angular motion. This motion causes the gyroscope to consume more current than an accelerometer. An example of MEMS gyroscope technology is that based on Coriolis acceleration [30]. As opposed to centrifugal acceleration, which is always present in rotation, Coriolis acceleration occurs whenever there is motion along the radius of the rotation.

Figure 3.3 shows the schematic of a single axis of a gyroscope which rotates clockwise, together with its semiconductor substrate. The gyroscope has capacitive fingers fabricated as part of its structure [20]. To this substrate a frame with a set of capacitive fingers is tethered with springs. In addition, a mass, also tethered to this frame with springs, is driven into mechanical resonance and constrained to move in one direction. On the left of the figure,

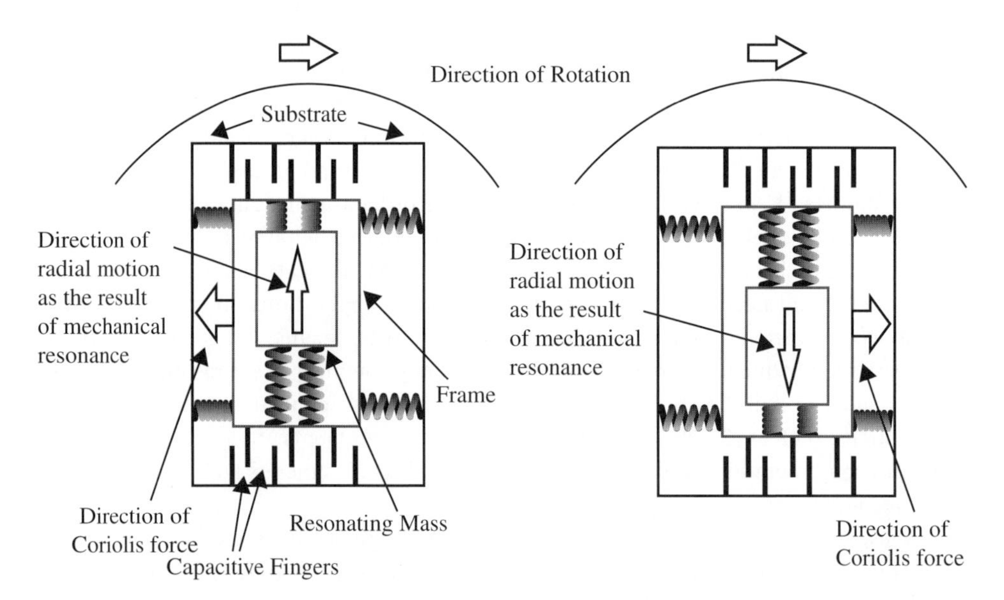


**Figure 3.3** A simple diagram of a gyroscope rotating clockwise. On the left, resonating mass is moving upwards, whereas, on the right, it is moving downwards. Direction of Coriolis force is also shown [20]. Source: Courtesy of Jarchi, D., Pope, J., Lee, T.K.M., Tamjidi, L., Mirzaei, A. and Sanei, S.

the mass is considered as it moves to the top of the figure along the radius of rotation. The Coriolis force acts on the frame, which deflects to the left as shown. On the right of the figure, as the mass resonates it moves to the bottom of the figure; in this case the Coriolis force causes a deflection to the right. The varying distance between the capacitive fingers is picked up as a voltage representing a measure of the angular speed. From [29], it can be shown that the displacement of the frame relative to the substrate is [20]:

$$\mathbf{D} = 2(\Omega \times \mathbf{v})M/k$$

where $\mathbf{\Omega}$ is the angular velocity of the gyroscope (towards the right) and $\mathbf{v}$ the velocity of the resonating mass along the radius of rotation, $M$ is mass, $k$ is the spring constant of the frame-substrate springs. By driving the resonating mass at a known frequency, its changing displacement $\mathbf{D}$ induces a voltage from which the angular speed can be derived reasonably linearly. However, as with MEMS devices, the problems of drift, noise, and other artefacts remain. An IMU can be a combination of both accelerometers and gyroscopes (and a magnetometer in a 9D version) that measures rate, angle, and direction of motion. IMUs can be attached to any part of the body and are effectively used in gait analysis. Recent IMUs have built-in magnetometers as well that sense gravity and can be used in fall detection.

### 3.2.3 Force Plates

Force plates measure vertical ground reaction forces (GRFs) applied by gait during walking. A force platform can be integrated under the moving belt of a treadmill or under the entire treadmill [31]. Force platforms are expensive and have some limitations as the plates are installed within a short space and need to remain stationary. These make their applications in free space difficult. Nevertheless, there are force or pressure sensors that can be accommodated inside or under shoes to monitor the walking steps of humans or help correct the posture of athletes.

### 3.2.4 Goniometer

A goniometer is used to track the angle changes and is used for angle measurement in gait analysis since it is flexible and can rotate proportionally to the joint angle being measured [32]. It is particularly useful for the analysis of ranges of motion. Using a goniometer, it is possible to determine the range of knee joint angular movement to monitor patients with knee injuries. An optical fibre-based goniometer has been introduced in [33].

### 3.2.5 Electromyography

Electromyography (EMG) is used to measure both physical (such as gait) and physiological activities (such as muscle diseases), as can be seen in Section 3.3. It represents the electrical potentials of neurons within the muscles and can be recorded using surface electrodes or by wires or needles inserted into the selected muscle of a lower extremity [34]. An understanding of muscle activity in gait can be obtained by processing the EMG signals. EMG provides feasibility of analysing relative contribution of the superficial muscles during movement. It records the activity of underlying motor units and plays an important role in clinical gait analysis for assessing walking performance of the people with gait impairments using muscle activity information, such as timing of the muscle activity and muscle strength.

### 3.2.6 Sensing Fabric

The goal in sensing fabric-based technology is the integration of sensors, communication components, and the processing elements into the fabric. The most common types are pressure sensors, including piezoelectric, piezoresistive, resistive, and capacitive sensors [35]. These sensors can be networked in a carpet to record the step pressure from both feet. In a more useful design, however, the pressure sensors and their associated electronics and wireless communication system, which does fit completely inside the shoe, is very demanding for the long-term monitoring and recordings of daily activities. Therefore, to make a valuable gait analysis platform, the sensing fabric technology has been directed towards the development of pressure-sensitive foot insoles with wireless communication capability [36].

## 3.3 Physiological Sensors

These sensors sense and capture the information emitted from the body inherently due to physiological or metabolic changes. These can be due to normal or abnormal human states. They can also be due to external effects such as viewing intriguing scenes, temperature change, and various physical activities.

### 3.3.1 Multichannel Measurement of the Nerves Electric Potentials

Electroencephalography (EEG), EMG, and electrocardiography (ECG) are probably the most common types of physiological measurement systems. They are widely available in the related clinical departments. In terms of operations, they measure, respectively, the electrical activities of nerves in the brain, muscles, and heart. Technically, low-noise differential amplifiers, with high amplification gains and large input impedance, are used to capture tiny variations in the voltage between each two electrodes or with respect to a common reference over the body. A typical three-stage differential amplifier is depicted in Section 3.5. Fundamental electronic components, such as operational amplifiers, resistors, capacitors, and diodes, are used to make this device. EEG and ECG (or even EMG) often involve a large number of channels. Nevertheless, due to their low bandwidth, the signals are often sampled in a low rate, which makes their real-time processing possible (Figure 3.4).

The EEG systems with their electrodes mounted over the cortex are called electrocorticography (ECoG) systems and those for measuring deep brain signals from deeper areas of the brain, are called intracranial EEG. In some applications, such as recording the neuron activities from the hippocampus, the measurement can be done by using multichannel electrodes inserted through human foramen ovale holes from both sides of the face. Figure 3.5a and b shows an ECoG and foramen ovale (FO) implant for deep brain source measurements.

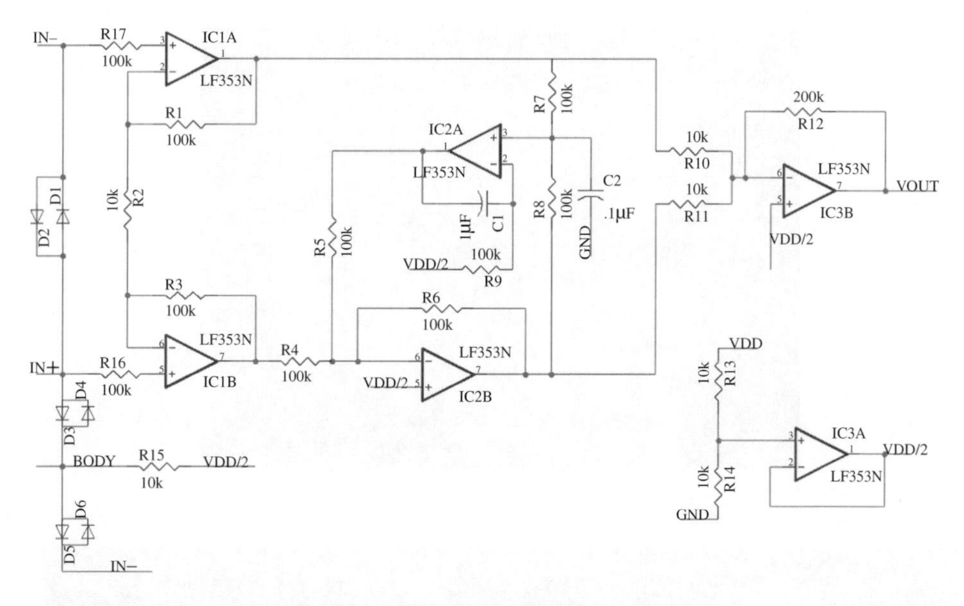


**Figure 3.4** A simple EEG differential amplifier used in EEG or EMG systems.

As another brain scanning modality, magnetoencephalography (MEG) is used to measure the magnetic fields produced by the electric currents in the brain. MEG and EEG look very similar. The advantage of MEG over EEG is its insensitivity against the changes in the tissue density, so it can be effectively used for more accurate brain source localisation. However, the system is complex, bulky, expensive, and not easily available for long-term patient monitoring, and requires a high maintenance cost.

ECG (also called EKG) is used to measure the electrical activity of the heart muscle nerves. The activity stems from pumping the blood through the right supraventricular down into the right ventricle and circling upward from the left ventricle to the left supraventricular and pumping into the arteries. Often 10–14 electrodes are used with reference to the arm, wrist or foot to capture the state of different heart sections.

The systems for measuring electrical activity of the body don't involve any time delay and therefore processing of EEG, MEG, EMG, and ECG doesn't suffer any fading, clutter, or time overlapping of the signals. On the other hand, the effective frequency range falls below 500 Hz, which requires low processing power and bandwidth.

There have been efforts to miniaturise these sensors with wireless capability. As an example, Figure 3.6 shows a new patch type ECG system which includes wireless connection to remote devices.

On the other hand, heart and lung sounds can be heard and recorded using electronic stethoscopes. They are sensitive microphones isolated from the surrounding environment when placed on the skin around the heart or back side of the body, so the ambient noise is greatly reduced. Most heart abnormalities as well as problems in the lung which cause wheezing or crackling can be easily heard and recorded through these systems. In some recent models, such as Modell 3200 by the Littman company (Figure 3.7), the stethoscope

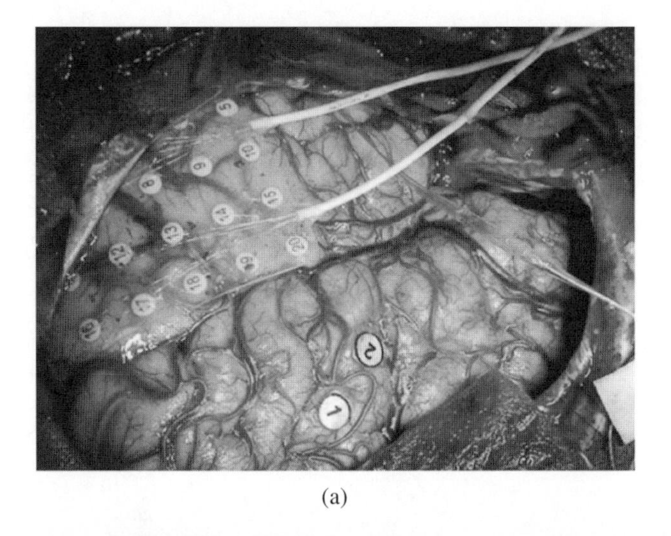

(a)

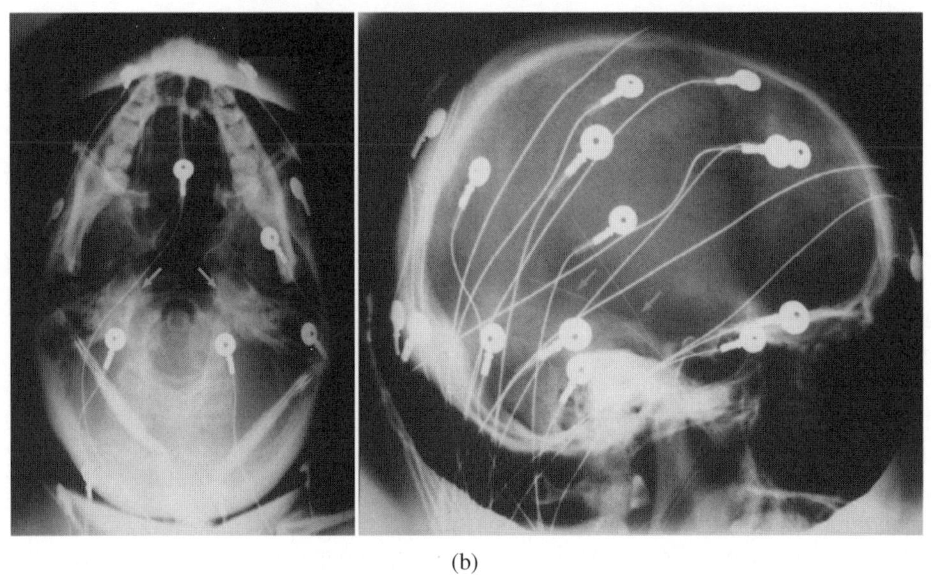

(b)

**Figure 3.5** (a) ECoG and (b) foramen ovale electrodes denoted by pointers. In this setup, a scalp EEG has also been used. (*See color plate section for color representation of this figure*)

can also transmit sound signals via Bluetooth to the remote device for observation, archiving, and analysis.

One of the challenging problems in assessing the recorded stethoscope data is separation of heart and lung sounds for automatic diagnosis. The difficulty stems from the fact that sound signals are subject to delay and for their separation a robust convolutive source separation technique is required. Although many solutions have been proposed by signal processing researchers [37–40], design of a clinically usable system which can effectively separate these two sound signals is still being researched.

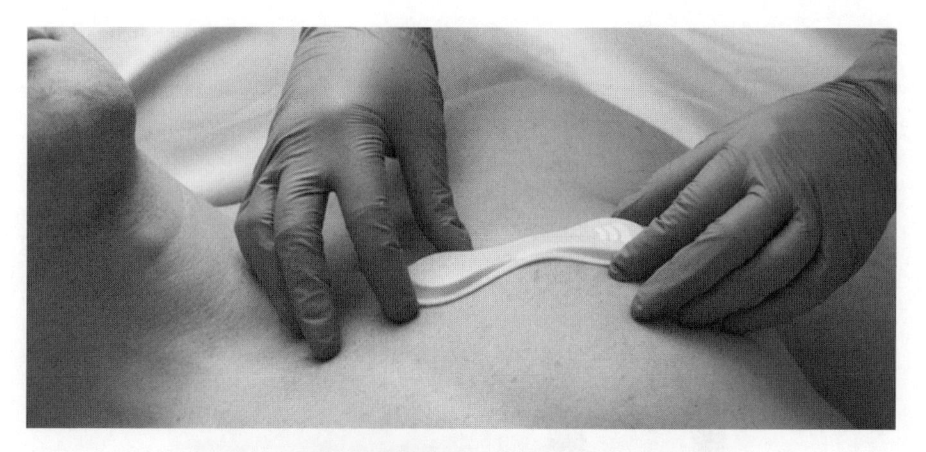

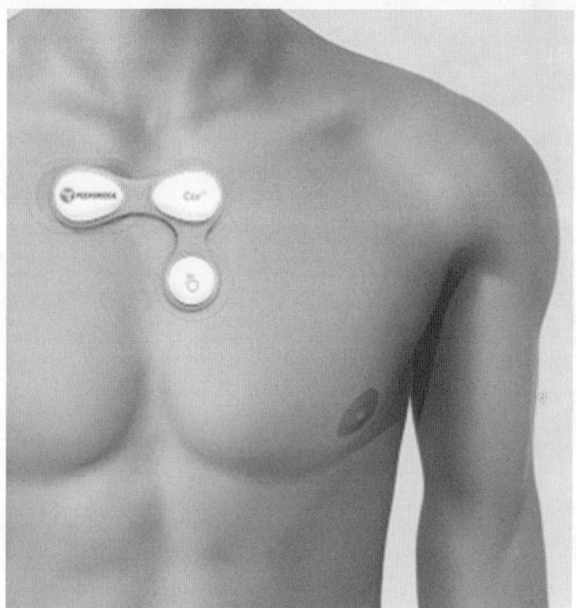

**Figure 3.6** New patch type ECG systems including electrodes and wireless connection.

### 3.3.2 Other Sensors

Oximeters are used to measure the blood oxygen saturation level ($SpO_2$). The traditional measure of $SpO_2$, called photoplethysmography, involves using two light beams in red and infrared wavelengths each sensitive to high and low blood oxygen levels respectively.

In principle, oximeters can operate in either transmissive or reflective mode. Those using an earlobe or finger measure the absorbed transmitted light through the tissue and others such as those used on the wrist work based on the reflective light by the tissue layers including the blood.

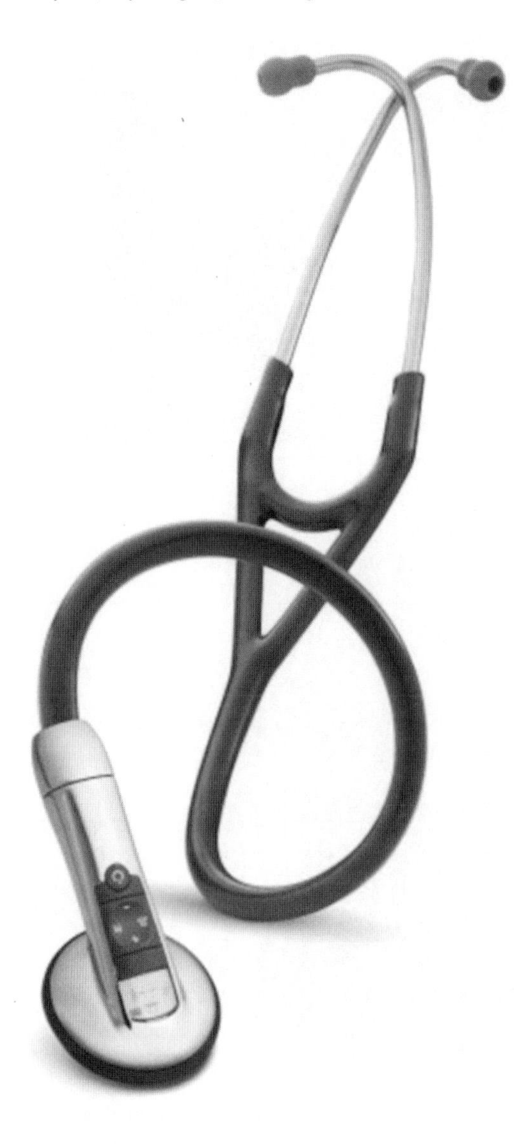

**Figure 3.7** Electronic stethoscope; Littmann 3200 model.

In general, the $SpO_2$ value is linearly dependent on the ratio between the transmitted infrared to red lights and decreases with this ratio. Figure 3.8 shows oximeters used in different modes of oximetry.

It is logical to accept that the oximeter reading is also an indicator of heart rate as the blood oxygen changes following the heart pumping cycle. Figure 3.9 shows two different oximeters with different recording modalities. The most important application of oximeters is in hospital intensive care units (ICUs) to ensure that the patient has normal breathing and heartbeat.

Amongst the other physiological measures, respiratory rate (RR) is important for the diagnosis of lung diseases and many other internal abnormalities. Therefore, it has an important role in healthcare and sport. In hospitals it is measured simply by the manual counting of

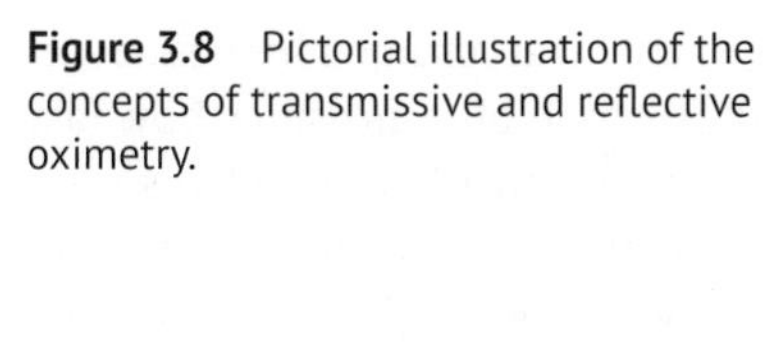

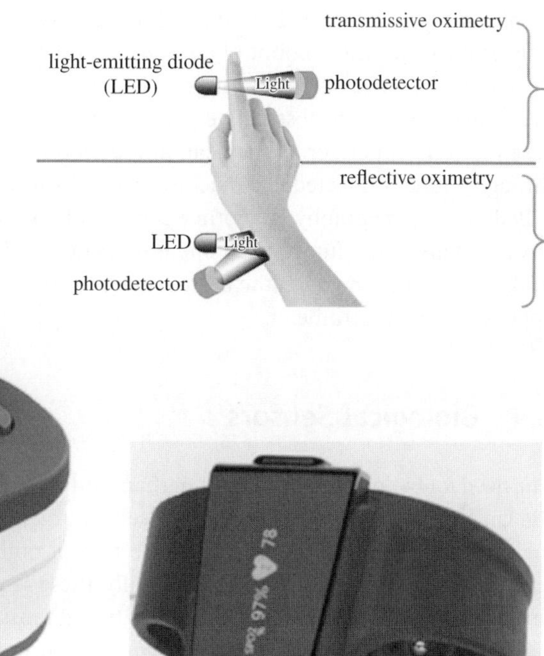


Figure 3.8 Pictorial illustration of the concepts of transmissive and reflective oximetry.

**Figure 3.9** Two different models of oximeters: (a) digital pulse oximeter used on the finger and (b) wireless wrist pulse oximeter, both from Viatom.

breaths, conventional impedance pneumography, or an optical technique called end-tidal carbon dioxide (et-$CO_2$).

The impedance pneumograph is a bioimpedance recorder for indirect measurement of respiration. Using superficial thoracic electrodes, the system measures respiratory volume and rate through the relationship between respiratory depth and thoracic impedance change [41].

In the et-$CO_2$ system, a device called a capnometer uses a small plastic tube inserted in the patient's mouth. The capnometer tube accumulates the expelled $CO_2$ in each breath. An infrared light is then emitted to the $CO_2$ from a moving source. Following absorption of infrared radiation overtime, a variation with breathing rate will be induced. The et-$CO_2$-based systems are used in hospitals mainly for patients in critical conditions often within ICUs as well as ambulances for emergency checking of patient states. The capnographs, which look approximately like square waveform in normal conditions, significantly change during intubation, after cardiac arrest, ventilation problem, shock, emphysema, or leaking alveoli in pneumothorax, hypoxia due to asthma or mechanical obstruction, poor lung compliance, obese, and pregnancy [42].

Spirometry used for respiration monitoring is also one of the most common lung function tests. It measures the amount of air one can inhale and exhale. It also measures how fast the air can be emptied out of lungs. It is used to help diagnose breathing problems such as asthma and chronic obstructive pulmonary disease (COPD). During the test, the subject breathes in as much air as they can and then quickly blows as much air out as possible through a tube connected to a machine called a spirometer.

Body plethysmography is another common lung function test. It measures how much air is actually in the lungs when one inhales deeply. It also checks how much air remains in the lungs after they breathe out as much as they can. A plethysmograph has the same applications as a spirometer.

## 3.4 Biological Sensors

The need for biological sensors was perhaps initially raised by the Department of Defense in the United States. In 2004 they announced an urgent need for the development of sensors for early-warning systems and the protection of military forces against potential chemical and biological attacks [43]. Traditionally, these sensors are particular types of chemical sensors and benefit from the high selectivity and sensitivity of biologically active materials.

Such sensors produce an electrical signal proportional to the concentration of a specific chemical or set of chemicals in the human body. Therefore, they should be biocompatible with and have no toxic effect on humans. However, current biosensors are more than a chemical sensor and are often used in biomedical applications.

The biosensors generally consist of a transducer, biological-recognition membrane in intimate contact with the transducer, and biologically active material, which is sensitive to the analyte molecule through a shape-specific recognition. Biosensors are generally divided into two categories: affinity-based and metabolic.

### 3.4.1 The Structures of Biological Sensors – The Principles

A biosensor consists of a biological sensing element and a transducer. The sensing elements can be organisms, tissues, cells, organelles, membranes, enzymes, receptors, antibodies, or nucleic acids. On the other hand, transducer types can be electrochemical, optical, calorimetric (thermal), or acoustic.

*Bio-affinity transducers* rely on very strong binding through which the transducer detects the bound receptor-analyte pair. The most common bio-affinity recognition processes are receptor-ligand binding and antibody-antigen binding.

*Metabolic biosensors*, on the other hand, sense the instances where the analytes are bound or the co-reactants are chemically altered, which in turn product molecules are formed, and the transducer detects the changes in concentration of the product molecules or co-reactants and consequently heat is released by the reaction. The common biometabolic processes are enzyme-substrate reactions and metabolism of specific molecules by organelles, tissues, and cells.

Biological recognition elements are immobilised on the surface of a transducer or in a membrane. This requires a bioreactor on top of a traditional transducer. The response of a biosensor is determined by four different factors: diffusion of the analyte, reaction products, co-reactants or interfering chemical species, and kinetics of the recognition process.

Various kinds of biosensors are mainly based on:

- chemically sensitive semiconductor devices;
- thermistors;
- chemically mediated electrodes;
- surface acoustic waves devices;
- piezoelectric microbalances;

and many other individual or hybrid systems such as MEMS. As an example, for a quartz crystal microbalance sensor depicted in Figure 3.10 [44]:

- an electrical AC voltage causes the resonator (i.e. the piezo layer) to oscillate;
- a target molecule binds with a receptor based on the lock-and-key principle;
- resonance frequency changes because of the weight change;
- frequency change is translated into an electrical signal and processed further.

Some examples of the transducers used in the design of biosensors are micro-electrodes, combined ion-selective field-effect transistors and micro-electrodes, fibre optodes and luminescence, thermistors and thermocouples, and surface acoustic wave (SAW) delay lines (often used as piezoelectric elements in sensors and modelled using transversal filters) and bulk acoustic wave microbalances.

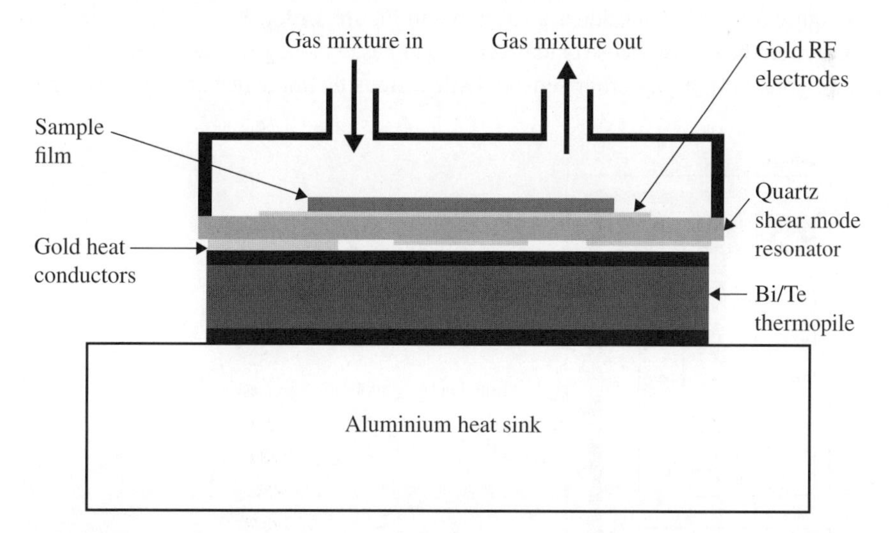


**Figure 3.10** A quartz crystal microbalance sensor; a thin film sample is coated on the top gold electrode of the quartz shear mode resonator (QCM). The QCM is then inserted between the thermopile and the sample chamber. The experiment consists of varying the composition of the gas mixture at constant temperature and observing changes in the resonant frequency and motional resistance of the QCM and the thermal power flowing between the QCM and the aluminium heat sink via the thermopile.

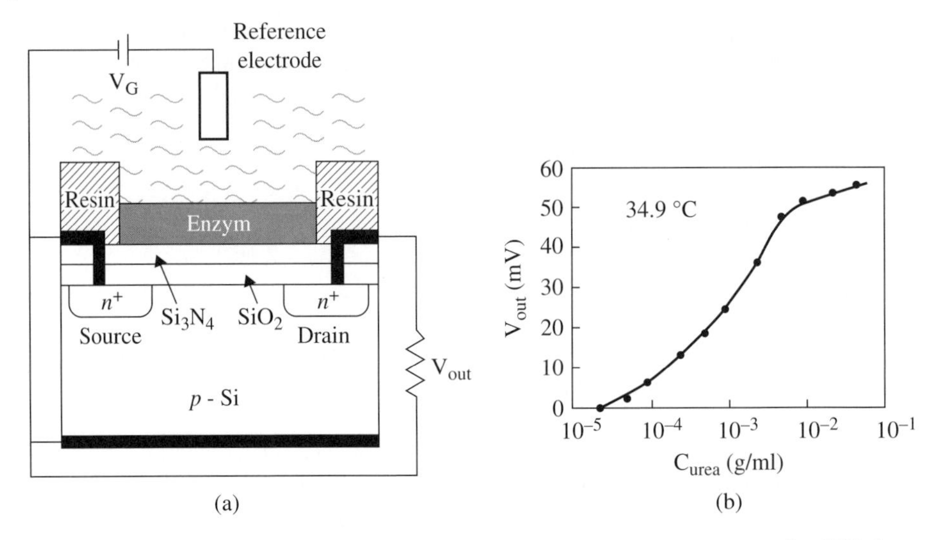


**Figure 3.11** (a) An EnzymFET and (b) its electrical output versus urea concentration [50]. Source: Courtesy of Middelhoek, S., and Audet, S.A.: Silicon Sensors, Academic Press Limited.

As an example, a schematic of a semiconductor sensor is illustrated in Figure 3.11. The back side is connected to a urea sensitive ISFET, often called an ENFET (EnzymFET), similar to that shown in Figure 3.11a. In Figure 3.11b the electrical voltage with respect to urea concentration is shown.

These sensors are carefully packaged using biocompatible materials. One of these types is shown in Figure 3.12. The backside is also shown in Figure 3.13.

Since 1962, when the early glucose oxidase electrode-based biosensor was manufactured by Clark and Lyons [45], many other sensors with various technologies and applications

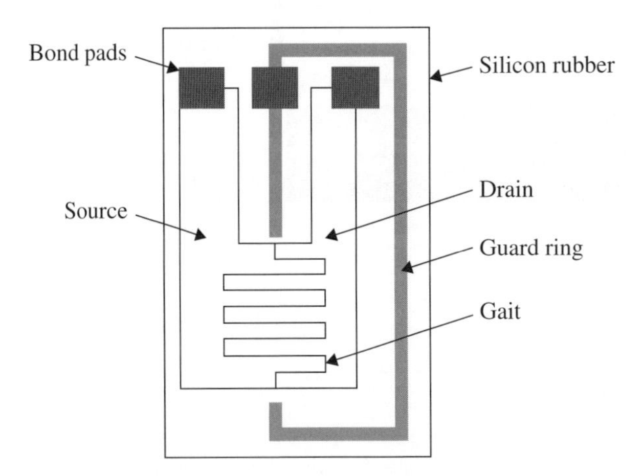


**Figure 3.12** A semiconductor biosensor schematic [51]. Source: Courtesy of John Wiley & Sons.

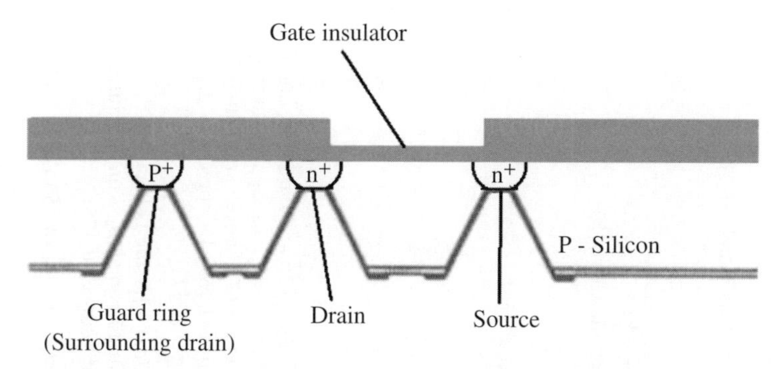


**Figure 3.13** A semiconductor biosensor schematic; backside contacts [51]. Source: Courtesy of John Wiley & Sons.

have been developed. Vigneshvar et al. [46] present a comprehensive study of biological sensors. Table 3.1 shows a large set of biosensors with their applications including the involved technology. Table 3.2 lists the uses of different biosensors for the diagnosis of various diseases [46].

### 3.4.2 Emerging Biosensor Technologies

There have been many emerging technologies recently, including those built in mobile handsets for measuring human biological metrics. As an example, a health app which can monitor people's glucose levels without breaking the skin has been developed recently. The so-called Epic app can help people find out if they develop diabetes and need to make lifestyle changes to avoid it. The app can also tell people about their respiration and blood oxygen saturation. SMBG (self-monitored blood glucose) is recommended for all people with diabetes [47, 48]. To alleviate the intrusiveness of the sensors, there is a great appetite among researchers to design contactless sensors. Some of these sensors are introduced in other chapters of this book.

## 3.5 Conclusions

Sensor technology devices have become more miniaturised, wearable, user-friendly, cost effective, less intrusive, and inclusive, and are often accessible to outpatients and individuals. Human vital signs as well as abnormalities can be captured invasively or noninvasively (and in some cases nonintrusively) by various sensor modalities, some packed together in one package. The new wireless technology in parallel with the advances in high-speed computing systems provides more accurate and accessible health-monitoring systems. Physical, physiological, biological, and other sensor types are fast developing and allow full body screening in all times. Emerging sensor technologies provide constant feedback to individuals and create a safer and healthier world.

**Table 3.1** Biosensors, their principle, applications, and bibliography.

| No. | Type | Principle | Applications | Bibliography |
|---|---|---|---|---|
| 1. | Glucose oxidase electrode-based biosensor | Electrochemistry using glucose oxidation | Analysis of glucose in biological sample | Clark and Lyons [45] |
| 2. | HbA1c biosensor | Electrochemistry using ferroceneboronic acid | Robust analytical method for measuring glycated haemoglobin | Wang et al. [52] |
| 3. | Uric acid biosensor | Electrochemistry | For detection of clinical abnormalities or diseases | Erden and Kilic [53] and Kim et al. [54] |
| 4. | Acetylcholinesterase inhibition-based biosensors | Electrochemistry | Understanding pesticidal impact | Pundir and Chauhan [55] |
| 5. | Piezoelectric biosensors | Electrochemistry | Detecting organophosphate and carbamate | Marrazza [56] |
| 6. | Microfabricated biosensor | Optical/visual biosensor using cytochrome P450 enzyme | For drug development | Schneider and Clark [57] |
| 7. | Hydrogel (polyacrylamide)-based biosensor | Optical/visual biosensor | Biomolecular immobilisation | Khimji et al. [58] |
| 8. | Silicon biosensor | Optical/visual/ fluorescence | Bioimaging, biosensing, and cancer therapy | Peng et al. [59] and Shen et al. [60] |
| 9. | Quartz crystal biosensor | Electromagnetic | For developing ultrahigh-sensitive detection of proteins in liquids | Ogi [61] |
| 10. | Nanomaterials-based biosensors | Electrochemical or optical/visual/fluorescence | For multifaceted applications including biomedicine, e.g. diagnostic tools | Li et al. [62], Kwon and Bard [63], Zhou et al. [64], Guo [65], Hutter and Maysinger [66], Ko et al. [67], Senveli and Tigli [68], Valentini et al. [69], Lamprecht et al. [70], and Sang et al. [71] |
| 11. | Genetically encoded or fluorescence-tagged biosensor | Fluorescence | For understanding biological process including various molecular systems inside the cell | Randriamampita and Lellouch [72], Oldach and Zhang [73], Kunzelmann et al. [74], and Wang et al. [75] |
| 12. | Microbial fuel-cell-based biosensors | Optical | To monitor biochemical oxygen demand and toxicity in the environment and heavy metal and pesticidal toxicity | Gutierrez et al. [76] and Sun et al. [77] |

**Table 3.2** Use of biosensors in disease diagnosis.

| No. | Biosensor(s) | Disease diagnosis or medical applications |
|---|---|---|
| 1. | Glucose oxidase electrode-based biosensor and HbA1c biosensor | Diabetes |
| 2. | Uric acid biosensor | Cardiovascular and general disease diagnosis |
| 3. | Microfabricated biosensor | Optical corrections |
| 4. | Hydrogel (polyacrylamide)-based biosensor | Regenerative medicine |
| 5. | Silicon biosensor | Cancer biomarker development and applications |
| 6. | Nanomaterials-based biosensors | For therapeutic applications |

## References


**1** Chen, S., Lach, J., Lo, B., and Yang, G.-Z. (2016). Toward pervasive gait analysis with wearable sensors: a systematic review. *IEEE Journal of Biomedical and Health Informatics* 20 (6): 1251–1537.

**2** Lee, T.K.M., Belkhatir, M., and Sanei, S. (2014). A comprehensive review of past and present vision-based techniques for gait recognition. *Multimedia Tools and Applications* 72 (3): 2833–2869.

**3** Lee, T.K.M., Sanei, S., and Belkhatir, M. (2011). Combining biometrics derived from different classes of nonlinear analyses of fronto-normal gait signals. *IARIA International Journal of Advances on Networks and Services* 4 (1–2): 232–243.

**4** Lee, T.K.M., Belkhatir, M., Lee, P.A., and Sanei, S. (2008). Nonlinear characterisation of fronto-normal gait for human recognition. In: *Advances in Multimedia Information Processing – PCM 2008*, Lecture Notes in Computer Science (eds. Y.-M.R. Huang et al.), 466–475. Berlin: Springer-Verlag.

**5** Elbaz, A., Mor, A., Segal, G. et al. (2016). Lower extremity kinematic profile of gait of patients after ankle fracture: a case-control study. *Journal of Foot and Ankle Surgery* 55 (5): 918–921.

**6** Ihlen, E.A., Weiss, A., Beck, Y. et al. (2016). A comparison study of local dynamic stability measures of daily life walking in older adult community-dwelling fallers and non-fallers. *Journal of Biomechanics* 49 (9): 1498–1503.

**7** Tadano, S., Takeda, R., Sasaki, K. et al. (2016). Gait characterization for osteoarthritis patients using wearable gait sensors (H-Gait systems). *Journal of Biomechanics* 49 (5): 684–690.

**8** Chini, G., Ranavolo, A., Draicchio, F. et al. (2017). Local stability of the trunk in patients with degenerative cerebellar ataxia during walking. *Cerebellum* 16 (1): 26–33.

**9** Gong, J., Lach, J., Qi, Y., and Goldman, M.D. (2015). Causal analysis of inertial body sensors for enhancing gait assessment separability towards multiple sclerosis diagnosis. In: *Proceedings of the 2015 IEEE 12th International Conference on Wearable and Implantable Body Sensor Networks*, 1–6. IEEE.

**10** Rapp, W., Brauner, T., Weber, L. et al. (2015). Improvement of walking speed and gait symmetry in older patients after hip arthroplasty: a prospective cohort study. *BMC Musculoskeletal Disorders* 16 (1): 291–298.

**11** Rampp, A., Barth, J., Schülein, S. et al. (2015). Inertial sensor-based stride parameter calculation from gait sequences in geriatric patients. *IEEE Transactions on Biomedical Engineering* 62 (4): 1089–1097.

**12** Kwasnicki, R.M., Hettiaratchy, S., Jarchi, D. et al. (2015). Assessing functional mobility after lower limb reconstruction: a psychometric evaluation of a sensor-based mobility score. *Annals of Surgery* 261 (4): 800–806.

**13** Pasluosta, C.F., Barth, J., Gassner, H. et al. (2015). Pull test estimation in Parkinson's disease patients using wearable sensor technology. In: *37th Annual International Conference of the IEEE Engineering in Medicine and Biology Society*, 3109–3112. IEEE.

**14** Mariani, B., Jimenez, M.C., Vingerhoets, F.J., and Aminian, K. (2013). Onshoe wearable sensors for gait and turning assessment of patients with Parkinson's disease. *IEEE Transactions on Biomedical Engineering* 60 (1): 155–158.

**15** Bagala, F., Klenk, J., Cappello, A. et al. (2013). Quantitative description of the lie-to-sit-to-stand-to-walk transfer by a single body-fixed sensor. *IEEE Transactions on Neural Systems and Rehabilitation Engineering* 21 (4): 624–633.

**16** Barth, J., Sünkel, M., Bergner, K. et al. (2012). Combined analysis of sensor data from hand and gait motor function improves automatic recognition of Parkinson's disease. In: *34th Annual International Conference of the IEEE Engineering in Medicine and Biology Society*, 5122–5125. IEEE.

**17** Benson, L.C., Clermont, C.A., Watari, R. et al. (2019). Automated accelerometer-based gait event detection during multiple running conditions. *Sensors (Basel)* 19 (7): 1–19.

**18** Li, R.T., Kling, S.R., Salata, M.J. et al. (2016). Wearable performance devices in sports medicine. *Sports Health* 8 (1): 74–78.

**19** Simon, S.R. (2004). Quantification of human motion: gait analysis-benefits and limitations to its application to clinical problems. *Journal of Biomechanics* 37 (12): 1869–1880.

**20** Jarchi, D., Pope, J., Lee, T.K.M. et al. (2018). A review on accelerometry based gait analysis and emerging clinical applications. *IEEE Reviews in Biomedical Engineering* 11: 177–194.

**21** Walter, P.L. (2006). The history of the accelerometer 1920s–1996: prologue and epilogue. *Sound & vibration* 41 (1): 84–92.

**22** Fennelly, J., Ding, S., Newton, J., and Zhao, Y. (2012). Thermal MEMS accelerometers fit many applications. *Sensor Magazine* 3: 18–20.

**23** Elble, R.J. (2005). Gravitational artifact in accelerometric measurements of tremor. *Clinical Neurophysiology* 116 (7): 1638–1643.

**24** Boser, B.E. and Howe, R.T. (1996). Surface micromachined accelerometers. *IEEE Journal of Solid-State Circuits* 31 (3): 366–375.

**25** Skog, I., Handel, P., Nilsson, J.O., and Rantakokko, J. (2010). Zero-velocity detection—an algorithm evaluation. *IEEE Transactions on Biomedical Engineering* 57 (11): 2657–2666.

**26** Kos, A., Tomazic, S., and Umek, A. (2016). Suitability of smartphone inertial sensors for real-time biofeedback applications. *Sensors (Basel)* 16 (3): 301.

**27** Mourcou, Q., Fleury, A., Franco, C. et al. (2015). Performance evaluation of smartphone inertial sensors measurement for range of motion. *Sensors (Basel)* 15 (9): 23168–23187.

**28** Kaiyu, T. and Malcolm, H.G. (1999). A practical gait analysis system using gyroscopes. *Medical Engineering & Physics* 21 (2): 87–94.

**29** Hestnes, E. (2016) Performance evaluation of smartphone inertial sensors measurement for range of motion. *NTNTU*, Master thesis, 2016.

**30** Geen, J. and Krakauer, D. (2003). New iMEMS angular-rate-sensing gyroscope. *Analog Dialogue* 37 (3): 1–4.

**31** Belli, A., Bui, P., Berger, A. et al. (2001). A treadmill ergometer for three-dimensional ground reaction forces measurement during walking. *Journal of Biomechanics* 34 (1): 105–112.

**32** Tesio, L., Monzani, M., Gatti, R., and Franchignoni, F. (1995). Flexible electrogoniometers: kinesiological advantages with respect to potentiometric goniometers. *Clinical Biomechanics* 10 (5): 275–277.

**33** Donno, M., Palange, E., Di Nicola, F. et al. (2008). A new flexible optical fiber goniometer for dynamic angular measurements: application to human joint movement monitoring. *IEEE Transactions on Instrumentation and Measurement* 57 (8): 1614–1620.

**34** Murley, G.S., Menz, H.B., and Landorf, K.B. (2009). Foot posture influences the electromyographic activity of selected lower limb muscles during gait. *Journal of Foot and Ankle Research* 2 (35) https://doi.org/10.1186/1757-1146-2-35.

**35** Hadi, A., Razak, A., Zayegh, A. et al. (2012). Foot plantar pressure measurement system: a review. *Sensors (Basel)* 12 (7): 9884–9912.

**36** De Rossi, S.M., Lenzi, T., Vitiello, N. et al. (2011). Development of an in-shoe pressure sensitive device for gait analysis. In: *33rd Annual International Conference of the IEEE Engineering in Medicine and Biology Society*. Boston, MA (30 August–3 September 2011), 5637–5640. IEEE.

**37** Tsalaile, T., Naqvi, S. M., Nazarpour, K., Sanei S., and Chambers, J. A. (2008) Blind source extraction of heart sound signals from lung sound recordings exploiting periodicity of the heart sound. 33rd IEEE International Conference on Acoustics, Speech and Signal Processing. Las Vegas (30 March–4 April 2008).

**38** Tsalaile, T., Sameni, R., Sanei, S. et al. (2009). Sequential blind source extraction for quasi-periodic signals with time-varying period. *IEEE Transaction on Biomedical Engineering* 56 (3): 646–655.

**39** Makkiabadi, B. Jarchi, D. and Sanei, S. (2012) A new time domain convolutive BSS of heart and lung sounds. *Proceedings of the IEEE International Conference on Acoustic, Speech, and Signal Processing, ICASSP*, Kyoto, Japan (25– March 2012).

**40** Ghaderi, F., Sanei, S., and McWhirter, J. (2010) Blind source extraction of cyclostationary sources with common cyclic frequencies. *Proceedings of the IEEE International Conference on Acoustics, Speech and Signal Processing, ICASSP*, Dallas, TX (14–19 March 2010).

**41** Pacela, A.F. (1966). Impedance pneumography: a survey of instrumentation techniques. *Medical & Biological Engineering* 4 (1): 1–15.

**42** Aminiahidashti, H., Shafiee, S., Zamani Kiasari, A., and Sazgar, M. (2018). Applications of end-tidal carbon dioxide (ETCO2) monitoring in emergency department; a narrative review. *Emergency (Tehran)* 6 (1): e5.

**43** Carrano, J. (2005). *Chemical and Biological Sensor Standards Study*, 1–30. Arlington, CA: DARPA report.

**44** Smith, A. L. (2005) Quartz crystal microbalance/heat conduction calorimetry. Online American Laboratory. https://americanlaboratory.com/914-Application-Notes/36163-Quartz-Crystal-Microbalance-Heat-Conduction-Calorimetry/ (accessed 25 November 2019).

**45** Clark, L.C. Jr., and Lyons, C. (1962). Electrode systems for continuous monitoring in cardiovascular surgery. *Annals of the New York Academy of Sciences* 102: 29–45.

**46** Vigneshvar, S., Sudhakumari, C.C., Senthilkumaran, B., and Prakash, H. (2016). Recent advances in biosensor technology for potential applications – an overview. *Frontiers in Bioengineering and Biotechnology* 4 (11) https://doi.org/10.3389/fbioe.2016.00011.

**47** Kazlauskaite, R., Soni, S., Evans, A.T. et al. (2009). Accuracy of self-monitored blood glucose in type 2 diabetes. *Diabetes Technology & Therapeutics* 11 (6): 385–392. https://doi.org/10.1089/dia.2008.0111.

**48** Weston, P. (2017) World's first diabetes app will be able to check glucose levels without drawing a drop of blood and will be able to reveal what a can of coke REALLY does to sugar levels. *MailOnline* (24 August). www.dailymail.co.uk/health/article-4817080/First-health-app-checks-glucose-levels-without-blood.html (accessed 26 November 2019).

**49** Analog Devices, Data Sheets (1996)ADXL50/ADXL05 Evaluation Modules. https://www.alldatasheet.com/datasheet-pdf/pdf/88616/AD/ADXL50.html (accessed 6 January 2020). Norwood, M.A., Alldatasheet.com.

**50** Middelhoek, S., Bellekom, A.A., Dauderstadt, U. et al. (1995). Silicon sensors. *Measurement Science and Technology* 6 (12): 1641.

**51** Sze, S.M. (1994). Biosensors. In: *Semiconductor Sensors* (ed. S.M. Sze). Wiley.

**52** Wang, B., Takahashi, S., Du, X., and Anzai, J. (2014). Electrochemical biosensors based on ferroceneboronic acid and its derivatives: a review. *Biosensors (Basel)* 4: 243–256.

**53** Erden, P.E. and Kilic, E. (2013). A review of enzymatic uric acid biosensors based on amperometric detection. *Talanta* 107: 312–323.

**54** Kim, J., Imani, S., de Araujo, W.R. et al. (2015). Wearable salivary uric acid mouthguard biosensor with integrated wireless electronics. *Biosensors & Bioelectronics* 74: 1061–1068.

**55** Pundir, C.S. and Chauhan, N. (2012). Acetylcholinesterase inhibition-based biosensors for pesticide determination: a review. *Analytical Biochemistry* 429: 19–31.

**56** Marrazza, G. (2014). Piezoelectric biosensors for organophosphate and carbamate pesticides: a review. *Biosensors (Basel)* 4: 301–317.

**57** Schneider, E. and Clark, D.S. (2013). Cytochrome P450 (CYP) enzymes and the development of CYP biosensors. *Biosensors & Bioelectronics* 39: 1–13.

**58** Khimji, I., Kelly, E.Y., Helwa, Y. et al. (2013). Visual optical biosensors based on DNA-functionalized polyacrylamide hydrogels. *Methods* 64: 292–298.

**59** Peng, F., Su, Y., Zhong, Y. et al. (2014). Silicon nanomaterials platform for bioimaging, biosensing, and cancer therapy. *Accounts of Chemical Research* 47: 612–623.

**60** Shen, M.Y., Li, B.R., and Li, Y.K. (2014). Silicon nanowire field-effect-transistor based biosensors: from sensitive to ultra-sensitive. *Biosensors & Bioelectronics* 60: 101–111.

**61** Ogi, H. (2013). Wireless-electrodeless quartz-crystal-microbalance biosensors for studying interactions among biomolecules: a review. *Proceedings of the Japan Academy. Series B, Physical and Biological Sciences* 89: 401–417.

**62** Li, M., Li, R., Li, C.M., and Wu, N. (2011). Electrochemical and optical biosensors based on nanomaterials and nanostructures: a review. *Frontiers in Bioscience (Scholar Edition)* 3: 1308–1331.

**63** Kwon, S.J. and Bard, A.J. (2012). DNA analysis by application of Pt nanoparticle electrochemical amplification with single label response. *Journal of the American Chemical Society* 134: 10777–10779.

**64** Zhou, Y., Chiu, C.W., and Liang, H. (2012). Interfacial structures and properties of organic materials for biosensors: an overview. *Sensors (Basel)* 12: 15036–15062.

**65** Guo, X. (2013). Single-molecule electrical biosensors based on single-walled carbon nanotubes. *Advanced Materials* 25: 3397–3408.

**66** Hutter, E. and Maysinger, D. (2013). Gold-nanoparticle-based biosensors for detection of enzyme activity. *Trends in Pharmacological Sciences* 34: 497–507.

**67** Ko, P.J., Ishikawa, R., Sohn, H., and Sandhu, A. (2013). Porous silicon platform for optical detection of functionalized magnetic particles biosensing. *Journal of Nanoscience and Nanotechnology* 13: 2451–2460.

**68** Senveli, S.U. and Tigli, O. (2013). Biosensors in the small scale: methods and technology trends. *IET Nanobiotechnology* 7: 7–21.

**69** Valentini, F., Galache, F.L., Tamburri, E., and Palleschi, G. (2013). Single walled carbon nanotubes/polypyrrole-GOx composite films to modify gold microelectrodes for glucose biosensors: study of the extended linearity. *Biosensors & Bioelectronics* 43: 75–78.

**70** Lamprecht, C., Hinterdorfer, P., and Ebner, A. (2014). Applications of biosensing atomic force microscopy in monitoring drug and nanoparticle delivery. *Expert Opinion on Drug Delivery* 11: 1237–1253.

**71** Sang, S., Wang, Y., Feng, Q. et al. (2015). Progress of new label-free techniques for biosensors: a review. *Critical Reviews in Biotechnology* 15: 1–17.

**72** Randriamampita, C. and Lellouch, A.C. (2014). Imaging early signaling events in T lymphocytes with fluorescent biosensors. *Biotechnology Journal* 9: 203–212.

**73** Oldach, L. and Zhang, J. (2014). Genetically encoded fluorescent biosensors for live-cell visualization of protein phosphorylation. *Chemistry & Biology* 21: 186–197.

**74** Kunzelmann, S., Solscheid, C., and Webb, M.R. (2014). Fluorescent biosensors: design and application to motor proteins. *Experientia Supplementum* 105: 25–47.

**75** Wang, S., Poon, G.M., and Wilson, W.D. (2015). Quantitative investigation of protein-nucleic acid interactions by biosensor surface plasmon resonance. *Methods in Molecular Biology* 1334: 313–332.

**76** Gutierrez, J.C., Amaro, F., and Martin-Gonzalez, A. (2015). Heavy metal whole-cell biosensors using eukaryotic microorganisms: an updated critical review. *Frontiers in Microbiology* 6 (48): 1–8.

**77** Sun, J.Z., Peter, K.G., Si, R.W. et al. (2015). Microbial fuel cell-based biosensors for environmental monitoring: a review. *Water Science and Technology* 71 (6): 801–809.

# 4

# Ambulatory and Popular Sensor Measurements

## 4.1 Introduction

Recognition and evaluation of physiological and biological states of human body including disease symptoms and variation in the related parameters may be obtained through the use of wearable or implanted sensors. Of these parameters, the vital signals play crucial roles in healthcare monitoring to predict the next state of a patient and to decide what treatment should be applied and how to avoid deterioration of the patient's situation. These data may also be used for patient management, rehabilitation, assistive technology, sport, and also human–machine interaction.

Primary vital signs include body temperature, heart rate (HR), respiratory rate (RR), and blood pressure (BP). There are other parameters which are recorded along the primary vital signs and can provide helpful, and in some cases vital, information in certain clinical settings. These signs are often grouped as 'fifth vital sign' or 'sixth vital sign'. The fifth vital signs may include blood oxygen saturation level ($SpO_2$), blood glucose level, menstrual cycle, or pain. The sixth vital signs include end-tidal $CO_2$, functional status, gait speed, and shortness of breath.

In this chapter, a number of vital signs including HR, RR, blood oxygen saturation level ($SpO_2$), BP, and blood glucose level are discussed in more details and recent techniques for their estimation are briefly explained.

## 4.2 Heart Rate

Electrocardiography (ECG) is a way of recording electrical activity of heart using mounted electrodes on the skin. From the ECG signals the HR parameters can be estimated. Numerous heart diseases such as coronary heart disease, arrhythmias, valvular heart disease, heart muscle disease, and congenital defects and their symptoms can be detected or monitored by detailed analysis of recorded ECG waveforms. Valvular heart diseases are characterised by abnormalities, stiffness, and damage to or a defect in one of the four heart valves, i.e. mitral, aortic, tricuspid, or pulmonary valves. In addition, by applying signal processing tools to the ECG, breathing rate, as a baseline change, can also be estimated.

Average HR can be easily computed by first detecting the R peaks which are the maximum amplitudes in the R waves. The R wave is the second upward deflection in each QRS wave

*Body Sensor Networking, Design and Algorithms,* First Edition. Saeid Sanei, Delaram Jarchi and Anthony G. Constantinides.

Companion Website: www.wiley.com/go/sanei/algorithm-design

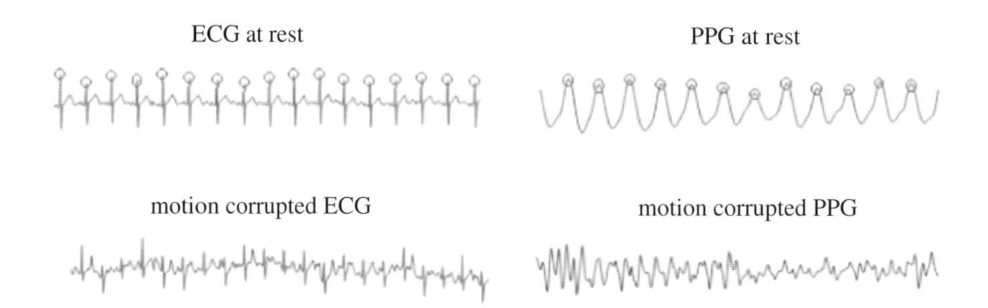


**Figure 4.1** Sample ECG and PPG signals are shown at rest and during motion. R peaks are shown by circles.

of the ECG. Heart rate variability (HRV) can be calculated by examining the time interval between two consecutive R peaks and estimating the changes of HR over time. HRV, also called instantaneous heat rate, is reciprocal to the timing of two consecutive R peaks multiplied by the sampling frequency to present beats per second. This can be further multiplied by 60 to provide beats per minute.

Photoplethysmography (PPG) sensors rely on light-base technology (introduced in the previous chapter) that can be used to estimate the average or instantaneous HR. ECG is usually used to provide reference data for HR for validating the outcome of PPG-based techniques and to estimate the HR parameters.

The pulse amplitude variability of PPGs is often investigated and compared with HR variabilities from ECGs. In one of the earliest PPG-based approaches, the variability of infrared PPG using spectral analysis was evaluated in various body locations such as the finger and earlobe [1]. Pulse amplitudes can be used to derive peak–peak (PP) intervals. Finger PPG has been used to measure the PP intervals where a high correlation between ECG and PPG exists within R–R intervals [2]. Figure 4.1 clearly shows the ECG and PPG correlation.

Using a tilt table test, the HRV recorded from finger PPG signals, called pulse rate variability (PRV), has been analysed [3]. During this test, following a head-up tilt, the subjects undertake a progressive orthostatic stress that induces changes in the modulation of HR. This results in an increased HR and a shrinkage of blood vessels in the legs.

In the first stage, the HRV derived from the ECG signals and PRV derived from the PPG signals are analysed with the help of time-frequency spectral and coherence analyses and a high correlation between them is demonstrated [3]. In another research attempt, time-frequency and entropy measurement algorithms were used to interpret the dynamics of PPGs for PRV [4]. A pulse frequency demodulation technique was performed in [5] to extract the instantaneous pulse rate from wrist PPGs. The earlobe PPGs recorded from healthy subjects are analysed in [6] and compared with the HRV estimated from the ECGs. The result of this study is important for certain applications such as monitoring Parkinson's patients where tremor and hyperkinesia particularly affect the PPGs recorded from finger/wrist. Frequency spectral analysis of finger PPG has revealed the possibility of evaluating peripheral circulatory control during haemodialysis [7].

### 4.2.1 HR During Physical Exercise

PPG signals are susceptible to a subject's motion where even a slight gap between the PPG sensor and the skin easily affects them. Owing to the correlation between PPG and other modality biomarkers, there has been some research to combine or fuse this information for more accurate diagnoses or better refining of PPG signals. In some recent studies, the effect of motion artefact has been reduced or removed from the motion contaminated PPG signals by exploiting simultaneously recorded acceleration signals. In [8] a real-time adaptive algorithm has been designed to estimate the HR recorded during an exercise without using any accelerometer signals. Adaptive filtering and time-frequency spectrum analysis are among the most popular techniques to enhance the time-frequency spectrum of motion contaminated PPG for better estimation of the average HR [9–13].

The average HR may be estimated from foot-worn PPG signals for various physical activities such as during fast bike exercise in [14]. IEEE Signal processing Cup 2015 dataset that includes simultaneous ECG, PPG, accelerometer recordings, and average HR as a gold standard are examined and used in [15]. In this research, a joint sparse spectrum reconstruction from simultaneously recorded accelerometer and PPG signals was proposed for estimation of average HR.

In a subsequent work, a general framework based on signal decomposition for denoising, sparse signal reconstruction for high-resolution spectrum estimation, and spectral peak tracking with verification, called TROIKA, was introduced. The HR estimation process includes three stages of signal decomposition for denoising, sparse signal reconstruction for spectrum estimation, and spectral peak tracking to estimate the average HR from a wrist PPG system during physical activities [16].

The spectrum of a selected accelerometer axis and motion corrupted PPG for subject #8 in the IEEE Signal processing Cup 2015 data is shown in Figures 4.2a and b, respectively. As shown in Figure 4.2b, the spectrum of PPG signals is corrupted by motion spectral components where some of them appear in the accelerometer spectrum. It has been shown that using two simultaneously recorded signal modalities, namely PPG and an accelerometer, the HR can be estimated [16]. As an effective approach, a normalised least mean square (NLMS) adaptive filter is employed to enhance the accelerometer signal and use it as a reference signal for PPG-based HR estimation. In this case the input to the adaptive filter is the accelerometer trace and the desired/target signal is the PPG signal. Theoretically, this is to adaptively estimate the motion free PPG by minimising the irrelevant spectral components measured by the accelerometer signal. Based on this, the PPG signal is modelled as:

$$p(t) = \tilde{p}(t) + m(t) + v(t) \tag{4.1}$$

where $\tilde{p}(t)$ is the motion interference-free PPG signal, $m(t)$ is the motion artefact, and $v(t)$ can be considered a residual sensor noise. In the modelled NLMS filter the motion artefact is assumed to be a linear function of the accelerometer signal:

$$m(t) = \mathbf{h}^{\mathrm{T}}(t)\mathbf{a}(t) \tag{4.2}$$

where $\mathbf{h}$ is an unknown transfer function for minimising the error to iteratively estimate $\mathbf{h}$. Hence, the error at each time point can be calculated as:

$$e(t) = p(t) - \mathbf{h}^{\mathrm{T}}(t)\mathbf{a}(t) \tag{4.3}$$

By minimising the error cost function, the filter weights $\mathbf{h}(t)$ are estimated using the following equation:

$$\mathbf{h}(t+1) = \mathbf{h}(t) + \frac{\mu(t)}{||\mathbf{a}(t)||^2}\mathbf{a}(t)e(t) \tag{4.4}$$

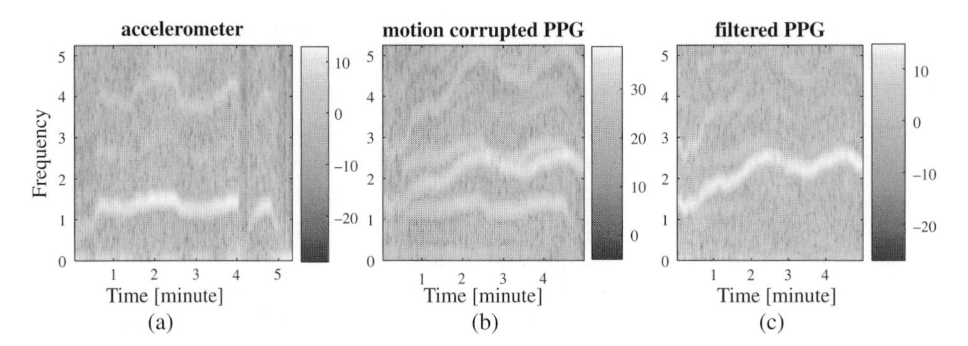


**Figure 4.2** (a) The spectrum of one accelerometer axis; (b) the spectrum of the raw PPG signals where the spectral components (similar to the spectral components of the simultaneously recorded accelerometer signal) related to motion are visible; (c) the spectrum of the PPG signal after applying adaptive filtering. (*See color plate section for color representation of this figure*)

where $\mu(t)$ is the step-size parameter and $\mathbf{a}(t)$ is a vector of length $L$ (filter order) of acceleration samples. As the result, the NLMS error output is the enhanced PPG signal. The spectrum of the filtered PPG using an NLMS is shown in Figure 4.2c. From this figure, the motion artefacts are suppressed and a dominant frequency trace around the HR frequency can be seen. A different NLMS filter can be constructed for each pair of the selected accelerometer axis and PPG signal. Such filters enhance the PPG spectrum [11]. To further motivate and stimulate this research, another public dataset has been created including simultaneous ECG, wrist PPG, accelerometer, and gyroscope recordings during fast/slow bike exercise [17].

HRV or instantaneous HR parameters are often estimated from ECG signals by constructing an RR interval time series. The PPG signals are used in [10] to estimate instantaneous HR. In this research, simultaneous accelerometer and PPG signals are used as inputs to an adaptive filter. Then, Hilbert transform has been employed to exploit the instantaneous frequency information in estimation of instantaneous HR from PPGs. In another study [18], a new time-varying spectral filtering algorithm for reconstruction of motion corrupted PPG signals during physical activities has been developed and applied to wrist (IEEE Signal processing Cup 2015 dataset) and forehead (Chon Lab dataset) PPG data. The power spectral densities of both PPG and accelerometer signals are created and compared to find the frequency peaks that originated from the motion artefact.

## 4.3 Respiration

Respiratory rate (RR), noted as number of breaths per minute (bpm), is one of the major and most important physiological parameters with significant applications in healthcare and fitness. Owing to its previous limited involvement in clinical diagnosis, RR has been noted as a neglected vital sign [19]. It is a quantification of early-warning scores and has been used to identify or predict the patient state in intensive care units (ICUs) [20]. Based on a study of 1025 emergency patients, an RR of greater than 20 bpm has been found to be directly associated with cardiopulmonary arrest within 72 hours and patient death within

30 days [21]. In another study including 1695 acute medical admissions, a mean RR of 27 bpm was associated to admission into the ICUs, cardiopulmonary arrest, or death within 24 hours [22].

Many recent research studies have focused on estimation of RR from wearable sensors, especially ECG and PPG signals, to enable automatic estimation of RR. One problem in validation of the RR from wearable sensors is the absence of a robust reference system to produce highly reliable ground truth data. The simplest way to measure RR is through manually counting the number of breaths. This method is cumbersome and cannot be used to produce a reference RR sequence, since the time intervals between respirations are not accurately recorded. As described in Chapter 3, the other methods include conventional impedance pneumography and an optical technique called end-tidal carbon dioxide (et-$CO_2$).

Impedance pneumography is a noninvasive approach to RR estimation. The conventional impedance pneumography method is highly susceptible to motion artefacts due to a subject's posture changes [23]. Therefore, it often produces erroneous results [24].

Using the et-$CO_2$-based method, a capnometer with its small plastic tube is inserted into the patient's mouth. The tube accumulates expelled $CO_2$ in each breath. The infrared radiation emits a light source that can move through the $CO_2$. Next, a variation with RR is induced following the absorption of infrared radiation over time. The et-$CO_2$-based systems have been used in hospitals mainly for patients in critical care units. However, the et-$CO_2$ system also suffers from limitations in terms of both accuracy and patient compliance especially in the clinical environment.

Figure 4.3a shows a sample of normal et-$CO_2$ capnometry output waveform, Figure 4.3b shows samples of capnography waveforms during advanced airway placement or intubation, and Figure 4.3c represents very common et-$CO_2$ capnography waveforms. Conventional signal processing and machine learning techniques can be used for the automatic recognition of various abnormalities in such waveforms.

Recently, there have been some efforts to automate unobtrusive RR estimation from wearable sensors. Both ECG and PPG signals have been used to estimate and validate RR. In [25], a benchmark dataset including both PPG and capnometry data was created during elective surgery and routine anaesthesia. For this dataset, the capnometry waveforms were used as a reference data for RR estimation. The recordings from PPG and impedance pneumography methods are also provided in the MIMIC-II dataset [26]. These include data from part of a larger cohort of ICU patients. For this dataset, the impedance pneumography waveforms are used as a reference data. These datasets – Capnobase and Medical Information Mart for Intensive Care (MIMIC) II – are used in various studies to validate the estimation of RR from PPG signals. A comprehensive review of the algorithms including Capnobase and MIMIC-II datasets has been recently published in [27, 28]. Most of these algorithms are very simple and easy to implement.

Given that the informative content of an ECG lies within a narrow band with low-frequency components, in one of the earliest approaches for estimation of RR, a bandpass filter with cut-off frequencies of 0.1 and 0.3 Hz, related to 6 and 18 bpm, respectively, was applied to the PPG signals [29]. Then, the number of breaths was obtained semi-automatically. The results are compared with a transthoracic impedance (TTI – which is a major determinant of transthoracic current flow in defibrillation) reference.

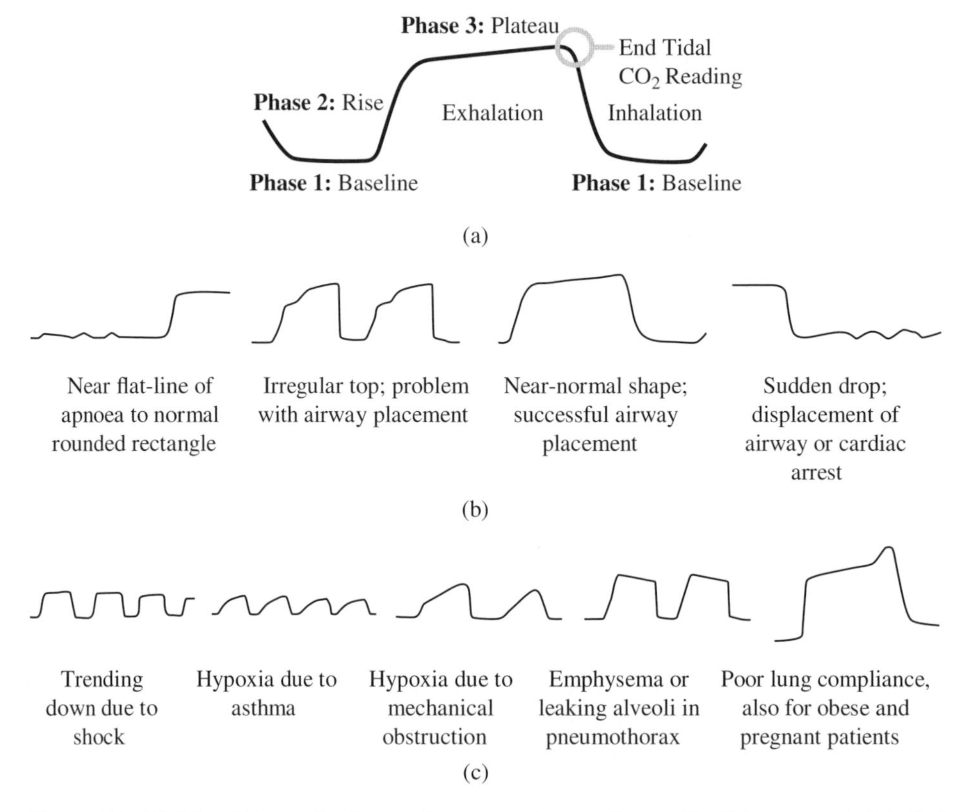


**Figure 4.3** (a) A breathing cycle of normal capnography waveform with all the segments labelled; (b) the waveforms during advanced airway placement or intubation; (c) a number of popular abnormal capnography waveforms.

A fundamental step in the estimation of RR from the ECG and PPG is extraction of a respiratory signal. A reasonable model for the generation of respiratory or cardiac signals involves three types of modulations representing respiratory-induced amplitude variation known as amplitude modulation (AM), respiratory-induced intensity variation called baseline wander (BW), and respiratory-induced frequency modulation (FM). These modulations are shown in Figure 4.4 and can be used to estimate RR. As shown in this figure, only BW is evident in the respiratory frequency ($\omega_r$) band. There are two main techniques to estimate RR from PPG or ECG signals including filter- or feature-based approaches.

In the filter-based technique, the input PPG/ECG signal is filtered to estimate the RR. For example in [30], the input signal is transformed using a continuous wavelet transform. This results in a time-frequency spectrum, with a dominant frequency at the HR frequency. This dominant frequency is tracked along time by simply finding the frequency component with the highest amplitude. Then, the amplitude and frequency of this dominant frequency are extracted, giving two time series. These two time series are then considered as the two respiratory signals (one relating to AM and one to FM). In [31], filtering has been performed using the centred-correntropy function (CCF) to extract the AM, FM, and BW components.

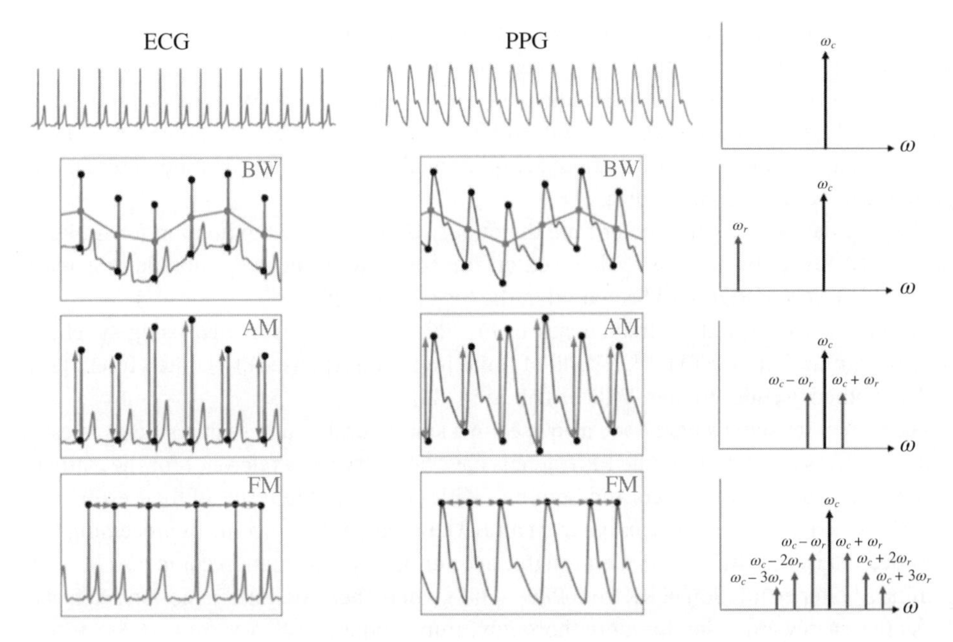


**Figure 4.4** ECG and PPG signals are shown at rest. Different respiratory induced modulations including BW, amplitude, and frequency modulations and their spectral characteristics are shown.

Employing a feature-based technique, the peaks in the signals are first detected. These peaks include Q and R peaks of the ECG signals (first column in Figure 4.4) or pulse peaks of the PPG signals (middle column in the Figure 4.4). These detected peaks allow derivation of FM, AM, and BW modulations and consequently estimation of RR. To derive a BW modulation, which is also called intensity modulation, the pulse peaks can be used to derive the BW. To derive the AM, the peaks and troughs of PPG pulses and the Q and R peaks of ECG signals need to be estimated. These parameters represent the height of each pulse. A time series including the height of all consecutive peaks produce the AM waveform. To derive FM, the timing between consecutive pulse peaks (for PPG) and R peaks (for ECG) are used to construct a time series representing frequency modulation.

After extraction of the respiratory modulations [32] from the PPG/ECG signals, a spectral analysis should be applied for identification of respiratory frequency indicating the breathing rate. RR is then estimated from the respiratory modulations, time-frequency representations including joint time-frequency approaches [33] and autoregressive (AR) modelling [34].

A considerable number of studies have proposed data fusion techniques to use multiple modulations [31–36]. Accordingly, several respiratory quality indexes (RQIs) have been defined in the literature with slightly different performances. FFT-RQI, AR-RQI, AutoCor-RQI, and Hjorth Parameter RQI are among the most popular ones. The FFT-RQI is based on Fourier transform and functions by calculating the area under the largest peak in the frequency domain compared to the rest of the signal. The AR-RQI works based on the autocorrelation function. The AR-RQI is calculated by finding the best-performing AR model order and selecting the largest pole for that model order to represent the AR-RQI.

The AutoCor-RQI is based on the autocorrelation function and seeks to find the area within the RR range where the signal best corresponds to itself; the higher the autocorrelation value is, the more likely it corresponds to an actual respiratory wave. Finally, the Hjorth Parameter RQI is calculated using the third Hjorth parameter, which is a measure of how sinusoidal a signal is where the periodicity of a signal is expected to correspond to a more pronounced respiratory waveform.

The RQI has been used to determine the quality of respiratory modulation for improving the RR fusion techniques [37]. Based on this, only the respiratory modulations with acceptable levels of RQI have been used in the fusion procedure.

Instantaneous respiratory frequencies from PPG have been estimated using synchrosqueezing transform (SST) [38], a bank of finite impulse response (FIR) notch filters [39] and smoothed pseudo Wigner–Ville distribution [40].

Accelerometer signals have been also used in a small number of research studies to estimate the RR [41–46]. In [46], the PPG signals have been used as a reference for the estimation of RR from accelerometers and both modalities compared in terms of their estimated RRs for patients following discharge from an ICU. Based on the outcome of processing the recorded dataset for this study, the available accelerometry information could also help in gaining a better estimation of RR from PPG signals where there are certain motion artefacts. This opens a new direction for more thorough future studies.

In a large clinical trial at the of Oxford, the data from patients following discharge from an ICU participated in a wearable sensor study called PICRAM (post-intensive care risk-adjusted and monitoring). The signals from ECG, PPG, and accelerometers were simultaneously recorded from a cohort of patients (Figure 4.5).

The estimated RRs from different signal modalities are shown in Figure 4.6. For an estimation of RR from the accelerometer using a devised signal quality index, the best accelerometer axis has been selected (Figure 4.6b). Then, the frequency spectrum has been analysed to estimate the dominant respiratory frequency and then presenting it in terms of bpm. For ECG- and PPG-based RR estimations, various respiratory modulations

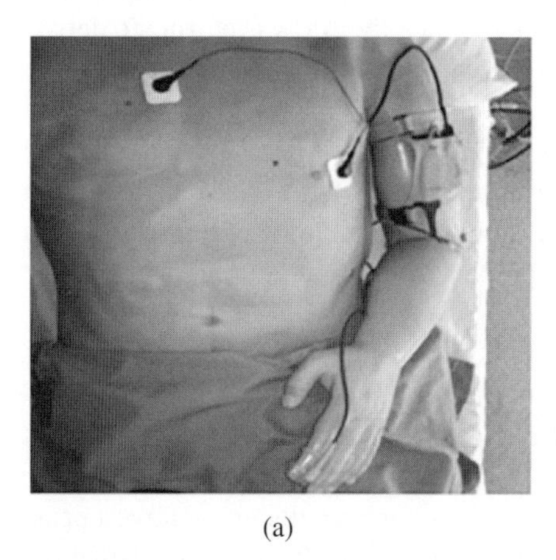

(a)

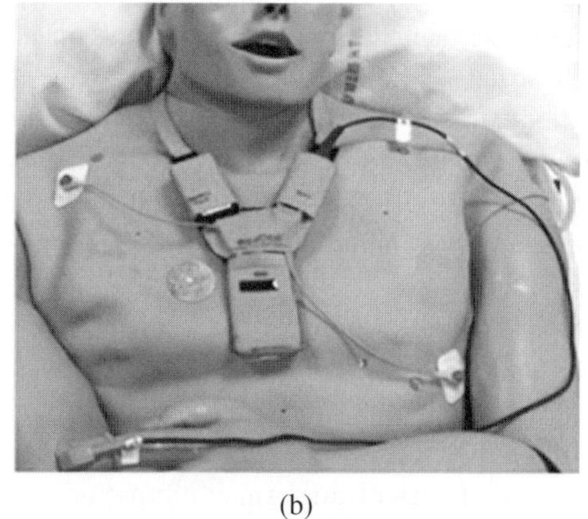

(b)

**Figure 4.5** A phantom subject equipped with ECG electrodes, pulse oximetry with finger probe. The accelerometer unit is placed either on the arm (a) or chest (b). In both figures the oximeter is worn on the finger and the ECG on the chest.

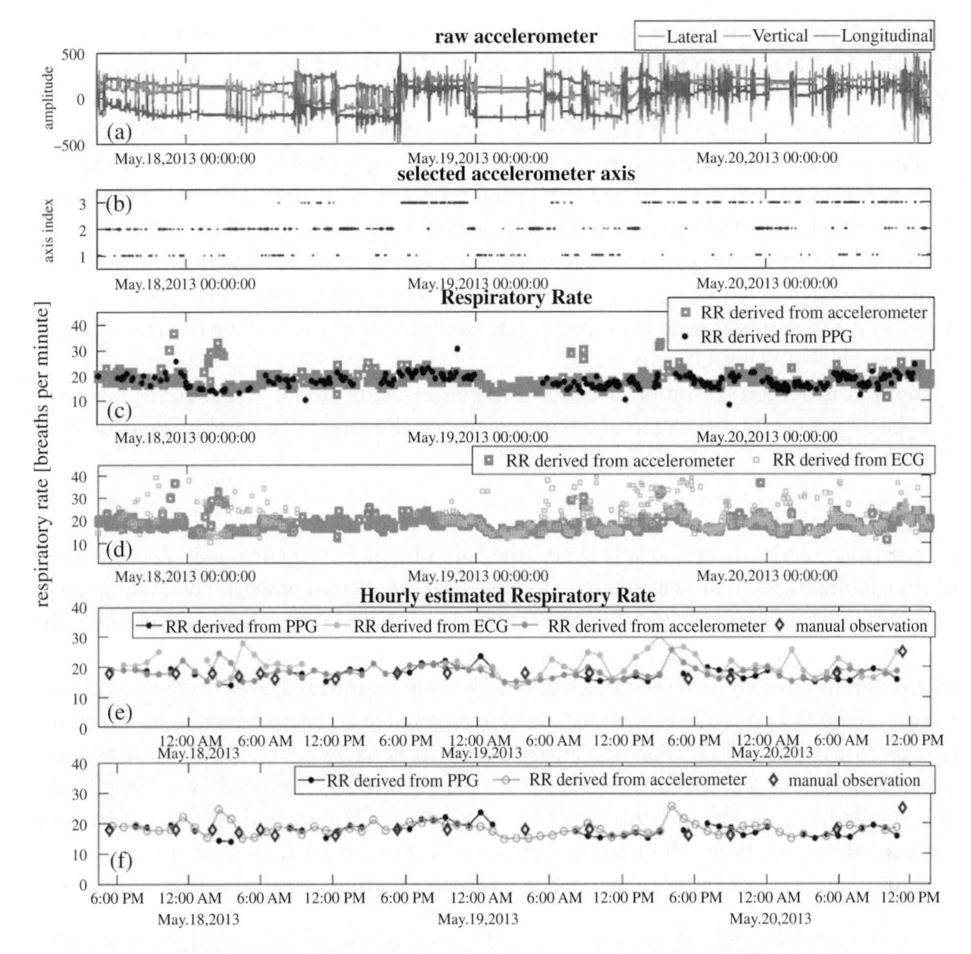


**Figure 4.6** Estimation of RR from simultaneous PPG, accelerometer, and ECG signals recorded from a patient following discharge from the ICU in the hospital ward (John Radcliff hospital, Oxford, 2013) for approximately two days. Raw acceleration data and selected accelerometer axis are shown in (a, b). There is a strong correlation between the PPG and the accelerometer based RR estimates (c, d). Averaged hourly RRs are also plotted (e, f). (*See color plate section for color representation of this figure*)

are used and the AR-based spectral analysis has been performed to fuse and estimate the RRs. Averaged hourly estimated RRs are shown in Figure 4.6e and f. There is a strong correlation between PPG and accelerometer-derived RRs and those by observations. However, the ECG-based RR estimates have provided partly correlated RR estimates. This is due to recording low-quality ECG signals over an extended period of time, which could be a result of electrode detachment, motion interferences, or patient sweating.

## 4.4 Blood Oxygen Saturation Level

The principle of the pulse oximeter was introduced in 1972 by Takuo Aoyagi [47], who showed that by applying red and infrared lights to human tissue and measuring the optical densities of the transmitted or reflected lights, the arterial blood oxygen saturation ($SaO_2$) could be estimated up to an acceptable accuracy. The OLV-5100 pulse oximeter, manufactured by Nihon Kohden Corporation, has been named as the world's first commercial ear-worn pulse oximeter to measure blood oxygen saturation level ($SpO_2$). In the transmission mode PPG, the probe is worn by the user on a finger [9, 48–50] or an earlobe [51]. Reflectance mode PPGs are embedded within wrist- or arm-worn devices. As shown in the previous chapter, the light illuminated through the light-emitting diode (LED) either is transmitted or reflected into a photodetector where the amount of transmitted or reflected light is estimated and used to derive the PPG signal. The quality of the PPG signal is usually better in the transmissive mode than the reflectance mode as the surface layers of the skin attenuate the light significantly.

The arterial oxygen saturation is the level of arterial haemoglobin oxygenation and can be calculated as the ratio of oxygenated haemoglobin to the total haemoglobin concentration in the blood, i.e.

$$SaO_2 = HbO_2/(HbO_2 + Hb) \tag{4.5}$$

where $HbO_2$ and Hb are, respectively, the amount of oxygenated and deoxygenated haemoglobins. $SaO_2$ can be estimated based on the Beer–Lambert law. Using pulse oximetry, an approximation to $SaO_2$ can be obtained noted as $SpO_2$, that corresponds to the percentage of oxygen bound in the blood. To estimate $SpO_2$, the lights from a pair of LEDs with different wavelengths (660 nm for red and 940 nm for infrared) are emitted and the difference in the absorption or reflection between oxygenated and deoxygenated haemoglobin is used to calculate $SpO_2$. For transmissive PPG, it is known that $HbO_2$ absorbs more infrared light and lesser red light than Hb (see Figure 4.7). For each wavelength, the transmitted light signals contain a direct current (DC) component and pulsatile alternating current (AC) component. These components are used in the calculation of $SpO_2$.

The $SpO_2$ values can be calculated using a 64-point Fourier transform [52]:

$$SpO_2 = 110 - 25 \times R \tag{4.6}$$

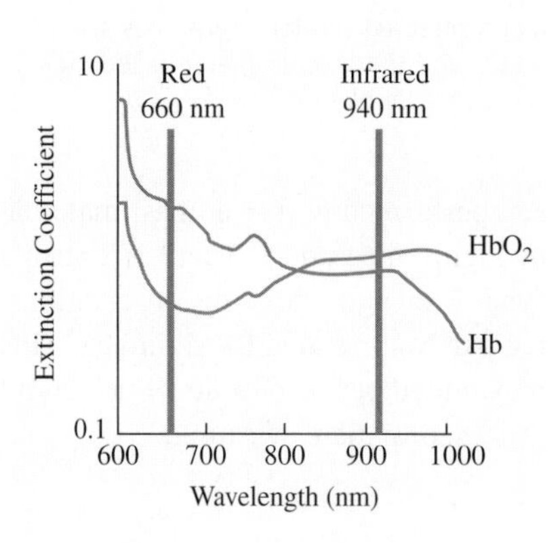


**Figure 4.7** Absorption spectral characteristics of oxygenated ($HbO_2$) and deoxygenated (Hb) haemoglobins with respect to the light wavelength.

where $R$ is the ratio between the infrared and red transmitted light intensity and can be calculated as:

$$R = \frac{\frac{dI_R(t)/dt}{I_R}}{\frac{dI_{IR}(t)/dt}{I_{IR}}} = \frac{\frac{I_R(t_2)-I_R(t_1)}{I_R(t_3)}}{\frac{I_{IR}(t_2)-I_{IR}(t_1)}{I_{IR}(t_3)}} = \frac{\frac{AC_R}{DC_R}}{\frac{AC_{IR}}{DC_{IR}}} \tag{4.7}$$

where $AC_R$ and $AC_{IR}$ represent, respectively, the signal variation at cardiac frequency for red and infrared. Similarly, $DC_R$ and $DC_{IR}$ represent, respectively, the average overall transmitted light intensity for red and infrared signals. In the above equation, $dI_R(t)/dt$ and $dI_{IR}(t)/dt$ are, respectively, the derivatives of AC components of red and infrared signals. As shown in Figure 4.8, two time instances ($t_1$ and $t_2$) are selected as the local minimum and maximum points of the waveform where their difference is noted as the AC value. As included in Eq. (4.7), a point ($t_3$) is selected between $t_1$ and $t_2$ as DC value. In Figure 4.8, these points are shown for infrared signal. Similar points should be extracted for the red signal, then, the $R$ value obtained from Eq. (4.7) is used to estimate the $SpO_2$ from Eq. (4.6). It is worth noting that the above method has been initially developed for transmissive PPG. Nevertheless, it is possible to use a similar approach for reflectance PPGs. In a recent work [53], reflective PPGs from a commercial sensor (WaveletHealth Inc., USA) have been used to validate the estimation of $SpO_2$. A commercial transmissive pulse oximeter with finger PPG probe has been used to provide the reference data.

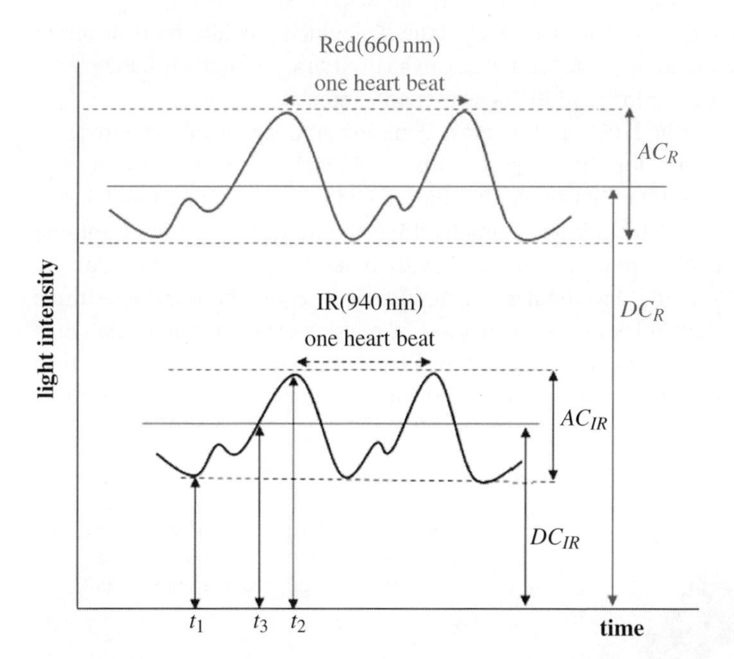


**Figure 4.8** Waveform of transmitted or reflected light at two wavelengths: 660 and 940 for red or infrared lights. The demonstrated AC and DC components are utilised to estimate the blood oxygen saturation level ($SpO_2$).

In some studies, the $SpO_2$ has been considered directly proportional to the $R$ ratio, i.e. $SpO_2 = k \times R$, rather than using Eq. (4.6). Based on this, constant $k$ is set during the calibration [48].

A phone oximeter consisting of a pulse oximeter with a smartphone to record $SpO_2$ and PPG has been used for identification of breathing in children with sleep disorder [54]. The PRV and $SpO_2$ patterns have been analysed and validated against polysomnography (PSG) as the gold standard. A motion-tolerant finger pulse oximeter, Onyx II, Model 9560 by Nonin Medical Inc. [8] and a finger transmittance PPG from Nonin Medical, Inc. [55] have been used for PPG recording to derive $SpO_2$ and HR for monitoring mobile patients in hospital.

## 4.5 Blood Pressure

Arterial BP is one of the vital signs and an important physiological parameter widely used for health monitoring of patients in a clinical setting. Continuous monitoring of BP is clinically important since large fluctuations in BP can be a warning sign of critical conditions. High BP or hypertension can damage heart, brain, kidney, arteries, and eyes, and cause sexual dysfunction.

Majority of BP measurements are performed noninvasively. However, an arterial catheter, which is a thin and empty tube, can be placed into an arterial site (wrist, groin, etc.) to measure the BP. This method is an invasive technique [56] which provides more accurate estimates than sphygmomanometer (BP cuff) and can be used as a gold standard. Figure 4.9 illustrates the invasive and noninvasive BP measurement systems.

Sphygmomanometry method [57] is the most common and applicable noninvasive technique and gold standard for BP measurements in clinical or home environments. It measures diastolic and systolic BP using cuff inflation and deflation. Other noninvasive methods which use BP cuff include oscillometry [58], auscultation [58], and volume clamping [59]. PPG signals alone or combined with other physiological data can be used to estimate BP. With the development of new devices, computational algorithms play increasingly important roles in the estimation of physiological parameters such as BP, as a crucial parameter for management and diagnosis of many abnormalities such as hypertension and hypotension particularly using less intrusive approaches, such as cuffless techniques.

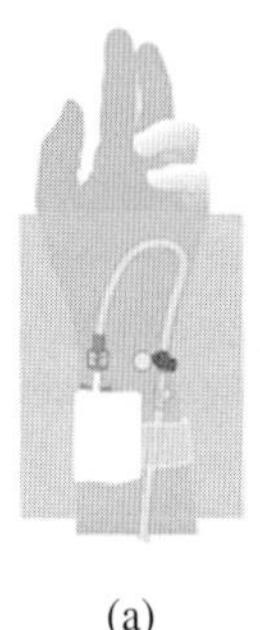

(a)

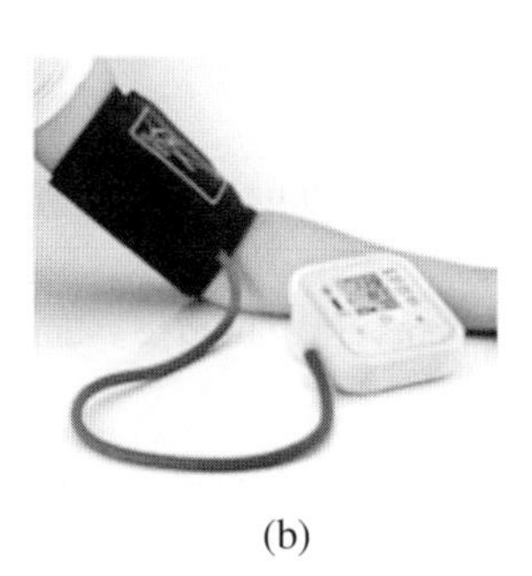

(b)

**Figure 4.9** (a) An invasive BP measurement using arterial catheter; (b) a noninvasive oscillometry cuff BP measurement system.

### 4.5.1 Cuffless Blood Pressure Measurement

The pulse transmit time (PTT) method is a well-known technique for cuffless BP measurement. Recent research studies have been directed towards designing accurate systems for cuffless BP monitoring using PTT. One advantage in using the PTT technique over the cuff-based BP method is that in cuffless BP monitoring using PTT the changes in BP can be tracked. PTT is denoted as the time delay taken by the arterial pulse to spread from heart to a peripheral site such as the ear, finger, or wrist. For example, it can be calculated by measuring the time interval started from the time point of an R peak in the ECG signal to an associated pulse peak of PPG signal recorded from the wrist. The characteristic pulse peak can be found using the first derivative of the PPG signal, as shown in Figure 4.10. Another useful parameter called pulse wave velocity (PWV) is proportional to reciprocal of the PTT multiplied by the distance between the arterial sites ($L$).

A number of studies have criticised the calculation of PTT for a reliable BP estimation [60, 61]. However, it has been found that PTT can be more useful for monitoring the variability of BP measurements where several comparative studies are also available. In another study a PPG intensive ratio has been proposed to improve the accuracy of the cuffless BP measurement systems [62]. This suggests monitoring the BP over extended periods of time and tracking its dynamics.

Recent studies have proposed machine learning platforms to extract the necessary features from various signals such as PPG, ECG, and accelerometer to provide more reliable estimates for systolic or diastolic BP. In this direction, researchers have started to explore the possibility of extracting systolic, diastolic, or even the whole waveform of BP using cuffless methods motivated by machine learning approaches. This relies on the measurement or estimation of a number of surrogates of BP levels from the time- and frequency-domain parameters derived from one or several physiological signals including PPG, ECG, and accelerometery. Additional information is often provided from questionnaires, which include peripheral variables such as age, sex, height, and weight. Then, machine learning algorithms are trained using the data that contain the reference

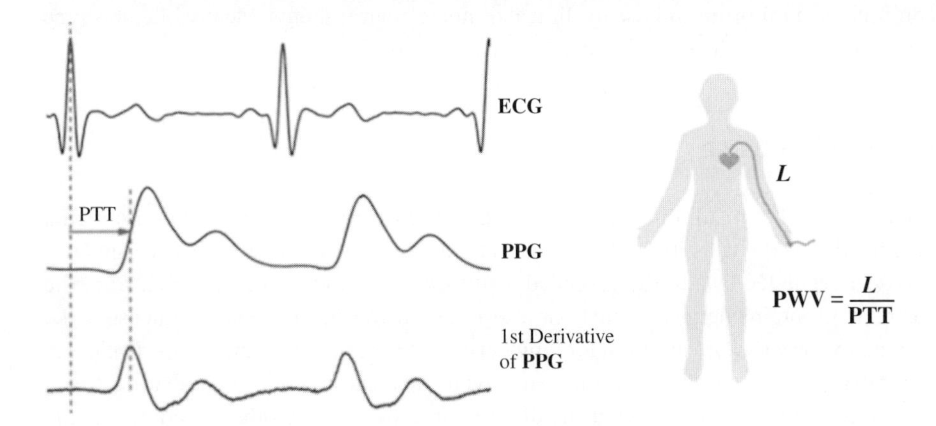


**Figure 4.10** The PTT can be calculated as the time between an ECG R peak and the next pulse peak located in the derivation of PPG within the same cardiac cycle. PWV can be approximated using PTT.

BP values. Finally, the algorithms are evaluated on the test datasets that have not been used during the training phase.

PPG-based BP devices known as cuffless devices use optical technology placed e.g. on the wrist [63], ear [64], or finger [65]. These devices are often embedded portable equipment or handphones. Consequently, the market for such devices is emerging. These wearable devices track vital signs via optical sensors and use beat-to-beat fluctuations together with mathematical models to compute diastolic and systolic BP values. The main advantage of cuffless devices is their ease of use and the potential for continuous monitoring. Unfortunately, this type of technology tends to be highly inaccurate. For example, in a recent review it was pointed out that nearly 77.5% of the individuals with hypertensive BP levels were wrongly assessed as normotensives. Although the cuffless BP technology is still new and not developed enough to be recommended for clinical use, it has been used in over 1000 clinical trials (www.clinicaltrials.gov) to assess the feasibility of its clinical use [66].

In some other studies [63], the researchers worked on a cuffless and noninvasive system for systolic BP and other haemodynamic variables using features from the PPG waveform and its second-order derivative (called acceleration plethysmogram). The features include the height of various peaks in the acceleration plethysmogram, the time elapsing from the rise of the signal within the same cardiac cycle, and the questionnaire-based variables such as the subject's height, weight, age, and sex.

Another cuffless, noninvasive system for BP monitoring has been developed using PPG technology based on a regression model and the restricted Boltzmann machine (RBM) [67]. The study has been extensively validated on 572 subjects against the USA Association for the Advancement of Medical Instrumentation protocol. Although this approach is very promising, in its present form it fails to meet the protocol requirements (standard deviation of differences between the new method and reference $> 8$ mmHg) and therefore its clinical use has not been recommended.

Many wearable devices track vital signs obtained using the signals from optical sensors and exploit beat-to-beat fluctuations together with computational or mathematical models to compute diastolic and systolic BP. The main advantage of cuffless devices is their ability for continuous monitoring and use in clinical or home environments. Meanwhile, this type of technology often produces highly inaccurate BP estimates.

## 4.6 Blood Glucose

An abnormal level of insulin in the body due to a malfunctioning pancreas not producing enough insulin or the cells in the body not using it adequately causes diabetes. The level of glucose is regulated by a hormone called insulin, by allowing cells to absorb it from the bloodstream to obtain energy or store it for future use. However, if the level of glucose in the blood remains very low or very high for long periods of time, it could cause hypoglycaemia or hyperglycaemia, respectively, leading to severe medical conditions, including tissue damage, kidney failure, blindness, heart disease, or stroke among others, and eventually death if left untreated.

To improve the quality of a diabetes patient's life, frequently measuring their glucose level is essential in order to adjust their insulin dose. This is traditionally made by electrochemical method which requires taking blood samples. To make the measurement more effective for severely diabetic patients, more frequent blood sampling needs to be made. This is not pleasant for the patients and causes pain and discomfort. Therefore, there has been a great interest in developing noninvasive glucose monitoring systems [68, 69]. The development of a noninvasive device for glucose measurement would be a life-changing factor for millions of patients around the world, allowing them to monitor their glucose level confidently and receive quick treatment if necessary. The market for such product covers approximately 450 million diabetes cases around the world, according to the World Health Organization.

A system which simultaneously records blood glucose and systolic and diastolic BP using PPG signals has been developed by Monte-Moreno [70]. Various machine learning approaches such as linear regression, neural network, support vector machine, regression trees, and random forest have been examined and used to train the model. The developed system was tested using a database from 410 individuals. For such a system there is no need for any personalised calibration. The producers claim that the system meets grade B criteria of the British Hypertension Society protocol. However, it does not meet the requirements for grade A due to the existence of 17 outliers with an error in measuring the BP greater than 15 mmHg.

For noninvasive glucose measurements, most of the systems developed previously are based on near-infrared spectroscopy and the majority have proved unsuccessful. Current developments try to exploit the characteristics of the glucose molecule at different frequencies in the spectrum, from DC and ultrasound to near-infrared and visible frequency bands. Yet, the most promising bands are near-infrared and infrared. Advanced technologies for glucose detection can be classified into four subgroups: optical, thermal, electrical, and nanotechnology methods mostly relying on the tissue near-infrared and infrared frequency characteristics [71]. Among them, optical methods and those based on near-infrared and infrared frequency characteristics are more in favour by the researchers [71].

Since the light absorption of glucose is two orders of magnitude below that of water, it can lead to unreliable measurements, and therefore recalibration either between individuals or over time becomes essential. Current systems suffer from lack of accuracy, usability, and patient compliance [72] limitations, which need to be improved in future studies.

Nevertheless, there are many cases where the intension is to check if the glucose level is within an acceptable range. In these cases the glucose level estimates don't need to be very accurate. Following this idea, sensor developers of the HELO Extense (NIR spectroscopy; World Global Network, Miami, FL, USA) [73] and the DermalAbyss (chemical fluorescence; MIT, Cambridge, MA, USA) [74] have been trying to design noninvasive glucose concentration level indicators.

## 4.7 Body Temperature

Body temperature represents the balance between heat production and heat loss. If the rate of heat generated equates to the rate of heat lost, the core body temperature (CBT) will be stable. The metabolising body cells generate heat in varying amounts. Therefore, body temperature is not evenly distributed across the body [75].

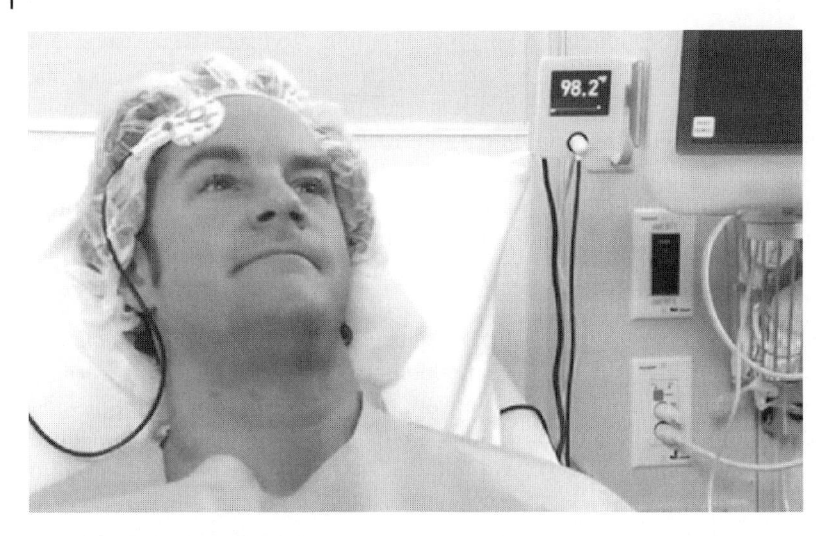


**Figure 4.11** Patient CBT monitoring using 3M™ SpotOn™ Temperature Monitoring System (http://go.3M.com/33NB).

CBT is another clinically important factor to evaluate the overall health state of a human body. A reliable measurement of the CBT is required for surgical patients to control their body temperature during surgery. The measurement of CBT may seem simple, but several issues affect the accuracy of the reading. These include the measurement site, the reliability of the instrument, and user technique. The measurement is more accurate when it is invasive (by inserting the thermometer inside the oesophagus, pulmonary artery, or urinary bladder).

Various devices using active or passive sensors have been developed to measure CBT noninvasively. When using active sensors such as in a zero-heat-flux thermometry measurement system (SpotOn™ by 3M™), as shown in Figure 4.11, a heating element is required [76]. This is not required when using passive sensors including single or dual heat flux technologies [77].

Most technologies for measuring the CBT have shown promising results. There are, however, some limitations such as the requirement for continuously warming the element when using active sensors. This is a barrier when the CBT is integrated within wearable systems and its implementation as a wearable device. It is also difficult to have an even thermal distribution when using passive sensors. To address these limitations, an ergonomic wearable sensor has been developed and has shown to produce a high accuracy when worn in two positions of forehead and behind the ears [78].

## 4.8 Commercial Sensors

In the following, several commercial sensors which have been used experimentally for validation studies or in certain clinical applications are briefly explained. The Simband smartwatch (Samsung Inc.) has been used in [79] to detect atrial fibrillation (AF). This is a wrist-worn device that includes eight PPG sensors, ECG, and accelerometer.

Its shimmer sensing platform records the data from ECG electrodes and has an ear-clip probe to record PPG signals. The shimmer device has been reconfigured to record PPG signals from the wrist [17]. A dataset has been created including PPGs, accelerometer, and gyroscopes data recorded during various conditions including walking, running, and easy/hard bike riding. The dataset contains reference data for HR estimates from chest-worn ECG electrodes.

A wrist-worn sensor known as Amigo that includes two-wavelength PPGs and simultaneously records accelerometer and gyroscope data has been developed by Wavelet Health. It has been used in recent research studies to detect AF from wrist PPGs with an accuracy of 91%. For validation of these measurements the ECG based on Xio Patch (iRhythm Tech. Inc.) was considered as the ground truth [80]. In another study, the sensor has also been used as a portable screening device for obstructive sleep apnoea and evaluated versus PSG system for the estimation of the HR, HRV, $SpO_2$, and RR parameters [81].

As a first study on analysis of patients with obstructive hypertrophic cardiomyopathy (OHCM) a wrist-worn sensor by Wavelet Health [82] has been used. It has been found that the combination of a wrist-worn biosensor and the associated machine learning algorithms can be employed to identify a signature of arterial blood flow in OHCM patients compared to those of unaffected controls. The findings are important to the noninvasively detected OHCM.

Currently, there are few commercially available products. The Somnotouch-NIBP [83] and ViSi Mobile (Sotera Wireless) [84] are among those that have been well received and commercialised. For example, Somnotouch-NIBP uses the finger PPG and three ECG leads that are connected to a wrist-mounted controller and provides systolic and diastolic BP measurements using PWV. The system has been validated by the European Society for Hypertension International Protocol. Sotera ViSi Mobile includes continuous noninvasive BP monitoring. It is determined on the beat-to-beat basis using PTT technology and calibrated using an automatic noninvasive BP device. It is a Food Standard Agency (FSA) approved device with clinical performance validated against ISO 81060-2. However, this standard has been set for cuff-based noninvasive sphygmomanometers [85].

## 4.9 Conclusions

Measurements given by wearable sensors have provided valuable input to patient monitoring systems. These data can estimate and detect changes in the physiological parameters of human bodies, often in unobtrusive ways. The development of optical sensors that can be embedded into smartwatches or wristbands has been a big step towards making outpatient continuous healthcare monitoring systems for everyday utilisation. One main challenge in designing such monitoring systems is the reliability of associated integrated algorithms to provide robust estimates of physiological parameters and accurate inference of the health status. Advances in machine learning and signal processing algorithms have provided a significantly important base to create highly reliable outputs for the estimation of physiological parameters. These factors can in turn be effectively used in healthcare monitoring systems and other applications. Although this chapter addresses a number of popular and useful wearable devices and technologies, there are many other sensors and platforms for measuring many other human vital factors. As an example, when people sweat, they unknowingly release chemicals that can noninvasively show the stress and the corresponding hormone level as well as concentration of glucose in the body. Researchers recently developed a new membrane that mitigates both issues that arise from direct dermal contact and sweat dilution for sweat biosensors. The membrane has been shown to perform very well and holds up to repeated use [86].

Another example is the instrumentation, called an electronic nose (eNose), designed to measure the level of smelly substances in the air. The basic concept of an electronic nose, or machine olfaction, is a measurement unit that generates complex multidimensional data for each measurement combined with a pattern recognition technique that interprets the complex data and relates it to a target value or class. The technologies which are feasible for application are QMB (quartz microbalance)/SAW (surface acoustic wave), conducting polymers, and metal-oxide sensors. A QMB is a quartz crystal with a chemically active surface, usually a polymer. When gas molecules adsorb to the surface, the mass changes and the resonant frequency of the crystal shifts. These minute shifts are measured with rather complex and expensive high frequency electronics. Small temperature variations result in similar frequency shifts thus dictating strict environmental temperature control. A variation of a QMB is a SAW sensor which also works on the principle of frequency shifts [87]. Rather than these sensors, there are a number of multisensor wearable systems – such as wearable EEG caps, head bands, glasses, ear plugs, or semi-crowns – which are becoming more acceptable by the users and likely to be popular soon.

## References


1. Bernardi, L., Radaelli, A., Solda, P.L. et al. (1996). Autonomic control of skin microvessels: assessment by power spectrum of photoplethysmographic waves. *Clinical Science* 90 (5): 345–355.
2. Selvaraj, N., Jaryal, A., Santhosh, J. et al. (2008). Assessment of heart rate variability derived from finger-tip photoplethysmography as compared to electrocardiography. *Journal of Medical Engineering & Technology* 32 (6): 479–484.
3. Gil, E., Orini, M., Bailon, R. et al. (2010). Photoplethysmography pulse rate variability as a surrogate measurement of heart rate variability during non-stationary conditions. *Physiological Measurement* 31 (9): 1271–1290.
4. Lu, S., Zhao, H., Ju, K. et al. (2008). Can photoplethysmography variability serve as an alternative approach to obtain heart rate variability information? *Journal of Clinical Monitoring and Computing* 22 (1): 23–29.
5. Hayano, J., Barros, A.K., Kamiya, A. et al. (2005). Assessment of pulse rate variability by the method of pulse frequency demodulation. *Biomedical Engineering Online* 4 (62) https://doi.org/10.1186/1475-925X-4-62.
6. Lu, G., Yang, F., Taylor, J.A., and Stein, J.F. (2009). A comparison of photoplethysmography and ECG recording to analyse heart rate variability in healthy subjects. *Journal of Medical Engineering & Technology* 33 (8): 634–641.
7. Javed, F., Middleton, P.M., Malouf, P. et al. (2010). Frequency spectrum analysis of finger photoplethysmographic waveform variability during haemodialysis. *Physiological Measurement* 31 (9): 1203–1216.

8 Yousefi, R., Nourani, M., Ostadabbas, S., and Panahi, I. (2014). A motion-tolerant adaptive algorithm for wearable photoplethysmographic biosensors. *IEEE Journal of Biomedical and Health Informatics* 18 (2): 670–681.

9 Fukushima, H., Kawanaka, H., Bhuiyan, M.S., and Oguri, K. (2012). Estimating heart rate using wrist-type photoplethysmography and acceleration sensor while running. In: *Annual International Conference of the IEEE Engineering in Medicine and Biology Society*, 2901–2904. IEEE.

10 Jarchi, D. and Casson, A. (2017). Towards photoplethysmography based estimation of instantaneous heart rate during physical activity. *IEEE Transactions on Biomedical Engineering* 64 (9): 2042–2053.

11 Schack, T., Sledz, C., Muma, M., and Zoubir, A.M. (2015). A new method for heart rate monitoring during physical exercise using photoplethysmographic signals. In: *23rd European Signal Processing Conference (EUSIPCO)*, 2666–2670. IEEE.

12 Boloursaz Mashhadi, M., Asadi, E., Eskandari, M. et al. (2016). Heart rate tracking using wrist-type photoplethysmographic (PPG) signals during physical exercise with simultaneous accelerometry. *IEEE Signal Processing Letters* 23 (2): 227–231.

13 Tautan, A.M., Young, A., Wentink, E., and Wiering, F. (2015). Characterization and reduction of motion artifacts in photoplethysmographic signals from a wrist-worn device. In: *Conference Proceedings: Annual International Conference of the IEEE Engineering in Medicine and Biology Society*, 6146–6149. IEEE.

14 Jarchi, D. and Casson, A. (2016). Estimation of heart rate from foot worn photoplethysmography sensors during fast bike exercise. In: *Conference Proceedings: Annual International Conference of the IEEE Engineering in Medicine and Biology Society*, 3155–2158. IEEE.

15 Zhang, Z. (2015). Photoplethysmography-based heart rate monitoring in physical activities via joint sparse spectrum reconstruction. *IEEE Transactions on Biomedical Engineering* 62 (8): 1902–1910.

16 Zhang, Z., Pi, Z., and Liu, B. (2015). TROIKA: a general framework for heart rate monitoring using wrist-type photoplethysmographic signals during intensive physical exercise. *IEEE Transactions on Biomedical Engineering* 62 (2): 522–531.

17 Jarchi, D. and Casson, A.J. (2017). Description of a database containing wrist PPG signals recorded during physical exercise with both accelerometer and gyroscope measures of motion. *Data* 2 (1) https://doi.org/10.3390/data2010001.

18 Salehizadeh, S.M.A., Dao, D., Bolkhovsky, J. et al. (2016). A novel time varying spectral filtering algorithm for reconstruction of motion artifact corrupted heart rate signals during intense physical activities using a wearable photoplethysmogram sensor. *Sensors (Basel)* 16 (1): 10.

19 Cretikos, M.A., Bellomo, R., Hillman, K. et al. (2008). Respiratory rate: the neglected vital sign. *The Medical Journal of Australia* 188 (11): 657–659.

20 Tarassenko, L., Clifton, D.A., Pinsky, M.R. et al. (2011). Centile-based early warning scores derived from statistical distributions of vital signs. *Resuscitation* 82 (8): 1013–1018.

21 Hong, W., Earnest, A., Sultana, P. et al. (2013). How accurate are vital signs in predicting clinical outcomes in critically ill emergency department patients. *European Journal of Emergency Medicine* 20 (1): 27–32.

**22** Subbe, C.P., Davies, R.G., Williams, E. et al. (2003). Effect of introducing the Modified Early Warning score on clinical outcomes, cardio-pulmonary arrests and intensive care utilisation in acute medical admissions. *Anaesthesia* 58 (8): 797–802.

**23** Yilmaz, T., Foster, R., and Hao, Y. (2010). Detecting vital signs with wearable wireless sensors. *Sensors (Basel)* 10 (12): 10837–10862.

**24** Goudra, B.G., Penugonda, L.C., Speck, R.M., and Sinha, A.C. (2013). Comparison of acoustic respiration rate, impedance pneumography and capnometry monitors for respiration rate accuracy and apnea detection during GI endoscopy anesthesia. *Open Journal of Anesthesiology* 3: 74–79.

**25** Karlen, W., Turner, M., Cooke, E. et al. (2010). Capnobase: signal database and tools to collect, share and annotate respiratory signals. In: *Proceedings of the Annual Meeting of the Society for Technology in Anesthesia*, 25. Society for Technology in Anesthesia.

**26** Saeed, M., Villarroel, M., Reisner, A.T. et al. (2011). Multiparameter intelligent monitoring in intensive care II (MIMIC-II): a public-access intensive care unit database. *Critical Care Medicine* 39 (5): 952–960.

**27** Charlton, P.H., Bonnici, T., Tarassenko, L. et al. (2016). An assessment of algorithms to estimate respiratory rate from the electrocardiogram and photoplethysmogram. *Physiological Measurement* 37 (4): 610–626.

**28** Charlton, P.H., Birrenkott, D.A., Bonnici, T. et al. (2017). Breathing rate estimation from the electrocardiogram and photoplethysmogram: a review. *IEEE Reviews in Biomedical Engineering* 11: 2–20.

**29** Nilsson, L., Johansson, A., and Kalman, S. (2000). Monitoring of respiratory rate in postoperative care using a new photoplethysmographic technique. *Journal of Clinical Monitoring and Computing* 16 (4): 309–315.

**30** Addison, P. and Watson, J. (2004). Secondary transform decoupling of shifted nonstationary signal modulation components: application to photoplethysmography. *International Journal of Wavelets, Multiresolution and Information Processing* 2: 43–57.

**31** Garde, A., Karlen, W., Ansermino, J.M., and Dumont, G.A. (2014). Estimating respiratory and heart rates from the correntropy spectral density of the photoplethysmogram. *PLoS One* 9 (1): e86427.

**32** Karlen, W., Raman, S., Ansermino, J.M., and Dumont, G.A. (2013). Multiparameter respiratory rate estimation from the photoplethysmogram. *IEEE Transactions on Biomedical Engineering* 60 (7): 1946–1953.

**33** Shelley, K.H., Awad, A.A., Stout, R.G., and Silverman, D.G. (2006). The use of joint time frequency analysis to quantify the effect of ventilation on the pulse oximeter waveform. *Journal of Clinical Monitoring and Computing* 20 (2): 81–87.

**34** Fleming, S.G. and Tarassenko, L. (2007). A comparison of signal processing techniques for the extraction of breathing rate from the photoplethysmogram. *International Journal of Biomedical Sciences* 2 (4): 232–236.

**35** Lazaro, J., Gil, E., Bailon, R. et al. (2013). Deriving respiration from photoplethysmographic pulse width. *Medical & Biological Engineering & Computing* 51 (1–2): 233–242.

**36** Pimentel, M.A.F., Charlton, P.H., and Clifton, D.A. (2015). Probabilistic estimation of respiratory rate from wearable sensors. In: *Wearable Electronics Sensors* (ed. S. Mukhopadhyay), 241–262. Springer.

**37** Birrenkott, D.A., Pimentel, M.A.F., Watkinson, P.J., and Clifton, D.A. (2016). Robust estimation of respiratory rate via ECG- and PPG-derived respiratory quality indices. In: *38th Annual International Conference of the IEEE Engineering in Medicine and Biology Society (EMBC)*, 676–679. IEEE.

**38** Dehkordi, P., Garde, A., Molavi, B. et al. (2015). Estimating instantaneous respiratory rate from the photoplethysmogram. In: *Conference Proceedings: Annual International Conference of the IEEE Engineering in Medicine and Biology Society*, 6150–6153. IEEE.

**39** Mirmohamadsadeghi, L., Fallet, S., Moser, V. et al. (2016). Real-time respiratory rate estimation using imaging photoplethysmography inter-beat intervals. In: *Computing in Cardiology Conference (CinC)*, 861–864. IEEE.

**40** Orini, M., Pelaez-Coca, M.D., Bailon, R., and Gil, E. (2011). Estimation of spontaneous respiratory rate from photoplethysmography by cross time-frequency analysis. In: *Computing in Cardiology (CinC)*, 661–664. IEEE.

**41** Lapi, S., Lavorini, F., Borgioli, G. et al. (2014). Respiratory rate assessments using a dual-accelerometer device. *Respiratory Physiology & Neurobiology* 191: 60–66.

**42** Jiang, P. and Zhu, R. (2016). Dual tri-axis accelerometers for monitoring physiological parameters of human body in sleep. In: *IEEE Sensors*, 1–3. IEEE.

**43** Morillo, D., Rojas Ojeda, J.L., Crespo Foix, L.F., and Jimenez, A. (2010). An accelerometer-based device for sleep apnea screening. *IEEE Transactions on Information Technology in Biomedicine* 14 (2): 491–499.

**44** Zhang, Z. and Yang, G.Z. (2015). Monitoring cardio-respiratory and posture movements during sleep: what can be achieved by a single motion sensor. In: *IEEE 12th International Conference on Wearable and Implantable Body Sensor Networks (BSN)*, 1–6. IEEE.

**45** Haescher, M., Matthies, D.J.C., Trimpop, J., and Urban, B. (2015). A study on measuring heart- and respiration-rate via wrist-worn accelerometer-based seismocardiography (SCG) in comparison to commonly applied technologies. In: *Proceedings of the 2nd international Workshop on Sensor-based Activity Recognition and Interaction*. ACM https://doi.org/10.1145/2790044.2790054.

**46** Jarchi, D., Rodgers, S.J., Tarassenko, L., and Clifton, D.A. (2018). Accelerometry-based estimation of respiratory rate for post-intensive care patient monitoring. *IEEE Sensors Journals* 18 (12): 4981–4989.

**47** Severinghaus, J.W. and Honda, Y. (1987). History of blood gas analysis: VII: pulse oximetry. *Journal of Clinical Monitoring* 3 (2): 135–138.

**48** Bagha, S. and Shaw, L. (2011). A real time analysis of PPG signal for measurement of $SpO_2$ and pulse rate. *International Journal of Computer Applications* 36 (11): 45–49.

**49** Preejith, S.P., Ravindran, A.S., Hajare, R. et al. (2016). A wrist worn $SpO_2$ monitor with custom finger probe for motion artifact removal. In: *38th Annual International Conference of the IEEE Engineering in Medicine and Biology Society (EMBC)*, 5777–5780. IEEE.

**50** Fu, Y. and Liu, J. (2015). System design for wearable blood oxygen saturation and pulse measurement device. *Procedia Manufacturing* 3: 1187–1194.

**51** Budidha, K. and Kyriacou, P.A. (2013). Development of an optical probe to investigate the suitability of measuring photoplethysmographs and blood oxygen saturation from the human auditory canal. In: *Annual International Conference of the IEEE Engineering in Medicine and Biology Society*, 1736–1739. IEEE.

**52** Webster, J.G. (1997). *Design of Pulse Oximeters*. Taylor & Francis.

**53** Jarchi, D., Salvi, D., Velardo, C. et al. (2018). Estimation of HRV and $SpO_2$ from Wrist-Worn Commercial Sensors for Clinical Settings. In: *IEEE 15th International Conference on Wearable and Implantable Body Sensor Networks (BSN)*, 144–147. IEEE.

**54** Garde, A., Dehkordi, P., Karlen, W. et al. (2014). Development of a screening tool for sleep disordered breathing in children using the phone oximeter. *PLoS One* 9 (11): e112959.

**55** Clifton, L., Clifton, D.A., Pimentel, M.A.F. et al. (2014). Predictive monitoring of mobile patients by combining clinical observations with data from wearable sensors. *IEEE Journal of Biomedical and Health Informatics* 18 (3): 722–730.

**56** McGhee, B.H. and Bridges, E.J. (2002). Monitoring arterial blood pressure: what you may not know. *Critical Care Nurse* 22 (2): 66–70.

**57** Perloff, D., Grim, C., Flack, J. et al. (1993). Human blood pressure determination by sphygmomanometry. *Circulation* 88 (5): 2460–2470.

**58** Alpert, B.S., Quinn, D., and Gallick, D. (2014). Oscillometric blood pressure: a review for clinicians. *Journal of the American Society of Hypertension* 8 (12): 930–938.

**59** Silke, B. and McAuley, D. (1998). Accuracy and precision of blood pressure determination with the Finapres: an overview using re-sampling statistic. *Journal of Human Hypertension* 12 (6): 403–409.

**60** Payne, R.A., Symeonides, C.N., Webb, D.J., and Maxwell, S.R. (2006). Pulse transit time measured from the ECG: an unreliable marker of beat-to-beat blood pressure. *Journal of Applied Physiology* 100 (1): 136–141.

**61** Naschitz, J.E., Bezobchuk, S., Mussafia-Priselac, R. et al. (2004). Pulse transit time by R-wave-gated infrared photoplethysmography: review of the literature and personal experience. *Journal of Clinical Monitoring and Computing* 18 (5–6): 333–342.

**62** Ding, X., Yan, B.P., Zhang, Y.T. et al. (2017). Pulse transit time based continuous cuffless blood pressure estimation: a new extension and a comprehensive evaluation. *Scientific Reports* 7 (1): 11554.

**63** Atomi, K., Kawanaka, H., Bhuiyan, M.S., and Oguri, K. (2017). Cuffless blood pressure estimation based on data-oriented continuous health monitoring system. *Computational and Mathematical Methods in Medicine* 2017: 1803485.

**64** Espina, J., Falck, T., Muehlsteff, J., and Aubert, X. (2006). Wireless body sensor network for continuous cuff-less blood pressure monitoring. In: *3rd IEEE/EMBS International Summer School on Medical Devices and Biosensors*, 11–15. IEEE.

**65** Zhang, Y.T., Poon, C.C.Y., Chan, C.h. et al. (2006). A health-shirt using e-textile materials for the continuous and cuffless monitoring of arterial blood pressure. In: *3rd IEEE/EMBS International Summer School on Medical Devices and Biosensors*, 86–89. IEEE.

**66** Goldberg, E.M. and Levy, P.D. (2016). New approaches to evaluating and monitoring blood pressure. *Current Hypertension Reports* 18 (6): 49.

**67** Ruiz-Rodríguez, J.C., Ruiz-Sanmartín, A., Ribas, V. et al. (2013). Innovative continuous non-invasive cuffless blood pressure monitoring based on photoplethysmography technology. *Intensive Care Medicine* 39 (9): 1618–1625.

**68** Zhang, Y., Zhu, J.M., Liang, Y.B. et al. (2017). Non-invasive blood glucose detection system based on conservation of energy method. *Physiological Measurement* 38 (2): 325–342.

**69** Jindal, G.D., Ananthakrishnan, T.S., Jain, R.K. et al. (2008). Noninvasive assessment of blood glucose by photo plethysmography. *IETE Journal of Research* 54 (3): 217–222.

**70** Monte-Moreno, E. (2011). Non-invasive estimate of blood glucose and blood pressure from a photoplethysmograph by means of machine learning techniques. *Artificial Intelligence in Medicine* 53 (2): 127–138.

**71** Gonzales, W.V., Mobashsher, A.T., and Abbosh, A. (2019). The progress of glucose monitoring—a review of invasive to minimally and non-invasive techniques, devices and sensors. *Sensors (Basel)* 19: 800. https://doi.org/10.3390/s19040800.

**72** Lin, T., Gal, A., Mayzel, Y. et al. (2017). Non-invasive glucose monitoring: a review of challenges and recent advances. *Current Trends in Biomedical Engineering & Biosciences* 6 (5): 1–8.

**73** Villena Gonzales, W., Toaha Mobashsher, A., and Abbosh, A. (2019). The progress of glucose monitoring—a review of invasive to minimally and non-invasive techniques, devices and sensors. *Sensors* 19 (4): 800. https://doi.org/10.3390/s19040800.

**74** Vega, K., Jiang, N., Liu, X. et al. (2017). The dermal abyss: interfacing with the skin by tattooing biosensors. In: *Proceedings of the 2017 ACM International Symposium on Wearable Computers (ISWC'17)*, vol. 11–15, 138–145. ACM.

**75** Childs, C. (2011). *Maintaining body temperature.* In: *Alexander's Nursing Practice* (eds. C. Brooker and M. Nicol), 617–633. Oxford: Elsevier.

**76** Iden, T., Horn, E.P., Bein, B. et al. (2015). Intraoperative temperature monitoring with zero heat flux technology (3M SpotOn sensor) in comparison with sublingual and nasopharyngeal temperature: an observational study. *European Journal of Anaesthesiology* 32 (6): 387–391.

**77** Feng, J., Zhou, C., He, C. et al. (2017). Development of an improved wearable device for core body temperature monitoring based on the dual heat flux principle. *Physiological Measurement* 38 (4): 652–666.

**78** Atallah, L., Ciuhu, C., Wang, C. et al. (2018). An ergonomic wearable core body temperature sensor. In: *IEEE 15th International Conference on Wearable and Implantable Body Sensor Networks (BSN)*, 70–73. IEEE.

**79** Nemati, S., Ghassemi, M.M., Ambai, V. et al. (2016). Monitoring and detecting atrial fibrillation using wearable technology. In: *Conference Proceedings: Annual International Conference of the IEEE Engineering in Medicine and Biology Society*, 3394–3397. IEEE.

**80** Quer, G., Nikzad, N., Lanka, S. et al. (2017). Preliminary evaluation of a wrist wearable heart rate sensor for the detection of undiagnosed atrial fibrillation in a real-world setting. *Circulation* 136: A18610.

**81** Steinberg, B.A., Yuceege, M., Mutlu, M. et al. (2017). Utility of a wristband device as a portable screening tool for obstructive sleep apnea. *Circulation* 136: A19059.

**82** Green, E.M., Mourik, R.V., Wolfus, C. et al. (2017). Machine learning detection of obstructive hypertrophic cardiomyopathy using a wearable biosensor. *Circulation* 136: A24031.

**83** Bilo, G., Zorzi, C., Ochoa Munera, J.E. et al. (2015). Validation of the Somnotouch-NIBP noninvasive continuous blood pressure monitor according to the European Society of

Hypertension International Protocol revision 2010. *Blood Pressure Monitoring* 20 (5): 291–294.

**84** Weenk, M., van Goor, H., Frietman, B. et al. (2017). Continuous monitoring of vital signs using wearable devices on the general ward: pilot study. *JMIR mHealth uHealth* 5 (7): e91.

**85** Ding, X.R., Zhao, N., Yang, G.Z. et al. (2016). Continuous blood pressure measurement from invasive to unobtrusive: celebration of 200th birth anniversary of Carl Ludwig. *IEEE Journal of Biomedical and Health Informatics* 20 (6): 1455–1465.

**86** Simmers, P., Yuan, Y., Sonner, Z., and Heikenfeld, J. (2018). Membrane isolation of repeated-use sweat stimulants for mitigating both direct dermal contact and sweat dilution. *Biomicrofluidics* 12 (3): 034101.

**87** Brattoli, M., de Gennaro, G., de Pinto, V. et al. (2011). Odour detection methods: olfactometry and chemical sensors. *Sensors (Basel)* 11 (5): 5290–5322.

# 5

# Polysomnography and Sleep Analysis

## 5.1 Introduction

Although most body sensor network (BSN) techniques are used for human monitoring during both awake and sleep situations, some types of measurements and methods are more suitable for monitoring humans during sleep. In this chapter some popular sleep monitoring systems and methods together with their analytical approaches are revised or presented and the applications explained in detail.

Human sleep monitoring is crucial for the diagnosis and management of certain medical conditions, such as sleep related disorders, which affect millions of people [1, 2]. A good state of mental and physical health is closely related to a high quality of sleep. There are various physiological parameters which can be used to detect or predict sleep related disorders and assess the quality of sleep. Wearable sensors can be used to measure these physiological parameters and use them in an inference system for monitoring, treatment, and the management of medical conditions.

There are various sleep disorders which stem from neurological, physiological, and even physical disabilities, abnormalities, or diseases. Obstructive sleep apnoea, upper airway resistance syndrome, periodic limb movement disorder, restless leg movement disorder, narcolepsy, rapid eye movement (REM) sleep behaviour, insomnia, circadian rhythm disorder, and consciousness disorder are probably the most common and important sleep abnormalities.

Currently, the use of wearable sensors has been limited to professional installations in clinical settings. Recent research aims to convert the sleep monitoring system into a home setting platform with a reduced number of sensors maintaining a comparable performance to that currently being used in clinics.

Here, the interest is in the methods used in sleep monitoring applications using simple and easily wearable body worn sensors including smartwatches and smartphones. In particular, the methods to identify various sleep stages, classify common body positions, and unobtrusively estimate respiratory rate (RR) are explored.

There is a direct link between physiological parameters and quality of sleep. The methods applied to estimate the physiological parameters in sleep monitoring systems can be integrated within specialised systems for the quantification of sleep quality related to humans' mental, physiological, or physical states.

*Body Sensor Networking, Design and Algorithms,* First Edition. Saeid Sanei, Delaram Jarchi and Anthony G. Constantinides.

Companion Website: www.wiley.com/go/sanei/algorithm-design

Signal processing and machine learning techniques play a crucial role in the analysis and quantification of sensor signals recorded during sleep. Deep neural networks (DNNs), for example, have shown much success in the classification of sleep signals. Many of these methods are explained in detail within this chapter.

## 5.2 Polysomnography

Using polysomnography (PSG), many parameters can be measured from the captured data and used to monitor sleep progress noninvasively. These parameters include heart rate and rhythms, blood oxygen level, RR and patterns, brain waves through electroencephalography (EEG), eye movement, body position, limb movement, and snoring sound. They are then quantified and used together with other measurable clinical evidences to characterise and identify the sleep stages as well as variety of sleep disorders. PSG is recognised as the gold standard for the measurement of sleep quality as well as a quantitative tool for diagnosis of sleep disorders in clinics. A PSG system is shown in Figure 5.1. From this figure, the sensors are placed on the human scalp and upper facial zones to record EEG and electrooculography (EOG) signals. The nose sensor has been used to measure air flow. The electromyography (EMG) electrodes are attached to the chin. Also, in the PSG system, the amount of force required to breathe is measured through the chest and abdominal fitted elastic belt sensors,

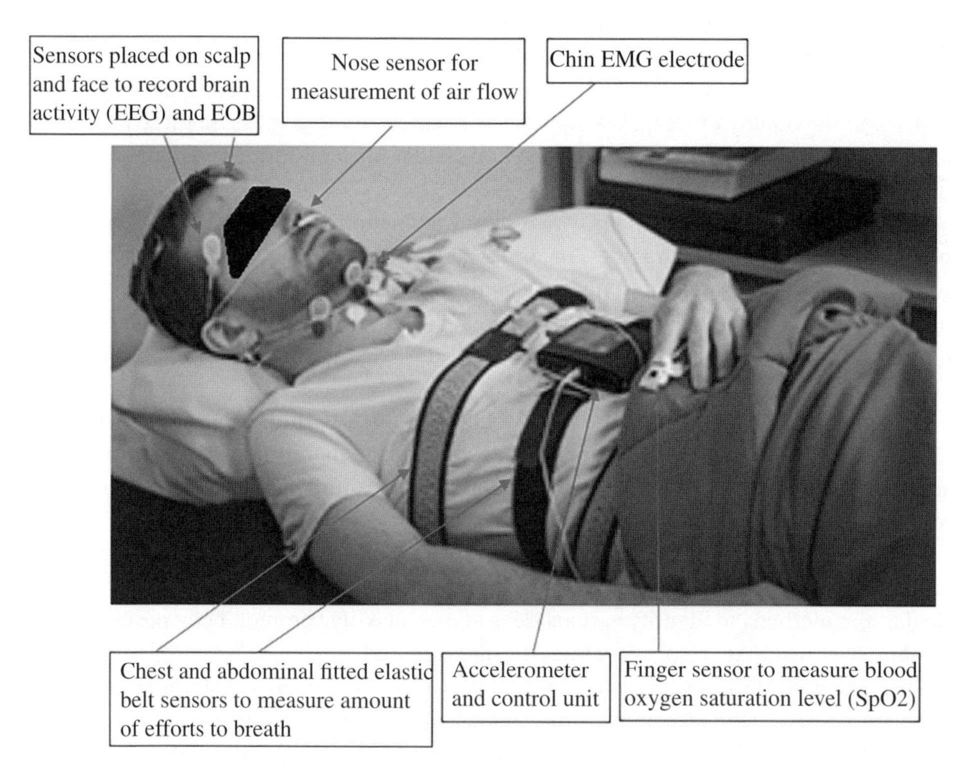


**Figure 5.1** A subject equipped with PSG monitoring system. Leg EMG electrodes may be connected and wired, too. The ECG electrodes are usually applied to the chest.

acceleration signals are recorded from the chest, and photoplethysmography signals are recorded from the finger.

The EEG signals can be processed and used to quantify or score the sleep stage. EOG signals are recorded to measure eye movement and determine the existence of the REM stage. EMG signals are recorded to monitor muscle activity, face twitches, and leg movements. These signals are also useful to determine the presence of REM stages and frequent leg movements, which can be related to restless leg syndrome (RLS).

The blood oxygen saturation level is measured using a finger-worn pulse oximeter. Sleep apnoea can be detected from a low $SpO_2$ value in the oximeter output. Using the technologies embedded in the PSG or actigraphy [3, 4], many systems have been designed for sleep monitoring, including the ZEO (http://www.myzeo.com/sleep), Fitbit (http://www.fitbit.com), and Sleeptracker (http://www.sleeptracker.com). Sleep monitoring using PSG has been performed in a number of studies [5–8].

To enable sleep state recognition, the multimodal PSG data have to be processed using advanced signal processing methods and the most descriptive features estimated accurately before applying them to a classifier for the identification of sleep stages or any mental, physiological, or physical abnormalities related to sleep.

## 5.3 Sleep Stage Classification

Before moving into processing or classifying sleep signals, we start with a study of sleep stages and how these stages can be recognised and scored. Then, EEG-based sleep stage classification techniques are explained. The steps followed by these techniques include pre-processing (if necessary) for removing the artefacts such as eye blinks, feature extraction mainly from single- or multichannel EEG data, and then application of a suitable classification algorithm. In addition, a separate section is devoted to the formulation of deep networks for sleep stage classification using PSG signals. These techniques are then used to perform continuous and automated sleep scoring.

### 5.3.1 Sleep Stages

For a normal human, in addition to the awake state, there are four stages of sleep: nonrapid eye movement (NREM) consisting of three sleep cycles called N1, N2, and N3, and REM [9]. For an adult during the night, the sleep cycles follow as N1 → N2 → N3 → N2 → REM. During N1, the low voltage EEG signals dominated by slow theta frequency patterns are observed. Other higher frequency activities such as beta may also appear at this stage. In the N2 stage, theta activities and sleep spindles or K-complexes are observed in the recorded EEG signals. As the sleep progresses into the N2 stage, high voltage EEG signals with slow wave activities appear and continue to the start of the N3 stage. In this stage, EEG activities in the delta band are usually observed. At the end of the N3 stage, there is a lighter NREM sleep episode which is also evident in the form of a series of body movements. Sleep spindles or K-complexes may continue from N2 to N3. Then, the initial REM stage starts in which the sawtooth waves appear. The cyclical patterns of NREM and REM stages are continued overnight with some variations in the length of different stages. A set of EEG recordings in different sleep stages is demonstrated in Figure 5.2.

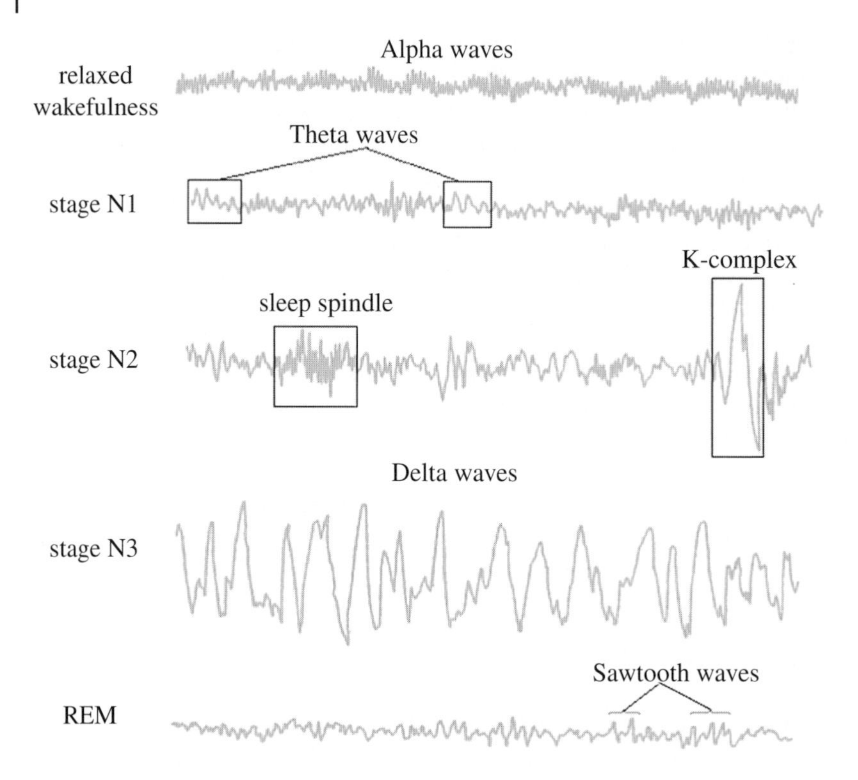


**Figure 5.2** An example of sleep EEG signals and waveforms that appear in different sleep stages.

### 5.3.2 EEG-Based Classification of Sleep Stages

EEG signals can be recorded from single or multiple channels. They are used in sleep monitoring applications mainly for the detection of various sleep events and more recently for the identification of sleep stages. Following data acquisition, the EEG signals are pre-processed using different techniques, mostly digital filtering (e.g. bandpass Butterworth filter). The EEG signals may also be manually or automatically scanned for the removal or suppression of artefacts such as eye blinks or eye movements. Blind source separation followed by a classifier such as that in [10] can be very effective approach for the removal of the eye-blink artefact.

Then, for each data epoch (the segmented EEG signal), most descriptive features are extracted through the application of suitable signal processing algorithms. These features are used by a classifier for epoch analysis and sleep stage classification. The features include time domain, frequency domain, time-frequency domain, and nonlinear features which are explained in the following sections.

#### 5.3.2.1 Time Domain Features

The basic time domain (temporal) features include mean, variance, SD, and median. The skewness [11] is a measure to determine the asymmetry of a probability distribution. The skewness is zero if the data are perfectly symmetrical. If the skewness is negative, it means the data are negatively (left) skewed, and if it is positive, it means the data are positively (right) skewed. Skewness of a zero-mean signal **x** having $N$ samples ($x_i$) is estimated as:

$$Skewness = \frac{1}{N}\sum_{i=1}^{N}\left(\frac{x_i}{\sigma}\right)^3 \tag{5.1}$$

where $\sigma$ is SD. Kurtosis [11] is a measure of how peaky or tailed the distribution of signal is. Kurtosis is defined as:

$$Kurtosis = \left[\frac{1}{N}\sum_{i=1}^{N}\left(\frac{x_i}{\sigma}\right)^4\right] - 3 \tag{5.2}$$

The Hjorth parameters [12] are another form of temporal parameters introduced by Bo Hjorth which include activity, mobility, and complexity. These parameters are computed using the variance of a target time-series or input signal **x**, and the first and second derivatives $\acute{X}$ and $\acute{\acute{X}}$, are defined as:

$$H_a = var(\mathbf{x}) \tag{5.3}$$

$$H_m = \sqrt{var(\mathbf{x})/var(\mathbf{x})} \tag{5.4}$$

$$H_c = \sqrt{var(\mathbf{x}) \times var(\acute{\mathbf{x}})/var(\acute{\mathbf{x}})^2} \tag{5.5}$$

where $H_a$, $H_m$, and $H_c$ represent activity, mobility, and complexity parameters, respectively. These parameters are applied to the PSG signals and have been shown to provide dynamic temporal information useful in sleep analysis mainly in the context of sleep stage classification [13, 14].

#### 5.3.2.2 Frequency Domain Features

Analysis of sleep EEGs in the frequency domain based on the extracted spectral features is very common. As an example, sleep signals can be scored by dividing the sleep EEGs into their typical frequency bands (e.g. delta, theta, alpha, beta) and analysing the spectral features. These features can be the ratios between signal powers in different bands. Non-parametric methods can be used to estimate the spectrum of the signal. These methods include Fourier transform [15], power spectral density (PSD), and spectrum estimation using the Welch method [16].

#### 5.3.2.3 Time-frequency Domain Features

Analysis of EEG signals that are generally nonstationary in a joint time and frequency domain is particularly important in sleep analysis [17]. Across various sleep events such as sleep spindles and arousal, sudden temporal changes in amplitude and frequency are visible in the joint time-frequency spectrum of the EEG waveforms. These changes can be further quantified to help in the identification of sleep events and the classification of sleep stages, so-called sleep scoring. The most commonly used methods to estimate the joint time-frequency spectrum include short-time Fourier transform (STFT), wavelet transform (WT), matching pursuits (MP), and empirical mode decomposition (EMD). These methods are briefly described in the following subsections.

#### 5.3.2.4 Short-time Fourier Transform

The simplest form of time-frequency analysis is STFT where the signal is uniformly segmented into overlapping short duration windows. Then, the fast Fourier transform (FFT) is applied to each window. For a signal $x(t)$, STFT is calculated as:

$$STFT\{x(t)\}(\tau, \omega) \equiv X(\tau, \omega) = \int_{-\infty}^{\infty} x(t)w(t-\tau)e^{-j\omega t}dt \tag{5.6}$$

where $w(t)$ is a window function (e.g. Hanning, Hamming, and Gaussian windows) that is centred around zero and $X(\tau, \omega)$ is the resulting Fourier transform of $x(t)w(t-\tau)$ that captures the phase and amplitude of the input signal over time and frequency. From Eq. (5.6), a two-dimensional representation (time and frequency) of the signal is obtained by applying STFT to the desired signal. In the cases where the squared magnitude of the STFT is used the method is noted as spectrogram. In practice we deal with signal samples and therefore, a discrete version of Eq. (5.6) is used where the integral changes to sum and $\omega t$ changes to $2\pi nk/N$, given that $N$ is the segment (or window) length.

#### 5.3.2.5 Wavelet Transform

The continuous wavelet transform (CWT) convolves the input signal with a set of functions produced using the specified mother wavelet [18–20]. The CWT of a signal $x(t)$ is expressed as:

$$X_w(a, b) = \frac{1}{|a|^{1/2}} \int_{-\infty}^{\infty} x(t)\overline{\psi}\left(\frac{t-b}{a}\right) dt \tag{5.7}$$

where $a$ is the scale parameter; $b$ is the translational (shifting) parameter; $\psi(t)$ is the mother wavelet, which is a continuous function in both time and the frequency domain; and $\overline{\psi}(t)$ represents its complex conjugate. The main objective in using the mother wavelet is to produce daughter wavelets which are simply the scaled and translated versions of the mother wavelet. Therefore, WT uses variable size windows to perform time-frequency decomposition of the input signal. The use of size adjustable windows in WT, where short duration wavelets represent higher frequencies and long duration wavelets represent low frequencies, enables the analysis of the temporal variation of a particular frequency component of the input signal. This is a major advantage of WT over the STFT, where fixed length windows are selected and the FFT is applied to each window. Discrete wavelet transform (DWT) can be applied to the input signal to perform denoising, which is not possible using the STFT. WT has been used in many different approaches for sleep analysis and automatic sleep stage identification [21–23].

#### 5.3.2.6 Matching Pursuit

MP is a signal dependent and adaptive technique in time-frequency representation of signals employing a large dictionary of functions which are not limited to sinusoids or wavelets as employed in Fourier transform or WT [24, 25]. Having a dictionary of basic functions, also called atoms, the operation of an MP algorithm is to find functions from the dictionary that their combination best represents the input signal. One of the shortcomings of an MP algorithm is its high computational cost. However, it generates a high resolution time-frequency pattern [26, 27] which is very desirable for the analysis of sleep EEG, particularly to estimate and recover the sparse activities such as sleep spindles or K-complexes [25, 28, 29].

In [29], singular spectrum analysis (SSA) has been applied for the decomposition of EEG signals to identify brain waves, K-complexes, and sleep spindles, where an MP approach has been used to represent time-frequency representations. SSA is a nonparametric technique which decomposes a single channel data into its constituent components through embedding, Hankelisation, singular value decomposition, and reconstruction stages [30].

#### 5.3.2.7 Empirical Mode Decomposition

EMD is a heuristic decomposition technique which is adaptive, data driven, and usually applied to nonstationary signals. An EMD algorithm uses an iterative sifting process to decompose the input signal into its main oscillatory modes, called intrinsic mode functions (IMFs). Hilbert transform [31] can then be applied to each extracted IMF to represent the instantaneous frequency and amplitude information. EMD has been used in limited research studies for the automatic identification of sleep states and the screening of obstructive sleep apnoea using ECG signals [32, 33].

#### 5.3.2.8 Nonlinear Features

EEG signals are assumed to be generated from stochastic processes which are generally nonlinear. Therefore, various nonlinear stochastic methods are used to model these signals. In the context of analysing sleep EEG waveforms, different complexity features are extracted and used to characterise the dynamical behaviour of EEGs in certain sleep stages. These features include fractal-based features and entropy measures.

Fractal dimension is a feature used to directly measure the complexity of a time-series by quantification of fractal dimensionalities (geometries that are self-similar on various scales). The fractal dimension can be a non-integer value. Different types of fractal dimensions include box counting, information, Higuchi, correlation, and Lyapunov dimensions.

Fractal dimensions can be applied to short segments of time-series data. Therefore, they are more appropriate for the detection of transient events in the EEG signals. There exist several algorithms for computation of fractal dimension which are also used in sleep EEG analysis.

Entropy is another nonlinear feature and an indicator of randomness, which has been demonstrated to be useful in identification of sleep disturbances [34, 35]. It is defined as:

$$Entropy = -K \sum_{i=1}^{N} p_i \ln(p_i) \tag{5.8}$$

where $K$ is a positive constant, $p_i$ is the probability of $i^{th}$ sample, and $N$ is the number of samples. Various entropy measures such as approximate entropy (ApEn) have been used to capture distinctive characteristics of EEG signals in different sleep stages [34, 35]. In some recent approaches more refined methods suitable for measuring the complexity in the biomedical signals have been defined. In a very recent approach [36] a multiscale fluctuation-based dispersion entropy (MFDE) has been developed as a robust approach at the presence of baseline wanders or trends in the data. MFDE has then been applied to focal and nonfocal EEGs; walking stride interval signals for young, elderly, and Parkinson's subjects; stride interval fluctuations for Huntington's disease and amyotrophic lateral sclerosis; EEGs for controls and Alzheimer's disease patients; and finally eye movement data for Parkinson's disease and ataxia. It proved to be a fast and accurate approach, especially where the mean value of a time-series considerably changes along the signal (e.g. eye movement data).

### 5.3.3 Classification Techniques

Following the construction of feature vectors using time domain (temporal), frequency domain, time-frequency domain or nonlinear/complexity features, the feature vectors (each feature vector corresponds to one epoch) can be classified into different sleep stages. Many classification techniques have been proposed in the literature for the identification of sleep stages. These techniques include $k$-nearest neighbour ($k$NN) [16], support vector machines (SVM) [37], random forest [38], linear discriminant analysis [39], and hidden Markov model (HMM) [40]. Most of these classifiers are able to classify sleep stages up to a very high accuracy as long as the features are optimal. For example, the features estimated using SSA followed by MP best describe the state of the signals and, therefore, an SVM approach can successfully perform their classification. Neural networks (NNs) are also very successful in many applications due to their nonlinearity and the flexibility in selecting their parameters (such as number of neurons and layers). Since DNNs, which are more scalable than NNs, have recently attracted much attention in the context of sleep analysis, a detailed explanation is provided in the following section.

In another approach for analysis of sleep EEG, for the first time, a tensor-based multiway SSA was designed and applied to the signals [41, 42]. It has been demonstrated that the method better exploits the signal nonstationarity and therefore more accurate scoring of the sleep stages is achieved. A quaternion SSA, on the other hand, better exploits the variability in the localisation of slow waves, spindles, and K-complexes within the brain and, therefore, the sleep waveform components are extracted more accurately [43]. Chapter 9 of this book covers more detailed explanations of various machine learning methods frequently used in BSNs.

#### 5.3.3.1 Using Neural Networks

NNs are designed based on biological neural systems although with a simpler structure and learning capabilities. They have been widely used in the design of intelligent systems for many applications, such as classification, pattern recognition, and feature selection. These networks have been used to process various types of physiological signals. As one example, they are used to analyse the EEG signals during cognitive load, sleep, or at rest [44, 45].

A simple feed-forward NN with two hidden layers is shown in Figure 5.3a. The NN consists of elementary units called neurons. Similar to a biological neuron, the neurons in the NN process the input data and produce an output. As can be seen from Figure 5.3b, each neuron multiplies the input values by their associated weights and then sums the weighted values up. Then, a specified bias is added and the final value is used as an input to an activation function to produce the neuron output. Different activation functions including linear, log-sigmoid, tan-sigmoid, or hard limiter can be used.

Consider a single layer NN (as shown in Figure 5.3b) with an input data vector as **a** with $n$ elements $(a_{1,} a_{2,} \dots, a_{N,})$ and associated weights **w** $(w_{1,} w_{2,} \dots, w_{N,})$. The dot product of **a**

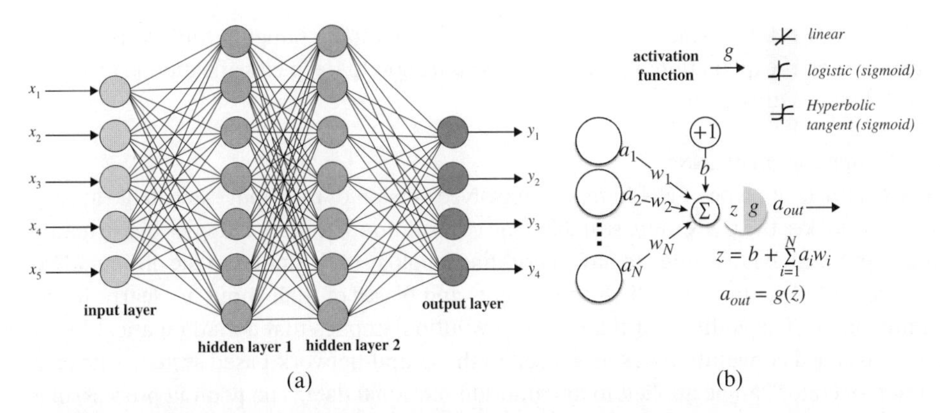


**Figure 5.3** (a) An NN with two hidden layers and (b) graphical model of a single layer NN.

and **w** is computed and a bias $b$ is added to the sum of products:

$$z = b + \sum_{i=1}^{N} a_i w_i \tag{5.9}$$

Then, the activation function **g** produces the neuron output as $y$ ($y = \mathbf{g}(z)$). The weights and biases of the neurons are learned in the training stage of the NN. In this stage, using the available training data and their class labels, the possible smallest error of differences between the desired and real outputs are used to adjust the weights and biases. Typically, mean squared error (MSE) is used to quantify the errors between the desired and real outputs.

In [46], an NN with two hidden layers is proposed to classify five sleep stages. The inputs to the NN are the filtered EEG, EMG, and EOG signals. To filter EEG signals, a bandpass filter at frequencies within 0.3–50 Hz has been used. A Butterworth filter may be used to filter the EOG signals as well as noise with cut-off frequencies of 0.5 and 100 Hz. Similarly, for the EMG signals, a Butterworth filter with cut-off frequencies of 40 and 4000 Hz is used for removing the disturbing signals. The NN outputs are associated with each of the five sleep stages.

Sleep signals are scored using NNs with features from single-channel EEG signals in [47]. In another study, by using EEG, EMG, and EOG recordings of five healthy subjects, sequential feature selection followed by the application of NNs is applied for the classification of various sleep stages [48]. Sequential feature selection algorithms are a family of greedy search algorithms utilised to reduce an initial high-dimensional feature space to a lower-dimensional feature subspace. Using these algorithms, a subset of features that is most relevant to the problem are automatically selected. Nevertheless, in the above work some of the features are selected manually.

Apart from the basic NNs, convolutional neural networks (CNNs) have been recently applied to the PSG data to analyse the sleep stages. An advantage of CNNs over using hand-crafted features and applying traditional classification is that the features are learnt from the data enabling more generalisation for new subjects or test data. On the other hand, the network is scalable and can be expanded without changing the feature space. Following a brief explanation of CNNs, their application to sleep classification is provided in the following subsections.

#### 5.3.3.2 Application of CNNs

A CNN is a specific type of NN in that a convolved version of the input is available at each layer. This makes the CNN more scalable. The convolution size often matches the structure of the input, which is assumed to have a grid-like topology [49]. Similar to the ordinary NN, CNNs consist of neurons in which the weights and biases of their input are learnt in the training phase. The addition of the term 'convolution' implies that a mathematical linear operation called convolution has been used in the neural-network-based system. The convolution operator can be applied to any multidimensional data. The pooling process after each convolution operation makes the best aggregate value of the number of samples. In general, three main layers in a CNN architecture include: convolutional later, pooling layer, and a fully connected layer. Often a large number of these layers exist in a CNN structure.

In the convolutional layer, a kernel needs to be constructed first. Then, the input data is convolved with the kernel to generate the output. This output is expected to provide an interesting feature of the input. Suppose that the input is a 2D image **I** (as a 2D matrix) and the kernel **K** is a small 2D matrix of size $m \times n$. The convolved image $S$ is produced by the following equation:

$$S(x,y) = (I * K)_{xy} = \sum_{i=1}^{m} \sum_{j=1}^{n} K_{ij} I_{x+i-1,y+j-1} \tag{5.10}$$

It has been well established that the use of CNNs improves the learning phase of the systems by having sparse weights or sparse connectivity using kernels of smaller size than the input data size. This is particularly useful with multidimensional data (with thousands or millions of entries) where using small kernels applied to a smaller region of the data (tens or hundreds of entries) allows meaningful features or edges to be detected. This makes the system more efficient as fewer parameters need to be retained. Consequently, it reduces the memory requirements of the model.

An example of a 1D kernel applied to a 1D signal (of size $W$) is provided in Figure 5.4. In this example, the kernel size is ($F = 3$), and it is applied to the 1D signal with two different

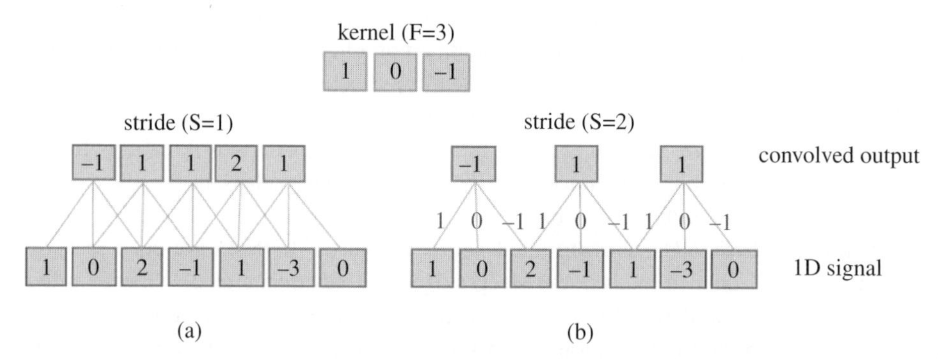


**Figure 5.4** One-dimensional convolution for (a) a stride length of 1 ($S = 1$) and (b) a stride length of 2 ($S = 2$).

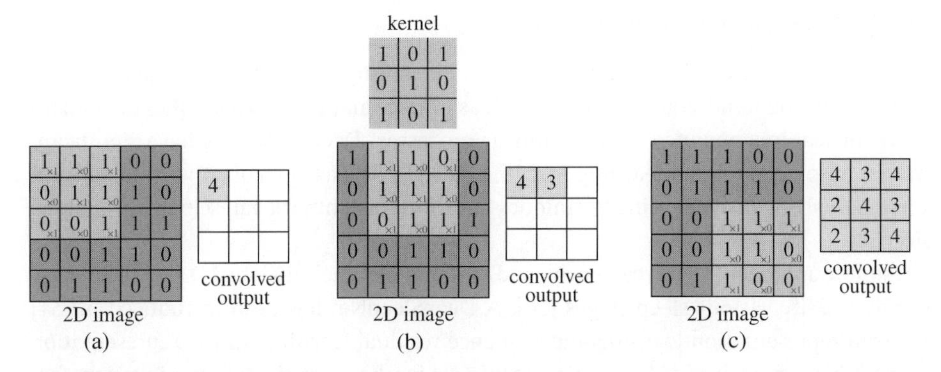


**Figure 5.5** Two-dimensional convolution: (a) first sliding window; (b) second sliding window; (c) the last sliding window.

stride lengths of ($S = 1$) and ($S = 2$) in Figure 5.4a and b, respectively. From Figure 5.4, the size of convolved output is reduced to $\left(\frac{W-F}{S}\right) + 1$. Another example for a 2D signal, such as an image with a 2D kernel, is demonstrated in Figure 5.5. In this figure the results of applying a 2D kernel of size ($F_1 \times F_2$) in three sliding windows (first, second, and last), with a stride length of 1 ($S = 1$), are provided. It can be easily followed that the size of input 2D signal has been reduced from ($w \times H$) to ($w_1 \times H_1$) where $w_1 = \left(\frac{w-F_1}{S}\right) + 1$, and $H_1 = \left(\frac{H-F_2}{S}\right) + 1$. Therefore, a 2D kernel of size ($3 \times 3$) has reduced the size of the input 2D image from ($5 \times 5$) into ($3 \times 3$) having a stride length of 1 ($S = 1$). In the case of applying $K$ filters, the output size will be $w_1 \times H_1 \times K$. In some convolutional layers a number of zeros are padded to the signals. Assume that the amount of zero-padding is $p$. Therefore, the output size considering $K$ filters would be $w_2 \times H_2 \times K$ where $w_2 = \left(\frac{w-F_1+2P}{S}\right) + 1$ and $H_2 = \left(\frac{H-F_2+2P}{S}\right) + 1$.

In the next stage, a nonlinear activation function such as rectified linear activation (RELU) is applied. Typically, for an input neuron $x$, $max(0, x)$ is used to threshold the data at zero. It is worth noting that applying an RELU does not change the network or layer dimension.

In the pooling layer, the output layer dimension is further decreased. Therefore, the pooling layer applies a down-sampling operation mainly along the spatial dimensions. Using one of the most common types of pooling, namely max pooling, a filter of size $F \times S$ is used to extract part of the input befor applying a maximum operation. Therefore, for an input size of $W_1 \times H_1 \times D_1$ the output size will be $W_2 \times H_2 \times D_2$, where $W_2 = \frac{W_1-F}{S} + 1, H_2 = \frac{H_1-F}{S} + 1$, and $D_2 = D_1$.

An average operation may be used instead of a maximum operation, though this is not very common. Application of zero padding in the max pooling stage is not usual either.

Finally, the neurons in the fully connected layer are connected to all neurons from the previous layer. The neuron's activations can be computed using a matrix multiplication operator following a bias of offset.

### 5.3.4 Sleep Stage Scoring Using CNN

The purpose of using CNN in automatic sleep stage scoring is to lessen the dependency of learning system on hand-crafted features such as time-domain features, frequency-domain features, or features obtained from nonlinear processes. Deep networks have significant power for automatic feature learning from the data. There has been ongoing research and interest in applying deep learning techniques to various benchmark datasets in the analysis of sleep stages [50–55].

In a simple approach, CNNs can be applied to single-channel EEGs as 1D signals to automatically classify various sleep stages [50]. A DeepSleepNet has been introduced in [51] to perform representation learning and sequence residual learning. In the representation learning, CNNs are trained to learn filters to be used for the extraction of time-invariant features using a single-channel EEG. Then, the bidirectional long short-term memory (LSTM) is trained for encoding the temporal information, mainly the sleep stage transitions, into the designed model. LSTM is a recurrent neural network (RNN) architecture used in deep learning. Unlike standard feedforward neural networks, LSTM has feedback connections and can be trained using the entire data sequence.

Analysis of raw 1D signals has many applications in single-channel systems, such as speech signals, and can be directly used as a CNN's input [56–60]. On the other hand, the time-frequency representation, for example using spectrogram of 1D signals, can be treated as an image or a 2D input to the CNNs. There are numerous studies in speech, acoustic [61–64], and activity recognition from accelerometer and gyroscope signals [65] that have applied CNNs to 2D input corresponding to time-frequency representations of the time-series data.

A CNN architecture for scoring sleep stages is demonstrated in Figure 5.6. From this figure, the input to the CNN architecture is a 1D EEG signal and two convolutional layers, each followed by a pooling layer, are devised. The CNN architecture includes two convolutional layers (C1, C2), two max-pooling layers (P1, P2), one stacking layer (converting 20 1D signals into a single 2D signal stack) after the first max pooling layer. Finally, a 5-class softmax layer has been added to enable scoring the sleep stages.

The outputs of CNN for a selected subject during night 1 for 120 epochs (1 epoch per 30 seconds) for a duration of approximately six hours are shown in Figure 5.7, where the manually scored hypnogram is also provided as a benchmark for comparison [50].

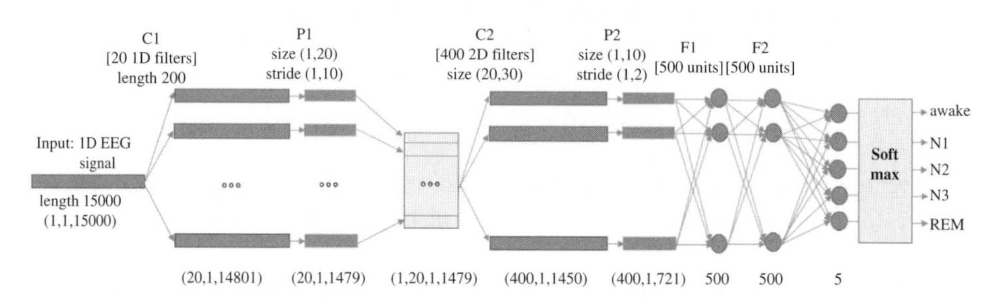


**Figure 5.6** CNN architecture for a single-channel sleep EEG signal [50]. Source: Courtesy of Zhang, J., and Wu, Y.

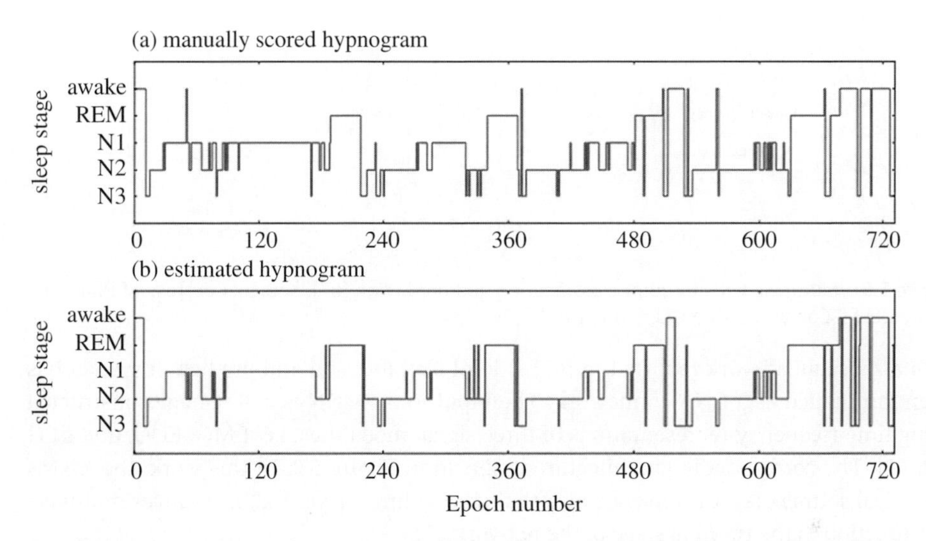


**Figure 5.7** The results of sleep stage classification using CNN architecture; top graph is the benchmark using hypnogram; bottom graph includes the estimated scores [50]. Source: Courtesy of Zhang, J., and Wu, Y.

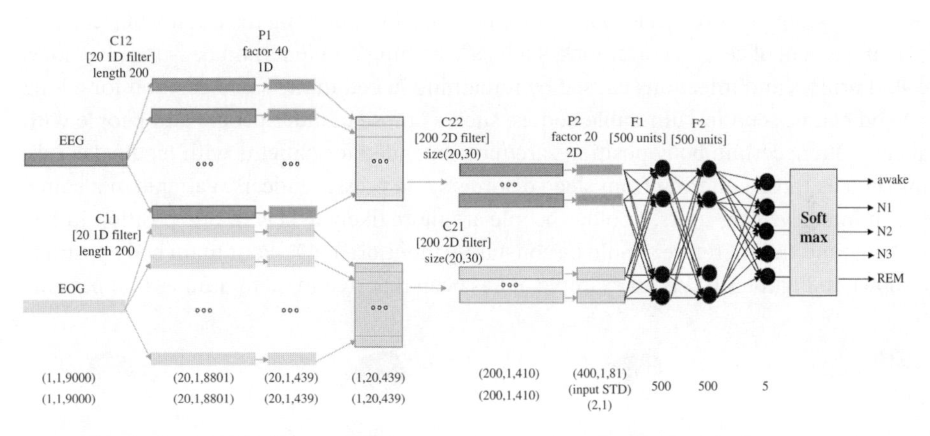


**Figure 5.8** A CNN architecture with EEG and EOG as its inputs [66]. Source: Courtesy of Mikkelsen, K., and De Vos, M.

Based on the results of the work in [50], the formulated CNN using single-channel EEG signals leads to a comparable performance with the state-of-the-art approaches from hand-engineered features. Followed by this work, the formulated CNN is slightly changed to include the EOG signals as another input [66]. Figure 5.8 shows the new architecture. In this architecture, each input data epoch has been rescaled to a standard deviation (SD) of 1. The two original SDs are used as an input to the F1 layer parallel to the output from P2 layer. Another variation in this CNN is that, instead of using one epoch per 30 seconds as in [50], 3 epochs with total duration of 90 seconds are used as input into the CNN where an improvement in the performance has been observed.

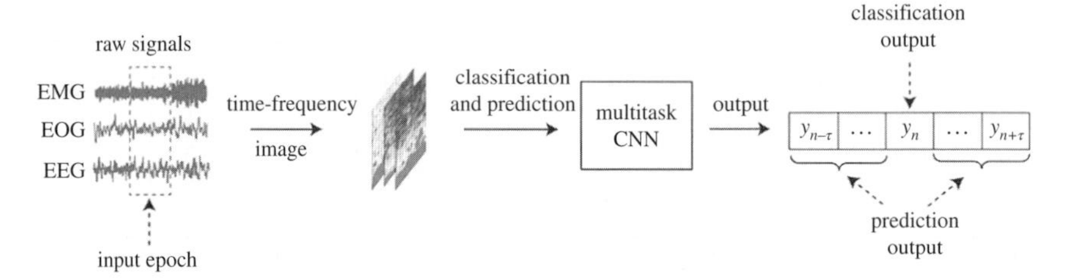


**Figure 5.9** Multitask CNN for joint classification and prediction [67]. Source: Courtesy of Phan, H., Andreotti, F., Cooray, N., Chen, O.Y., and De Vos, M.

In a subsequent work reported in [67], a joint classification and prediction system has been formulated as a CNN framework. The input to the CNN is a 2D image constructed using time-frequency representations of three signal modalities, i.e. EMG, EOG, and EEG [67, 68]. The corresponding architecture is shown in Figure 5.9. In this work, the CNN's canonical softmax layer is replaced by a multitask softmax layer that introduces multitask loss function in the training stage of the network.

## 5.4 Monitoring Movements and Body Position During Sleep

Monitoring sleep posture has important applications in improving the quality of sleep and the management of certain conditions, such as breathing problems and bedsores. Bedsores are skin bruises and infections caused by remaining in bed in the same position for a long time and can be seen in vulnerable bodies, such as those of older people and people with diabetes. Often, certain body positions are recommended for patients with medical conditions such as those suffering from sleep disorders and pressure ulcers. Patients in a coma, those in intensive care units, or older people are more likely to stay in one position in bed for a long time and therefore should be constantly monitored [69]. Four main body postures are shown in Figure 5.10. As an example, it has been reported in many studies that patients

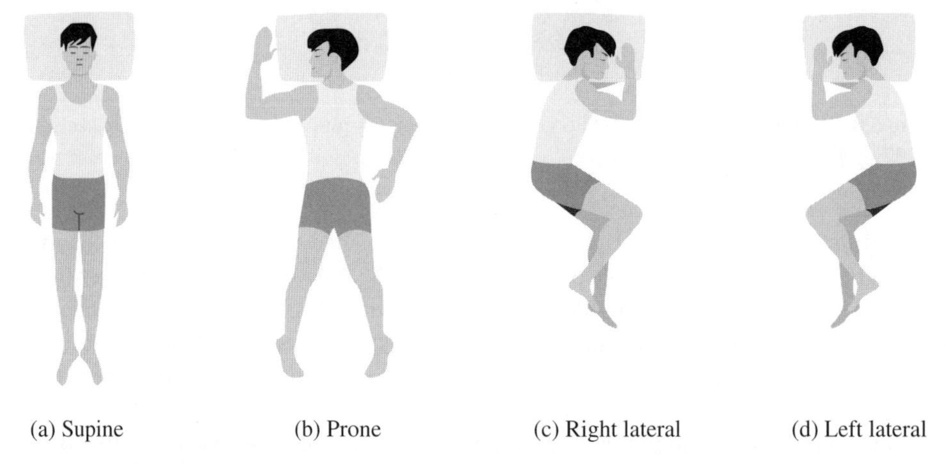


**Figure 5.10** Body postures for a subject lying down: (a) facing upwards; (b) facing downwards; (c) facing right; (d) facing left.

with sleep apnoea such as obstructive sleep apnoea (OSA) should avoid sleeping in a supine position [70–72]. This position has shown to be more associated with the occurrence of OSA than other postures. Hence, detection of various body postures and their changes throughout sleep is crucial to prevent patients with OSA sleeping in a supine position.

The body position during sleep can be captured by video cameras and classified automatically after the images are processed and the distinctive features extracted. An effective approach using tensor factorisation followed deep learning classifier optimised by transfer learning for classification of infrared images captured by a Kinect system is proposed in [73]. Ten subjects each in twelve different sleep postures participated in the experiment. It was shown that, even when using a blanket during sleep, the postures can be classified with an acceptable accuracy. However, the video capturing of humans is not always acceptable, owing to security and privacy.

Systems such as depth cameras [74], integrated pressure sensors like dense pressure-sensitive bedsheet/mattress [75], and wearable sensors [76–79] are used in research studies to monitor body postures. The body postures are estimated using ECG signals from 12 capacitively coupled electrodes where a conductive textile sheet [76] has been used as a reference electrode for the measurement of the electrical potential. The textile sheet has been attached to the lower part of the bed below the leg of the subject. The coupled electrodes are used to classify the body postures into the main four postures of supine, prone, right, or left lateral shown in Figure 5.10. Based on the work in [76], similar QRS morphologies are observed for the same group of body postures. An example is provided in Figure 5.11.

Based on this work, various classifiers – such as linear discriminant analysis (LDA), SVM, and artificial neural network (ANN) – are trained and their performances for classification of body postures are examined and compared. Most applied classifiers achieved a high level of accuracy (over 97%) with some differences in their performances.

In another research, a single chest-worn triaxial accelerometer has been used to identify the static or dynamic movements during sleep. The aim is to detect and quantify posture and cardiorespiratory parameters during sleep and distinguish static postures, associated with small variations in the acceleration signals, from dynamic movements associated with posture changes. Once the static movements are recognised, the subject's posture is further

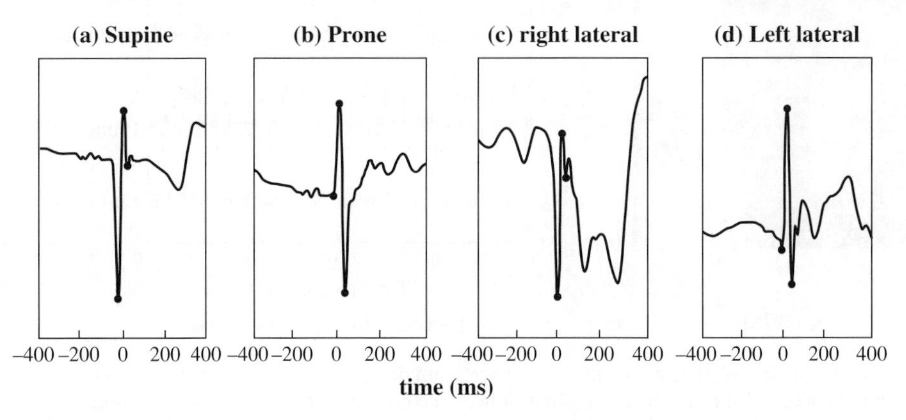


**Figure 5.11** QRS morphologies for different body postures using coupled electrodes [76]. Source: Courtesy of Lee, H.J., Hwang, S.H., Lee, S.M., Lim, Y.G., and Park, K.S.

classified into one of the four common body postures (shown in Figure 5.10). The LDA classification has been used where the feature inputs are obtained using the average of accelerations from three axes [77].

A wireless identification and sensing platform (WISP) as a low-power and low-cost device for continuous monitoring of static body postures [78] and the transitions between postures [79] has been designed. The WISP is placed over the subject's sternum, as shown in Figure 5.12a. The acceleration signals from the y-axis for various body positions are shown in Figure 5.12b. The noise level of accelerations and SDs of the accelerations in three axes are used to detect the body movements and transitions to a new posture [79]. Recognition of body postures is shown to be comparable with the use of pressure sensors. These sensors have also been used to detect entry to and exit from the bed [78]. It has been demonstrated that this postural transition often associated with the occurrence of fall mainly in the elderly care units can, therefore, be used for fall detection of vulnerable patients in hospital environments.

Built-in sensors from smartwatches or smartphones are also used in various sleep analysis applications. Acceleration signals given by the build-in accelerometer of a smartwatch have been used for the estimation of both body position and RR in sleep monitoring applications [80]. Based on the measurements from the developed SleepMonitor device, the acceleration data are collected from the smartwatch worn on the subject's wrist. Then, two approaches are followed and implemented on Android Wear-based smartwatches. The outcome of this research has shown a good performance for monitoring respiration and body positions. For an estimation of the RR, FFT has been applied to each acceleration axis since the frequency of the respiration is assumed to be reflected in the fluctuations caused by repeated inhalation and exhalation. Then, a Kalman-filtering-based approach is designed for sensor fusion of estimated RR from three axes.

For an estimation of body positions, various features including mean, SD, 20th percentile, median, 80th percentile of each accelerometer axis, and the covariance and the

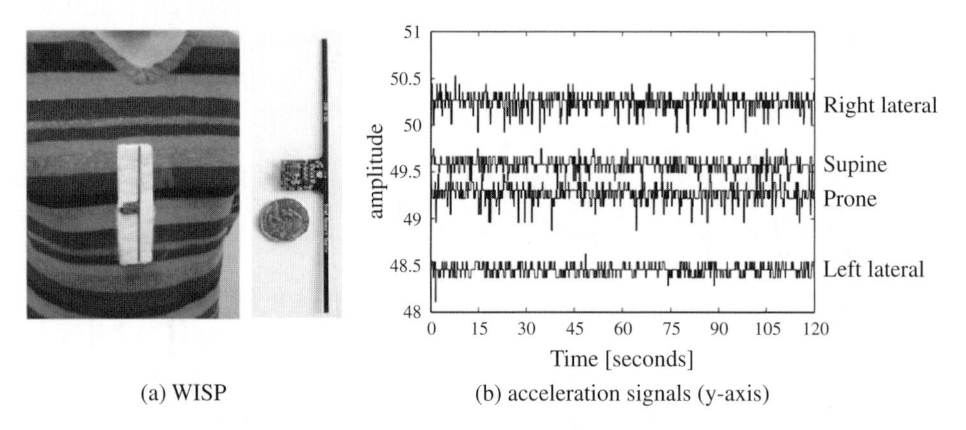


**Figure 5.12** (a) The WISP is a 145 mm × 20 mm × 2 mm device which needs to be attached to the sternum; (b) acceleration signals from y-axis of the accelerometer in four common postures.

correlation between selected accelerometer axes have been used. Then, four classification algorithms – namely Bayesian network, naïve Bayes, decision tree, and random forest – have been employed for the estimation of the four common body positions.

In other studies, acoustic signals have been analysed to quantify sleep quality using two established indices [81, 82] given by the actigraphy and Pittsburgh Sleep Quality Index [83]. Based on the data from an iSleep system in [81] the acoustic signals given by the built-in microphones of a smartphone have been used to detect various events such as body movement, snore, and cough. A decision-tree algorithm has been devised for the classification of different events using the features extracted from the acoustic signals. The features include root mean square, ratio of low-band to high-band energies, and variance.

A system of sleep stage detection based on actigraphy called Sleep Hunter has been proposed which enables the prediction of transitions between sleep stages by a smartphone [82]. Human sleep quality has been investigated which provides a smart call service, using artificial intelligence, to wake up the subjects while they are in the light sleep stage. Using a Sleep Hunter device, the data from microphone, accelerometer, light sensor, and clock, and personal information, are fused to exploit acoustic, vibration, illumination, sleep duration, and personal information in sleep analysis.

The performance of the Sleep Hunter for sleep stage identification has been compared with Jawbone Up and Sleep As Android, which are two representative actigraphy-based products used in monitoring sleep stages [82]. It has been concluded that the performance of Sleep Hunter is superior to that of existing actigraphy-based system for the detection of sleep stages.

## 5.5 Conclusions

The continuous monitoring of sleep has certain applications for the management and monitoring of sleep disorders associated with the mental, physical, and physical state of humans. It also helps improve the quality of sleep for patients in care units or those recovering from surgery. Wearable sensors have been widely employed in sleep monitoring systems. While there is evidence to support the use of wearable sensors in sleep analysis, current research needs to be directed towards making reliable and unobtrusive systems ensuring both patient compliance and accuracy of the designed system for long-term and home-based monitoring. The new miniaturised embedding devices, microcontrollers, and microelectronic circuits will also bring the overall cost of the equipment down, making them accessible to more people. In addition, more work is underway to better make sense of the sleep data from a minimum number of, and minimally intrusive, sensors. Last but not least, owing to the variation of the human body's metabolism from awake to sleep, a great deal of research is being carried out to examine the effect of taking medicine during sleep (e.g. through intravenous injection) compared to that during the day. Such a medication regimen may (or may not) have significant impact on treating various diseases. Therefore, better and more advanced sensing of biological and physiological changes of the human body during sleep will soon be a hot research area.

## References


1 Krieger, A.C. (ed.) (2017). *Social and Economic Dimensions of Sleep Disorders, an Issue of Sleep Medicine Clinics*. Elsevier.

2 Redmond, S.J. and Heneghan, C. (2006). Cardiorespiratory-based sleep staging in subjects with obstructive sleep apnea. *IEEE Transactions on Biomedical Engineering* 53 (3): 485–496.

3 Ancoli-Israel, S., Cole, R., Alessi, C. et al. (2003). The role of actigraphy in the study of sleep and circadian rhythms. *Sleep* 26 (3): 342–392.

4 Penzel, T. and Conradt, R. (2000). Computer based sleep recording and analysis. *Sleep Medicine Reviews* 4 (2): 131–148.

5 Accardo, J.A., Shults, J., Leonard, M.B. et al. (2010). Differences in overnight polysomnography scores using the adult and pediatric criteria for respiratory events in adolescents. *Sleep* 33 (10): 1333–1339.

6 Malhotra, A., Younes, M., Kuna, S.T. et al. (2013). Performance of an automated polysomnography scoring system versus computer-assisted manual scoring. *Sleep* 36 (4): 573–582.

7 Younes, M., Younes, M., Mu, B., and Giannouli, E. (2016). Accuracy of automatic polysomnography scoring using frontal electrodes. *Journal of Clinical Sleep Medicine* 12 (5): 735–746.

8 Procházka, A., Kuchynka, J., Vyšata, O. et al. (2018). Sleep scoring using polysomnography data features. *Signal, Image and Video Processing* 12 (6): 1–9.

9 Khalighi, S., Sousa, T., Pires, G., and Nunes, U. (2013). Automatic sleep staging: a computer assisted approach for optimal combination of features and polysomnographic channels. *Expert Systems with Applications* 40 (17): 7046–7059.

10 Shoker, L., Sanei, S., and Chambers, J. (2005). Artifact removal from electroencephalograms using a hybrid BSS-SVM algorithm. *IEEE Signal Processing Letters* 12 (10): 721–724.

11 Šušmáková, K. and Krakovská, A. (2008). Discrimination ability of individual measures used in sleep stages classification. *Artificial Intelligence in Medicine* 44 (3): 261–277.

12 Hjorth, B. (1970). EEG analysis based on time domain properties. *Electroencephalography and Clinical Neurophysiology* 29 (3): 306–310.

13 Koley, B. and Dey, D. (2012). An ensemble system for automatic sleep stage classification using single channel EEG signal. *Computers in Biology and Medicine* 42 (12): 1186–1195.

14 Herrera, L.J., Fernandes, C.M., Mora, A.M. et al. (2013). Combination of heterogeneous EEG feature extraction methods and stacked sequential learning for sleep stage classification. *International Journal of Neural Systems* 23 (3): 1350012.

15 Liang, S.-F., Kuo, C.-E., Hu, Y.-H., and Cheng, Y.-S. (2012). A rule-based automatic sleep staging method. *Journal of Neuroscience Methods* 205 (1): 169–176.

16 Güneş, S., Polat, K., and Yosunkaya, Ş. (2010). Efficient sleep stage recognition system based on EEG signal using *k*-means clustering based feature weighting. *Expert Systems with Applications* 37 (12): 7922–7928.

**17** Motamedi-Fakhr, S., Moshrefi-Torbati, M., Hill, M. et al. (2014). Signal processing techniques applied to human sleep EEG signals: a review. *Biomedical Signal Processing and Control* 10: 21–33.

**18** Chui, C.K. (1992). *An Introduction to Wavelets.* Academic Press.

**19** Addison, P.S. (2002). *The Illustrated Wavelet Transform Handbook.* Taylor & Francis.

**20** Durka, P.J. (2003). From wavelets to adaptive approximations: time–frequency parametrization of EEG. *Biomedical Engineering Online* 2 (1): 1–30.

**21** Ebrahim, F., Mikaeili, M., Estrada, E., and Nazeran, H. (2008). Automatic sleep stage classification based on EEG signals by using neural networks and wavelet packet coefficients. In: *Conference Proceedings: Annual International Conference of the IEEE Engineering in Medicine and Biology Society*, 1151–1154. IEEE.

**22** Schlüter, T. and Conrad, S. (2012). An approach for automatic sleep stage scoring and apnea-hypopnea detection. *Frontiers of Computer Science* 6 (2): 230–241.

**23** Jobert, M., Tismer, C., Poiseau, E., and Schulz, H. (1994). Wavelets: a new tool in sleep biosignal analysis. *Journal of Sleep Research* 3 (4): 223–232.

**24** Mallat, S.G. and Zhang, Z.F. (1993). Matching pursuits with time–frequency dictionaries. *IEEE Transactions on Signal Processing* 41 (12): 3397–3415.

**25** Durka, P.J., Ircha, D., and Blinowska, K.J. (2001). Stochastic time–frequency dictionaries for matching pursuit. *IEEE Transactions on Signal Processing* 49 (3): 507–510.

**26** Zygierewicz, J., Blinowska, K.J., Durka, P.J. et al. (1999). High resolution study of sleep spindles. *Clinical Neurophysiology* 110 (12): 2136–2147.

**27** Blinowska, K.J. and Durka, P.J. (2001). Unbiased high resolution method of EEG analysis in time–frequency space. *Acta Neurobiologiae Experimentalis* 61 (3): 157–174.

**28** Huupponen, E., De Clercq, W., Gomez-Herrero, G. et al. (2006). Determination of dominant simulated spindle frequency with different methods. *Journal of Neuroscience Methods* 156 (1–2): 275–283.

**29** Mahvash Mohammadi, S., Kouchaki, S., Ghavami, M., and Sanei, S. (2016). Improving time-frequency domain sleep EEG classification via singular spectrum analysis. *Journal of Neuroscience Methods* 273: 96–106.

**30** Sanei, S. and Hassani, H. (2015). *Singular Spectrum Analysis of Biomedical Signals.* CRC Press.

**31** Huang, N.E., Shen, Z., Long, S.R. et al. (1998). The empirical mode decomposition and the Hilbert spectrum for nonlinear and non-stationary time series analysis. *Proceedings of the Royal Society of London A* 454: 903–995.

**32** Hassan, A.R. and Bhuiyan, M.I.H. (2017). Automated identification of sleep states from EEG signals by means of ensemble empirical mode decomposition and random under sampling boosting. *Computer Methods and Programs in Biomedicine* 140: 201–210.

**33** Mendez, M.O., Corthout, J., Van Huffel, S. et al. (2010). Automatic screening of obstructive sleep apnea from the ECG based on empirical mode decomposition and wavelet analysis. *Physiological Measurement* 31 (3): 273–289.

**34** Jiayi, G., Peng, Z., Xin, Z., and Mingshi, W. (2007). Sample entropy analysis of sleep EEG under different stages. In: *IEEE/ICME International Conference on Complex Medical Engineering*, 1499–1502. IEEE.

**35** Acharya, R., Faust, O., Kannathal, N. et al. (2005). Non-linear analysis of EEG signals at various sleep stage. *Computer Methods and Programs in Biomedicine* 80 (1): 37–45.

**36** Azami, H., Arnold, S.E., Sanei, S. et al. (2019). Multiscale fluctuation-based dispersion entropy and its applications to neurological disease. *IEEE Access* 7 (1): 68718–68733.

**37** Wu, H., Talmon, R., and Lo, Y. (2015). Assess sleep stage by modern signal processing techniques. *IEEE Transactions on Biomedical Engineering* 62 (4): 1159–1168.

**38** Fraiwan, L., Lweesy, K., Khasawneh, N. et al. (2012). Automated sleep stage identification system based on time–frequency analysis of a single EEG channel and random forest classifier. *Computer Methods and Programs in Biomedicine* 108 (1): 10–19.

**39** Fraiwan, L., Lweesy, K., Khasawneh, N. et al. (2010). Classification of sleep stages using multi-wavelet time frequency entropy and LDA. *Methods of Information in Medicine* 49 (3): 230–237.

**40** Flexerand, A., Dorffner, G., Sykacek, P., and Rezek, I. (2002). An automatic, continuous and probabilistic sleep stager based on a hidden Markov model. *Applied Artificial Intelligence* 16: 199–207.

**41** Kouchaki, S., Sanei, S., Arbon, E.L., and Dijk, D. (2015). Tensor based singular spectrum analysis for automatic scoring of sleep EEG. *IEEE Transactions on Neural Systems and Rehabilitation Engineering* 23 (1): 1–9.

**42** Kouchaki, S., Eftaxias, K., and Sanei, S. (2015). An adaptive filtering approach using supervised SSA for identification of sleep stages from EEG. *Journal of Frontiers in Biomedical Technologies* 1 (4): 233–239.

**43** Enshaeifar, S., Kouchaki, S., Took, C.C., and Sanei, S. (2016). Quaternion singular spectrum analysis of electroencephalogram with application in sleep analysis. *IEEE Transactions on Neural Systems and Rehabilitation Engineering* 24 (1): 57–67.

**44** Peters, B.O., Pfurtscheller, G., and Flyvbjerg, H. (2001). Automatic differentiation of multichannel EEG signals. *IEEE Transactions on Biomedical Engineering* 48 (1): 111–116.

**45** Gevins, A. and Smith, M.E. (1999). Detecting transient cognitive impairment with EEG pattern recognition methods. *Aviation, Space, and Environmental Medicine* 70 (10): 1018–1024.

**46** Tagluk, M.E., Sezgin, N., and Akin, M. (2010). Estimation of sleep stages by an artificial neural network employing EEG, EMG and EOG. *Journal of Medical Systems* 34 (4): 717–725.

**47** Ronzhina, M., Janoušek, O., Kolářová, J. et al. (2012). Sleep scoring using artificial neural networks. *Sleep Medicine Reviews* 16 (3): 251–263.

**48** Ozsen, S. (2013). Classification of sleep stages using class-dependent sequential feature selection and artificial neural network. *Neural Computing and Applications* 23 (5): 1239–1250.

**49** Goodfellow, I., Bengio, Y., and Courville, A. (2016). *Deep Learning*. Cambridge, MA: MIT Press.

**50** Zhang, J. and Wu, Y. (2017). A new method for automatic sleep stage classification. *IEEE Transactions on Biomedical Circuits and Systems* 11 (5): 1097–1110.

**51** Supratak, A., Dong, H., Wu, C., and Guo, Y. (2017). DeepSleepNet: a model for automatic sleep stage scoring based on raw single-channel EEG. *IEEE Transactions on Neural Systems and Rehabilitation Engineering* 25 (11): 1998–2008.

**52** Stephansen, J.B., Olesen, A.N., Olsen, M. et al. (2018). Neural network analysis of sleep stages enables efficient diagnosis of narcolepsy. *Nature Communications* 9 (1): 5229.

**53** Tsinalis, O., Matthews, M., Guo Y., and Zafeiriou S. (2017) Automatic sleep stage scoring with single-channel EEG using convolutional neural networks. arXiv:1610.01683.

**54** Phan, H., Andreotti, F., Cooray, N. et al. (2018). Automatic sleep stage classification: Learning sequential features with attention-based recurrent neural networks. https://doi.org/10.1109/EMBC.2018.8512480. In: *40th Annual Intl. Conf. of the IEEE Engineering in Medicine and Biology Society (EMBC)*, 1452–1455. IEEE.

**55** Dong, H., Supratak, A., Pan, W. et al. (2018). Mixed neural network approach for temporal sleep stage classification. *IEEE Transactions on Neural Systems and Rehabilitation Engineering* 26 (2): 324–333.

**56** Goncharova, I.I., McFarland, D.J., Vaughan, T.M., and Wolpaw, J.R. (2003). EMG contamination of EEG: spectral and topographical characteristics. *Clinical Neurophysiology* 114 (9): 1580–1593.

**57** Palaz, D., Collobert, R., and Magimai-Doss, M. (2014). Estimating phoneme class conditional probabilities from raw speech signal using convolutional neural networks. *Interspeech*: 1766–1770.

**58** Swietojanski, P., Ghoshal, A., and Renals, S. (2014). Convolutional neural networks for distant speech recognition. *IEEE Signal Processing Letters* 21 (9): 1120–1124.

**59** Palaz, D., Magimai.-Doss, M., and Collobert, R. (2015). Convolutional neural networks-based continuous speech recognition using raw speech signal. In: *IEEE International Conference on Acoustics, Speech and Signal Processing (ICASSP)*, 4295–4299. IEEE.

**60** Hoshen, Y., Weiss, R.J., and Wilson, K.W. (2015). Speech acoustic modeling from raw multichannel waveforms. In: *IEEE International Conference on Acoustics, Speech and Signal Processing (ICASSP)*, 4624–4628. IEEE.

**61** Huang, J., Li, J., and Gong, Y. (2015). An analysis of convolutional neural networks for speech recognition. In: *IEEE International Conference on Acoustics, Speech and Signal Processing (ICASSP)*, 4989–4993. IEEE.

**62** Dieleman, S. and Schrauwen, B. (2014). End-to-end learning for music audio. In: *IEEE International Conference on Acoustics, Speech and Signal Processing (ICASSP)*, 6964–6968. IEEE.

**63** Sainath, T.N., Mohamed, A., Kingsbury, B., and Ramabhadran, B. (2013). Deep convolutional neural networks for LVCSR. In: *IEEE International Conference on Acoustics, Speech and Signal Processing*, 8614–8618. IEEE.

**64** Zhang, H., McLoughlin, I., and Song, Y. (2015). Robust sound event recognition using convolutional neural networks. In: *IEEE International Conference on Acoustics, Speech and Signal Processing (ICASSP)*, 559–563. IEEE.

**65** Ravi, D., Wong, C., Lo, B., and Yang, G.-Z. (2016). Deep learning for human activity recognition: a resource efficient implementation on low-power devices. In: *IEEE 13th International Conference on Wearable and Implantable Body Sensor Networks (BSN)*, 71–76. IEEE.

**66** Mikkelsen, K., and De Vos, M. (2018) Personalizing deep learning models for automatic sleep staging. arXiv:1801.02645.

**67** Phan, H., Andreotti, F., Cooray, N. et al. (2019). Joint classification and prediction CNN framework for automatic sleep stage classification. *IEEE Transactions on Biomedical Engineering* 66 (5): 1285–1296.

**68** Phan, H., Andreotti, F., Cooray, N. et al. (2018). DNN filter bank improves 1-max pooling CNN for automatic sleep stage classification. In: *Conference Proceedings of the IEEE Engineering in Medicine and Biology Society*, 453–456. IEEE.

**69** Hayn, D., Falgenhauer, M., Morak, J. et al. (2015). An eHealth system for pressure ulcer risk assessment based on accelerometer and pressure data. *Journal of Sensors* 2015: 1–8.

**70** Akita, Y., Kawakatsu, K., Hattori, C. et al. (2003). Posture of patients with sleep apnea during sleep. *Acta Oto-Laryngologica Supplementum* 123 (543): 41–45.

**71** Oksenberg, A., Khamaysi, I., Silverberg, D.S., and Tarasiuk, A. (2000). Association of body position with severity of apneic events in patients with severe nonpositional obstructive sleep apnea. *Chest* 118 (4): 1018–1024.

**72** Oksenberg, A., Silverberg, D.S., Arons, E., and Radwan, H. (1997). Positional vs. nonpositional obstructive sleep apnea patients: anthropomorphic, nocturnal polysomnographic, and multiple sleep latency test data. *Chest* 112 (3): 629–639.

**73** Mahvash Mohammadi, S., Kouchaki, S., Sanei, S., Dijk, D.-J., Hilton, A. and Wells, K. (2019) Tensor Factorisation and Transfer Learning for Sleep Pose Detection. *Proceedings of 2019 European Signal Processing Conference (EUSIPCO).* IEEE1–5.

**74** Grimm, T., Martinez, M., Benz, A., and Stiefelhagen, R. (2016). Sleep position classification from a depth camera using bed aligned maps. In: *23rd International Conference on Pattern Recognition (ICPR)*, 319–324. IEEE.

**75** Liu, J.J., Xu, W., Huang, M.-C. et al. (2013). A dense pressure sensitive bedsheet design for unobtrusive sleep posture monitoring. In: *IEEE International Conference on Pervasive Computing and Communications (PerCom)*, 207–215. IEEE.

**76** Lee, H.J., Hwang, S.H., Lee, S.M. et al. (2013). Estimation of body postures on bed using unconstrained ECG measurements. *IEEE Journal of Biomedical and Health Informatics* 17 (6): 985–993.

**77** Zhang, Z. and Yang, G.-Z. (2015). Monitoring cardio-respiratory and posture movements during sleep: what can be achieved by a single motion sensor. In: *IEEE 12th International Conference on Wearable and Implantable Body Sensor Networks (BSN)*, 1–6. IEEE.

**78** Ranasinghe, D.C., Shinmoto Torres, R.L., Hill, K., and Visvanathan, R. (2013). Low cost and batteryless sensor-enabled radio frequency identification tag based approaches to identify patient bed entry and exit posture transitions. *Gait & Posture* 39 (1): 118–123.

**79** Hoque, E., Dickerson, R.F., and Stankovic, J.A. (2010). Monitoring body positions and movements during sleep using WISP. In: *Wireless Health '10*, 44–53. ACM.

**80** Sun, X., Qiu, L., Wu, Y. et al. (2017). SleepMonitor: monitoring respiratory rate and body position during sleep using smartwatch. *Proceedings of the ACM on Interactive, Mobile, Wearable and Ubiquitous Technologies* 1 (3): 104.

**81** Hao, T., Xing, G., and Zhou, G. (2013). iSleep: Unobtrusive sleep quality monitoring using smartphones. In: *Proceedings of the 11th ACM Conference on Embedded Networked Sensor Systems, SenSys*, 1–14. ACM.

**82** Gu, W., Shangguan, L., Yang, Z., and Liu, Y. (2016). Sleep hunter: towards fine grained sleep stage tracking with smartphones. *IEEE Transactions on Mobile Computing* 15 (6): 1514–1527.

**83** Buysse, D.J., Reynolds, C.F., Monk, T.H. et al. (1989). The Pittsburgh sleep quality index: a new instrument for psychiatric practice and research. *Psychiatry Research* 28 (2): 193–213.

# 6

# Noninvasive, Intrusive, and Nonintrusive Measurements

## 6.1 Introduction

Longing for a more effective, easier-to-use, and publicly acceptable monitoring of human physical and physiological activities? Recently there has been an increasing interest in developing noninvasive sensor technology. Great strides have undoubtedly been made in the design of noninvasive monitoring techniques for recording many human vital signs such as brain, heart, muscle, breathing, and oxygenation signals. At the same time, continuous use of wearables and the applications for vulnerable people has encouraged even more nonintrusive sensor designs to be made. For example, exploiting the photonic property of optical fibres and the use of phase-sensitive optical fibre interferometers has allowed a mat to be designed for patients to lie or sleep on. Breath and heart beat induce slight strain changes on the mat resulting in the variation of light propagating within the fibre. This in turn can be analysed using signal processing tools and algorithms to detect breathing and heart beat waveforms. In the following sections some of these systems are introduced and explained.

## 6.2 Noninvasive Monitoring

Basically, three prerequisites are necessary for most clinical applications: (i) a simple and inexpensive method for collecting biological samples with minimal discomfort; (ii) determination of specific biomarkers associated with health or disease; and (iii) an accurate, portable, and easy-to-use technology for disease diagnosis and health screening [1]. Numerous sensor systems have been designed to record human vital signs. Most traditional systems are first developed in an invasive way. In parallel with technological advances, ongoing research has been focussed on replacing invasive techniques allowing preventive and cost-effective personal medicine on a large scale. A great deal of information that used to be available through invasive techniques may be derived from noninvasive measurements. This is mainly due to modernising medical equipment and the integration of advanced signal processing and artificial intelligence algorithms to these systems.

In addition to making noninvasive and less-intrusive systems for patients, the long-term monitoring of patients mainly in their home environments eliminates the need of a 'white-coat', thus reducing costs and the time spent on hospital visits. Therefore, the

*Body Sensor Networking, Design and Algorithms,* First Edition. Saeid Sanei, Delaram Jarchi and Anthony G. Constantinides.

Companion Website: www.wiley.com/go/sanei/algorithm-design

long-term monitoring of physiological parameters such as vital signs has become a goal in healthcare monitoring, preventive treatments, and personalised medicine. As an example, the adoption of invasive electromyography (EMG) has largely been reduced by virtue of new signal processing methods which can easily separate the interfering signals from the surface EMG. A new adaptive line enhancer based on singular spectrum analysis (SSA) has been designed and used to perfectly remove the electrocardiogram (ECG) interference from the hand surface EMG signals [2]. This method is currently used for the online restoration of surface EMG signals in many hospitals around the world.

Another good example is noninvasive glucose monitoring. Diabetes affects and kills several million people worldwide every year. A noninvasive painless method of glucose testing would greatly improve compliance and glucose control while reducing the number of in-patient admissions in hospitals with better patient and cost management. GlucoWatch Biographer has been on the market since 1998. In many experiments, such as that reported in [3], a good agreement between its results and those of the HemoCue blood glucose analyser has been shown.

Saliva, also known as the 'mirror of the body', is clinically of great interest as a bio-medium for clinical diagnostics. Its unique properties, such as noninvasive accessibility and the presence of various disease biomarkers, make it particularly attractive for disease diagnosis and monitoring [4, 5]. Saliva can be easily collected by individuals at or away from clinics with modest instruction and in comfort. Changes in saliva are believed to indicate the wellness of an individual. There are a large number of diagnostic analytes present in saliva, including glucose [6, 7], steroid hormones [8], and even the HIV antibody [8, 9].

Zhang et al. [10] developed a disposable saliva nano biosensor using salivary biomarkers to enable accurate, low-cost, and continuous glucose monitoring. Here, glucose oxidase, gold nanoparticles, chitosan, poly(allylamine), acetate buffer solution, D-(+)-glucose, phosphate buffered saline, and other materials as well as a number of tools, devices, and equipment have been used in this research. Eight clinical trials were carried out on two healthy individuals after consuming regular food and having an intake of standard glucose beverage. High accuracy was reported when compared with ultraviolet (UV) spectrophotometer. In addition, by measuring subjects' salivary glucose and blood glucose in parallel, they discovered that the two generated profiles share the same fluctuation trend while the correlation between them is subject dependent pronounced in a time lag between the two peak glucose values from blood and saliva. By further adjustment of the time of taking saliva samples, it has been reported that this method can indeed replace the invasive techniques for monitoring diabetes.

Wireless wearable sensors have attracted great attention in making such systems. Various wearable systems and their clinical applications are explored in Chapters 3 and 4. New devices such as DiaMonTech (DMT) or Dexcom G6 are used to noninvasively measure glucose by resting the finger on its finger pad (sensor). This technology is based on the absorption of laser light by glucose molecules. Their wearable wristband device, DMT band, continuously measures the blood glucose and includes data logging and alarms for too high or too low glucose values. Figure 6.1 shows a number of these wearable devices.

Two different types of monitoring systems, including contactless and implantable sensors, are reviewed in the following sections. Both these systems have the potential to make a significant impact on certain clinical applications.

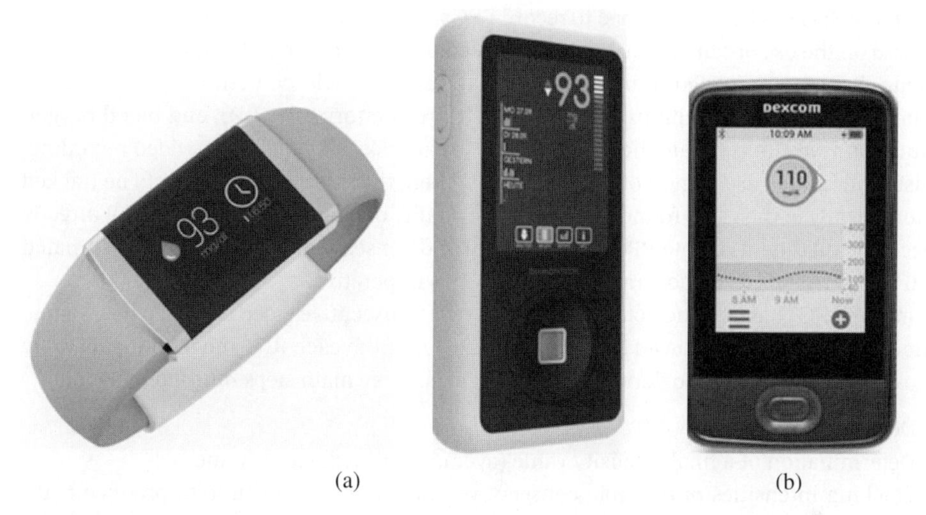


**Figure 6.1** (a) DiaMonTech (DMT) and (b) Dexcom G6 devices for noninvasive glucose measurement.

## 6.3 Contactless Monitoring

### 6.3.1 Remote Photoplethysmography

Photoplethysmography (PPG) is an optical-based technology which uses light to measure the changes in microvascular blood volume [11, 12]. As the pioneers who contributed to establishing the PPG technique, Hertzman and colleagues initially introduced the term 'photoplethysmography' (*plethysmos* means 'enlargement' in Greek) in the 1930s and suggested that it represented volumetric change in the dermal vasculature.

PPG signal recording requires physical contact of the PPG sensors with the skin. The most common places used for placing the PPG include the finger, ear, and wrist. The human skin has been found to be continuously changing in colour after each heart beat which may not be visible to the naked eye. By magnification or enhancing the contrast of video images, subtle changes in the colour of the human face are detected over time [13]. As an example [13], in four video frames from a subject's face demonstrated in Figure 6.2 the colour changes can be viewed.

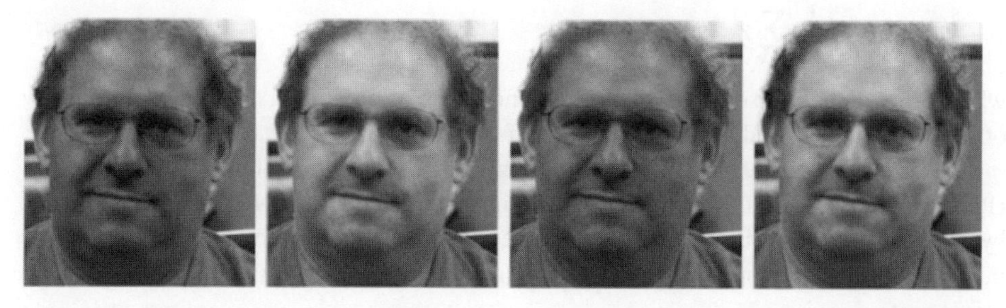

**Figure 6.2** Amplification of colour changes in four video frames from a subject's face [13]. Source: Courtesy of Wu, H.-Y., Rubinstein, M., Shih, E., Guttag, J., Durand, F., and Freeman, W.T. (*See color plate section for color representation of this figure*)

Remote PPG has been developed to record PPGs without any physical contact. Its concept is based on the use of camera images to detect variations of colour in the human skin. These colour changes are related to blood flow variations and could be used to derive PPGs in estimating vital signs, including heart rate (HR), respiratory rate (RR), and blood oxygen saturation level. To estimate these signs, each suitable skin regions in the video recording, must be identified as a region of interest (ROI). Then, the selected ROI needs to be tracked automatically in the video frame sequence. The quality of image in the selected ROI directly affects the quality of remote PPG (rPPG) signal and consequently the validity of estimated vital signs. Therefore, an accurate ROI detection is imperative.

Once the ROI has been detected and tracked over consecutive video frames, a new signal, namely rPPG, can be produced by extracting a feature from each ROI and tracking its value across video recordings. To derive the rPPG signals, three main steps must be followed:

1. Detection of the ROI.
2. Determination of a final intensity value (average) for each video frame.
3. Tracking intensities of multiple consecutive video frames over time to produce rPPG time-series.

These steps are explained in detail below.

#### 6.3.1.1 Derivation of Remote PPG

An RGB image is represented by three channels (red, green, and blue). To find the face regions containing useful physiological information, common face detection algorithms such as simple conventional image processing can be applied for face detection and feature point tracking. As an example, the skin classifier developed in [14] can be applied to an RGB image. Based on the work in [15], a classifier has been constructed using the Bayesian decision rule [16] to detect the skin and nonskin colour pictures. Then, through skin registration the ROIs have been highlighted to achieve a stabilised segmentation contour. An adaptation technique [15] is applied where a large intensity variation has been detected in the last ROIs or in the cases where the ROI disappears. Following the detection of ROIs, a channel has been selected. In various studies green channels provide a higher quality of PPG signals than red or blue channels [15]. Then, for each frame, the average intensity of pixel values is calculated using the selected channel (e.g. green). The estimated average intensities over consecutive video frames are used to derive an rPPG time-series.

The rPPG is derived as:

$$rPPG_n = \frac{1}{|R|} \sum_{i \in R} \gamma_n^i \tag{6.1}$$

where $R$ is the set of pixels in the ROI, $|R|$ denotes the number of pixels, $\gamma_n^i$ is the intensity of pixel $i$ in frame $n$, and $rPPG_n$ is the value of rPPG signals at frame $n$ [17].

It has been found that the size of ROI can influence the estimation of vital signals from rPPG signals. As an example, the increase in signal-to-noise ratio (SNR) has been observed by an increase in the size of ROI [17]. However, larger ROIs can create other problems including nonrelevant or undesired components such as HR, breathing, or those related to movement.

The rPPG time-series can further be processed and subjected to spectral analysis for HR, RR, or blood oxygen saturation level estimation. One popular spectral analysis method is based on autoregressive (AR) modelling which has been very useful especially in estimation of respiratory frequency or rate [18].

### 6.3.2 Spectral Analysis Using Autoregressive Modelling

AR models have widely been used for analysis of electroencephalography (EEG) signals [19] such as for the estimation of brain connectivity through multivariate AR, EEG prediction, modelling, classification, feature extractions, and compression. In an AR model, the current value of a time-series is predicted as a linearly weighted sum of the preceding $p$ values, where $p$ stands for the model order. An AR model is described as:

$$x(n) = -\sum_{i=1}^{p} a_i x(n-k) + e(n) \tag{6.2}$$

where $x(n)$ is the $n^{th}$ value of input signal $x$, $a_i$ is the $i^{th}$ coefficient, and $e(n)$ is the error which is assumed to be white Gaussian noise with a normal distribution of zero mean variance of $\sigma^2$. The model order $p$ must be much smaller than the length of the input signal $x$. Setting of model order $p$ is important since it can affect the system's performance. Taking a larger prediction order, less correlated components belonging to noise and artefacts will be included in the prediction, whereas low orders smooth the signal and reject fast fluctuating components. A correct model order greatly depends on the signal and noise subspaces. Several methods, such as the Akaike information criterion, may be employed for model order estimation.

If the AR model parameters are estimated optimally, the error $e(n)$ is white and has a flat spectrum. Therefore, the spectrum of $X(z)/E(z)$ represents the spectrum of $X(z)$. Based on this assumption and by exploiting the $z$-transform, the transfer function which maps the input to the output can be obtained as:

$$H(z) = \frac{1}{\sum_{k=1}^{p} a_k z^{-k}} \tag{6.3}$$

Factorising $H(z)$ denominator into a product of $p$ elements leads to:

$$H(z) = \frac{z^p}{(z - z_1)(z - z_2)\dots(z - z_p)} \tag{6.4}$$

where, $z_1, z_2, \dots, z_p$ are the poles of $H(z)$ which are complex-valued in general and are related to spectral peaks in the power spectrum of the input signal. The phase angle of each pole can be used to find the spectral peak frequency using:

$$\theta = 2\pi f \Delta t \tag{6.5}$$

where $\theta$ is in radians and $\Delta t$ is the sampling interval. The poles may contain physiological information, be related to artefacts such as artificial light flicker, or any component outside the ROI. The poles, which are unrelated to the physiological information, must be suppressed and removed. A method which reconstructs a new transfer function for removing/cancelling undesired poles corresponding to the aliased components from the denominator of the transfer function in Eq. (6.3) or (6.4) is proposed in [20]. The new transfer function represents the reflected light intensity from the ROI [20]. The pole cancellation procedure is summarised in the following steps:

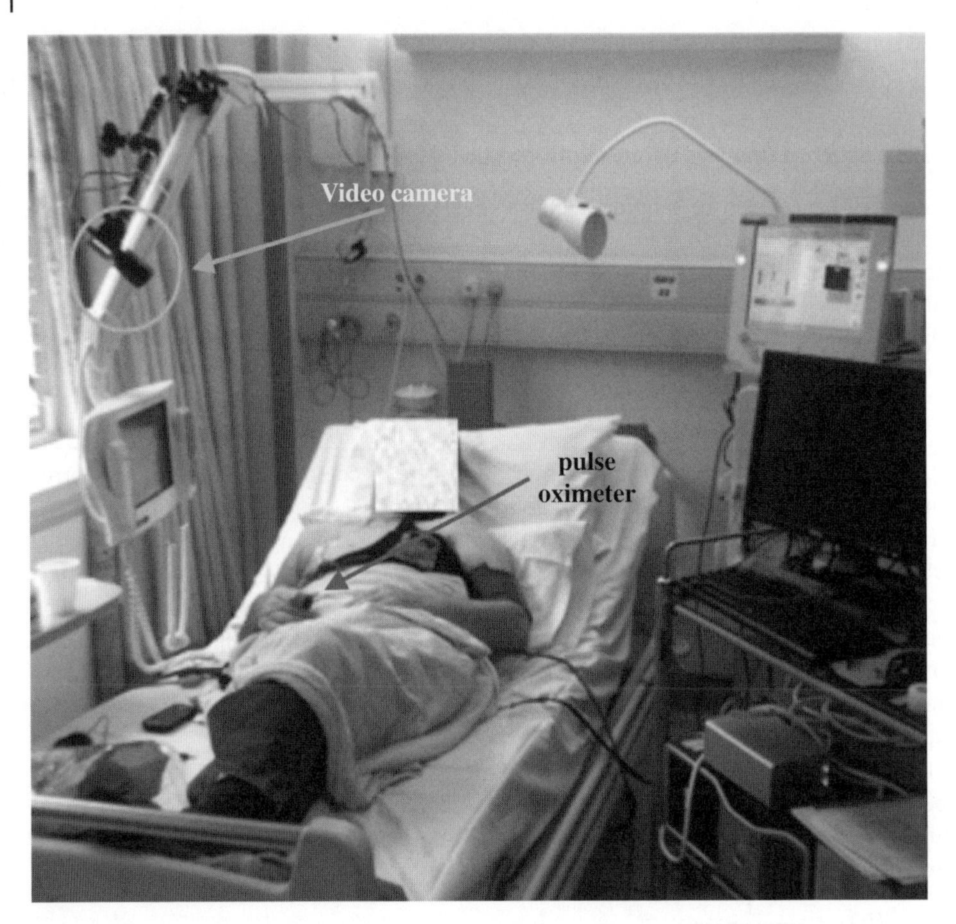


**Figure 6.3** Recording of video frames using a digital video camera, shown in red circle, for a patient at the Oxford Kidney Unit [20]. A pulse oximeter is worn on the patient's finger. Source: Courtesy of Tarassenko, L., Villarroel, M., Guazzi, A., Jorge, J., Clifton, D.A., and Pugh, C. (*See color plate section for color representation of this figure*)

1. Identify an ROI within the subject's face. This place can be the forehead or cheek. A patient being monitored at the Oxford Kidney Unit during a four-hour dialysis session is shown in Figure 6.3. In this figure, the video camera and finger-worn pulse oximeter are shown. The video camera produces an image including the patient's face where an appropriate ROI within the face must be selected. Analysis of the selected ROI is expected to provide the relevant physiological information, such as HR, RR, and blood oxygen saturation level. This ROI is denoted as the selected region of interest ($ROI_s$).
2. Select a reference region of interest outside the subject's face. This place can include the background wall. Analysis of this ROI will not result in obtaining relevant physiological information; however, it is used to discard poles unrelated to the desired information from the subject. This ROI is denoted as the reference region of interest ($ROI_r$). Figure 6.4 shows an example of both $ROI_s$ and $ROI_r$ including their derived signals from the green channel.

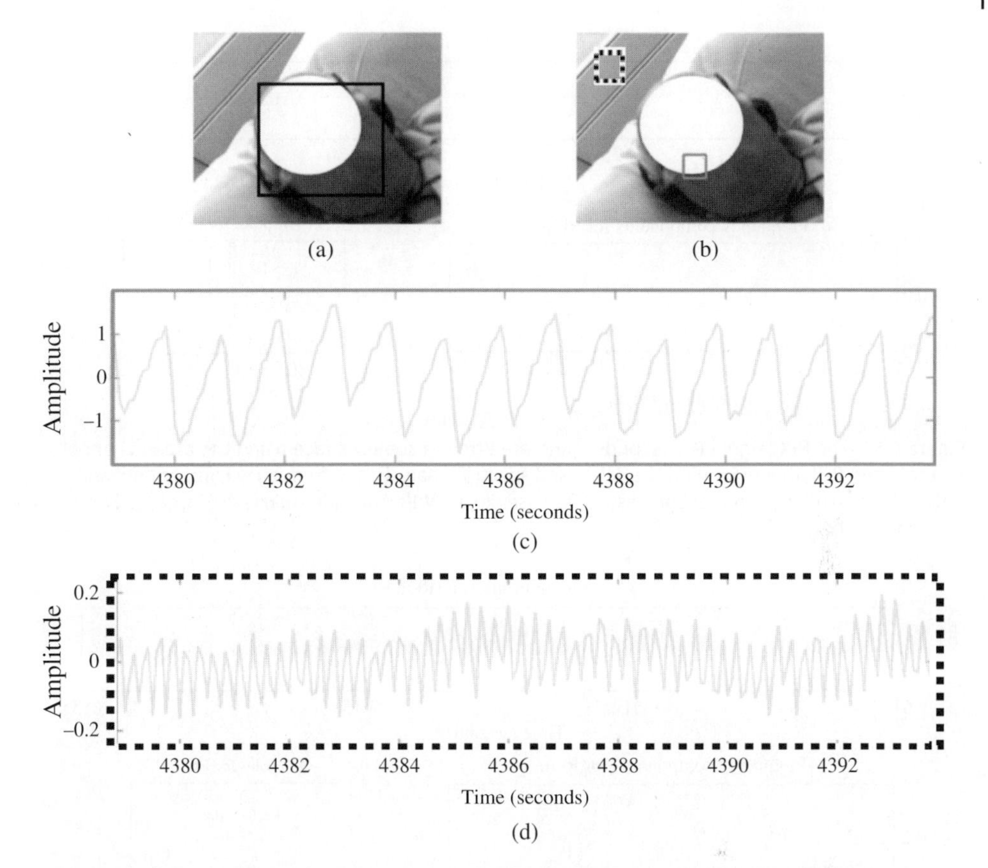


**Figure 6.4** Region of interest: (a) the face registration has been performed and presented as a black square; (b) selected region of interest containing physiological information ($ROI_s$) in smaller square, the selected $ROI_r$ has been used to discard information unrelated to the patient's physiology; (c) 15 seconds of $ROI_s$ time-series extracted from the green channel; (d) 15 seconds of $ROI_r$ time-series extracted from the green channel [20]. Source: Courtesy of Tarassenko, L., Villarroel, M., Guazzi, A., Jorge, J., Clifton, D.A., and Pugh, C.

3. Derive the reconstructed time-series signal, using spatial averaging of intensities for each video frame using both $ROI_s$ and $ROI_r$.
4. Fit the AR model to the time-series derived from $ROI_s$ and calculate poles. An example of $ROI_s$ and calculated poles using AR model is shown in Figure 6.4.
5. Fit the AR model to the time-series derived from $ROI_r$ and calculate poles. An example of $ROI_r$ and the estimated poles using AR model is shown in Figure 6.5.
6. Identify the poles in the $ROI_r$ and $ROI_s$ which are within $m$ degrees of each other. These poles and their complex conjugates are cancelled in the transfer function for the $ROI_s$. As an example, a flickering light leads to the generation of poles in both $ROI_s$ and $ROI_r$ at the same frequencies. Considering the examples in Figures 6.5 and 6.6, after pole cancellation, the remained frequency components and AR poles are shown in Figure 6.6. Here, $Ps_2$, $Ps_3$, and $Ps_4$ are the poles which appeared for both $ROI_s$ and the background ($ROI_r$) and do not correspond to the patient's physiology, and so they are cancelled.

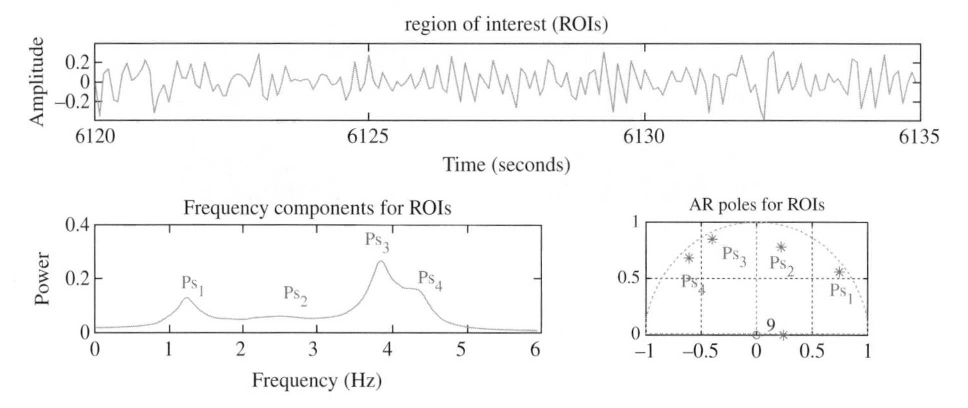


**Figure 6.5** The PPG signal (15 seconds) generated from a subject's face using the green channel and selected ROI. By performing the AR-based spectral analysis, the frequency components and poles are shown [20]. Source: Courtesy of Tarassenko, L., Villarroel, M., Guazzi, A., Jorge, J., Clifton, D.A., and Pugh, C.

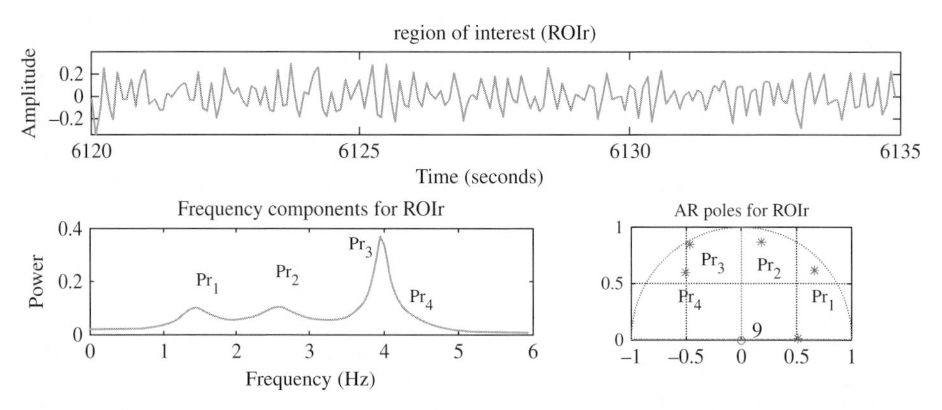


**Figure 6.6** The PPG signal (15 seconds) generated from a background behind subject using a green channel and selected ROI. By performing the AR-based spectral analysis, the frequency components and poles are shown [20]. Source: Courtesy of Tarassenko, L., Villarroel, M., Guazzi, A., Jorge, J., Clifton, D.A., and Pugh, C.

A reconstructed signal using the remaining pole at the cardiac frequency is shown in Figure 6.7.

In the following, it is explained that estimation of RR, HR, and blood oxygen saturation level can be also achieved using a similar AR algorithm.

### 6.3.3 Estimation of Physiological Parameters Using Remote PPG

#### 6.3.3.1 Heart Rate Estimation

Employing a remote PPG or noncontact PPG imaging method for HR estimation has demonstrated that a flickering light, such as fluorescent light, produces frequencies around the cardiac frequencies. Therefore, unnecessary poles have to be cancelled. Following construction of the time-series using the spatial averaging of intensities for each video

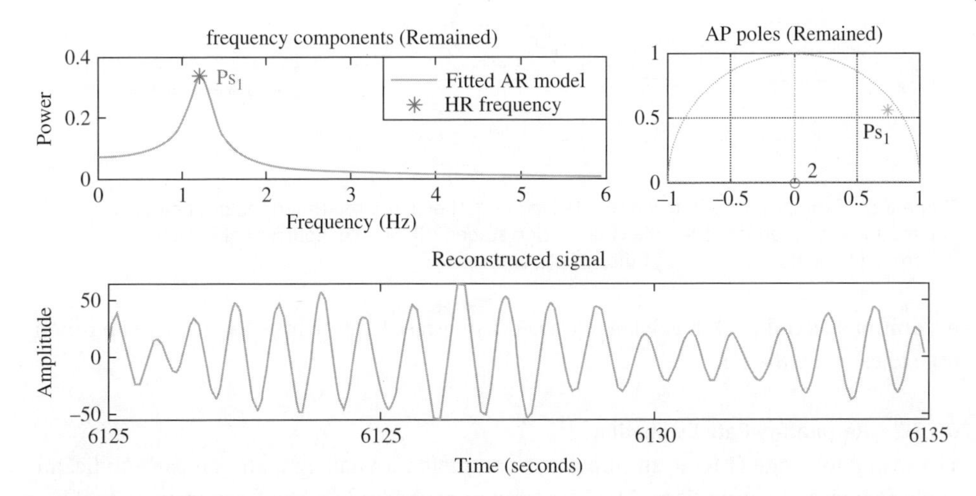


**Figure 6.7** The frequency component and zero-pole plot using an AR model are shown after pole cancellation. The reconstructed signal at cardiac frequency after cancelling the pole is also illustrated [20]. Source: Courtesy of Tarassenko, L., Villarroel, M., Guazzi, A., Jorge, J., Clifton, D.A., and Pugh, C.

frame, using both $ROI_s$ and $ROI_r$, the time-series signals are detrended. It is worth mentioning that the reconstructed time-series using $ROI_s$ is the desired remote PPG signal. Signal windows of 15 seconds' duration are then selected, the bandpass filtered, and the AR model fitted to each window of the reconstructed times-series. For each 15-second window and video sampling rate of 12 frames per second, 180 samples are obtained from the time-series. These samples are then used to estimate the AR model coefficients. A suitable model order must be selected for dominant cardiac frequency estimation. In [20], a model order of 9 has been selected.

Following calculation of the AR model coefficients and the corresponding poles, the pole cancellation has been carried out by comparing the poles obtained from the AR model fitted to the reconstructed time-series (rPPG) in the $ROI_s$, and those from the AR model fitted to the reconstructed time-series in the $ROI_r$. Further to pole cancellation, a new transfer function is constructed to present a new AR model representing $ROI_k$. The pole with the highest magnitude in the AR model for $ROI_k$ corresponding to a frequency between zero and half of sampling frequency is then selected as the HR pole. The phase of this pole ($\theta$ in radians) can be used to estimate the HR in beats per minute by multiplying $\theta$ by $60\frac{f_s}{2\pi}$, where $f_s$ is the sampling frequency measured in Hz. From a pole-zero plot, the radius of the HR pole is related to the HR component amplitude. For HR estimation, usually a sliding window with a width of at least twice the HR cycle is used. In practice a sliding window of 15 seconds with 14 seconds' overlap (sliding each window by one second) is used to estimate the HR for every one-second interval. The procedure for the proposed approach is demonstrated in Figure 6.8 [20].

In other studies, various spectral analysis algorithms are proposed for HR estimation from rPPG. In [21], Hilbert transform has been used for beat-to-beat detection and the estimation of HR variability using rPPG derived from a subject's forehead using camera images. In [15],

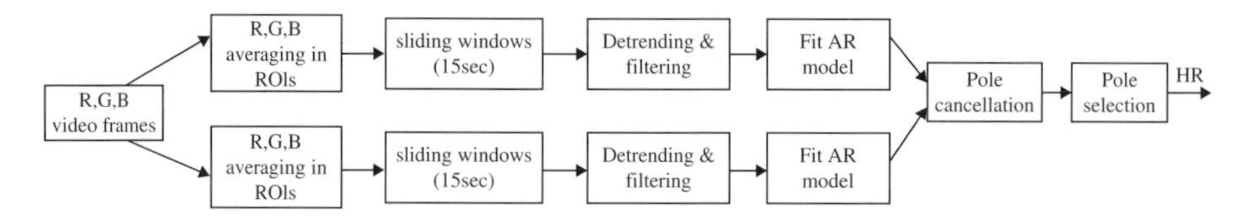


**Figure 6.8** Estimation of HR using the derived rPPG. Poles not related to the physiological information are ignored in the pole cancellation stage [20]. Source: Courtesy of Tarassenko, L., Villarroel, M., Guazzi, A., Jorge, J., Clifton, D.A., and Pugh, C.

maximum spectral peak has been extracted using fast Fourier transform (FFT) and then used to estimate the HR.

#### 6.3.3.2 Respiratory Rate Estimation

The respiratory rate (RR) is an important parameter, a vital sign, and an early indicator of physiological deteriorations. The RR component resides in a low-frequency range. Also, it has been demonstrated that the ROI size for RR estimation is typically smaller when compared to that of HR estimation [20]. Similar HR estimation procedures – such as the use of bandpass filtering, sliding windows, and fitting AR models – can be applied to RR estimation but with some differences. The upper cut-off frequency for the rPPG bandpass filter must be lowered to 0.7 Hz. This cut-off frequency corresponds to a RR of 42 breaths per minute as a reasonable normal breathing rate. This cut-off frequency rejects the heart beat frequency component of approximately 1 Hz (60 beats per minutes). Furthermore, no pole cancellation (as in heart beat estimation) is required since the RR component appears in very low frequencies.

Signals of 30-second windows, with a frame rate of 12 frames per second are bandpass filtered to extract the low-frequency respiratory component as proposed by [20], down-sampled to 2 Hz to increase their angular resolution, and each window is separately modelled with an AR model order of 7. Finally, the pole with the lowest angle in 0.1–0.7 Hz frequency range and the magnitude greater than 95% of the highest-magnitude pole is selected. The selected pole is used to estimate the respiratory frequency. The diagram of the proposed approach is demonstrated in Figure 6.9 [20].

Realtime vision-based respiration monitoring is also performed in [22], where the video camera is directed at the subject's chest and abdominal areas at different angles. In [23], the respiration signals are extracted by detecting changes due to the chest movement. Additional signal processing techniques are also proposed in order to detect nonrespiratory motion detection such as when a subject is changing their position in bed or waving their arms in front of their chest. The proposed motion detector algorithm labels the images as 'moving' and 'stationary' states. PPG signal is generated using the 'moving' state images. Contactless monitoring of the RR is performed by using occlusion of a dot pattern [24].

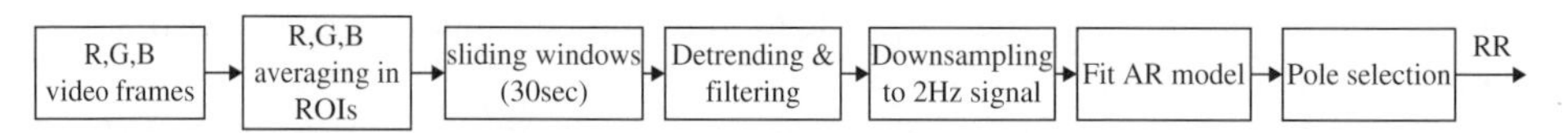


**Figure 6.9** Estimation of RR using derived rPPG [20]. Source: Courtesy of Tarassenko, L., Villarroel, M., Guazzi, A., Jorge, J., Clifton, D.A., and Pugh, C.

To do this, a dotted pattern is mounted on one of the two sides of the bed and the camera placed on the opposite side to monitor the dots. This is based on the concept of dot pattern occlusion during breathing. For example, during inhalation, there is an increase in the volume that occludes the background and vice versa during exhalation.

#### 6.3.3.3 Blood Oxygen Saturation Level Estimation

Recalling from Chapter 4, the $SpO_2$ level can be estimated using the Beer–Lambert law, sometimes called the 'ratio of ratios' method. That is, for each wavelength, such as red (660 nm) or infrared (940 nm), direct current (DC) and pulsatile alternating current (AC) components are used to calculate the $SpO_2$ level [25] as:

$$SpO2 = k_1 - k_2 \times R \tag{6.6}$$

where $k_1$ and $k_2$ are constant coefficients obtained during the calibration process.

$$R = \frac{\frac{dI_R(t)/dt}{I_R}}{\frac{dI_{IR}(t)/dt}{I_{IR}}} = \frac{\frac{I_R(t_2) - I_R(t_1)}{I_R(t_3)}}{\frac{I_{IR}(t_2) - I_{IR}(t_1)}{I_{IR}(t_3)}} = \frac{\frac{AC_R}{DC_R}}{\frac{AC_{IR}}{DC_{IR}}} \tag{6.7}$$

Several limitations are reported in [20], for example an overlap in broadband spectral responses of the three colour sensors for RGB cameras. Therefore, only windows of the estimated signals in agreement with the reference HR are selected for subsequent DC and AC component estimations. To obtain the DC component, a 10-second moving average filter has been used, while for derivation of the AC component the peak-to-trough heights are averaged in every 10-second window. As with other estimates such as HR and RR, the ROI has been obtained from the patient's face. Then, $k_1$ and $k_2$ are estimated using the available ground truth data for $SpO_2$ from a commercial finger-worn pulse oximeter and a best-fit linear equation. Once $k_1$ and $k_2$ are obtained, the $SpO_2$ values can be derived using Eqs. (6.6) and (6.7). The block diagram of the proposed approach is demonstrated in Figure 6.10 following the approach in [20].

In another work, Monte Carlo simulation of light transport is applied to digital RGB images to find the relative concentration of oxygen saturation [26]. In [27], a camera targets the subject's left inner arm skin through a custom-built apochromatic lens; 3$\lambda$-LED-ringlight ($\lambda_1$ = 660 nm, $\lambda_2$ = 810 nm, $\lambda_3$ = 940 nm). This method has the potential to derive rPPG signals using different wavelengths to apply the 'ratios of ratio' method to estimate SpO2. Elsewhere, two cameras are used to generate videos of human faces. Two narrow-band filters with wavelengths of 660 and 520 nm are mounted to the camera to obtain the rPPGs [28].

Face images have been used to estimate the $SpO_2$where the validation has been performed using a commercial wearable oximeter [29]. In another study, an oxygen desaturation of subjects has been performed in order to generate variable $SpO_2$ between 80 and 100% [30].

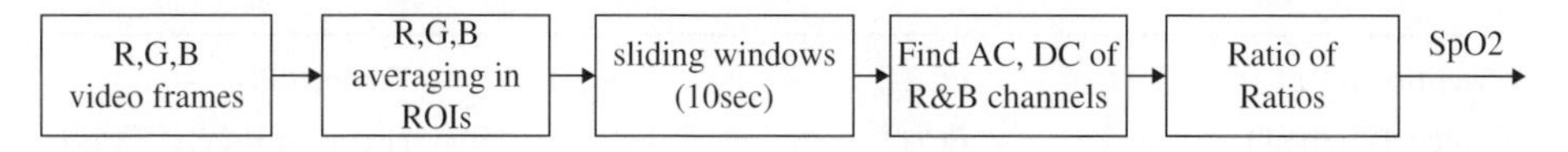


**Figure 6.10** Estimation of blood oxygen saturation level ($SpO_2$) using derived rPPG [20]. Source: Courtesy of Tarassenko, L., Villarroel, M., Guazzi, A., Jorge, J., Clifton, D.A., and Pugh, C.

Continuous noncontact monitoring of vital signs including HR, RR, and blood oxygen saturation level in neonatal intensive care unit has been performed in [31].

#### 6.3.3.4 Pulse Transmit Time Estimation

In Chapter 4, pulse transmit time (PTT) has been introduced which could be used to estimate blood pressure (BP) [32, 33]. PTT in a conventional way has been estimated by measuring the time delay between two cardiac-related synchronised signals. As described in Chapter 4, the two synchronous signals are typically the ECG signal (attached to the chest) and PPG recorded from the finger. Therefore, the conventional PTT (cPTT) is denoted when the time delay between ECG and PPG signals is considered. To evaluate the capability of video recordings to estimate PTT and consequently the blood pressure or changes in blood pressure, the first step is to replace the ECG or finger PPG recordings using rPPG-based methods. Based on the definition in [17], remote pulse transmit time (rPTT) has been used when the time delay is measured between ECG and rPPG (from a video camera). In addition, the differential PTT (dPTT) has been denoted where the two rPPGs are recorded from two different body sites, then the time delay or PTT is calculated between these two different sites from PPGs. Therefore, a camera-based PTT can be obtained using the two methods of rPTT and dPTT. The notations for different PTT-based methods are provided in Table 6.1.

In some other research works, estimation of PTT using video recordings has been performed under different scenarios. The PTT could have been simply obtained by an extension of the rPPG methods, for example using the study in [34] to derive rPPG and then PTT. However, explicit measurement of the rPTT/dPTT has been introduced in 2014 and 2015 by several researchers [35–37]. The dPTT using rPPG signals has been measured from the mouth and palm using 10 subjects in [35]. In [36], the time difference of the pulse peaks is calculated using rPPGs derived from the wrist and ankle. For this study, 10 patients were participated in the experiments that lasted for only 30 seconds where a strong correlation between their dPTT and blood pressure was reported. This correlation was comparable to the correlation obtained using cPTT methods in the related literature, i.e. the correlation between cPTT and BP.

The study in [37] investigated the dPTT under normal and stressed conditions, and higher average dPTT values under stressed conditions were found. Therefore, based on the findings, dPTT can be used as a marker for stress. However, this work used short video recordings from 15 subjects and therefore the results do not sufficiently support the claim. For this study, rPPGs were recorded from the face and hand. In [38], dPTT and rPTT were derived using two rPPGs and a combination of ECG and rPPG, respectively. The rPPGs

**Table 6.1** Description and notation of the different PTT-based methods.

| Method | First signal (proximal site) | Second signal (distal site) |
|---|---|---|
| Conventional PTT (cPTT) | ECG | Finger PPG |
| Remote PTT (rPTT) | ECG | $rPPG_1$ or $rPPG_2$ |
| Differential PTT (dPTT) | $rPPG_1$ | $rPPG_2$ |

were recorded from the forehead and palm. This study found a slightly better correlation between rPTT and BP than that obtained between dPTT and BP.

However, the limitation of the study was the small size of its participant cohort and short duration of the recordings, which made it difficult to infer significant conclusions. In other studies, dPTT has been estimated using rPPGs derived from the head and palm [39], and rPTT estimated using ECG and rPPG from the palm where the subjects were asked to press their palm against a glass surface [40].

The major limitations of all the mentioned studies are the short video recordings and small number of subjects. For a reliable estimation of PTT from video cameras, longer video durations and a larger number of subjects are required. The relationship between BP and cPTT has been investigated in many studies, as detailed in Chapter 4. However, the relationship between BP and rPTT has been investigated much less. In order to properly evaluate the accuracy of the rPTT-based methods to measure PTT and then BP, in addition to having longer video recordings and more subjects, there are other important factors. One factor is the selection of a good location to record rPPG where the skin is visible to video cameras. For example, the face, forehead, and cheek are more visible to cameras. Other locations, such as the hand, depend on whether the subjects have worn clothes. Another important factor is the presence of the same blood supply between the two selected body sites to properly estimate the PTT. As an example, there are different blood branches supplied to the face where the blood does not flow directly from the cheek to the forehead. Therefore, it is not completely clear where the pulse should arrive first. This essentially affects the estimation of dPTT where the blood branching complicates the measurement of dPTT.

As with the cPTT methods, the changes in BP have been better detected than an estimation of actual raw BP values. Considering the limitations of the camera-based methods to estimate rPTT/dPTT, the main objective in most relevant methods is to look for the changes in camera-derived PTT where the changes in BP are detected.

Estimation of the PTT from video recordings are summarised in the following steps:

1. Pre-processing of the video recordings.
2. Selection of the ROI.
3. Derivation of the rPPG signal.
4. Processing of the rPPG signals.
5. Calculation of the rPTT/dPTT.

These steps are explained in connection with the proposed approaches in [17]. Unlike in previous attempts, in this research longer duration (up to 40 minutes) videos and a relatively large number of subjects (43 subjects) have been used. The subjects were under hypoxic condition to induce changes in their physiological parameters.

#### 6.3.3.5 Video Pre-processing

The video data which are in a form of $M \times N \times 3 \times T$ are recorded representing $T$ frames of an $M \times N$ video pixel with three colour channels: red, green, and blue. Then, a strong object detection technique (e.g. the Viola–Jones object detection framework in [41]) must be applied to detect the subject's face within the camera video frame.

Usually, the Viola–Jones object detector is applied to the first frame where the area of interest (AOI) of size $M_a \times N_a$ pixels is found. For subsequent frames, the centroid of the previously detected AOI has been used to re-initialise the new AOI around a new centroid which could vary around a certain threshold. The AOI is split into a grid each of its elements containing a $12 \times 12$ pixel area of the frame. For each element, the pixels are summed and down-sampled, thus the original AOI of size $M_a \times N_a$ is converted into $M_a/12 \times N_a/12$. These new smaller frames obtained from each colour channel are combined to form a final matrix of size $M_a/12 \times N_a/12 \times 3 \times T$. Based on this video image compression using a resolution of $12 \times 12$, the size of video files is significantly reduced. However, careful consideration must be made not to use smaller grid elements that could lead to the generation of rPPG signals with a low SNR.

#### 6.3.3.6 Selection of Regions of Interest

Several ROIs (such as the forehead, cheek, and neck) are manually detected. Additional ROIs are selected to evaluate the local variations of dPTT/rPTT around the neck and forehead regions. In [17], a small element size ($1 \times 1$) was used to produce a better localised rPPG signal. This corresponds to $12 \times 12$ pixels in the original video frame. ROI can be considered static or dynamic. For dynamic ROIs, image tracking techniques should be used to warrant the ROI moves with the subjects. Static ROIs are easy to handle since the image tracking techniques are no longer required. However, the subjects need to be informed to remain still and minimise their movements during the experiments.

#### 6.3.3.7 Derivation of the rPPG Signal

Following determination of the ROI, the rPPG signal needs to be produced by spatial averaging. This is explained in Section 6.3.1. Each video frame contributes one value to the rPPG signal. Then, over time, the rPPG time-series is obtained. This time-series has a sampling rate equal to the video frame rate.

#### 6.3.3.8 Processing rPPG Signals

Filtering the rPPG signal to estimate the PTT needs to be carefully performed to remove noise and artefacts while preserving clinically important signal information. This is due to the fact that pulse peaks located on rPPG used to estimate the PTT may be affected by filtering the rPPG. Although it is very important to use an appropriate filter to process the rPPG signals, a small shift in the pulse peaks locations is acceptable. Since the main objective is to detect the changes in BP, the relative changes in the PTT remain unaltered if the filtering process causes a slight shift in the location of pulse peaks.

In [17], it is suggested to select a finite impulse response (FIR) bandpass filter to maintain the cardiac frequencies for resting HR where the cut-off frequencies are selected using the 1st and 99th percentiles of HR estimated from a set of healthy adults. Once the rPPG signal is filtered, the pulse peaks must be located to estimate the PTT. These pulse peaks must be estimated as accurately as possible.

The resolution of rPPG is similar to the frame rate of the video. For example, the sampling rate of the rPPG is 15 in [20] and 16 in [17]. These are much less than a sampling rate of 256 Hz as used in conventional PPG recorded using commercial devices. The time difference between the detected peaks and the actual peak can produce a big error in the calculation of the PTT in the event of the sampling rate being low. Resampling the rPPG signal into a higher rate seems necessary. It is suggested in [17] to resample the 16 Hz rPPG into

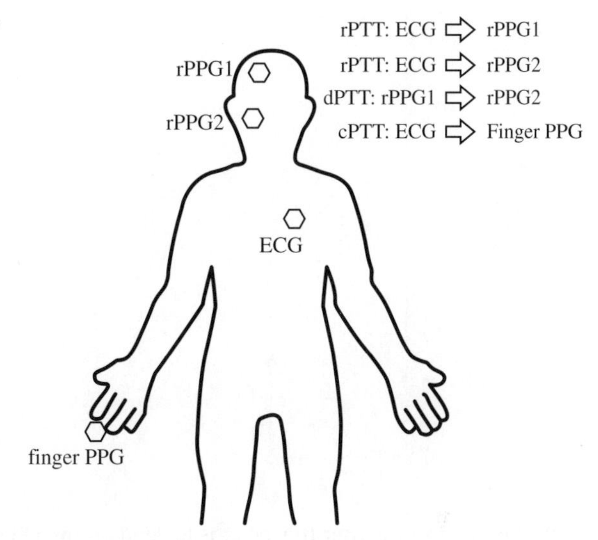


**Figure 6.11** Demonstration of body sites to record ECG, finger PPG, and rPPGs and various approaches for PTT estimation (rPTT, dPTT, cPTT).

1 kHz signal using piecewise cubic spline interpolation. One assumption using this interpolation is that the rPPG waveform is a near-sinusoid one. This procedure has been shown to be effective in producing more accurate pulse peaks for rPPG improving the estimation of PTT.

#### 6.3.3.9 Calculation of rPTT/dPTT

Once the rPPG signal is resampled into a desired sampling rate, it can be used along with ECG, or another processed rPPG signal (from a different site) to estimate rPTT/dPTT. In Figure 6.11, a schematic has been provided to specify the positions for the PTT on the body. The synchronisation of ECG and one rPPG or two rPPGs is an important factor in deriving the PTT signal. One goal in evaluating remote PTT monitoring is to compare the measured PTT using ECG and conventional PPG (finger PPG), to the one using ECG and rPPG (from a camera). To be able to do such comparison synchronisation of the finger PPG and rPPG is important. The finger PPG and rPPG are both derived from heart beats; they are expected to have small phase differences due to their different measurement sites. In [17], a time lag using cross-correlation calculation is performed as:

$$\phi = argmax \int_{-\infty}^{\infty} f^*(\tau)g(t+\tau)d\tau \tag{6.8}$$

where $\phi$ is the lag, $f^*$ is the complex conjugate of finger PPG signal, $g$ is the rPPG signal derived from the camera, and $\tau$ is the time shift. The lag with maximum cross-correlation has been used to synchronise the PPG and rPPG.

Using Eq. (6.8), it is helpful to synchronise the reference HR and HR estimated from the rPPG obtained from the cross-correlation function [17]. Once the lag with maximum cross-correlation has been found, the timestamps of the camera and the reference data can be synchronised. Interpolation of the signals into similar timestamps is also required.

Following the synchronisation and interpolation steps, the R peaks need to be located on the ECG signals and the pulse peaks/troughs on the rPPG signal. Then, rPTT can be simply calculated. For a calculation of the rPTT, first an R peak is detected on the ECG signal,

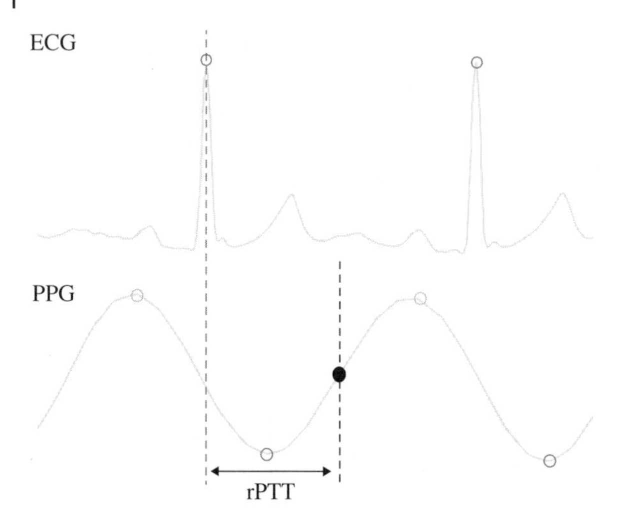


**Figure 6.12** The ECG R peaks are denoted as small circles in the top signal. The bold black circle presents the detected peak on PPG signal where its amplitude is approximately half of the amplitudes of the first PPG trough after the ECG R peak and the following PPG pulse peak. The interval is equivalent to rPTT

then the first trough after this peak is located on the rPPG signal. In the next step, the first peak on the rPPG after the rPPG trough is detected. An amplitude is calculated, which is half of the amplitude of the selected peak and trough on rPPG. The sample with the closest distance to the corresponding calculated amplitude (halfway between the peak and trough amplitudes) has been identified and its index saved. The number of samples from the ECG R peak and the selected index on rPPG is converted to a time delay by dividing that by the sampling rate. In Figure 6.12, the ECG and PPG signals and the required peaks used in determination of PTT are illustrated.

## 6.4 Implantable Sensor Systems

The wearable sensors are either invasive or unobtrusive. Recent research has been directed towards the development of robust systems for unobtrusive monitoring such as using the wearable sensors or contactless monitoring. On the other hand, there is a limited but increasing amount of research on implantable devices for certain clinical applications. For these systems to be launched on the market, not only clinical studies must be performed but also new regulations and standards for their use must be set, introduced, and reported properly and in a timely manner [42, 43].

Heart failure detection and continuous BP monitoring are two main targets for implantable devices. Monitoring BP is very important since a rise in BP increases strain to the vessels and can damage the cardiovascular system [43]. Long-term monitoring of BP is crucial to find anomalies in the trend of BP and consequently to find the right therapy. Invasive BP monitoring systems are only usable for acute clinical settings. They also induce risks such as inflammations to the patients. Typical standard noninvasive systems, such as sphygmomanometry, can be used for the short duration of BP measurements. Implantable systems are designed to solve the existing problems with invasive and non-invasive measurements and to provide long-term monitoring of patients without limiting and obstructing their lives during their daily activities.

The development of BP monitoring systems is addressed in various studies [44–47]. These studies can mainly be used for monitoring hypertension in patients. In [42], a peel-away

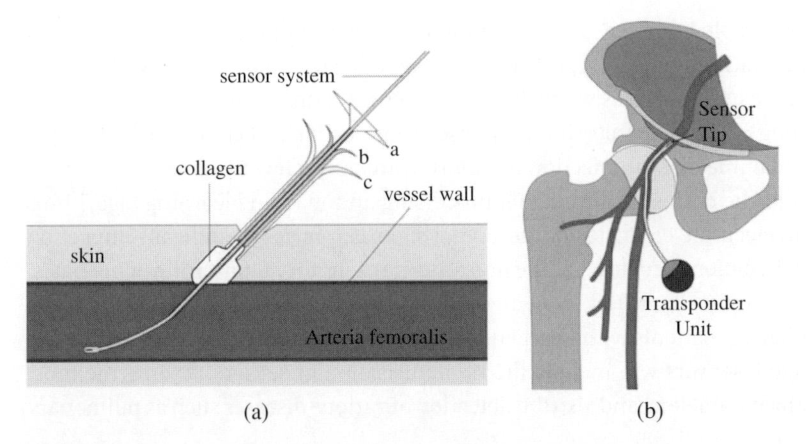


**Figure 6.13** (a) Functional principle of the peel-away sheath introducer set; (b) an implanted sensor system in a human femoral artery [42]. Source: Courtesy of Cleven, N.J., Müntjes, J.A., Fassbender, H. et al.

sheath introducer set was developed to help the sensor's system implantation procedure. A sheath was used to prevent bleeding during surgery and once the punctured artery was sealed securely the sheath was removed. Figure 6.13, shows the implanted sensor in a human femoral artery and its function principle.

Many implanted systems have been developed for cardiovascular applications. These systems, introduced in [48–52], need to be clinically validated.

A large-scale study demonstrated that by using remote monitoring of heart failure the earlier detection of critical events was achieved, thus prompting timely intervention [53]. It has also been reported that by managing heart failure patients using remote monitoring of implanted defibrillators, and thus having access to their alerts and data reviews, reduced the number of patients admitted to emergency departments while scores for quality of life increased [54]. Implantations of devices for cardiac resynchronisation therapy have shown to be effective in monitoring arrhythmia in heart failure patients [55]. In another study, an implantable ECG sensor was developed and its performance tested using a laboratory pig [56].

Implantable sensors can be employed for many medical applications, such as hydrocephalus monitoring, glaucoma monitoring, cystometry, and prosthetic replacement, and are implanted in brain, eye, urinary tract/bladder, and big joints. Implantable sensors are developed for the detection of glaucoma [57] where a wireless intraocular pressure transducer is implanted into the human eye [58]. Intracranial pressure has been a target for the development of implantable pressure sensors [59–62].

Among many others, cortical EEG electrodes and subdural foramen ovale brain electrodes are used in some clinical departments to scan deep brain activities for some brain abnormalities, such as seizure.

## 6.5 Conclusions

Certain systems have been developed for the long-term remote monitoring of patients using contactless monitoring and implantable sensors. In this chapter, following an introduction to noninvasive wearable body sensors, estimation of three important vital signs (HR, RR, and blood oxygen saturation level) has been investigated using camera videos. These are only a few examples of nonintrusive health monitoring systems. The use of optical fibre mats for breathing and HR monitoring [12], laser cameras for detection of facial movements, and thermal imaging for infection monitoring are only a few to name.

Improvement of the corresponding signal processing and machine learning algorithms to achieve highly accurate estimates of these vital signs can have a significant impact on monitoring certain patients, such as for the noncontact respiratory monitoring of neonates.

Patient remote monitoring using implanted devices has shown significant applications for the management of patients with heart failure and hypertension. Future research in the field of implantable sensors will include the miniaturisation of sensors, finding the most important sensing parameters, and also the detection of various diseases such as pulmonary disease.

Since wearable technology is becoming more user-friendly, significant growth is expected in the development of noninvasive wearable devices for patients, athletes, and even those members of the public who wish to be under active surveillance.

## References


1 Zhang, L., Xiao, H., and Wong, D.T. (2009). Salivary biomarkers for clinical applications. *Molecular Diagnosis & Therapy* 13 (4): 245–259.

2 Sanei, S., Lee, T., and Abolghasemi, V. (2012). A new adaptive line enhancer based on singular spectrum analysis. *IEEE Transactions on Biomedical Engineering* 59 (2): 428–434.

3 Tamada, J.A., Garg, S., Jovanovic, L. et al. (1999). Noninvasive glucose monitoring: comprehensive clinical results. *Journal of the American Medical Association* 282 (19): 1839–1844.

4 Lamy, E. and Mau, M. (2012). Saliva proteomics as an emerging, non-invasive tool to study livestock physiology, nutrition and diseases. *Journal of Proteomics* 75: 4251–4258.

5 Wong, D.T. (2006). Salivary diagnostics powered by nanotechnologies, proteomics and genomics. *Journal of the American Dental Association (1939)* 137: 313–321.

6 Yamaguchi, M., Mitsumori, M., and Kano, Y. (1998). Development of noninvasive procedure for monitoring blood glucose levels using saliva. In: *Proceedings of the 20th Annual International Conference of the IEEE Engineering in Medicine and Biology Society*, 1763–1766. IEEE.

7 Panchbhai, A.S., Degwekar, S.S., and Bhowte, R.R. (2010). Estimation of salivary glucose, salivary amylase, salivary total protein and salivary flow rate in diabetics in India. *Journal of Oral Science* 52: 359–368.

8 Forde, M.D., Koka, S., Eckert, S.E. et al. (2006). Systemic assessments utilizing saliva: part 1. General considerations and current assessments. *The International Journal of Prosthodontics* 19 (1): 43–52.

9 Reynolds, S.J. and Muwonga, J. (2004). OraQuick advance rapid HIV-1/2 antibody test. *Expert Review of Molecular Diagnostics* 4: 587–591.

**10** Zhang, W., Du, Y., and Wang, M.L. (2015). Noninvasive glucose monitoring using saliva nano-biosensor. *Sensing and Bio-Sensing Research* 4: 23–29.

**11** Sun, Y. and Thakor, N. (2016). Photoplethysmography revisited: from contact to non-contact, from point to imaging. *IEEE Transactions on Biomedical Engineering* 63 (3): 463–477.

**12** Yu, C., Xu, W. Zhang, N., Yu, C. (2017) Non-invasive smart health monitoring system based on optical fiber interferometers. *16th International Conference on Optical Communications and Networks (ICOCN)*, Wuzhen, China.

**13** Wu, H.-Y., Rubinstein, M., Shih, E. et al. (2012). Eulerian video magnification for revealing subtle changes in the world. *ACM Transactions on Graphics* 31 (4) https://doi.org/10.1145/2185520.2185561.

**14** Jones, M.J. and Rehg, J.M. (2002). Statistical color models with application to skin detection. *International Journal of Computer Vision* 46 (1): 81–96.

**15** Trumpp, A., Lohr, J., Wedekind, D. et al. (2018). Camera-based photoplethysmography in an intraoperative setting. *Biomedical Engineering Online* 17 (1): 1.

**16** Duda, R.O., Hart, P.E., and Stork, D.G. (2001). Maximum Likelihood and Bayesian Parameter Estimation. In: *Pattern Classification*, 2e, 84–160. New York: Wiley.

**17** Daly, J. (2016) Video camera monitoring to detect changes in haemodynamics. PhD thesis, University of Oxford.

**18** Fleming, S. and Tarassenko, L. (2007). A comparison of signal processing techniques for the extraction of breathing rate from the photoplethysmogram. *International Journal of Biomedical Sciences* 2: 232–236.

**19** Pardey, J., Roberts, S., and Tarassenko, L. (1996). A review of parametric modelling techniques for EEG analysis. *Medical Engineering & Physics* 18 (1): 2–11.

**20** Tarassenko, L., Villarroel, M., Guazzi, A. et al. (2014). Non-contact video-based vital sign monitoring using ambient light and auto-regressive models. *Physiological Measurement* 35 (5): 807–831.

**21** Blöcher, T., Schneider, J., Schinle, M., and Stork, W. (2017). An online PPGI approach for camera based heart rate monitoring using beat-to-beat detection. In: *IEEE Sensors Applications Symposium (SAS)*, 1–6. IEEE.

**22** Tan, K.S., Saatchi, R., Elphick, H., and Burke, D. (2010). Real-time vision based respiration monitoring system. In: *2010 7th International Symposium on Communication Systems, Networks & Digital Signal Processing (CSNDSP 2010)*, 770–774. IEEE.

**23** Bartula, M., Tigges, T., and Muehlsteff, J. (2013). Camera-based system for contactless monitoring of respiration. In: *Conference Proceedings: Annual International Conference of the IEEE Engineering in Medicine and Biology Society*, 2672–2675. IEEE.

**24** Makkapati, V.V. and Rambhatla, S.S. (2016). Remote monitoring of camera based respiration rate estimated by using occlusion of dot pattern. In: *IEEE International Conference on Advanced Networks and Telecommunications Systems (ANTS)*, 1–5. IEEE.

**25** Webster, G. (1997). *Design of Pulse Oximeters.* Taylor & Francis.

**26** Nishidate, I., Sasaoka, K., Yuasa, T. et al. (2008). Visualising of skin chromophore concentrations by use of RGB images. *Optics Letters* 33 (19): 2263–2265.

**27** Wieringa, F.P., Mastik, F., and van der Steen, A.F. (2005). Contactless multiple wavelength photoplethysmographic imaging: a first step toward 'Sp$O_2$ camera' technology. *Annals of Biomedical Engineering* 33 (8): 1034–1041.

28 Kong, L., Zhao, Y., Dong, L. et al. (2013). Non-contact detection of oxygen saturation based on visible light imaging device using ambient light. *Optics Express* 21 (15): 17464–17471.

29 Bal, U. (2014). Non-contact estimation of heart rate and oxygen saturation using ambient light. *Biomedical Optics Express* 6 (1): 86–97.

30 Guazzi, A.R., Villarroel, M., Jorge, J. et al. (2015). Non-contact measurement of oxygen saturation with an RGB camera. *Biomedical Optics Express* 6 (9): 3320–3338.

31 Villarroel, M., Guazzi, A., Jorge, J. et al. (2014). Continuous non-contact vital sign monitoring in neonatal intensive care unit. *Healthcare Technology Letters* 1 (3): 87–91.

32 Wang, R., Jia, W., Mao, Z.H. et al. (2014). Cuff-free blood pressure estimation using pulse transit time and heart rate. In: *International Conference on Signal Processing Proceedings*, 115–118. IEEE.

33 Gao, M., Olivier, N.B., and Mukkamala, R. (2016). Comparison of noninvasive pulse transit time estimates as markers of blood pressure using invasive pulse transit time measurements as a reference. *Physiological Reports* 4 (10): e12768.

34 Verkruysse, W., Svaasand, L.O., and Nelson, J.S. (2008). Remote plethysmographic imaging using ambient light. *Optics Express* 16 (26): 21434–21445.

35 Shao, D., Yang, Y., Liu, C. et al. (2014). Noncontact monitoring breathing pattern, exhalation flow rate and pulse transit time. *IEEE Transactions on Biomedical Engineering* 61 (11): 2760–2767.

36 Murakami, K., Yoshioka, M., and Ozawa, J. (2015). Non-contact pulse transit time measurement using imaging camera, and its relation to blood pressure. In: *14th IAPR International Conference on Machine Vision Applications (MVA)*, 414–417. IEEE.

37 Kaur, B., Tarbox, E., Cissel, M. et al. (2015). Remotely detected differential pulse transit time as a stress indicator. In: *Independent Component Analyses, Compressive Sampling, Large Data Analyses (LDA), Neural Networks, Biosystems, and Nanoengineering XIII*, vol. 9496. SPIE https://spie.org/Publications/Proceedings/Paper/10.1117/12.2177886?SSO=1 (accessed 6 January 2020).

38 Secerbegovic, A., Bergsland, J., Halvorsen, P.S. et al. (2016). Blood pressure estimation using video plethysmography. In: *IEEE 13th International Symposium on Biomedical Imaging (ISBI)*, 461–464. IEEE.

39 Jeong, I.C. and Finkelstein, J. (2016). Introducing contactless blood pressure assessment using a high speed video camera. *Journal of Medical Systems* 40: 77.

40 Volynsky, M.A., Mamontov, O.V., Sidorov, I.S., and Kamshilin, A.A. (2016). Pulse wave transit time measured by imaging photoplethysmography in upper extremities. *Journal of Physics: Conference Series* 737 (1): 1–5.

41 Viola, P. and Jones, M. (2001). Rapid object detection using a boosted cascade of simple features. In: *IEEE Computer Society Conference on Computer Vision and Pattern Recognition, CVPR*. IEEE https://doi.org/10.1109/CVPR.2001.990517.

42 Cleven, N.J., Müntjes, J.A., Fassbender, H. et al. (2012). A novel fully implantable wireless sensor system for monitoring hypertension patients. *IEEE Transactions on Biomedical Engineering* 59 (11): 3124–3130.

43 Clausen, I. and Glott, T. (2014). Development of clinically relevant implantable pressure sensors: perspectives and challenges. *Sensors (Basel)* 14 (9): 17686–17702.

**44** Cibula, E., Donlagic, D., and Stropnik, C. (2002). Miniature fiber optic pressure sensor for medical applications. In: *Proceedings of IEEE Sensors*, 711–714. IEEE.

**45** Chenga, X., Xue, X., Ma, Y. et al. (2016). Implantable and self-powered blood pressure monitoring based on a piezoelectric thinfilm: simulated, *in vitro* and *in vivo* studies. *Nano Energy* 22: 453–460.

**46** Murphy, O.H., Bahmanyar, M.R., Borghi, A. et al. (2013). Continuous in vivo blood pressure measurements using a fully implantable wireless saw sensor. *Biomedical Microdevices* 15 (5): 737–749.

**47** Schnakenberg, U., Kruger, C., Pfeffer, J.G. et al. (2004). Intravascular pressure monitoring system. *Sensors and Actuators A: Physical* 110 (1–3): 61–67.

**48** Merchant, F.M., Dec, G.W., and Singh, J.P. (2010). Implantable sensors for heart failure. *Circulation: Arrhythmia and Electrophysiology* 3 (6): 657–667.

**49** Hoppe, U.C., Vanderheyden, M., Sievert, H. et al. (2009). Chronic monitoring of pulmonary artery pressure in patients with severe heart failure: multicentre experience of the monitoring pulmonary artery pressure by implantable device responding to ultrasonic signal (PAPIRUS) II study. *Heart* 95 (13): 1091–1097.

**50** Bio-Medicine. Issys Inc. Titan wireless implantable hemodynamic monitor. (2020). https://mems-iss.com/intelligent-titan-wireless-implantable-hemodynamic-monitor/ (accessed 6 January 2020).

**51** ClinicalTrials.gov (2019). Left atrial pressure monitoring to optimize heart failure therapy (laptop-hf). http://clinicaltrials.gov/ct2/show/NCT01121107 (accessed 25 November 2019).

**52** Pour-Ghaz, I., Hana, D., Raja, J. et al. (2019). CardioMEMS: where we are and where can we go? *Annals of Translational Medicine* 7 (17): 418.

**53** Varma, N., Epstein, A.E., Irimpen, A. et al. (2010). Efficacy and safety of automatic remote monitoring for implantable cardioverter-defibrillator follow-up: the Lumos-T Safely Reduces Routine Office Device Follow-up (TRUST) trial. *Circulation* 122 (4): 325–332.

**54** Landolina, M., Perego, G.B., Lunati, M. et al. (2012). Remote monitoring reduces healthcare use and improves quality of care in heart failure patients with implantable defibrillators: the evolution of management strategies of heart failure patients with implantable defibrillators (EVOLVO) study. *Circulation* 125 (24): 2985–2992.

**55** Shanmugam, N., Boerdlein, A., Proff, J. et al. (2012). Detection of atrial high-rate events by continuous home monitoring: clinical significance in the heart failure– cardiac resynchronization therapy population. *Europace* 14 (2): 230–237.

**56** Lee, J.-H. (2016). Miniaturized human insertable cardiac monitoring system with wireless power transmission technique. *Journal of Sensors* 2016: 5374574.

**57** Chen, P.J., Rodger, D.C., Agrawal, R. et al. (2007). Implantable micromechanical parylene-based pressure sensors for unpowered intraocular pressure sensing. *Journal of Micromechanics and Microengineering* 17 (10): 1931–1938.

**58** Melki, S., Todani, A., and Cherfan, G. (2014). An implantable intraocular pressure transducer: initial safety outcomes. *JAMA Ophthalmology* 132 (10): 1221–1225.

**59** Frischholz, M., Sarmento, L., Wenzel, M. et al. (2007). Telemetric implantable pressure sensor for short- and long-term monitoring of intracranial pressure. In: *Proceedings of*

*2007 Annual International Conference of the IEEE Engineering in Medicine and Biology Society*, 514–518. IEEE.

**60** Hierold, C., Clasbrummel, B., Behrend, D. et al. (1999). Low power integrated pressure sensor system for medical applications. *Sensors and Actuators A: Physical* 73 (1–2): 58–67.

**61** Rai, P. and Varadan, V.K. (2010). Organic electronics based pressure sensor towards intracranial pressure monitoring. In: *Nanosensors, Biosensors, and Info-Tech Sensors and Systems*, vol. 7646 (ed. V.K. Varadan). Bellingham, WA: SPIE https://doi.org/10.1117/12.849188.

**62** Yoon, H.J., Jung, J.M., Jeong, J.S., and Yang, S.S. (2004). Micro devices for a cerebrospinal fluid (CSF) shunt system. *Sensors and Actuators A: Physical* 110 (1–3): 68–76.

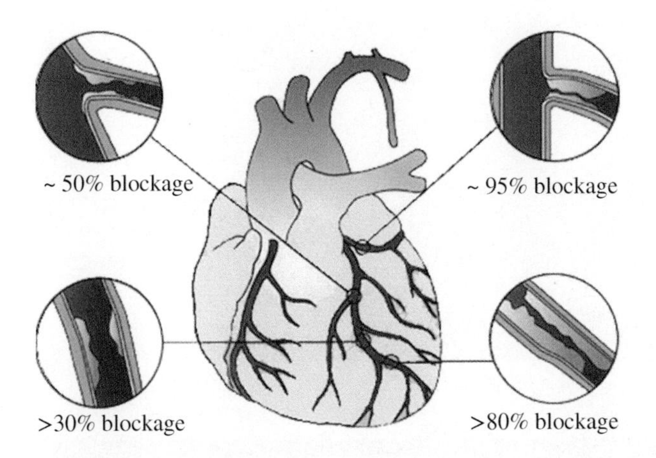


**Figure 2.1** Schematic of some possible blocking of heart arteries with different severities.

**Figure 2.2** Different brain sensory zones.

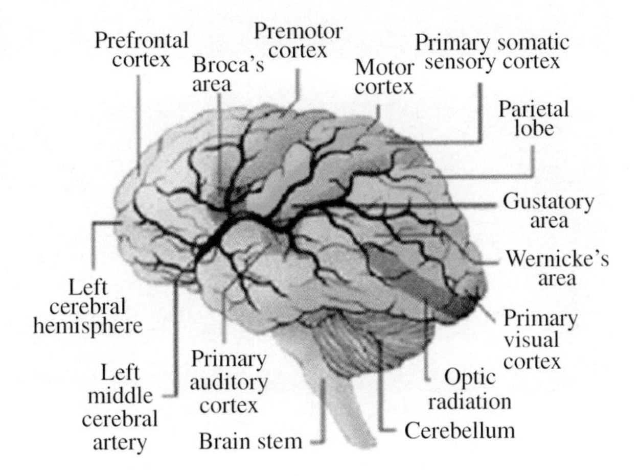


*Body Sensor Networking, Design and Algorithms*, First Edition. Saeid Sanei, Delaram Jarchi and Anthony G. Constantinides.

Companion Website: www.wiley.com/go/sanei/algorithm-design

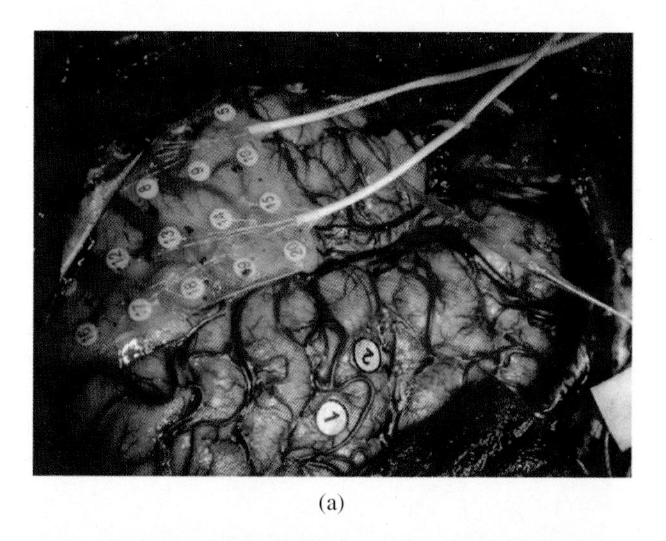

(a)

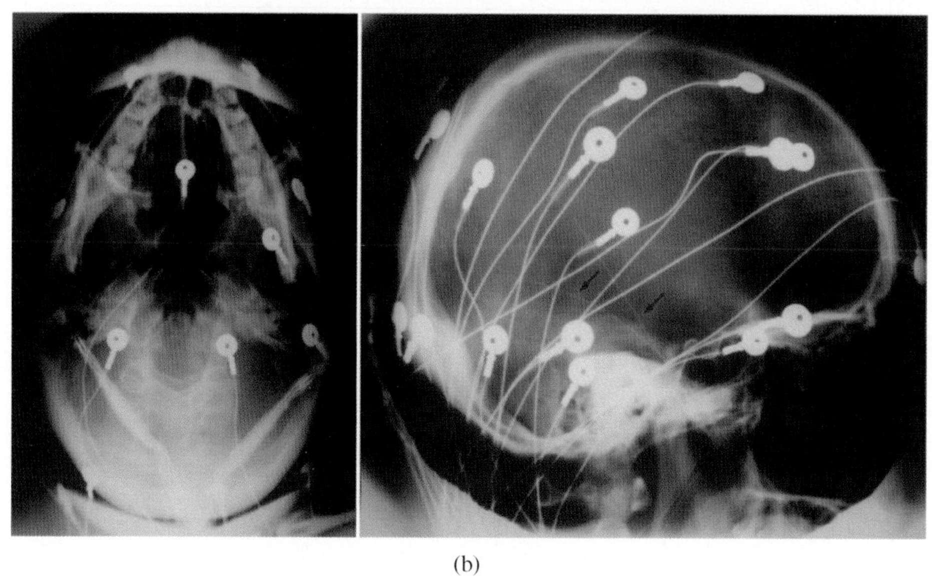

(b)

**Figure 3.5** (a) ECoG and (b) foramen ovale electrodes denoted by pointers. In this setup, a scalp EEG has also been used.

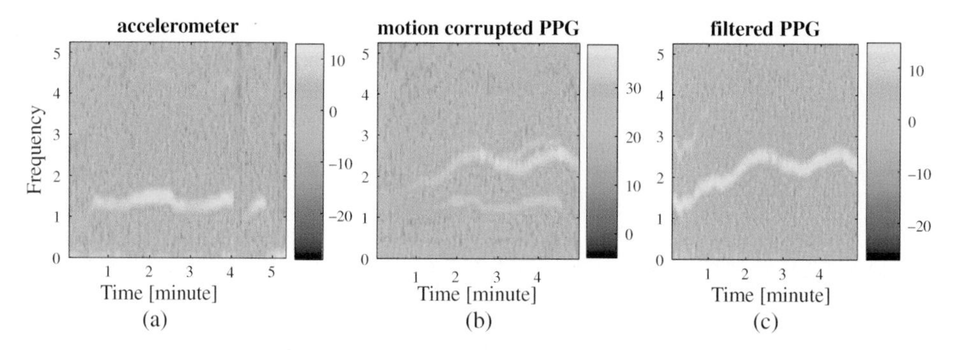


**Figure 4.2** (a) The spectrum of one accelerometer axis; (b) the spectrum of the raw PPG signals where the spectral components (similar to the spectral components of the simultaneously recorded accelerometer signal) related to motion are visible; (c) the spectrum of the PPG signal after applying adaptive filtering.

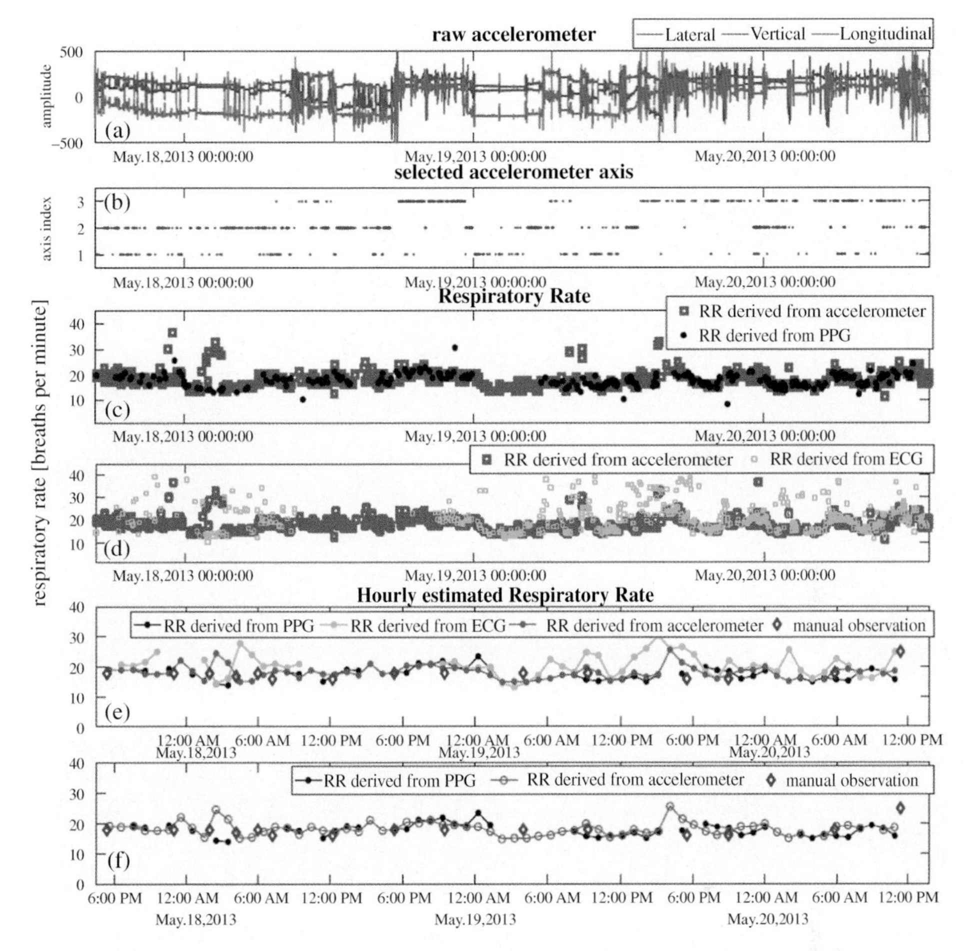


**Figure 4.6** Estimation of RR from simultaneous PPG, accelerometer, and ECG signals recorded from a patient following discharge from the ICU in the hospital ward (John Radcliff hospital, Oxford, 2013) for approximately two days. Raw acceleration data and selected accelerometer axis are shown in (a, b). There is a strong correlation between the PPG and the accelerometer based RR estimates (c, d). Averaged hourly RRs are also plotted (e, f).

**Figure 6.2** Amplification of colour changes in four video frames from a subject's face [13]. Source: Courtesy of Wu, H.-Y., Rubinstein, M., Shih, E., Guttag, J., Durand, F., and Freeman, W.T.

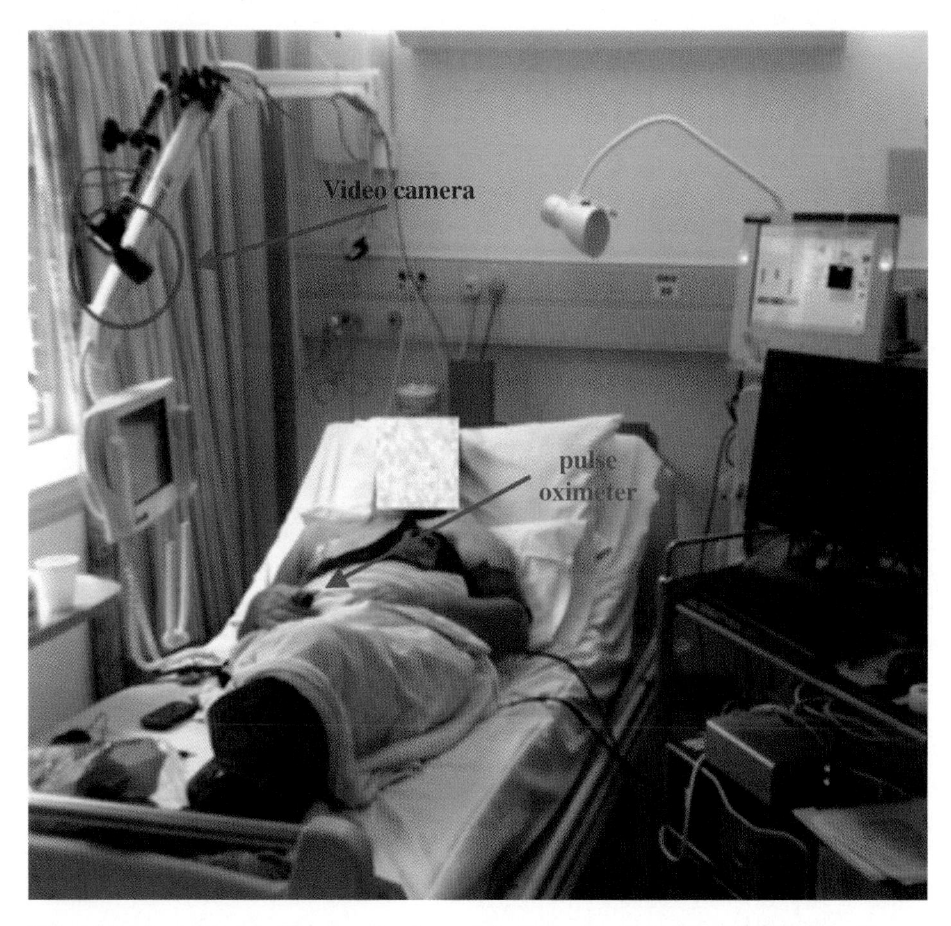


**Figure 6.3** Recording of video frames using a digital video camera, shown in red circle, for a patient at the Oxford Kidney Unit [20]. A pulse oximeter is worn on the patient's finger. Source: Courtesy of Tarassenko, L., Villarroel, M., Guazzi, A., Jorge, J., Clifton, D.A., and Pugh, C.

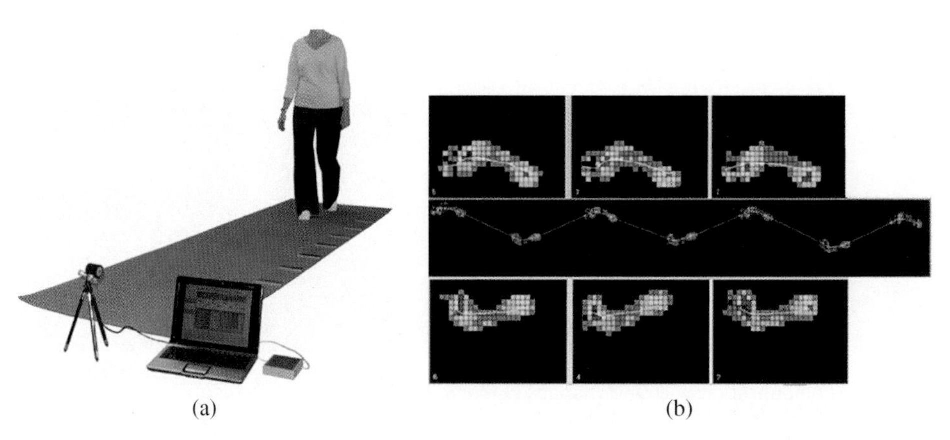


**Figure 7.6** (a) A subject walking on a GAITRite walkway (Source: taken from www.emsphysio.co.uk/product/gaitrite-platinum/); (b) an example of dynamic pressure map during walking using GAITRite.

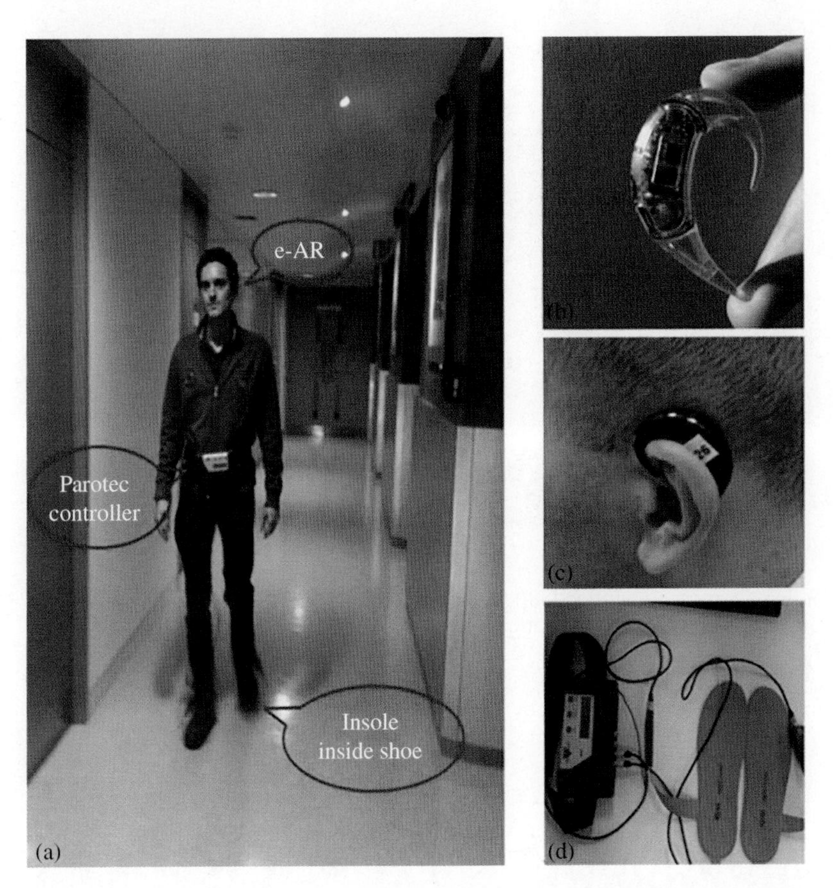


**Figure 7.12** A subject wearing a single ear-worn accelerometer (e-AR) sensor and an in-shoe foot plantar measurement (Parotec) sensor [38]. Source: Courtesy of Jarchi, D., Lo, B., Ieong, E., Nathwani, D., and Yang, G.-Z.

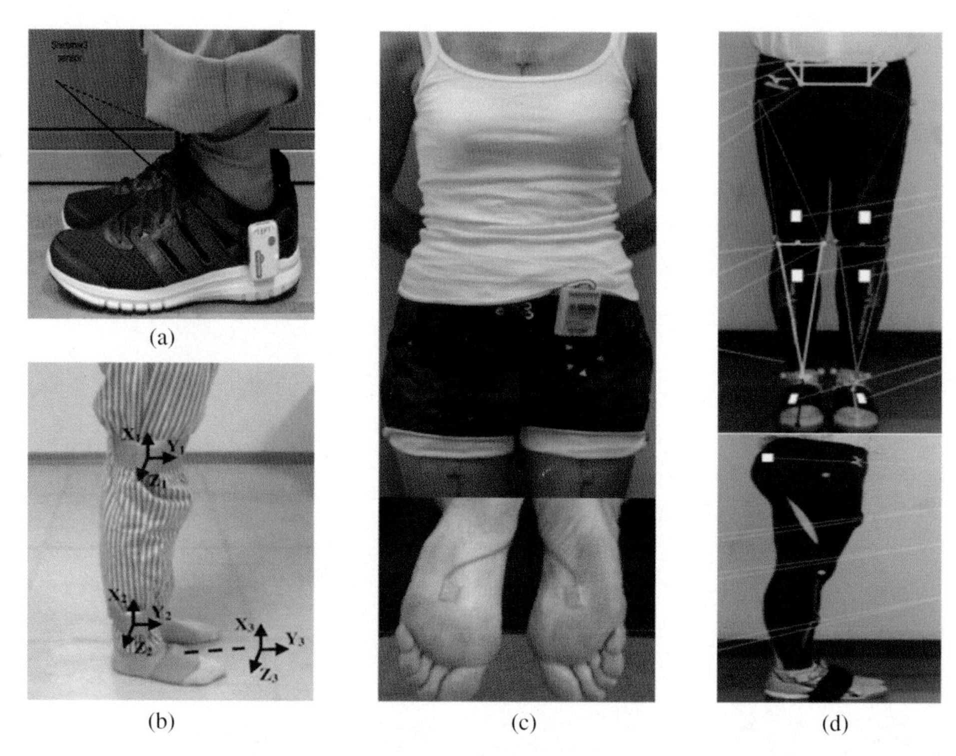


**Figure 7.13** (a) Two-sensor configuration [38]; (b) three-sensor configuration [45]; (c) five-sensor configuration [50]; (d) seven-sensor configuration [52]. Source: Courtesy of Jarchi, D., Lo, B., Ieong, E., Nathwani, D., and Yang, G.-Z.

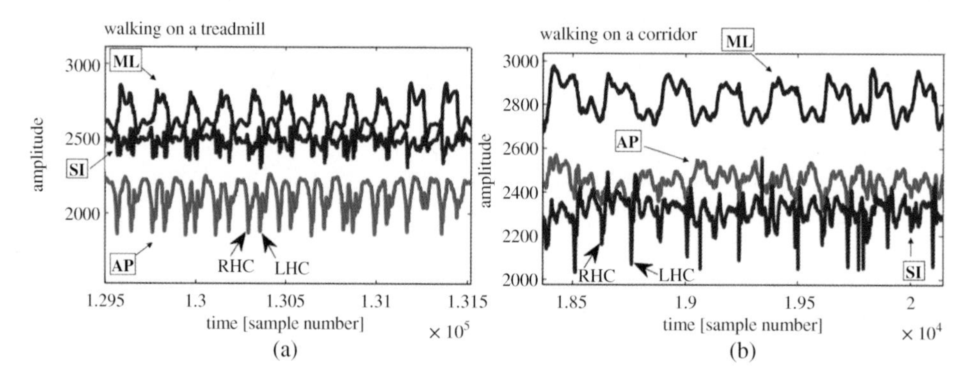


**Figure 7.14** (a) Output acceleration signals for a subject walking on a treadmill; (b) output accelerations for a subject walking outside the gait lab in a corridor [54]. AP, ML, and SI refer, respectively, to anterior–posterior, mediolateral, and superior–inferior axes. Source: Courtesy of Jarchi, D., Lo, B., Wong, C., Ieong, E., Nathwani, D., and Yang, G.-Z.

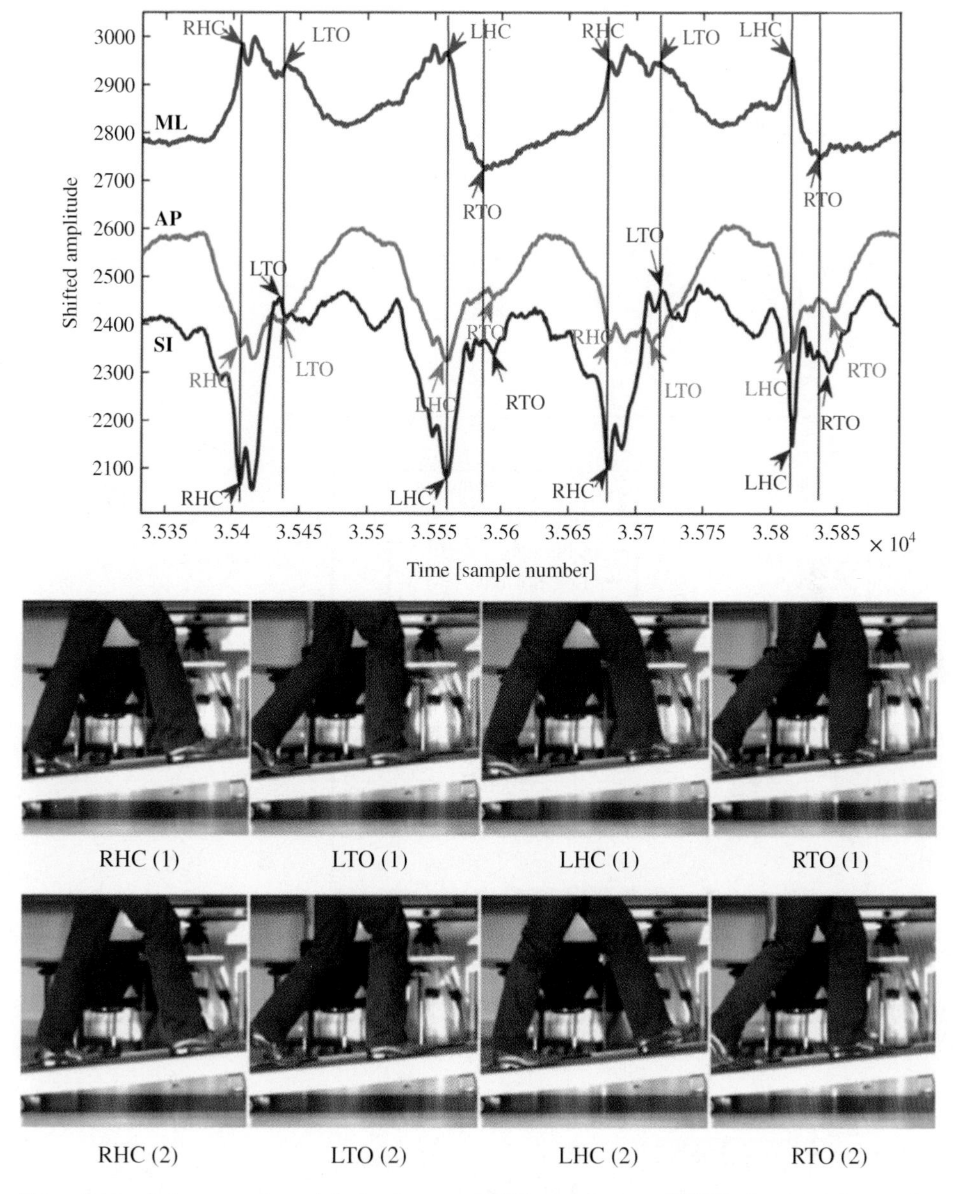


**Figure 7.15** Gait events, from right heel contact (RHC) to right toe-off (RTO) through left toe-off (LTO), and left heel contact (LHC) are located on the acceleration signals (top plot) from the ear-worn sensor, with the help of produced images given by a high-speed camera [55]. Source: Courtesy of Jarchi, D., Wong, C., Kwasnicki, R., Heller, B., Tew, G., and Yang, G.-Z.

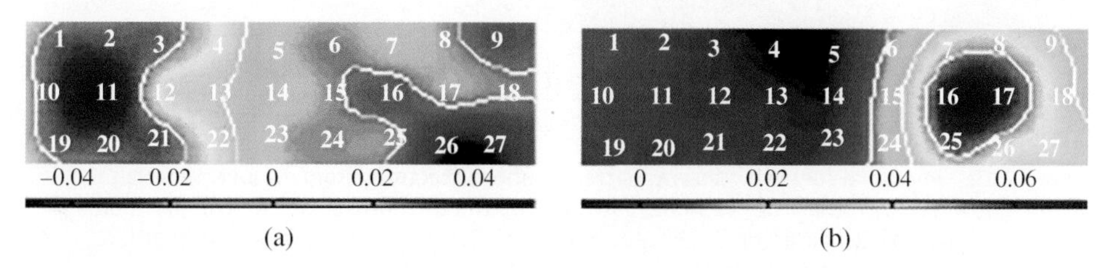


**Figure 9.12** CSPs related to right-hand movement (a) and left-hand movement (b). The EEG channels are indicated by numbers that correspond to three rows of channels (electrodes) within the central and centroparietal regions.

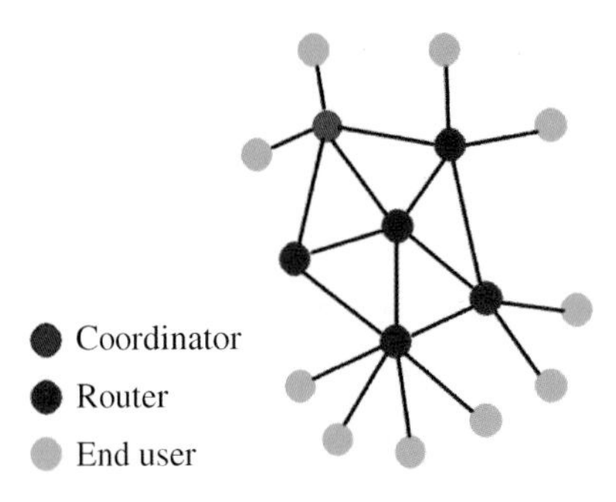


**Figure 11.4** A typical ZigBee network topology.

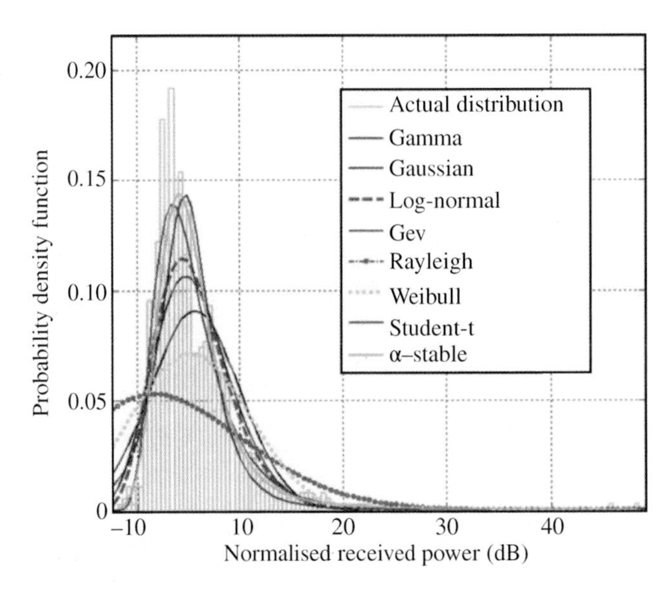


**Figure 11.5** The pdfs representing on-body channel-gain agglomerate from everyday activity of 10 subjects and 540 kHz bandwidth at 2360 MHz [70]. Source: Courtesy of El-Sallabi, H., Aldosari, A., and Abbasi, Q.H.

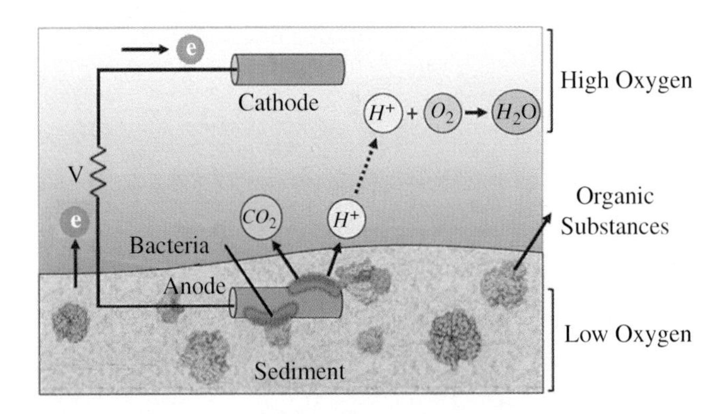


**Figure 12.10** Microbial fuel cell concept: bacteria remove electrons from organic material; the electrons flow through an anode–cathode, battery-like, structure where the two compartments are located in mediums with different $O_2$ concentration [14]. Source: Courtesy of Anwar Bhatti, N., Alizai, M.H., Syed, A.A., and Mottola, L.

# 7

# Single and Multiple Sensor Networking for Gait Analysis

## 7.1 Introduction

Systematic evaluation of bipedal locomotion or gait analysis continues to be a major clinical tool in exercise physiology, biomechanical, pathological, and rehabilitative assessments. Although in some studies 'gait' refers to a general state of moving body parts, the focus for most gait analysis techniques is mainly on human leg movement during walking. A considerable number of platforms have been developed and deployed for applications, including laboratory-based and mobile/portable gait analysis. Force-plates and video cameras are mainly used inside gait laboratories, while wearable sensors such as accelerometers or gyroscopes can be deployed in a free environment outside the laboratories. Therefore, wearable sensors are more useful but may not be as accurate as the indoor measurement system. However, there is always the possibility of validating the performance of wearable sensors for gait analysis using the standard gait platforms using force-plates or cameras inside controlled laboratories. In-shoe foot plantar measurement systems also can be used for validation of wearable systems outside the laboratories.

In this chapter first a brief introduction of gait events, features, and parameters is provided. Next, several important standard gait analysis systems are introduced. Then, exemplar single and multiple wearable sensing platforms are presented. Finally, some of the effective and popular algorithms for the estimation of gait events and parameters are described.

## 7.2 Gait Events and Parameters

### 7.2.1 Gait Events

Gait or stride cycles have two main phases of stance and swing [1] which are described and subdivided into various states in detail in the following sections.

1. *Stance phase*. During this phase, the foot remains in contact with the ground. The stance phase consists of initial contact, loading response, midstance, terminal stance, and pre-swing. These states can be defined as follows:
   a. *Initial contact*, also called heel strike (HS) or heel contact (HC), is the moment the foot makes contact with the ground.

*Body Sensor Networking, Design and Algorithms,* First Edition. Saeid Sanei, Delaram Jarchi and Anthony G. Constantinides.

Companion Website: www.wiley.com/go/sanei/algorithm-design

   b. *Loading response*. Succeeding the initial contact of the foot, the loading response phase begins immediately and it will be continued until the lift of the contralateral limb for the swing phase.
   c. *Mid-stance*. Following the lift of the contralateral limb from the ground to the point where the body weight is aligned with the forefoot, the mid-stance starts.
   d. *Terminal stance*. After heel rising in the frontal plane, the terminal stance period begins and will be continued just before the initial contact of the contralateral limb with the ground.
   e. *Pre-swing*. This period starts from the initial contact of the contralateral limb with the ground and ends with the lift of the ipsilateral limb from the ground.
2. *Swing phase*. During this phase, the foot is not in touch with the ground. The swing phase consists of the initial swing, the mid-swing, and the terminal swing:
   a. *Initial swing*. This phase, which is also called toe-off (TO), starts from lifting the foot off the ground until the knee flexion is increased to its maximum position.
   b. *Mid-swing*. This phase starts immediately after knee flexion and ends when the tibia is vertical (VL).
   c. *Terminal swing*. This phase begins following the VL tibia position prior to the initial contact with the ground.

### 7.2.2 Gait Parameters

There are various parameters, also called features, used for characterising gait. These parameters are often used to quantify different gait phases. Temporal, spatial, kinetic, and kinematic parameters are amongst the most descriptive features to evaluate the gait phases.

#### 7.2.2.1 Temporal Gait Parameters

The temporal gait parameters are timing-based parameters which characterise the gait events or properties based on their temporal behaviour. They can be obtained by finding the timing of the gait events or applying statistical approaches to find important gait features which could be useful for certain clinical applications. In the following the temporal gait parameters are briefly explained.

*Acceleration amplitude variability*. Standard deviations (SDs) of the acceleration signals segmented for each step are averaged over all the steps to represent the mean variability of the acceleration signal during the entire walking period. The steps are identified by locating the acceleration peaks in the VL acceleration axis. Segmented accelerations are averaged and normalised by time to estimate the $mean \pm SD$. Then, the SDs are averaged to produce mean amplitude variability [2].

*Cadence*. Number of steps per minute (i.e. steps/min) [3] or the total number of steps ($M$) divided by the walking time ($S$) [4] or number of strides per minute [5]. The cadence is derived as [6]:

$$c = 60 \times \frac{M}{S} = 60 \times \frac{\frac{N}{n}}{\frac{N}{f}} = 60 \times \frac{f}{n} \tag{7.1}$$

where $N$ is the number of samples for a $D$ metre walk ($D$ is the completed distance of walking in metres), $n$ is the number of samples per dominant period, $f$ is the sampling frequency in hertz, and $c$ is the cadence (steps/min). Having the values for $D, f, n$, and $S$ ($S$ is the time in seconds to walk $D$ metres), it is possible to estimate other parameters, e.g. $N = Sf$ or $n = \frac{N}{M}$. Therefore, the equation to estimate the cadence is obtained using only two parameters: $f$ and $n$ [6].

*Cycle frequency*. The fundamental frequency obtained by applying discrete Fourier transform (DFT) where it has been found that stabilised walking generates periodic patterns in the acceleration signals [7].

*Double support duration*. The time duration of phase of support on both feet. It can be presented as a percentage of gait cycle where both feet are in contact with the ground [8].

*Foot symmetry*. The step duration presented as a percentage of gait cycle [9].

*Gait cycle time*. The time duration between two consecutive HS events.

*Gait irregularity*. The average SDs of the right and left step times [10, 11] that correlates with the variability in successive steps of the same foot.

*Gait variability*. The SD of gait parameters or their coefficient of variation (CV) defined as the ratio of SD to the mean (i.e. $CV = \frac{SD}{mean}$) [4, 12] which is based on stride to stride fluctuations.

*Harmonic ratio*. The ratio between sum of even and sum of odd (($\Sigma$ even harmonic)/($\Sigma$ odd harmonic)) and sum of odd and sum of even (($\Sigma$ odd harmonic)/($\Sigma$ even harmonic)) harmonic amplitudes calculated using DFT [2] for the anterior–posterior (AP), vertical (VT), and mediolateral (ML) axes, respectively. This parameter presents the smoothness and rhythmicity of acceleration patterns. The even harmonics for the AP and VT axes are related to the in-phase components of acceleration signal, while the odd harmonics correspond to the out-of-phase components that are minimum for a normal gait [2].

*Interstride acceleration variability*. It is estimated by using calculated autocorrelation coefficients for each axis [5].

*Normalised speed*. The speed normalised with respect to the subject's height [9].

*Root mean square*. Calculated root mean square (RMS) of the acceleration magnitudes [2]. This parameter indicates the dispersion of accelerometry data relative to zero, as opposed to SD, which is a measure of dispersion relative to the mean acceleration. In the case of having zero acceleration mean, the RMS has the same concept as SD. This parameter is used to express the average magnitude of the accelerations for each axis [2].

*Stance duration*. The timing from HS and TO of the same foot that is usually expressed as a percentage of gait cycle.

*Step asymmetry*. The ratio of the difference between mean step time of each foot to the combined mean step time of both feet [11, 13], or the difference in consecutive step times [10] which is equivalent to $\frac{A_{d1}}{A_{d2}}$, where $A_{d1}$ and $A_{d2}$ are obtained by calculating the autocorrelation coefficients at first dominant period and second dominant period of the acceleration signal (VL or AP), respectively [6].

*Step duration*. The time duration between the heel contacts of the opposite feet (between ipsilateral and contralateral HSs) [9].

*Step frequency*. Half of the fundamental frequency calculated using DFT. It can be scaled with respect to the subject's height to utilise the influence of the height on step frequency. This has shown to be useful when comparing male and female adult populations [14].

*Step regularity*. Defined as $A_{d1}$: calculated as autocorrelation coefficient at first dominant period of the VL or AP accelerations [6]. A bigger value for $A_{d1}$ corresponds to a higher step regularity. It has been concluded that a periodic signal like acceleration signals produces autocorrelation coefficients which have peak values for the lags corresponding to the periodicity of the signal, called the dominant period [6].

*Step timing variability*. SD of time between consecutive heel contacts considering whole duration of walking [3].

*Single support duration*. Percentage of gait cycle during which only one foot is in contact with the floor or it can be expressed as the duration of the phase of support on one foot.

*Stride duration*. The time between two consecutive HSs of the same foot.

*Stride frequency*. Number of cycles per second (hertz).

*Stride regularity*. Defined as $A_{d2}$: calculated autocorrelation coefficient at second dominant period of the VL or AP accelerations [6].

*Stride symmetry*. $\frac{A_{d1}}{A_{d2}} \times 100$ [6, 15].

*Swing duration*. The time from TO to HS of the same foot that can be expressed as percentage of gait cycle.

*Walking intensity*. Calculated from the integral of the modulus accelerometer output [5].

*Walking time*. Duration of walk which is measured using a stopwatch, as in [16].

#### 7.2.2.2 Spatial Gait Parameters

Spatial gait parameters are noted as distance-related parameters. They are presented in the following manner:

*Step length*. Travelled distance in metres divided by the number of completed steps, i.e. distance (m)/number of completed steps [2]. The distance is considered between the ipsilateral and contralateral HSs [9]. In the most common way having the trunk accelerations, an inverted pendulum model is used to estimate step length [9, 16, 17].

*Step width*. ML distance between the heels in double support phase [18]. In a simple way it was registered from footprints in [3].

*Step width variability*. SD of the step width [3].

*Stride length*. The distance between two successive HSs of the same foot or average of speed (m/s)/stride frequency (in hertz) [7, 19]. In [14], the stride has been scaled for the subject's height. Convolutional neural networks have recently been used in [20] for the estimation of stride length from accelerometer and gyroscope signals.

*Stride velocity*. Ratio of the stride length to stride time [8] or the mean value of foot velocity in forward direction during gait cycle [12].

*Walking distance*. Multiplication of mean step length over a specified duration by the number of steps.

*Walk ratio*. Defined as the average step length (in centimetres) divided by cadence (steps/min) [2]. This parameter provides a link between the amplitude and frequency of rhythmic foot movements during walking [2].

*Walking (gait) speed*. Ratio of distance and the walking time [4]. It has been measured in m/s using a chronometer [7].

*Walking velocity*. (Distance covered)/(number of data points × sampling period) [2] which is described in terms of m/s.

*Lateral foot placement* (LFP) is defined as the distance between the heel position and the orthogonal projection of this heel position (of the same foot) on the average walking path [21].

#### 7.2.2.3 Kinetic Gait Parameters

By using kinetic gait parameters, it is possible to measure both magnitude and direction of external forces on lower limbs in different gait phases. The ground reaction force (GRF) is an important external force produced by the ground which is considered the reaction to the force the body exerts on the ground. Force-plates are usually used for measurement of the GRFs. The GRF as a force can be described using Newton's second law, i.e.

$$F = ma \tag{7.2}$$

where $m$ is the subject's body mass and $a$ is the VL acceleration of the body's centre of gravity. The GRF typically represents the effects of body's centre of gravity and movement or acceleration in three reference planes (VL, ML, and AP). The recorded GRF during stance phase from heel contact of one foot to the TO of the same foot is shown in Figure 7.1. From this figure, the initial heel contact happens shortly before the first maximum peak of the GRF and the TO event after the second maximum peak of the GRF where the GRF reaches zero.

#### 7.2.2.4 Kinematic Gait Parameters

Kinematic gait analysis mainly involves analysis of mobility of lower extremity joints [22]. The joint angles of the lower extremity, such as hip, knee, and ankle, are considered as kinematic gait parameters (see Table 7.1). They are used in various clinical settings for the diagnosis of many diseases, such as gait impairments and their management through the right choice of rehabilitation techniques. Wearable sensors such as inertial measurement units (IMUs) can be placed on various joints to measure the ranges of motion (RoMs) of each joint, which corresponds to joint flexibility [24] (see Figure 7.2).

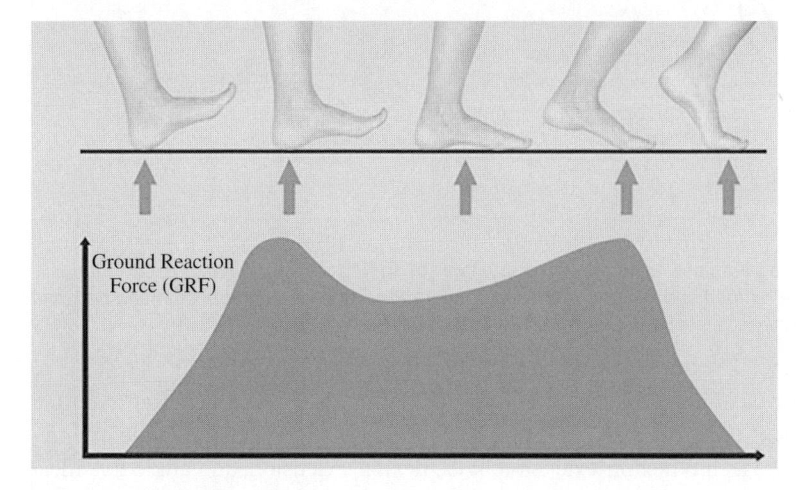


**Figure 7.1** GRF measured from a subject walking on a force-plate during stance phase. Source: Courtesy of MASS4D INC.

**Table 7.1** Average RoMs for different hip, knee, and ankle movements according to American Academy of Orthopaedic Surgeons [23].

| Movement | Normal RoM |
|---|---|
| Hip flexion | 0–100∼125° |
| Hip extension | 0–10∼30° |
| Knee flexion | 0–135∼150° |
| Knee extension[a)] | 0–10∼30° |
| Ankle dorsiflexion | 0–20° |
| Ankle plantarflexion | 0–40∼50° |

a) Any hyperextension is noted where the degree goes beyond 0 and is measured in negative degrees.
Source: Courtesy of American Academy of Orthopaedic Surgeons.

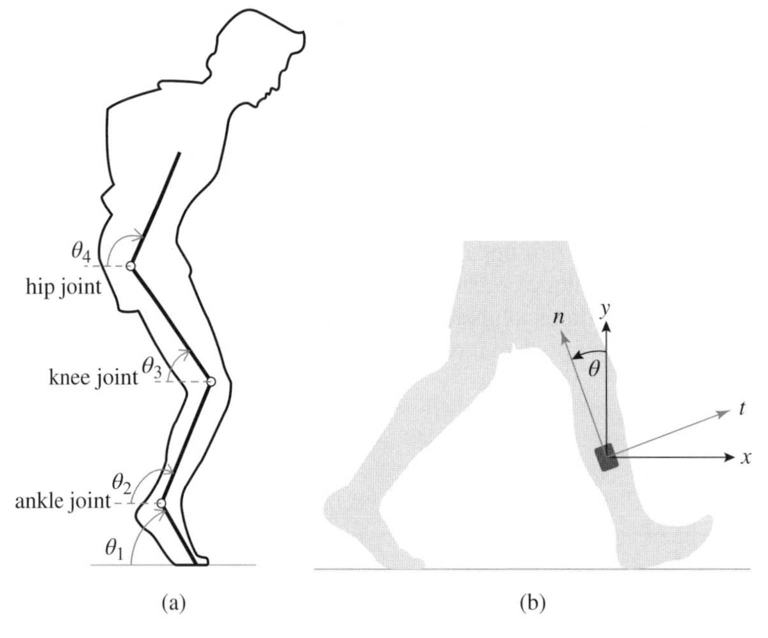


**Figure 7.2** (a) Limb joints of a lower human body part; (b) an IMU sensor attached to the shank; the sensor axes are defined along *n* and *t* directions, while the world coordinate is defined by *x* and *y* axes [24]. Source: Courtesy of Yang, S., Mohr, C., and Li, Q.

A reduced joint flexibility which can be reflected in a reduced RoM has been reported to contribute to gait impairments [25]. Three RoMs for the hip, knee, and ankle are presented in Figure 7.2 as hip flexion/extension, knee flexion/extension, and ankle dorsiflexion/plantarflexion, respectively (Figure 7.3).

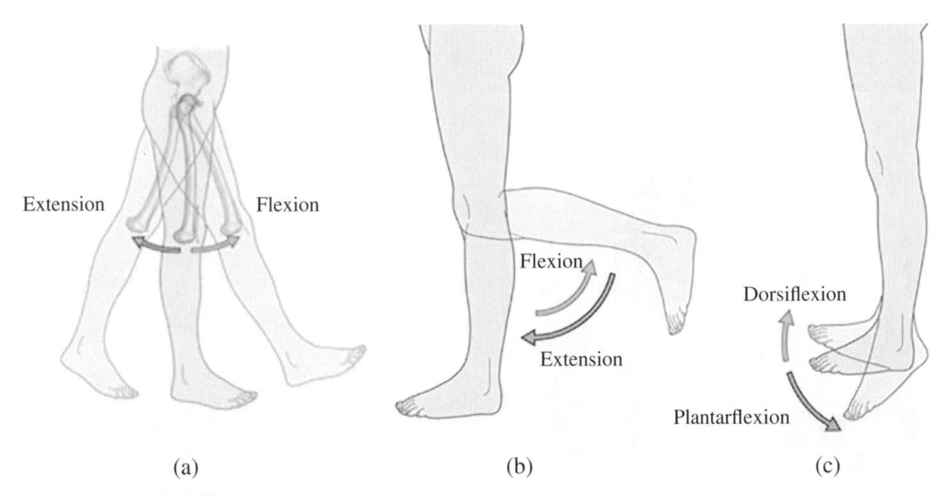


**Figure 7.3** Hip, knee, and ankle movements: (a) hip flexion and extension; (b) knee flexion and extension; (c) ankle dorsiflexion and plantarflexion (https://clinicalgate.com).

## 7.3 Standard Gait Measurement Systems

### 7.3.1 Foot Plantar Pressure System

Development of an in-shoe foot plantar system has gained importance for mobile gait analysis outside of the laboratory. Example systems include F-Scan (Tekscan, South Boston, MA, USA), and Parotec (Paromed, Neubeuern, Bavaria, Germany) [26]. These systems include an insole embedded with multiple pressure sensors to generate force signals which can be used to determine gait events and derive various gait parameters. Example outputs of the Tekscan system are shown in Figures 7.4 and 7.5. The generated force signals over time (green/red curves for the left/right foot) are shown in the right while the sensor pressure distribution maps are visualised for the left/right foot at one instant of time in the left. The time instant is denoted as a VL dash line in the force vs. time plot. Wearable sensors are usually miniaturised and cheap. Hence, it is convenient to use them for mobile gait analysis while in-shoe foot plantar system is more expensive and mostly bulky. However, plantar systems measure and analyse the gait more accurately and therefore are usually used as a reference platform to validate the wearable sensors. For certain patient populations where a detailed analysis of foot pressure is required, these in-shoe foot plantar systems could be more ideal compared to wearable sensors.

### 7.3.2 Force-plate Measurement System

Force-plates or force platforms measure GRFs produced by a subject walking on them. Laboratory/portable walkways or treadmills can be instrumented with force platforms to analyse the gait. The GAITRite system (CIR Systems Inc. Clifton, NJ, USA) consists of a portable pressure sensitive walkway embedded with pressure sensors. The subjects can walk along a walkway (a carpet 8.3 m long and 0.98 m wide forming an electrical walkway) and have their

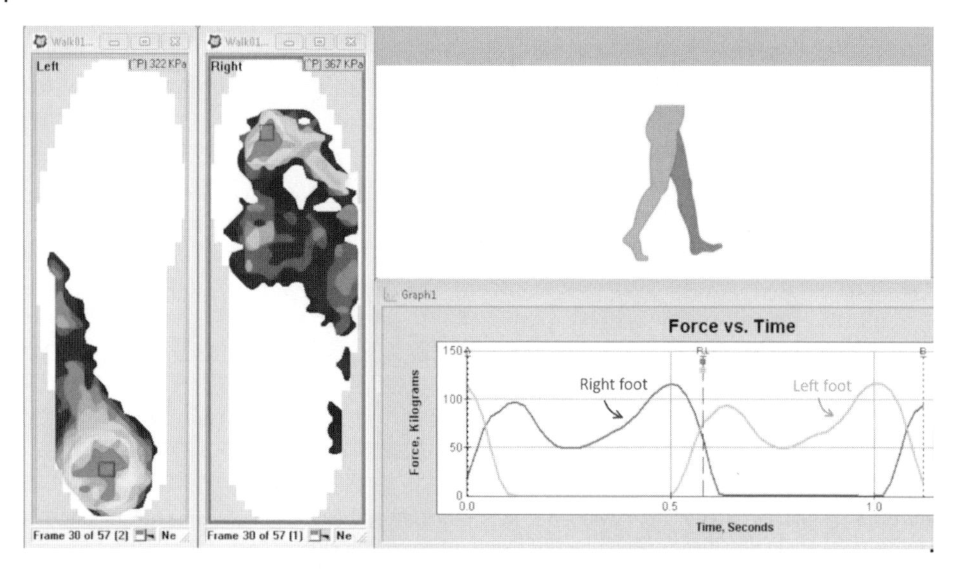


**Figure 7.4** Force signal generated for the left foot – curve with the nonzero amplitudes (~20 to ~100) from 0.5 to 1 second – and right foot – curve with the nonzero amplitudes (~20 to ~100) from 0 to 0.5 second – over time. The pressure maps for both shoes are shown in the left (F-scan system https://www.tekscan.com/).

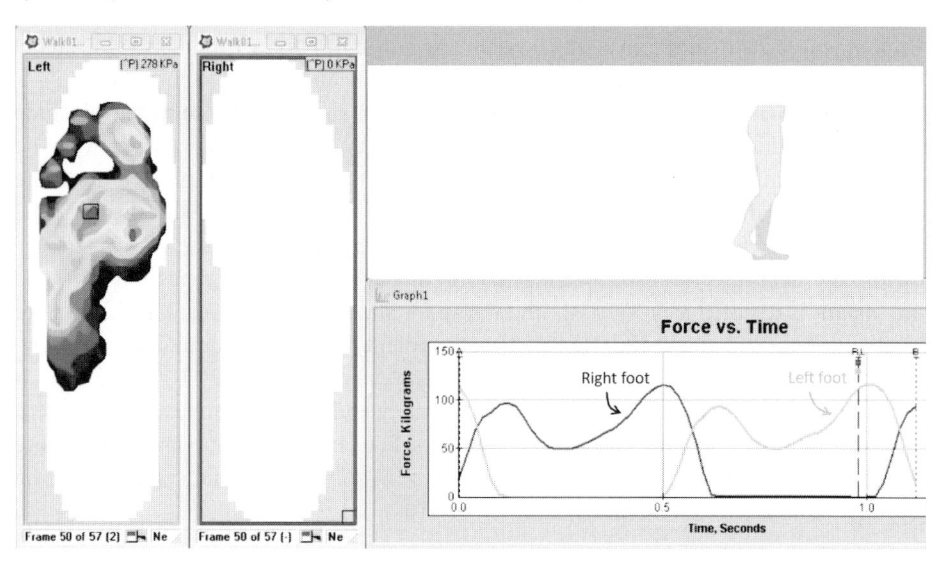


**Figure 7.5** Force signal generated for the left foot (green curve) and right foot (red curve) over time and pressure maps similar to Figures 4 and 5 (F-scan system https://www.tekscan.com/).

gaits characterised without any need for wires or markers (Figure 7.6). The active area covered by the carpet is about 7.32 m long and 0.61 m, where there is a distance of about 7.7 mm between each sensor. Gait parameters such as walking speed, cadence, step length, single, and double support durations and stride widths can be estimated using GAITRite [27–31]. This system scans the pressure sensors constantly to provide the geometry of 2D footprints

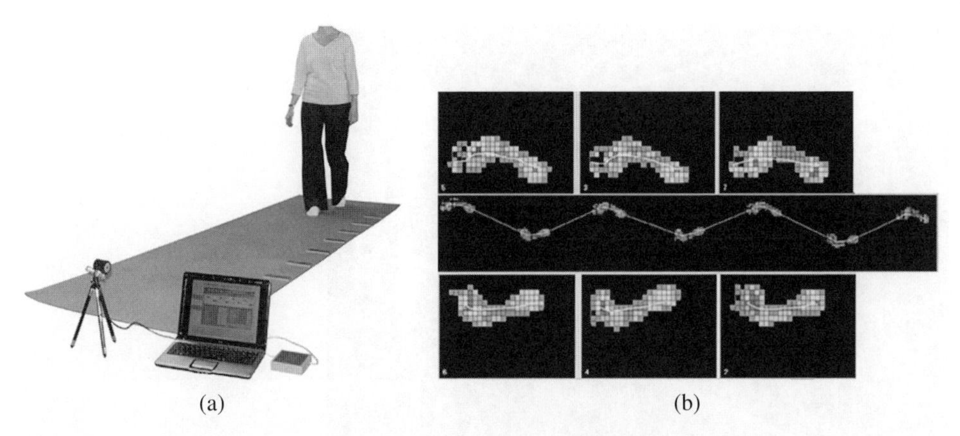

(a) (b)

**Figure 7.6** (a) A subject walking on a GAITRite walkway (Source: taken from www.emsphysio.co.uk/product/gaitrite-platinum/); (b) an example of dynamic pressure map during walking using GAITRite. (*See color plate section for color representation of this figure*)

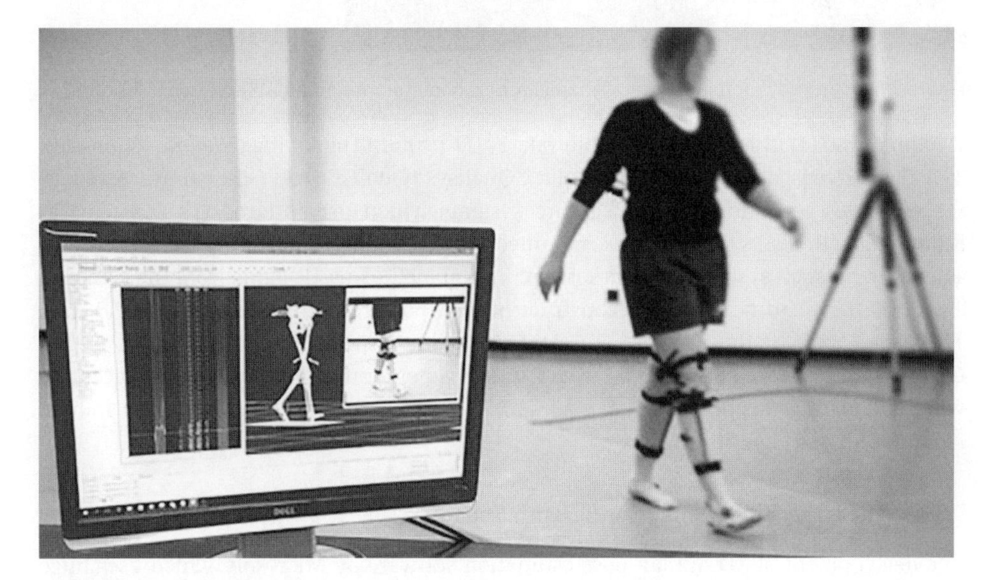

**Figure 7.7** Camera-based motion analysis (https://codamotion.com) and force-plates. Source: Courtesy of Codamotion.

which help form the dynamic pressure mapping (Figure 7.7). A grid of about 13 424 sensors is used and the measurements from these sensors can be used to find the location of activated sensors and the timestamps of activation and deactivation. An example output is shown in Figure 7.6. Gait analysis configuration using force-plates and camera based motion analysis systems are shown in Figure 7.7.

### 7.3.3 Optical Motion Capture Systems

Vicon Motion Capture System (Vicon Motion Systems, Los Angeles, USA) [32], Cartesian Optoelectronic Dynamic Anthropometer (CODA) motion analysis system (Charnwood

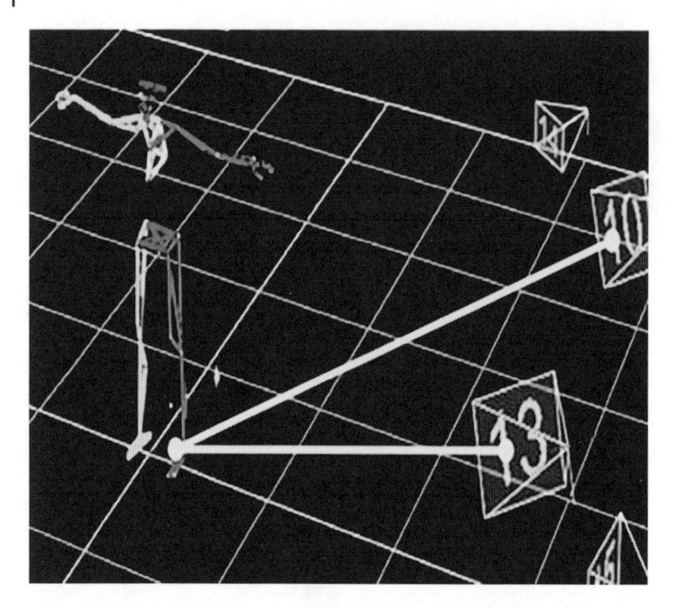

**Figure 7.8** Cartesian Optoelectronic Dynamic Anthropometer motion analysis system.

Dynamics Ltd., Rothley, UK) [33], Optotrak [34] (Northern Digital Inc., Waterloo, Canada), Tracklab (Freedspace, Victoria, Australia), Qualisys (Qualisys AB, Gothenburg, Sweden) are examples of optical motion capture systems which use infrared (IR) cameras to illuminate retro-reflective markers, mounted on the body, for capturing their positions. Therefore, in such systems the markers are illuminated using IR lights from the cameras. These markers can be used to reconstruct human posture and motion (Figure 7.8–7.9). The coordinates of the markers are estimated with respect to a pre-calibrated origin. Optical systems have shown to be useful in the generation of motion and very accurate when estimating position and angular rates. Nevertheless, they are limited to indoor, laboratory-based experiments.

### 7.3.4 Microsoft Kinect Image and Depth Sensors

The development of 3D human pose estimation software by Microsoft, which uses huge datasets including numerous variations, has provided a rich source for an accurate estimation of body position, including 3D full body motion registration. The use of depth images has also removed the need for calibration procedures and placing markers as required by other systems, such as Optotrak.

The 3D positions from 25 body positions including head, neck, spine shoulder, spine mid, spine base, and right and left shoulders, elbow, wrist, hand, thumb, hand tip, hip, knee, ankle, and foot have been provided by Microsoft Kinect. The sampling rate has been set as 30 samples per second. Using the Kinect system RGB-D camera [35], depth images are produced for accurately tracking body points. Microsoft Kinect v2 provides depth images with higher resolutions and an improved body point tracking than those achieved by its predecessor, Microsoft Kinect v1. In addition, there is a wider field of view for larger distances leading to higher depth-image qualities for Microsoft Kinect v2 [36]. As shown in Figure 7.10, the four Kinect v2 sensors are placed on tripods alongside a walkway of 10 m

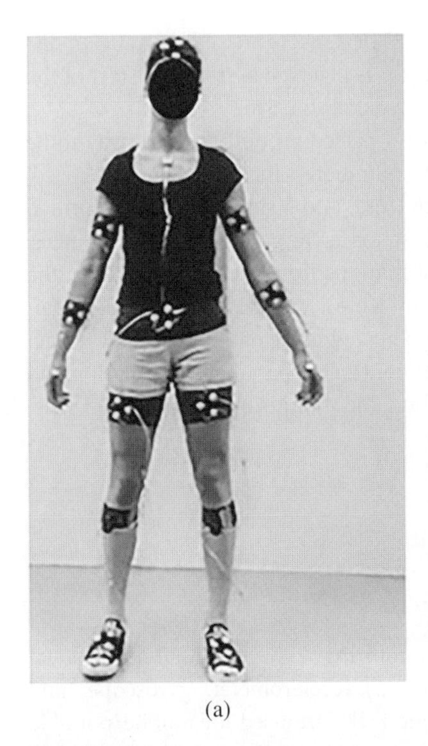
(a)

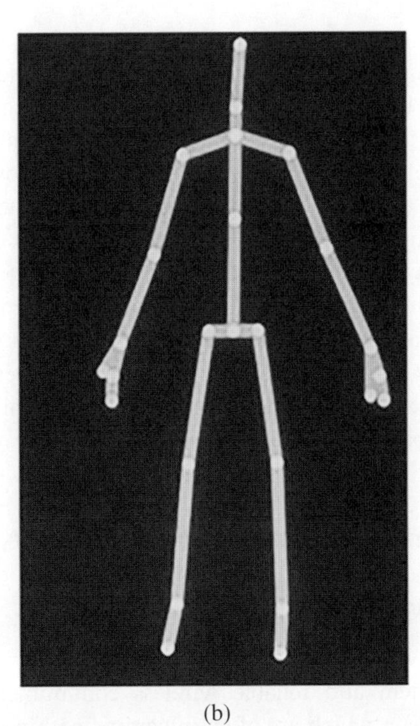
(b)

**Figure 7.9** (a) The markers based on the Optotrak system for a selected subject, often used as a gold standard [34]; (b) derived body points of subject in a using the human-pose estimation algorithm of Kinect v2 [34]. Source: Courtesy of Geerse, D.J., Coolen, B.H., and Roerdink, M.

**Figure 7.10** RGB-D camera setup for multi-Kinect v2 setup.

each covering 0.5 m including a height of 0.75 m [34]. The sensors are placed in 0.5 m distance from the left side of the walkway. The first sensor is placed 4 m from the start of the walkway. The distance between the other three sensors is set to 2.5 m.

An Optotrak system, including five cameras, is also placed around the walkway to be used as a reference. Then, a number of spatiotemporal parameters including walking speed, cadence, step length, stride length, step width, step time, and stride time are quantified and compared using multi-Kinect v2 and Optotrak systems as the gold standard references. Based on a high level of agreement between the estimations from these two systems, the results of the study in [34] suggests the use of Microsoft Kinect v2 system for automatic gait assessments for a 10-minute walking test (10MWT). The downside of the Microsoft Kinect technology is that it is limited to indoor-based experiments. Gait parameter estimation and motion tracking using Microsoft Kinect for recognition and monitoring of Parkinson's disease (PD) patients have been investigated in [37].

## 7.4 Wearable Sensors for Gait Analysis

### 7.4.1 Single Sensor Platforms

There are two primary objectives for outdoor gait monitoring: to have a small multimodal sensor system and reliable wireless communication. Accelerometer, gyroscope, and magnetometer which are often integrated within one IMU, are used for mobile/portable gait analysis. Ear, thorax, shoe, shank, heel, knee, thigh, and pelvis are the body sites often used for the placement of wearable sensors. Among single sensor platforms, a trunk mounted accelerometer (Figure 7.11) has been used in some studies [4] where the main gait events, such as HS and TO, and temporal/spatial gait parameters are estimated. Kinematic gait parameters are usually estimated using gyroscope/IMU placed on lower body parts. A recently designed and prototyped ear-worn sensor including a single accelerometer has been used in a number of studies for the estimation of temporal gait parameters and investigating gait patterns for normal and orthopaedic patients. The sensor has been used outside the gait laboratory for monitoring the rehabilitation of an orthopaedic patient in a hospital corridor. As seen in Figure 7.12, the ear-worn sensor is used for gait analysis and

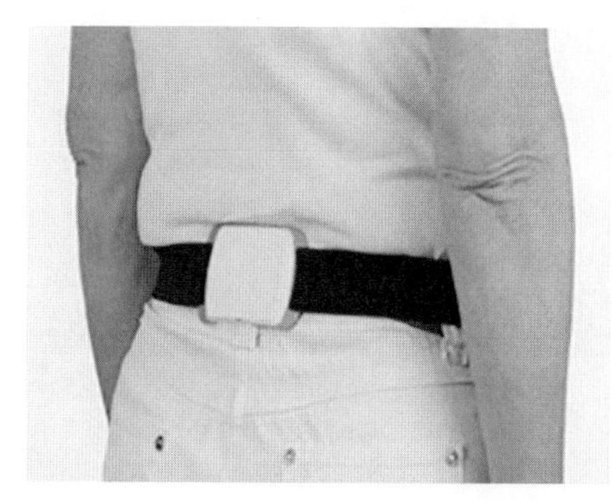

**Figure 7.11** Trunk mounted accelerometer (Dynaport® MiniMod system) used in clinical studies.

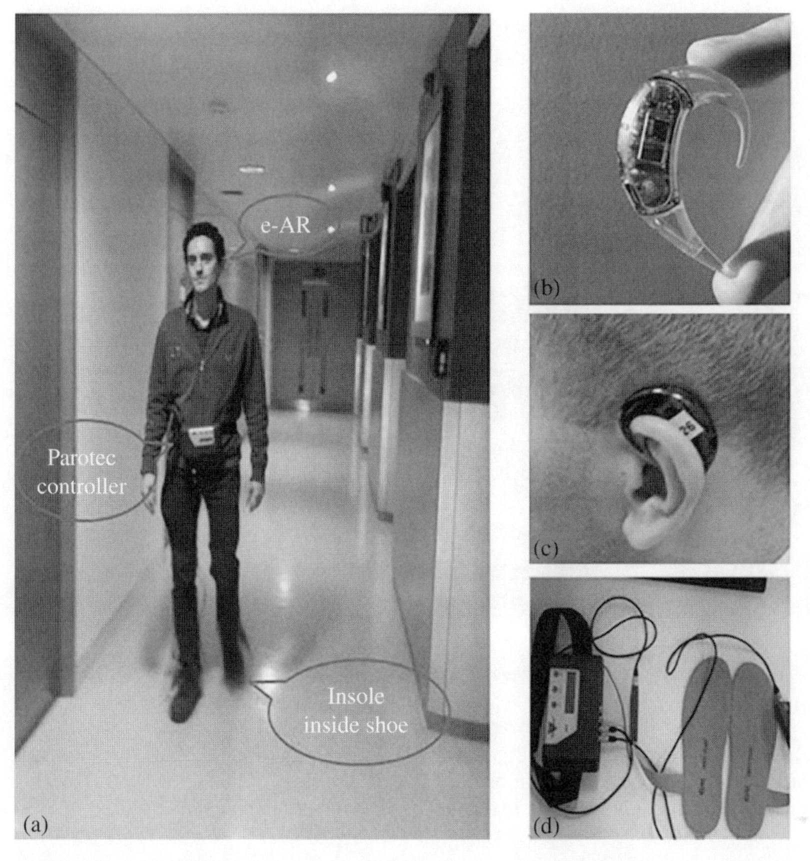


**Figure 7.12** A subject wearing a single ear-worn accelerometer (e-AR) sensor and an in-shoe foot plantar measurement (Parotec) sensor [38]. Source: Courtesy of Jarchi, D., Lo, B., Ieong, E., Nathwani, D., and Yang, G.-Z. (*See color plate section for color representation of this figure*)

validated using the Parotec systems. Smartphones are also used in recent studies for gait pattern analysis [39] and monitoring patients with chronic diseases [40].

### 7.4.2 Multiple Sensor Platforms

An IMU unit can be considered as a single sensor unit consisting of an accelerometer, a gyroscope, and a magnetometer. Multiple accelerometers, gyroscopes, or IMU units are placed in different parts of the body and used in various studies to achieve an improved performance that can be crucial for certain clinical applications. Multiple sensor platforms can have a significant impact on monitoring specific patient groups where many gait analysis systems are not applicable to normal gait. In addition, various walking conditions such as stair climbing, treadmill exercise, or outdoor walking produce different gait patterns for which further modifications to the analysis algorithms become inevitable.

As examples of two-sensor configurations, IMUs are placed on the left and right shoes to validate foot-worn IMUs for the gait measurement of children with cerebral palsy. They are also used for the mobile gait analysis of PD patients [38, 41]. Triaxial/biaxial

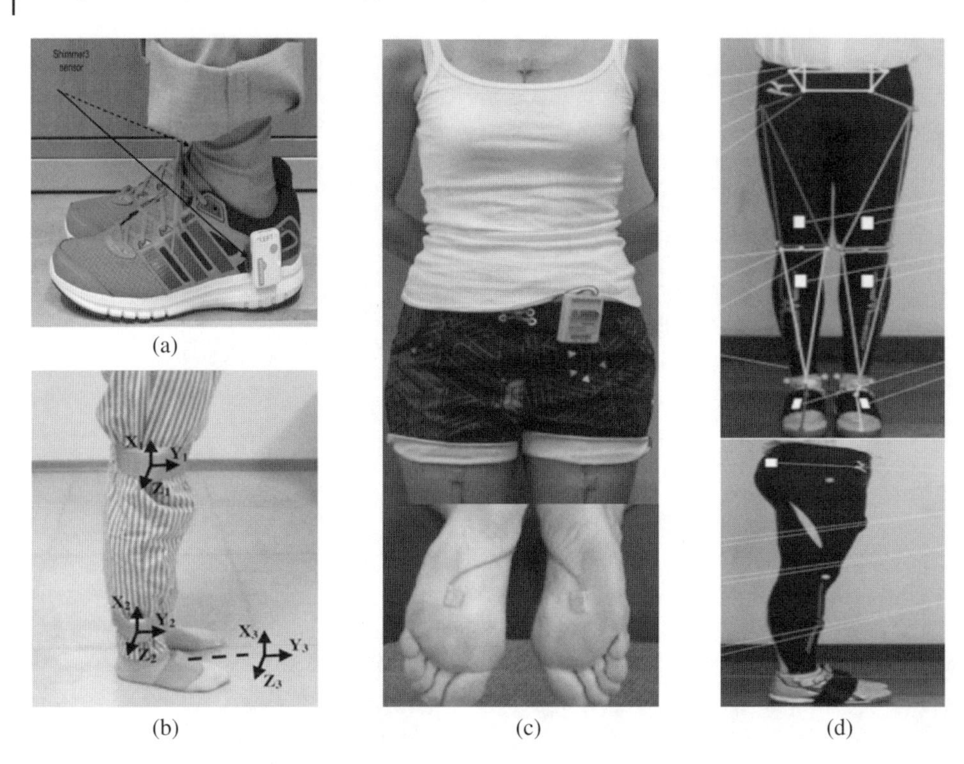


**Figure 7.13** (a) Two-sensor configuration [38]; (b) three-sensor configuration [45]; (c) five-sensor configuration [50]; (d) seven-sensor configuration [52]. Source: Courtesy of Jarchi, D., Lo, B., Ieong, E., Nathwani, D., and Yang, G.-Z. (*See color plate section for color representation of this figure*)

accelerometers are placed on the head/trunk joints for elderly gait analysis [42], ankle/knee joint for the evaluation of hemiplegic gait and control of an intelligent knee prosthesis of above-knee amputees [43], and the foot/shank joint for the detection of main gait events for spinal-cord-injured patients [44]. Three-sensor configurations have been used in various studies. Placement of the IMUs on the shank/thigh/foot to discriminate gait symptoms in a group of hemiplegia and asymptotic subjects [45] is only one example. In another study the IMUs were placed on the trunk and the left and right shoes [46] for the analysis of gait and balance in patients with Alzheimer's disease. A four-sensor configuration was used for the quantification of gait asymmetries and the freezing of gait in patients with PD from just before the metatarsophalangeal joint with hallux and above lateral malleolus [47] and for the analysis of lower limb osteoarthritis from the head, lower back, and left and right feet [48]. A five-sensor configuration was used in [49, 50] for the gait analysis of patients with cerebral palsy and during level walking/stair climbing, respectively. A six-sensor configuration was used for the gait analysis of patients with multiple sclerosis [51]. A seven-sensor configuration was used in [52] for the characterisation of osteoarthritis gait. Examples of multiple sensor configuration are shown in Figure 7.13.

## 7.5 Gait Analysis Algorithms Based on Accelerometer/Gyroscope

### 7.5.1 Estimation of Gait Events

Gait events can be identified from the signals recorded by wearable sensors simply by using peak detection and template matching methods. The timings of gait events need to be determined. The local peaks in accelerometer and gyroscope signals recorded from different parts of the body are detected to find the initial foot contacts to use them to determine the footsteps. The Pan-Tompkins method, which was initially developed for the detection of QRS complexes in electrocardiography (ECG) signals, was used to detect steps from accelerometer signals [53]. The steps can be detected by template matching applied to foot acceleration signals too [53]. Based on this method, the accelerometer signals should be first segmented into several 10-second data blocks. These signals need to be filtered by a lowpass filter with a cut-off frequency of 20 Hz. Adaptive templates representing a typical step cycle must be constructed to detect steps when there is a good match between the signal and the template. Since the morphology of steps changes over time, the temporal template must be updated. Steps are identified as peaks in the accelerations by comparing adaptive templates to the recorded acceleration signals [53]. A dual-axis peak-detection method has been developed in [53] where the steps have been identified as the negative peak when both AP and VL acceleration signals coincide.

The autocorrelation method is proposed by Moe-Nilssen and Helbostad [6] using an accelerometer placed on the back of the lower waist. The autocorrelation coefficient can be calculated by:

$$A_{unbiased} = \frac{1}{N-\mid m\mid}\sum_{i=1}^{N-|m|} x_i x_{i+m} \tag{7.3}$$

where $x_i$, $(i = 1, 2, \ldots, N - \mid m|)$, is the acceleration data, $N$ is the number of samples, and $m$ is the time lag phase shift parameter. The autocorrelation coefficients with increasing time lags are computed. The peak values on the autocorrelation signal represent the lags equivalent to the periodicity of the signal. The peak values of the autocorrelation coefficient sequence at the first and second dominant periods correspond to one step and one stride, respectively.

In [54, 55], the acceleration signals from an e-AR sensor have been used to estimate the gait events. From Figure 7.14, the acceleration signals are more regular and periodic while a subject is walking on the treadmill (TM in Figure 7.14) than walking along a corridor (Figure 7.14). A high-speed camera was used to locate the gait events on acceleration signals recorded from the ear in [55]. The gait events, from right heel contact (RHC) to right toe-off (RTO) through left toe-off (LTO) and left heel contact (LHC) are located on the acceleration signals (top plot) from the ear-worn sensor in Figure 7.15 where simultaneous video camera images helped to better identify the gait events.

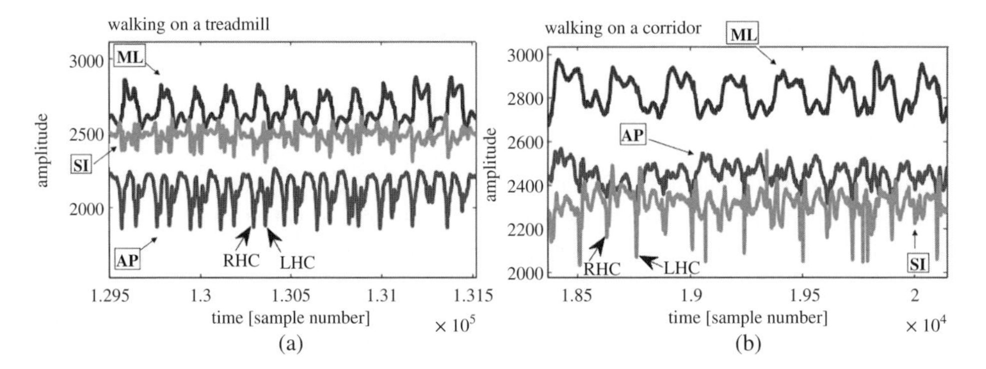


**Figure 7.14** (a) Output acceleration signals for a subject walking on a treadmill; (b) output accelerations for a subject walking outside the gait lab in a corridor [54]. AP, ML, and SI refer, respectively, to anterior–posterior, mediolateral, and superior–inferior axes. Source: Courtesy of Jarchi, D., Lo, B., Wong, C., Ieong, E., Nathwani, D., and Yang, G.-Z. (*See color plate section for color representation of this figure*)

### 7.5.2 Estimation of Gait Parameters

Temporal gait parameters can be obtained following the estimation of gait events. For other parameters – such as kinematic, kinetic, and spatial parameters – the associated models usually need to be constructed. Two important related parameters include orientation and displacement. In the following subsections, first, estimation of orientation including a short note on using gyroscopes for assessing displacement is described. Then, several methods developed for estimation of step length, as an important spatial parameter, are explained.

#### 7.5.2.1 Estimation of Orientation

Two methods (Euler angles and quaternions) are popular for the estimation of orientation from the gait signals. For simplicity's sake, we start by formulating the orientations using a Euler angle [56, 57]. The inertial frame is shown in Figure 7.16 presenting roll, pitch, and yaw angles as the rotations along x, y, and z axes, respectively.

The 3D rotation matrixes for rotating the frame along x, y, or z axes can be obtained as [56]:

$$R(\phi) = \begin{bmatrix} 1 & 0 & 0 \\ 0 & \cos(\phi) & \sin(\phi) \\ 0 & -\sin(\phi) & \cos(\phi) \end{bmatrix}, R(\theta) = \begin{bmatrix} \cos(\theta) & 0 & -\sin(\theta) \\ 0 & 1 & 0 \\ \sin(\theta) & 0 & \cos(\theta) \end{bmatrix},$$

$$R(\Psi) = \begin{bmatrix} \cos(\Psi) & \sin(\Psi) & 0 \\ -\sin(\Psi) & \cos(\Psi) & 0 \\ 0 & 0 & 1 \end{bmatrix} \tag{7.4}$$

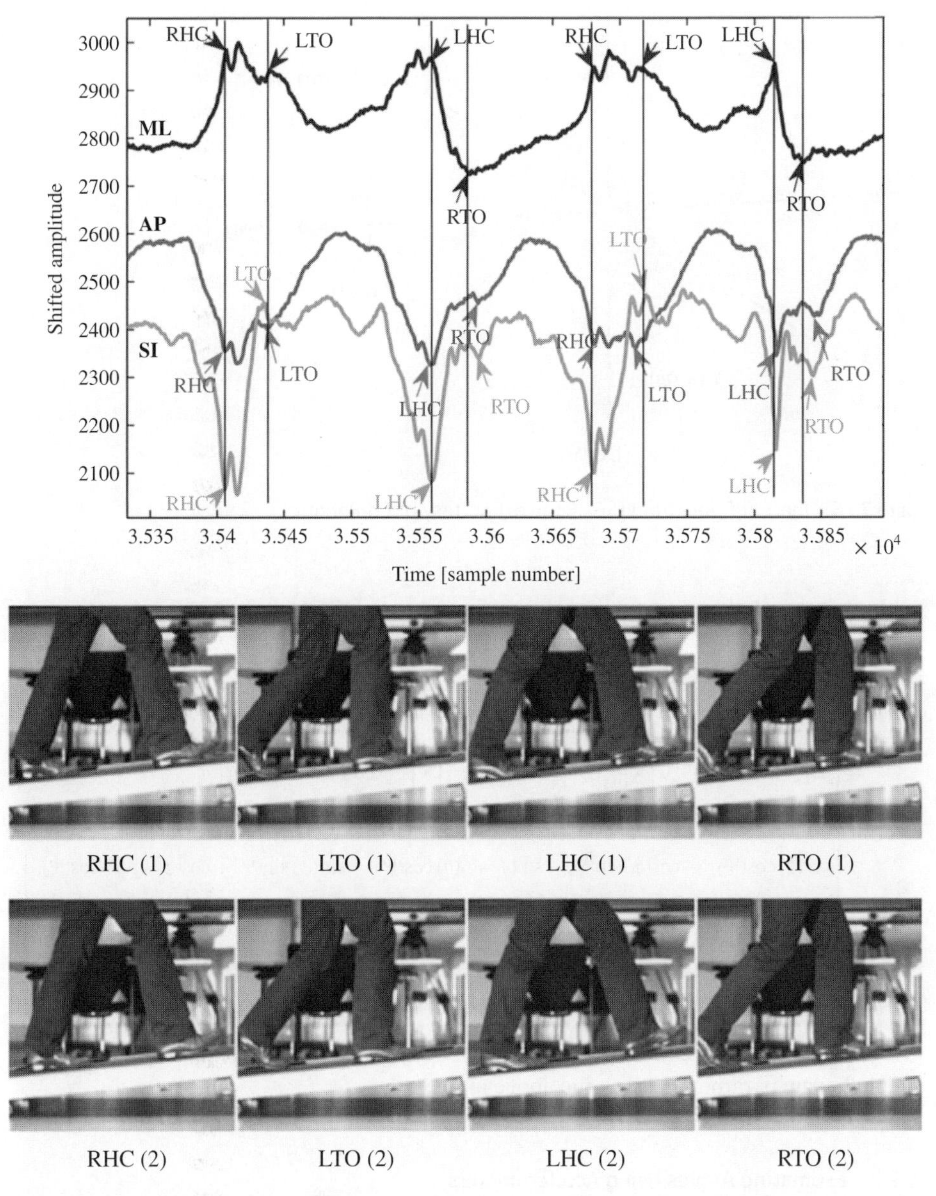


**Figure 7.15** Gait events, from right heel contact (RHC) to right toe-off (RTO) through left toe-off (LTO), and left heel contact (LHC) are located on the acceleration signals (top plot) from the ear-worn sensor, with the help of produced images given by a high-speed camera [55]. Source: Courtesy of Jarchi, D., Wong, C., Kwasnicki, R., Heller, B., Tew, G., and Yang, G.-Z. (*See color plate section for color representation of this figure*)

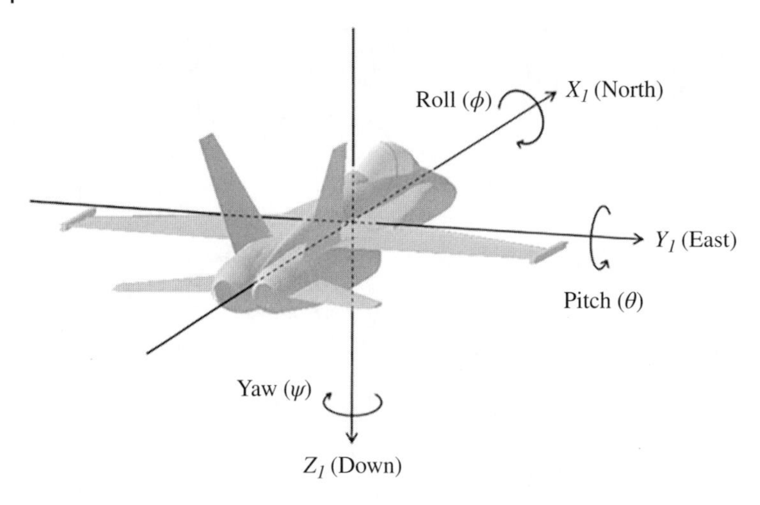


**Figure 7.16** Inertial frame from [56]. Source: Courtesy of Ch Robotics.

A sequence of rotations by yaw, pitch, and roll is calculated as:

$$
\begin{aligned}
&R(-\Psi) \times R(-\theta) \times R(-\phi)\\
&= \begin{bmatrix} \cos(\Psi) & -\sin(\Psi) & 0 \\ \sin(\Psi) & \cos(\Psi) & 0 \\ 0 & 0 & 1 \end{bmatrix} \times \begin{bmatrix} \cos(\theta) & 0 & \sin(\theta) \\ 0 & 1 & 0 \\ -\sin(\theta) & 0 & \cos(\theta) \end{bmatrix} \times \begin{bmatrix} 1 & 0 & 0 \\ 0 & \cos(\phi) & -\sin(\phi) \\ 0 & \sin(\phi) & \cos(\phi) \end{bmatrix}\\
&= \begin{bmatrix} \cos(\theta)\cos(\Psi) & -\cos(\phi)\sin(\Psi) + \sin(\phi)\sin(\theta)\cos(\Psi) & \sin(\phi)\sin(\Psi) + \cos(\phi)\sin(\theta)\cos(\Psi) \\ \cos(\theta)\sin(\Psi) & \cos(\phi)\cos(\Psi) + \sin(\phi)\sin(\theta)\sin(\Psi) & -\sin(\phi)\cos(\Psi) + \cos(\phi)\sin(\theta)\sin(\Psi) \\ -\sin(\theta) & \sin(\phi)\cos(\theta) & \cos(\phi)\cos(\theta) \end{bmatrix}
\end{aligned}
\tag{7.5}
$$

The motion x–y–z coordinate can then be converted to the new system with roll–pitch–yaw parameters using the above matrix.

#### 7.5.2.2 Estimating Angles Using Accelerometers

Roll, pitch, and yaw angles can be estimated using acceleration signals. Consider the inertial frame and a stationary object where the gravity is presented by a vector $[0\ 0\ \mathrm{mg}]^{\mathrm{T}}$, which means that the gravity only affects the $z$ axis and there is no force on the $x$ or $y$ axes. To rotate the gravity vector into the frame of the sensors, the following rotation matrix should be constructed:

$$
\mathbf{R} = R(\phi) \times R(\theta) = \begin{bmatrix} 1 & 0 & 0 \\ 0 & \cos(\phi) & \sin(\phi) \\ 0 & -\sin(\phi) & \cos(\phi) \end{bmatrix} \times \begin{bmatrix} \cos(\theta) & 0 & -\sin(\theta) \\ 0 & 1 & 0 \\ \sin(\theta) & 0 & \cos(\theta) \end{bmatrix}
$$

$$\mathbf{R} = \begin{bmatrix} \cos(\theta) & 0 & -\sin(\theta) \\ \sin(\phi)\sin(\theta) & \cos(\phi) & \sin(\phi)\cos(\theta) \\ \cos(\phi)\sin(\theta) & -\sin(\phi) & \cos(\phi)\cos(\theta) \end{bmatrix} \tag{7.6}$$

Then, after multiplication of R matrix by the gravity vector, we obtain the accelerometer signals in the sensor frame as:

$$F_g = R \times \begin{bmatrix} 0 \\ 0 \\ mg \end{bmatrix} = \begin{bmatrix} \cos(\theta) & 0 & -\sin(\theta) \\ \sin(\phi)\sin(\theta) & \cos(\phi) & \sin(\phi)\cos(\theta) \\ \cos(\phi)\sin(\theta) & -\sin(\phi) & \cos(\phi)\cos(\theta) \end{bmatrix} \times \begin{bmatrix} 0 \\ 0 \\ mg \end{bmatrix}$$
$$= \begin{bmatrix} -mg\sin(\theta) \\ mg\sin(\phi)\cos(\theta) \\ mg\cos(\phi)\cos(\theta) \end{bmatrix} \tag{7.7}$$

where $m$ is the body mass. Following Newton's second law i.e. $a_m = \frac{F_g}{m}$, the acceleration vector is achieved easily as:

$$a_m = \begin{bmatrix} a_{mx} \\ a_{my} \\ a_{mz} \end{bmatrix} = \begin{bmatrix} g\sin(\theta) \\ -g\sin(\phi)\cos(\theta) \\ -g\cos(\phi)\cos(\theta) \end{bmatrix} \tag{7.8}$$

Therefore, it is straightforward to estimate pitch and roll angles from the acceleration signals as:

$$\widehat{\theta}_{acc} = sin^{-1}\left(\frac{a_{mx}}{g}\right), \widehat{\phi}_{acc} = tan^{-1}\left(\frac{a_{my}}{a_{mz}}\right) \tag{7.9}$$

The pitch angle can also be estimated as [58, 59]:

$$\widehat{\theta}_{acc} = tan^{-1}\left(\frac{a_{mx}}{\sqrt{{a_{my}}^2 + {a_{mz}}^2}}\right) \tag{7.10}$$

However, this is a simple case where the only force applied on the accelerometer is gravity. In the cases where other forces (external, vibration, etc.) are applied through different angles, the roll, pitch, and yaw angles will be in more complex forms.

#### 7.5.2.3 Estimating Angles Using Gyroscopes

Gyroscope measurements are typically unaffected by accelerations imposed by external forces [59]. Gyroscopes are considered good alternatives for angle estimation, especially in situations where the accelerometers cannot be used for orientation estimation. Using Euler integration, the angles can be derived as:

$$\widehat{\theta}^{+} = \widehat{\theta} + T\dot{\theta} \tag{7.11}$$

$$\widehat{\phi}^{+} = \widehat{\phi} + T\dot{\phi} \tag{7.12}$$

where, $\widehat{\theta}^{+}$ and $\widehat{\theta}$ are current and previous estimates of pitch, $\widehat{\phi}^{+}$ and $\widehat{\phi}$ are the current and previous estimates of roll, and $T$ is the sampling interval. Using this approach, the estimated angles are robust against external forces. Nevertheless, any small error in the measurements given by the gyroscope can produce a drift due to integration of the error. In addition, initial angles should be known in advance. In order to compensate or correct the drift caused by the integration of the gyroscope measurements, the information from accelerometers can be used. In the following section we can see how, by fusing the information from both accelerometer and gyroscope modalities, this problem is solved and the angle estimation process improved.

#### 7.5.2.4 Fusing Accelerometer and Gyroscope Data

The angles are first estimated using Eqs. (7.11) and (7.12) from the gyroscope measurements. Then, a weighted difference between the angles estimated using accelerometer and gyroscope is added to each angle measured from the gyroscope only to estimate the corrected angles:

$$\hat{\theta} = \hat{\theta}^{+} + L(\hat{\theta}_{acc} - \hat{\theta}^{+}) \tag{7.13}$$

$$\hat{\phi} = \hat{\phi}^{+} + L(\hat{\phi}_{acc} - \hat{\phi}^{+}) \tag{7.14}$$

$L$ is a constant between 0 and 1. For $L = 1$, the estimated angles are purely related to the acceleration measurements. Otherwise, for $L = 0$, they are purely dependent on the gyroscope data.

#### 7.5.2.5 Quaternion Based Estimation of Orientation

A quaternion is used to represent a rotation in the 3D space of a frame $R$ with respect to a frame $B$. It consists of a vector part in the form of a $3 \times 1$ column vector and a scalar. A quaternion can be represented in the following form:

$${}^{B}_{R}\boldsymbol{q} = [q_0\; q_1\; q_2\; q_3]^T = \left[cos\frac{\propto}{2}\; e_x sin\frac{\propto}{2}\; e_y sin\frac{\propto}{2}\; e_z sin\frac{\propto}{2}\right]^T \tag{7.15}$$

where $\boldsymbol{e} = [e_x\; e_y\; e_z]^T$ is a unit vector representing the rotation matrix and $\propto$ is the rotation angle. A conjugate quaternion describes the inverse quaternion as:

$${}^{B}_{R}\boldsymbol{q}^{*} = {}^{R}_{B}\boldsymbol{q} = [q_0 - q_1 - q_2 - q_3]^T \tag{7.16}$$

A sequence of rotations can be presented by simply using quaternion multiplications. Quaternion multiplication for two quaternions $\boldsymbol{p}$ and $\boldsymbol{q}$ is defined as:

$$\boldsymbol{p} \otimes \boldsymbol{q} = \begin{bmatrix} p_0q_0 - p_1q_1 - p_2q_2 - p_3q_3 \\ p_0q_1 + p_1q_0 + p_2q_3 - p_3q_2 \\ p_0q_2 - p_1q_3 + p_2q_0 + p_3q_1 \\ p_0q_3 + p_1q_2 - p_2q_1 + p_3q_0 \end{bmatrix} \tag{7.17}$$

where $\otimes$ represents quaternion multiplication. In order to rotate a 3D vector $\boldsymbol{v}$ in the reference frame $R$ into the frame $B$, unit quaternions can be used as:

$${}^{B}\boldsymbol{v}_q = {}^{B}_{R}\boldsymbol{q} \otimes {}^{R}\boldsymbol{v}_q \otimes {}^{B}_{R}\boldsymbol{q}^{*} \tag{7.18}$$

Both ${}^{B}\boldsymbol{v}_q$ and ${}^{R}\boldsymbol{v}_q$ vectors can be written in the form of quaternion as:

$$\boldsymbol{v}_q = [0\; \boldsymbol{v}]^T = [0\; v_x\; v_y\; v_z]^T \tag{7.19}$$

The rotation presented in the above equation can be re-expressed in matrix form as:

$$^{B}\boldsymbol{v} = R(^{B}_{R}\boldsymbol{q})^{R}\boldsymbol{v} \tag{7.20}$$

where $R(^{B}_{R}\boldsymbol{q})$ is the 3D rotation matrix presented in the quaternion form as:

$$R(^{B}_{R}\boldsymbol{q}) = \begin{bmatrix} q_0^2 + q_1^2 - q_2^2 - q_3^2 & 2(q_1q_2 - q_0q_3) & 2(q_1q_3 + q_0q_2) \\ 2(q_1q_2 + q_0q_3) & q_0^2 - q_1^2 + q_2^2 - q_3^2 & 2(q_2q_3 - q_0q_1) \\ 2(q_1q_3 - q_0q_2) & 2(q_2q_3 + q_0q_1) & q_0^2 - q_1^2 - q_2^2 + q_3^2 \end{bmatrix} \tag{7.21}$$

The gyroscope measurements can be modelled based on the sum of angular rate $\omega_{G,\,t}$, the bias $b_{G,\,t}$, and white noise $w_{G,\,t}$ [60]:

$$y_{G,t} = \omega_{G,t} + b_{G,t} + w_{G,t} \tag{7.22}$$

The gyroscope bias variation over time can be modelled using a first-order Markov process with Gaussian noise and covariance matrix:

$$b_{G,t} = b_{G,t-1} + u_{bG,t} \tag{7.23}$$

Acceleration signals can be modelled as the sum of motion acceleration ($a_t$), gravitational acceleration ($g_t$), bias ($b_{A,\,t}$), and a white Gaussian noise ($w_{A,\,t}$) as:

$$y_{A,t} = a_t + g_t + b_{A,t} + w_{A,t} \tag{7.24}$$

Based on the signal model used for gyroscope measurements, the predicted orientation at time step $t$ of a body frame with respect to the reference frame (according to global GPS) can be represented as [60, 61]:

$$\hat{q}_t = \exp\left(\frac{1}{2}\Omega(\acute{y}_{G,t})\Delta t\right) q_{t-1} \tag{7.25}$$

where $\Delta t$ is the sampling interval, $\acute{y}_{G,t} = y_{G,t} - b_{G,t}$, $q_{t-1}$ presents the quaternion updated at time $t-1$, and $\Omega(\acute{y}_{G,t})$ is a skew symmetric matrix:

$$\Omega(\acute{y}_{G,t}) = \begin{bmatrix} -[\acute{y}_{G,t}\times] & \acute{y}_{G,t} \\ -(\acute{y}_{G,t})^T & 0 \end{bmatrix} \tag{7.26}$$

where $[\acute{y}_{G,t}\times]$ represents cross-product matrix (here, it is applied to $\acute{y}_{G,t}$).

The motion acceleration can be integrated before predicting the velocity at time step $t$:

$$a_t^G = C(\hat{q}_t)\acute{y}_{A,t} - g_0 \tag{7.27}$$

$$C(\hat{q}_t) = (\hat{q}_t^2 - e_t^T e_t)I_3 + 2e_t^T e_t - 2\hat{q}_t[e_t\times] \tag{7.28}$$

$$\acute{y}_{A,t} = y_{A,t} - b_{A,t} \tag{7.29}$$

This can then be used to predict the velocity as:

$$\hat{v}_t = v_{t-1} + a_t^G \Delta t \tag{7.30}$$

which can be used for prediction of displacement as:

$$p_t = p_{t-1} + \hat{v}_t \Delta t \tag{7.31}$$

It is also a challenging task to estimate motion acceleration from the accelerometer measurements due to bias and errors that result from integration. In [60] an error state vector complementary Kalman filter (CKF) has been designed for orientation and displacement estimation using three-axis gyroscope, accelerometer, and magnetometer measurements. The error state process has been modelled by the CKF as:

$$\delta x_t = \mathbf{F}_t \delta x_{t-1} + u_t \tag{7.32}$$

where $u_t$ is the process noise with covariance matrix $Q = E(u_t u_t^T)$ and $\mathbf{F}_t$ is the state transition matrix, which needs to be determined. The measurement model is defined by:

$$\delta z_t = \mathbf{O}_t \delta x_t + n_t \tag{7.33}$$

where $\delta x_t$ is the measurement error, $n_t$ is the measurement noise assuming to be Gaussian with covariance matrix $R = E(n_t n_t^T)$, and $\mathbf{O}_t$ is the measurement matrix, which needs to be determined.

For velocity and displacement, the errors can be considered as:

$$\delta v_t = \hat{v}_t - v_t \text{ or, } v_t = \hat{v}_t - \delta v_t \tag{7.34}$$

$$\delta p_t = \hat{p}_t - p_t \text{ or, } p_t = \hat{p}_t - \delta p_t \tag{7.35}$$

In fact, the orientation error cannot be simply defined as an arithmetic difference between the estimated and actual orientations. However, it can be expressed as a quaternion that yields the true quaternion when combined with the estimated quaternion [60]:

$$\delta p_t = q_t \otimes \hat{q}_t^{-1}, \text{or } q_t = \hat{q}_t \otimes \delta p_t \tag{7.36}$$

A simplified error state vector $\delta x_t = [\delta q_t^T, \delta b_{G,t}^T]^T$ has been used to estimate the orientation where the state transition and measurement matrices are obtained as:

$$\mathbf{F}_t = \begin{bmatrix} \mathbf{I}_3 + \Delta t[\acute{y}_{G,t}\times] & -\frac{1}{2}\Delta t\mathbf{I}_3 \\ \mathbf{0} & \mathbf{I}_3 \end{bmatrix} \tag{7.37}$$

and

$$\mathbf{O}_t = \begin{bmatrix} 2[\hat{z}_{A,t}\times] & \mathbf{0} \\ 2[\hat{z}_{M,t}\times] & \mathbf{0} \end{bmatrix} \tag{7.38}$$

where $\mathbf{0}$ is a 3×3 matrix of zeros, $\mathbf{I}_3$ is a 3×3 identity matrix, and $\hat{z}_{A,t}$ and $\hat{z}_{M,t}$ are the normalised gravitational acceleration and the earth magnetic vector in the sensor frame [60], respectively.

#### 7.5.2.6 Step Length Estimation

A large number of algorithms for the estimation of step/stride length are based on double integration [62–64]. These approaches are a class of zero-velocity update point (ZUPT) methods which aim at re-initialising the integration process to reduce the integration drift. For this purpose, either acceleration signals or these signals in combination with the gyroscope data are used. They are also prone to drift integration and measurement errors. In some studies the magnetometer information has also been used [64]. Kalman filter and quaternion [61, 62] based approaches are proposed to estimate orientations where the integration with respect to time has been done directly. In [65] a Kalman filter has been formulated to estimate orientation, velocity, and position from the accelerometer and gyroscope measurements. The other approaches, referred to as model based, are explained as follows:

In [66] the step length has been modelled based on a linear combination of step frequency and variance of acceleration signals from an IMU placed on the hip during each step by:

$$\mathbf{SL}_s = \alpha.f_s + \beta.\sigma_s^2 + \gamma \tag{7.39}$$

where $\alpha$, $\beta$, and $\gamma$ are model parameters that should be learnt in the training stage, $f_s$ is the step frequency, and $\sigma_s^2$ is the accelerometer signal variance which can be obtained using the following equations:

$$f_s = \frac{1}{t_k - t_{k-1}} \tag{7.40}$$

$$\sigma_s^2 = \frac{1}{n-1}\sum_{k=1}^{n}(a_k - \overline{a})^2 \tag{7.41}$$

where $n$ is the number of samples, $\overline{a}$ is the average acceleration during one step, $a_k$ is the accelerations at $k^{th}$ sample, and $t_k$, $t_{k-1}$ are the timings of two consecutive detected steps.

The remaining methods are a kind of parametric methods as they use variation of accelerometer data, gait speed, walking frequency, or other parameters to estimate the step length. In one of the earliest approaches for estimation of step length (Zijlstra's method), an inverted pendulum model has been used [67] to find a relationship between VL displacement of the body centre of mass (CoM) and signals recorded by a single accelerometer attached to the CoM. The mathematical formulation to estimate the step length is expressed as:

$$\mathbf{SL}_s = 2K\sqrt{2hL - h^2} \tag{7.42}$$

where, $K$ is the individual correction factor, $L$ is the leg length of the individual calculated from the position of the sensor to heel, and $h$ is the VL displacement of CoM during each step. It has been suggested to find the value of $h$ by double integration of the VL accelerations between two consecutive initial foot contacts. In addition, $K$ can be obtained using the ratio between the mean reference step length and the mean estimated step length during a training session as:

$$K = \frac{mean\,(\mathbf{SL}_r)}{mean\,(\mathbf{SL}_s)} \tag{7.43}$$

where $\mathbf{SL}_r$ and $\mathbf{SL}_s$ present, respectively, the reference and estimated step lengths. In another study, it has been proposed to use a generic correction factor of 1.24 instead of individual correction factor to eliminate the training stage.

The Weinberg algorithm [68] reflects a strong relationship between step length and the difference between maximum and minimum VL acceleration during each step recorded from the accelerometer placed at the CoM or waist using the following equation:

$$\mathbf{SL}_s = K\sqrt[4]{\max(a_v) - \min(a_v)} \tag{7.44}$$

where $\max(a_v)$ and $\min(a_v)$ represent, respectively, the maximum and minimum of VL accelerations. $K$ is a user-specific constant similar to that in Zijlstra's method. It can be learnt in a training stage for each subject. As an example, $K$ has been assigned to various values ranging from 0.5 to 0.57 in [69] for different subjects.

In other studies, a mobile phone's accelerometer sensor placed in the right trouser pocket of the user has been used to estimate the step length [70]. A simple linear relationship between step length and swing speed has been formulated by using a foot-mounted gyroscope [71].

## 7.6 Conclusions

In this chapter, a number of standard and wearable platforms for gait analysis have been introduced. The use of standard platforms which include video cameras or force-plates are usually limited to laboratory-based assessments of the gait, while the foot plantar pressure systems can be used outside gait laboratories. All of these standard platforms are used as a reference for validation of gait event/parameter estimation from wearable sensors. Manipulation of the measurements from gait sensors is of great interest to ensure achieving meaningful data from these sensors. On the other hand, with the advancements in the evolution of wearable sensors, the development of signal processing and machine learning algorithms for the estimation of major gait events/parameters is crucial. These parameters can be used for the diagnosis and management of certain gait impairments in clinical studies.

## References


1 Perry, J. and Burnfield, J. (2010). *Gait Analysis: Normal and Pathological Function*, 2e. SLACK Incorporated.

2 Menz, H.B., Lord, S.R., and Fitzpatrick, R.C. (2003). Acceleration patterns of the head and pelvis when walking on level and irregular surfaces. *Gait & Posture* 18 (1): 35–46.

3 Moe-Nilssen, R. and Helbostad, J.L. (2005). Interstride trunk acceleration variability but not step width variability can differentiate between fit and frail older adults. *Gait & Posture* 21 (2): 164–170.

4 Hartmann, A., Luzi, S., Murer, K. et al. (2009). Concurrent validity of a trunk tri-axial accelerometer system for gait analysis in older adults. *Gait & Posture* 29 (3): 444–448.

5 Annegarn, J., Spruit, M.A., Savelberg, H.H. et al. (2012). Differences in walking pattern during 6-min walk test between patients with COPD and healthy subjects. *PLoS One* 7 (5): e37329.

6 Moe-Nilssen, R. and Helbostad, J.L. (2004). Estimation of gait cycle characteristics by trunk accelerometry. *Journal of Biomechanics* 37 (1): 121–126.

7 Auvinet, B., Bileckot, R., Alix, A.S. et al. (2006). Gait disorders in patients with fibromyalgia. *Joint, Bone, Spine* 73 (5): 543–546.

8 Doheny, E.P., Greene, B.R., Foran, T. et al. (2012). Diurnal variations in the outcomes of instrumented gait and quiet standing balance assessments and their association with falls history. *Physiological Measurement* 33 (3): 361–373.

9 Bugan, F., Benedetti, M.G., Casadio, G. et al. (2012). Estimation of spatial-temporal gait parameters in level walking based on a single accelerometer: validation on normal subjects by standard gait analysis. *Computer Methods and Programs in Biomedicine* 108 (1): 129–137.

10 Senden, R., Grimm, B., Heyligers, I.C. et al. (2009). Acceleration-based gait test for healthy subjects: reliability and reference data. *Gait & Posture* 30 (2): 192–196.

**11** Senden, R., Grimm, B., Meijer, K. et al. (2011). The importance to including objective functional outcomes in the clinical follow up of total knee arthroplasty patients. *The Knee* 18 (5): 306–311.

**12** Aminian, K., Mariani, B., Paraschiv-Ionescu, A. et al. (2011). Foot worn inertial sensors for gait assessment and rehabilitation based on motorized shoes. In: *Conference Proceedings of the IEEE Engineering in Medicine and Biology Society*, 5820–5823. IEEE.

**13** Dalton, A., Khalil, H., Busse, M. et al. (2013). Analysis of gait and balance through a single triaxial accelerometer in presymptomatic and symptomatic Huntington's disease. *Gait & Posture* 37 (1): 49–54.

**14** Auvinet, B., Berrut, G., Touzard, C. et al. (2002). Reference data for normal subjects obtained with an accelerometric device. *Gait & Posture* 16 (2): 124–134.

**15** Terrier, P., Deriaz, O., Meichtry, A., and Luthi, F. (2009). Prescription footwear for severe injuries of foot and ankle: effect on regularity and symmetry of the gait assessed by trunk accelerometry. *Gait & Posture* 30 (4): 492–496.

**16** Brandes, M., Zijlstra, W., Heikens, S. et al. (2006). Accelerometry based assessment of gait parameters in children. *Gait & Posture* 24 (4): 482–486.

**17** Zijlstra, W. (2004). Assessment of spatio-temporal parameters during unconstrained walking. *European Journal of Applied Physiology* 92 (1–2): 39–44.

**18** Helbostad, J.L. and Moe-Nilssen, R. (2003). The effect of gait speed on lateral balance control during walking in healthy elderly. *Gait & Posture* 18 (2): 27–36.

**19** Gillain, S., Warzee, E., Lekeu, F. et al. (2009). The value of instrumental gait analysis in elderly healthy, MCI or Alzheimer's disease subjects and a comparison with other clinical tests used in single and dual-task conditions. *Annals of Physical and Rehabilitation Medicine* 52 (6): 453–474.

**20** Hannink, J., Kautz, T., Pasluosta, C. F., Barth, J., Schülein, S., Gaßmann, K.-G., Klucken, J., and Eskofier, B. M. (2017). Stride Length Estimation with Deep Learning. arXiv:1609.03321v3.

**21** Martin Schepers, H., van Asseldonk, E.H., Baten, C.T., and Veltink, P.H. (2010). Ambulatory estimation of foot placement during walking using inertial sensors. *Journal of Biomechanics* 43 (16): 3138–3143.

**22** Levangie, P.K. and Norkin, C.C. (2005). *Joint Structure and Function: A Comprehensive Analysis*. Philadelphia, PA: F. A. Davis co.

**23** Roaas, A. and Andersson, G. (1982). Normal range of motion of the hip, knee and ankle joints in male subjects, 30–40 years of age. *Acta Orthopaedica* 53 (2): 205–208.

**24** Yang, S., Mohr, C., and Li, Q. (2011). Ambulatory running speed estimation using an inertial sensor. *Gait & Posture* 34 (4): 462–466.

**25** Sharma, B.S.M., Vidhya, S., and Kumar, N. (2017). System for measurement of joint range of motion using inertial sensors. *Biomedical Research* 28 (8): 3699–3704.

**26** Paromed Medizintechnik GmbH (1997). *Parotec System Instruction Manual*. Neubeuern, Germany: Paromed.

**27** Bilney, B., Morris, M., and Webster, K. (2003). Concurrent related validity of the GAITRite walkway system for quantification of the spatial and temporal parameters of gait. *Gait & Posture* 17 (1): 68–74.

**28** Selby-Silverstein, L. and Besser, M. (1999). Accuracy of the GAITRite® system for measuring temporal-spatial parameters of gait. *Physical Therapy* 79: S59.

**29** Cutlip, R.G., Mancinelli, C., Huber, F., and Di, J. (2000). Pasquale evaluation of an instrumented walkway for measurement of the kinematic parameters of gait. *Gait & Posture* 12 (2): 134–138.

**30** Menz, H.B., Latt, M.D., Tiedemann, A. et al. (2004). Reliability of the GAITRite walkway system for the quantification of temporo-spatial parameters of gait in young and older people. *Gait & Posture* 20 (1): 20–25.

**31** Titianova, E.B., Mateev, P.S., and Tarkka, I.M. (2004). Footprint analysis of gait using a pressure sensor system. *Journal of Electromyography and Kinesiology* 14 (2): 275–281.

**32** Windolf, M., Götzen, N., and Morlock, M. (2008). Systematic accuracy and precision analysis of video motion capturing systems: exemplified on the Vicon-460 system. *Journal of Biomechanics* 41 (12): 2776–2280.

**33** Gao, Z., Song, H., Ren, F. et al. (2017). Reliability and validity of CODA motion analysis system for measuring cervical range of motion in patients with cervical spondylosis and anterior cervical fusion. *Experimental and Therapeutic Medicine* 14 (6): 5371–5378.

**34** Geerse, D.J., Coolen, B.H., and Roerdink, M. (2015). Kinematic validation of a multi-Kinect v2 instrumented 10-meter walkway for quantitative gait assessments. *PLoS One* 10 (10): e0139913.

**35** Shotton, J., Fitzgibbon, A., Cook, M. et al. (2011). Real-time human pose recognition in parts from single depth images. In: *The 24th IEEE Conference on Computer Vision and Pattern Recognition*, 1297–1304. Colorado Springs, CO: CVPR.

**36** Microsoft (2020) Kinect for Windows features. https://developer.microsoft.com/en-us/windows/kinect (accessed 6 January 2020).

**37** Ťupa, O., Procházka, A., Vyšata, O. et al. (2015). Motion tracking and gait feature estimation for recognising Parkinson's disease using MS Kinect. *Biomedical Engineering Online* 14: 97.

**38** Jarchi, D., Lo, B., Ieong, E. et al. (2014). Validation of the e-AR sensor for gait event detection using the Parotec foot insole with application to post-operative recovery monitoring. In: *11th International Conference on Wearable and Implantable Body Sensor Networks*, 127–131. IEEE.

**39** Yang, M., Zheng, H., Wang, H. et al. (2012). Assessing the utility of smart mobile phones in gait pattern analysis. *Health and Technology* 2 (1): 81–88.

**40** Juen, J., Cheng, Q., Prieto-Centurion, V. et al. (2014). Health monitors for chronic disease by gait analysis with mobile phones. *Telemedicine Journal and E-Health* 20 (11): 1035–1041.

**41** Kluge, F., Gaßner, H., Hannink, J. et al. (2017). Towards mobile gait analysis: concurrent validity and test-retest reliability of an inertial measurement system for the assessment of spatio-temporal gait parameters. *Sensors (Basel)* 17 (7).

**42** Kavanagh, J.J., Barrett, R.S., and Morrison, S. (2004). Upper body accelerations during walking in healthy young and elderly men. *Gait & Posture* 20 (3): 291–298.

**43** Torrealba, R.R., Cappelletto, J., Fermin-Leon, L. et al. (2010). Statistics-based technique for automated detection of gait events from accelerometer signals. *Electronics Letters* 46 (22): 1483–1485.

**44** Jasiewicz, J.M., Allum, J.H., Middleton, J.W. et al. (2006). Gait event detection using linear accelerometers or angular velocity transducers in able-bodied and spinal-cord injured individuals. *Gait & Posture* 24 (4): 502–509.

**45** Guo, Y., Wu, D., Liu, G. et al. (2012). A low- cost body inertial-sensing network for practical gait discrimination of hemiplegia patients. *Telemedicine Journal and E-Health* 18 (10): 748–754.

**46** Hsu, Y.L., Chung, P.C., Wang, W.H. et al. (2014). Gait and balance analysis for patients with Alzheimer's disease using an inertial-sensor-based wearable instrument. *IEEE Journal of Biomedical and Health Informatics* 18 (6): 1822–1830.

**47** Stamatakis, J., Crmers, J., Maquet, D. et al. (2011). Gait feature extraction in Parkinson's disease using low-cost accelerometers. In: *Proceedings of the Annual International Conference of the IEEE Engineering in Medicine and Biology Society*, 7900–7903. IEEE.

**48** Barrois, R., Gregory, T., Oudre, L. et al. (2016). An automated recording method in clinical consultation to rate the limp in lower limb osteoarthritis. *PLoS One* 11 (10).

**49** Gorelick, M.L., Bizzini, M., Maffiuletti, N.A. et al. (2009). Test-retest reliability of the IDEEA system in the quantification of step parameters during walking and stair climbing. *Clinical Physiology and Functional Imaging* 29 (4): 271–276.

**50** Mackey, A.H., Stott, N.S., and Walt, S.E. (2008). Reliability and validity of an activity monitor (IDEEA) in the determination of temporal-spatial gait parameters in individuals with cerebral palsy. *Gait & Posture* 28 (4): 634–639.

**51** Spain, R.I., St George, R.J., Salarian, A. et al. (2012). Body-worn motion sensors detect balance and gait deficits in people with multiple sclerosis who have normal walking speed. *Gait & Posture* 35 (4): 573–578.

**52** Tadano, S., Takeda, R., Sasaki, K. et al. (2016). Gait characterization for osteoarthritis patients using wearable gait sensors (h-gait systems). *Journal of Biomechanics* 49 (5): 684–690.

**53** Ying, H., Silex, C., Schnitzer, A. et al. (2007). Automatic step detection in the accelerometer signal. In: *4th International Work on Wearable and Implantable Wireless Sensor Network*, 80–85. IEEE.

**54** Jarchi, D., Lo, B., Wong, C. et al. (2016). Gait analysis from a single ear-worn sensor: reliability and clinical evaluation for orthopaedic patients. *IEEE Transactions on Neural Systems and Rehabilitation Engineering* 24 (8): 882–892.

**55** Jarchi, D., Wong, C., Kwasnicki, R. et al. (2014). Gait parameter estimation from a miniaturized ear-worn sensor using singular spectrum analysis and longest common subsequence. *IEEE Transactions on Biomedical Engineering* 61 (4): 1261–1273.

**56** CH Robotics (2020) Understanding Euler Angles. http://www.chrobotics.com/library/understanding-euler-angles (accessed 6 January 2020).

**57** Abhayasinghe, K. N. (2016). Human gait modelling with step estimation and phase classification utilising a single thigh mounted IMU for vision impaired indoor navigation. PhD Thesis. Curtin University.

**58** Pedley, M. (2014). Tilt sensing using a three-axis accelerometer. Freescale. https://cache.freescale.com/files/sensors/doc/app_note/AN3461.pdf (accessed 25 November 2019).

**59** Rampp, A., Barth, J., Schülein, S. et al. (2014). Inertial sensor based stride parameter calculation from gait sequences in geriatric patients. *IEEE Transactions on Biomedical Engineering* 62 (4): 1089–1097.

**60** Meng, X., Zhang, Z.Q., Wu, J.K., and Wong, W.C. (2013). Hierarchical information fusion for global displacement estimation in microsensor motion capture. *IEEE Transactions on Biomedical Engineering* 60 (7): 2052–2063.

61 Choukroun, D., Bar-Itzhack, I., and Oshman, Y. (2006). Novel quaternion Kalman filter. *IEEE Transactions on Aerospace and Electronic Systems* 42 (1): 174–190.
62 Mariani, B., Hoskovec, C., Rochat, S. et al. (2010). 3D gait assessment in young and elderly subjects using foot-worn inertial sensors. *Journal of Biomechanics* 43 (15): 2999–3006.
63 Rebula, J.R., Ojeda, L.V., Adamczyk, P.G., and Kuo, A.D. (2013). Measurement of foot placement and its variability with inertial sensors. *Gait & Posture* 38 (4): 974–980.
64 Trojaniello, D., Cereatti, A., Pelosin, E. et al. (2014). Estimation of step-by-step spatio-temporal parameters of normal and impaired gait using shank-mounted magneto-inertial sensors: application to elderly, hemiparetic, parkinsonian and choreic gait. *Journal of Neuroengineering and Rehabilitation* 11: 152.
65 Ferrari, A., Ginis, P., Hardegger, M. et al. (2016). A mobile Kalman-filter based solution for the real-time estimation of spatio-temporal gait parameters. *IEEE Transactions on Neural Systems and Rehabilitation Engineering* 24 (7): 764–773.
66 Shin, S.H. and Park, C.G. (2011). Adaptive step length estimation algorithm using optimal parameters and movement status awareness. *Medical Engineering & Physics* 33 (9): 1064–1071.
67 Zijlstra, W. and Hof, A.L. (1997). Displacement of the pelvis during human walking: experimental data and model predictions. *Gait & Posture* 6 (3): 249–262.
68 Weinberg, H. (2002). Using the ADXL202 in pedometer and personal navigation applications. Analog Devices, Norwood, MA. https://www.analog.com/media/en/technical-documentation/application-notes/513772624AN602.pdf (accessed 25 November 2019).
69 Fang, L., Antsaklis, P.J., Montestruque, L.A. et al. (2005). Design of a wireless assisted pedestrian dead reckoning system: the NavMote experience. *IEEE Transactions on Instrumentation and Measurement* 54 (6): 2342–2358.
70 Bylemans, I., Weyn, M., and Klepal, M. (2009). Mobile phone-based displacement estimation for opportunistic localisation systems. In: *Third International Conference on Mobile Ubiquitous Computing Systems, Services and Technologies*, 113–118. IEEE.
71 Zhang, Z., Wong, C., Lo, B., and Yang, G.-Z. (2015) A simple and robust stride length estimation method using foot-mounted micro-gyroscopes. https://pdfs.semanticscholar.org/3e2e/6f83c8a36bf2ed041bf74f15fe57e200c6ab.pdf (accessed 6 January 2020).

# 8

# Popular Health Monitoring Systems

## 8.1 Introduction

Numerous healthcare technologies have been developed to assist clinicians in health monitoring, improve patient care with timely and effective interventions, and transform traditional systems. These technologies are intended to be used in both clinical settings, such as inside hospital environments, and at home. In addition, many portable and wearable devices are used for health monitoring by patients suffering from breathing (e.g. asthma), heart rate (HR) variability (e.g. stroke), paralysis (e.g. amputees), and diabetes; as well as athletes; drivers; pilots; and others to prevent them from danger or help them remain fit, alert, and healthy. Recent body sensor platforms are often equipped with wireless communication technologies which can be exploited for real-time and continuous monitoring of patients. In this chapter, first, a brief overview of the technology for data acquisition is provided. Then, popular physiological health monitoring systems (HMSs) are described in various contexts such as detecting patient deterioration, ambient assisted living (AAL), movement tracking, and fall detection/prevention, monitoring patients with chronic obstructive pulmonary disease (COPD), as well as monitoring patients with dementia and those suffering from Parkinson's disease (PD).

## 8.2 Technology for Data Acquisition

Body area network (BAN) is a term used to refer to a short-range wireless network which consists of sensors or devices placed on or around the body [1]. The data communication is limited to a short distance, for example a range over just few metres. Both wearable and implanted devices often use a BAN for data communications. There are different types of communication techniques in terms of the medium they use, ranging from wired systems to short-range radio communication and sophisticated digital wireless technologies. Recently, the use of wireless communication systems has been increased in the health and medical applications using body sensor networks (BSN) [2, 3]. Wireless technologies include but are not limited to Wi-Fi, Zigbee, and Bluetooth. The wireless body area network (WBAN) [4–6] architecture is shown in Figure 8.1. The WBAN architecture can contain several physiological sensors mounted on or around the human body. Then, the data communication is performed using wireless technology where it provides a functional tool

*Body Sensor Networking, Design and Algorithms*, First Edition. Saeid Sanei, Delaram Jarchi and Anthony G. Constantinides.

Companion Website: www.wiley.com/go/sanei/algorithm-design

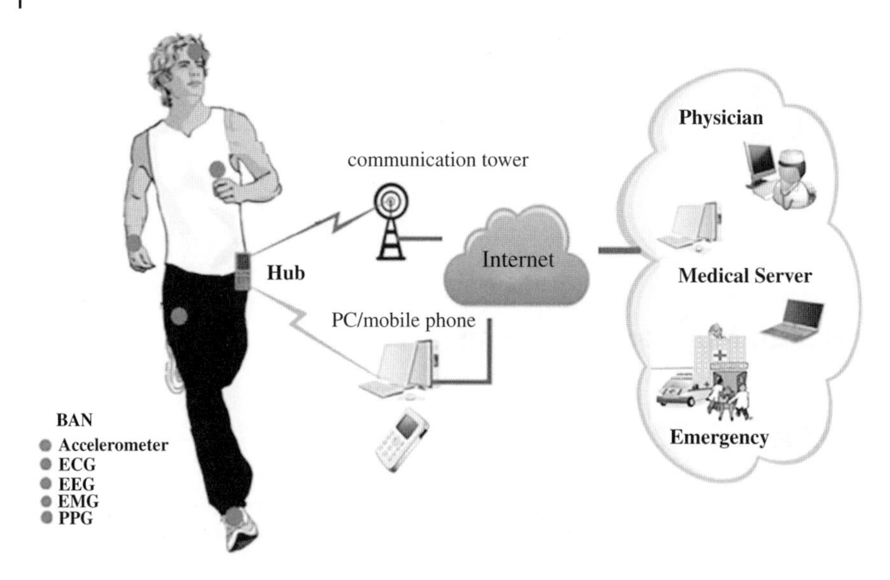


**Figure 8.1** Architecture of WBAN.

without affecting the subject's mobility. The wireless sensors used in medical applications are expected to meet the following requirements of: wearability, reliability, security, and interoperability [7].

Some of the sensors and devices used in HMSs are listed in Table 8.1 [3]. The purpose for each sensor/device is also provided in this table. The devices are categorised into personal sensor network (PSN), BSN, and multimedia devices. PSNs, which are also called environmental sensors, are usually used to record contextual data associated with the subjects and their environment. PSN devices can be placed in a free-living environment or mounted on various objects, such as a bed, chair, sofa, door, or table. 'BSN' refers to wearable sensors which can be embedded into any body-related accessory, such as cloth, watch, belt, or glasses, or directly attached to the body. Accelerometer, gyroscope, and magnetometer are popular sensors for movement detection. Other sensors, such as electrocardiogram (ECG) and photoplethysmogram (PPG), are used for physiological health monitoring like measurement of HR and blood oxygen saturation level ($SpO_2$). In addition, several multimedia devices, as listed in Table 8.1, can be used as interface to the HMS. Such devices, including cameras, microphones, speakers, telephones, or TV sets, can provide a platform to exchange the data between the user/subject and the system.

## 8.3 Physiological Health Monitoring Technologies

### 8.3.1 Predicting Patient Deterioration

The vital signs are continuously measured in clinics or hospitals, especially those from patients inside intensive care units (ICUs) and recorded on the patient's clinical chart

**Table 8.1** Sensors/devices used in health monitoring applications [3].

| Category | Sensor/device | Purpose |
|---|---|---|
| Personal sensor network (PSN) | Radio frequency identification (RFID) | Objects/person identification |
| | Passive infrared (PIR) | Motion/movement detection |
| | Light | Light/intensity of light |
| | Temperature | Room temperature measurement |
| | Humidity | Room humidity measurement |
| | Contact switches | Door status detection (open/close) |
| | Ultrasonic | Posture analysis and location tracking |
| Body sensor network (BSN) | Electrocardiography (ECG) | Cardiac monitoring |
| | Photoplethysmography (PPG)/pulse oximetry | HR analysis and measurement of blood oxygen saturation level |
| | Electrooculography (EOG) | Eye movement monitoring |
| | Electroencephalography (EEG) | Brainwave analysis |
| | Electromyography (EMG) | Muscle activity monitoring |
| | BP monitor | BP measurement |
| | Electro-olfactography | Measurement of olfactory bulb sensitivity to odour |
| | Global positioning system (GPS) | Motion detection and location tracking |
| | Gyroscope | Orientation estimation and motion detection |
| | Accelerometer | Fall detection, daily activity monitoring, posture analysis |
| | SKT | Skin temperature |
| Multimedia devices (MD) | Microphone | Voice detection |
| | Cameras | Movement and gesture recognition |
| | Speakers | Alerts and instructions |
| | TV | Visual information |

Source: Courtesy of Mshali, H., Lemlouma, T., Moloney, M., and Magoni, D.

record. They are used to detect patient deterioration which can lead to death. Six vital signs included in the national early-warning score (NEWS) [8] are listed as:

1. heart rate
2. respiratory rate
3. oxygen saturation
4. temperature
5. systolic blood pressure
6. level of consciousness.

Using a simple scoring system, a number is assigned to each measured physiological parameter (vital sign). These scores are aggregated and the final score which reflects the severity of abnormality in the subject has been associated with a clinical risk. The NEWS is a popular score which has been used to identify patients at high risk for further assessment and appropriate medical interventions. Table 8.2 lists the vital signs and their scores based on the initial proposed NEWS system. For each patient, the scores are summed and the aggregated NEWS score is used to identify the overall clinical risk, as seen in Table 8.3.

Any measured vital sign associated with the score of 3 demonstrates extreme variation in a single vital sign (see the corresponding two columns in Table 8.2). Such extreme variation should not be ignored and implies an urgent clinical assessment. Five key vital signs including HR, respiratory rate (RR), $SpO_2$, temperature, and systolic blood pressure (BP) have been used in many clinical settings, for example for patients with long-term conditions. These include older people with respiratory health conditions. The latest version of the NEWS was announced in December 2017 [9].

The problems in paper-based systems for recording early-warning scores lead to missing the early detection and worsening trend of a patient's situation in the hospital. An

**Table 8.2** Vital signs and their scores based on the NEWS system.

| Physiological parameters | 3 | 2 | 1 | 0 | 1 | 2 | 3 |
|---|---|---|---|---|---|---|---|
| HR (beats per minute) | $\leq 40$ | | 41–50 | 51–90 | 91–110 | 111–130 | $\geq 131$ |
| RR (breaths per minute) | $\leq 8$ | | 9–11 | 12–20 | | 21–24 | $\geq 25$ |
| $SpO_2$ (%) | $\leq 91$ | 92–93 | 94–95 | $\geq 96$ | | | |
| Temperature (C) | $\leq 35$ | | 35.1–36.0 | 36.1–38.0 | 38.1–39.0 | $\geq 39.1$ | |
| Systolic BP (mmHg) | $\leq 90$ | 19–100 | 101–110 | 111–219 | | | $\geq 220$ |
| Any supplemental Oxygen | | Yes | | No | | | |
| Level of consciousness | | | | A | | | V, P, or U |

**Table 8.3** Clinical risk associated with the final aggregate NEWS score.

| NEWS score | Clinical risk |
|---|---|
| Aggregate 1–4 | Low |
| Aggregate 5–6<br>Or if there is a RED score (3) | Medium |
| Aggregate 7 or more | High |

evidence-based system for identification of patient deterioration has been proposed in [10]. The data have been recorded from over 64 000 hours in the UK and USA and used to produce system alerts to clinicians whenever any abnormal ranges of data is detected. Development of the corresponding system presented as system for electronic notes documentation (SEND) has been used by four hospitals within the Oxford University Hospitals NHS Foundation Trust. Using the SEND system, clinicians are able to view the data on any tablet within the hospital. This results in saving time and improved patient care. It has been found that in one year following the introduction of SEND system, there was a 10% reduction in the cardiac arrest rate. There are several number of early warning score (EWS) based systems introduced in the literature. The performances of 22 EWS systems are compared in [11] using a performance measure called area under the receiver-operating characteristic curve (AUC).

Anomaly or novelty detection approaches have been widely used for the identification of unusual, out of range, or abnormal data and have applications in HMSs, for example to detect patient deterioration. Conventional pattern classification approaches usually provide classification of two or more classes, while the problem of novelty detection is represented within the framework of one-class classification. The positive class related to the normal patterns is usually very well sampled; however, the abnormal patterns are relatively low in the frequency of occurrences. The normal patterns are then used in the training phase. Then, a threshold is applied to the novelty scores using a model of normality. The threshold represents a binary decision process [12–14].

As an example, the studies in [15, 16] propose a novelty detection-based approach where a Parzen window is used to construct the distribution of vital sign data. The approach was later used in [17] for patient monitoring in a clinical trial including 336 patients where an early warning of critical physiological event could be produced.

Earlier HMSs were developed in hospital settings, while recent systems are aimed at being deployed in home environments. Smart HMSs which are able to monitor physiological health are helpful for the management of patients with chronic conditions such as diabetes, hypertension, COPD, and cardiovascular disease. These systems often need to be real time and provide a continuous remote monitoring of human physiological parameters.

A general architecture of an HMS is provided in Figure 8.2. In such a system important vital signs are estimated. These signs include HR, RR, BP, $SpO_2$, and blood glucose. Statistical methods and machine learning approaches are applied to detect abnormal conditions and immediately provide the required care and services. There are a number of challenges for the development of smart home monitoring platforms, including reliable inference tools exploiting statistical and machine-learning-based methods, networking infrastructure, and energy requirement and harvesting of wearable devices.

A randomised trial was performed in [18] to evaluate telemonitoring of heart failure patients where vital signs information were transmitted daily to a central nursing station. The results suggest that the telemonitoring operation can be used for an improved ambulatory management of heart failure patients which can result in fewer visits to the emergency department. On the other hand, it is concluded in [19] that there is no significant difference in the number of hospitalisations, emergency department visits, number of deaths between interventions (telehealth case management), and the control group population (case management).

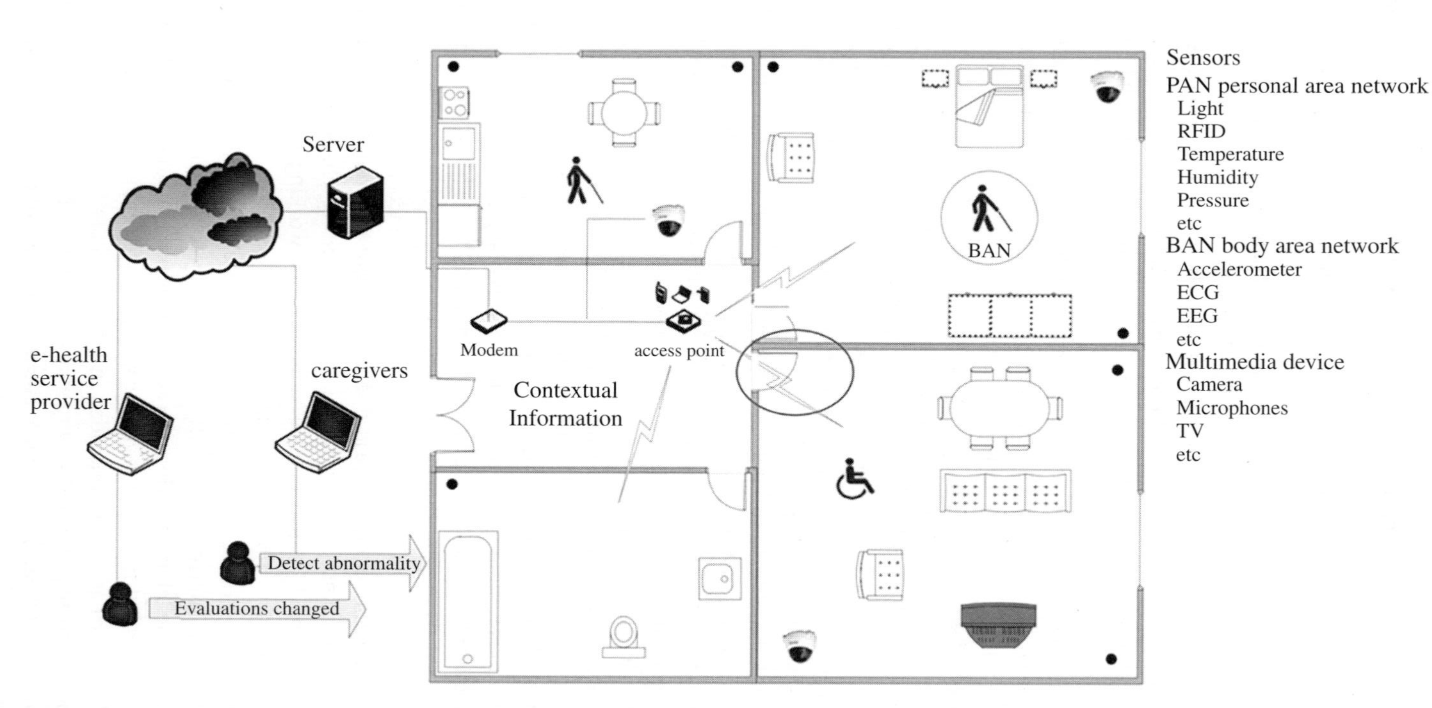


**Figure 8.2** General architecture of an HMS. A rapid intervention is expected to be performed in the case of detecting any abnormality. A regular and continuous evaluation of a human's condition has been done to detect any changes in health condition [3]. Source: Courtesy of Mshali, H., Lemlouma, T., Moloney, M., and Magoni, D.

### 8.3.2 Ambient Assisted Living: Monitoring Daily Living Activities

Human activity and behavioural patterns in real-life settings are quantified using activity recognition methods. One application of activity recognition is automatic human behaviour monitoring, especially for older populations and in care homes. Owing to the diverse and complex nature of the human's behaviour, monitoring activities of daily living (ADLs) has been a challenging problem. There can be several activities being performed simultaneously. Both wireless area and body sensors are widely used for activity recognition with the help of advances in data processing, computational intelligence, and statistical techniques. These techniques include very popular methods such as hidden Markov models (HMMs) [20–22], often used for the estimation of the state of the system; Bayesian network [23–25]; neural networks, as a nonlinear and effective classifier, [26]; support vector machines (SVM), which is optimal for separable cases [27–29]; multiclass logistic regression [30]; decision tree [31–33]; and rule-based fuzzy logic ontologies [34–37]. In some earlier studies, a randomised control trial was performed for monitoring ADLs of frail older people [38, 39]. A decrease in costs was achieved in [38].

The raw data from wearable accelerometer sensors was used for human activity recognition in [20] incorporating HMM in the context of multiple regression denoted as MHMMR (Multiple Hidden Markov Model Regression). Based on the proposed method, automatic activity recognition is performed by a number of wearable accelerometers placed on chest, thigh, and ankle. The ambulatory activities were classified into three groups: dynamic, static, and transitional. The dynamic activities include walking, stairs down and stairs up. The static activities include standing, sitting, and lying. Transitional activities include changing position from sitting to standing, standing to sitting, sitting to lying, and lying to sitting. These activities can include transition from one way of sitting to another, too.

An automatic activity recognition system called hybrid reasoning for context-aware activity recognition (COSAR) has been developed incorporating both statistical and ontological reasoning [30]. The activity recognition system has been aimed for general populations, not specifically older adults, and it has been shown that the accuracy of this system is increased by including ontological reasoning [30].

In [29], accelerometer and gyroscope sensors that were placed on the wrist and foot of the subjects were used to identify six activities, including standing, walking, writing, smoking, and jogging.

One important direction in monitoring ADLs is to search for abnormal events or movements. This concept is noted as anomaly detection, which aims at finding a pattern in the data that does not comply with normal or expected behaviour. In the context of activity recognition, the boundary between a normal and abnormal behaviour is not specified in detail. This is due to the fact that human behaviour continually changes and the anomalies may vary based on the subject's situation [40]. For example, some activities or behaviour could be normal for one subject but abnormal for the other [40]. In addition, the noise created by sensors or any interference by electronic systems can often be mistaken for real anomalies. Large-scale data collection and training is usually required for more effective anomaly detection. Prediction of health conditions well ahead of their occurrences is another important feature of an activity-based HMS. For this, usually a set of long-term and historical data is required so that the predictive and analytical methods can be applied to understand the subject's normal behaviour.

One of the main objectives in the development of AAL is to utilise the information and communication technology [41] to assist elderly population and support them for doing their daily life activities [41]. Development of ambient intelligent tools [42] for AAL has focused on improvement of the quality of older people's life using modern smart technology and also an ability to live independently. Typical environment exploiting the AAL technologies includes smart homes, hospital wards used for long-term care, and rehabilitation centres [42]. A cloud-oriented context-aware middleware (CoCaMAAL) model for monitoring ADL in AAL is proposed in [43]. Based on this work, machine learning and statistical methods are applied for the detection of behavioural changes and abnormalities. In addition, BSNs are employed in AAL where cloud computing has been exploited to provide real-time service for assisted living [44].

### 8.3.3 Monitoring Chronic Obstructive Pulmonary Disease Patients

COPD is a condition that leads to the hospitalisation of patients mainly due to exacerbations and is the third cause of death worldwide by 2020 [45]. Timely detection and prevention of exacerbation using telemedicine [46] helps reduce the number of hospitalisations, and thus reduce cost as well as implement timely interventions. In a comprehensive study followed by clinical trials, the vital signs ($SpO_2$, HR, near-body temperature, overall physical activity) of patients were monitored for of nine months [46]. A wristband sensor integrated with Bluetooth including sensors for measuring HR, physical activity, near-body temperature, and galvanic skin response was coupled with a commercial pulse-oximeter (Nonin Medical Inc., Plymouth, MN, USA) that was used to collect the data. A commercial cellular telephone was used to connect to the wristband using Bluetooth connection, to receive the data from the wristband and send the data to the monitoring system.

The study in [46] suggested that, using a multiparametric remote monitoring system, the COPD patients had a lower rate of exacerbation compared to those patients who only received follow-up using the standard care. On the other hand, a study in [47] showed that in-home remote monitoring of COPD has not reduced hospitalisation of the patients or improved the quality of patients' lives. Long-term management of COPD patients using wearable sensors has been a target in recent years and is explored in various studies [48–50].

A personalised HMS has been implemented for continuous measurement of vital signs of older people to monitor their wellness condition and predict potential health risks. Daily noninvasive measurement of these signs has been performed with the aid of a commercial station-based all-in-one health monitor equipment (TeleMedCare Health Monitor, TeleMedCare, Sydney, Australia) [51–54]. The subjects were asked to wear a commercial wearable tracker (Sony SmartBand 2, Sony Corporation, Tokyo, Japan). An integrated system is proposed to monitor the wellness condition of older people in [55]. The system includes a computer-aided decision support for clinicians and community nurses by means of which they can easily monitor and analyse an older person's overall activity and vital signs using a wearable wellness tracker and an all-in-one satiation-based monitoring device. Multimodality recordings have been used and a personalised scheme for forecasting older people's one-day-ahead wellness condition has been proposed via data integration and statistical learning. Using this platform, a pilot study at a nursing home in Hong Kong was performed and the implementation of the system demonstrated. In this study a health index was introduced to quantify the general health condition of older people.

### 8.3.4 Movement Tracking and Fall Detection/Prevention

Among their wide applications, accelerometers can be used to measure acceleration and perform fall detection and posture analysis. On the other hand, gyroscopes measure orientation and are used for motion detection in general. Motion detection and location tracking can be also carried out using a global positioning system (GPS). A fall detection system for older people is introduced in [56, 57] using accelerometers and gyroscopes. In an open or mobile environment, GPS can be used to monitor location-based activities. For example, it is possible to predict movement based on the GPS data logs [58]. In addition to wearable sensors, such as accelerometers or gyroscopes, passive infrared sensors are used for detecting a subject presence [59] or falling down [60]. Vision-based approaches are sometimes used in video tracking systems to detect inactivity and falls and for 3D motion reconstruction [61]. These methods are, however, subject to breach of privacy, and therefore not allowed for all applications.

A monitoring system for older subjects called the ANGELAH (assisting elders at home) is been proposed in [61]. The basis for the ANGELAH framework is a context-awareness and group-based collaboration to guarantee older people's safety, detect risky and dangerous situations, and inform emergency response groups. The response groups include a group of caregivers or volunteers present in the vicinity to help in the case of emergencies. The sensors and actuators used in ANGELAH include a video camera, sound sensors, radio frequency identification (RFID) readers and appliances such as microphones and smart door locks.

To deploy an RFID system in a smart home, active and passive tags are affixed to objects and readers worn by the subject [62]. RFID is utilised to identify subjects and objects of the smart home environment [62]. The other sensors include pressure for using in a smart chair [63] and ultrasonic for developing an ultrasonic 3D tag system which locates ultrasonic tags in real time and deploys the system in a care home to monitor the position of older people [64]. Environmental sensors are deployed in different places of a smart home to detect light, temperature, and humidity as well as to monitor environmental conditions and identify ADLs.

Movement path variability (tortuosity) has been quantified using a commercially available ultra-wideband radio real-time location sensor network including a spatial resolution of approximately 20 cm. Fractal dimension technique has been applied to measure tortuosity. An improved performance in fall detection has shown to be linked with the evaluation of everyday movements in combination with other known risk factors for falls [65]. An object balance testing using a mechanical apparatus was demonstrated to be effective for the remote monitoring of balance abilities which could potentially prevent falls [66].

Although the most useful techniques in fall detection involve free-space monitoring of the subjects, there are methods which use video information and processing for this purpose. These approaches include vision-based approaches employing depth cameras [67, 68].

For example, depth camera images are used in [67] where body part tracking has been performed using the generated 3D trajectory. The 3D body joints trajectory has been extracted using an improved randomised decision tree algorithm. This trajectory has been used as an input to an SVM to detect fall.

In [69], MEMS sensor data was used to record inertial information and the input to the fall detection algorithm has been acceleration and orientation (yaw, pitch, and roll) of user's body. Based on the results of the study in [69], the best performance of the algorithm has been obtained when the sensor is worn on the waist (Figure 8.3). Hence, some ADL signs can be misdetected as falling when the sensor is not placed on the waist.

### 8.3.5 Monitoring Patients with Dementia

Dementia is a progressive (degenerative) neurological disorder which affects nearly 50 million people worldwide, and 10 million new cases are diagnosed every year [70]. Currently, there is no treatment found for dementia to stop its progression. Nevertheless, there are ways to improve the lives of people suffering from dementia. These include optimising their activity, physical health, cognition, and sharing their information to the carers to provide necessary assistance and support. With recent advancements in the Internet of Things (IoT) technology, it is possible to monitor the health status of people with dementia remotely. There are a number of mental health conditions such as cognitive decline in older

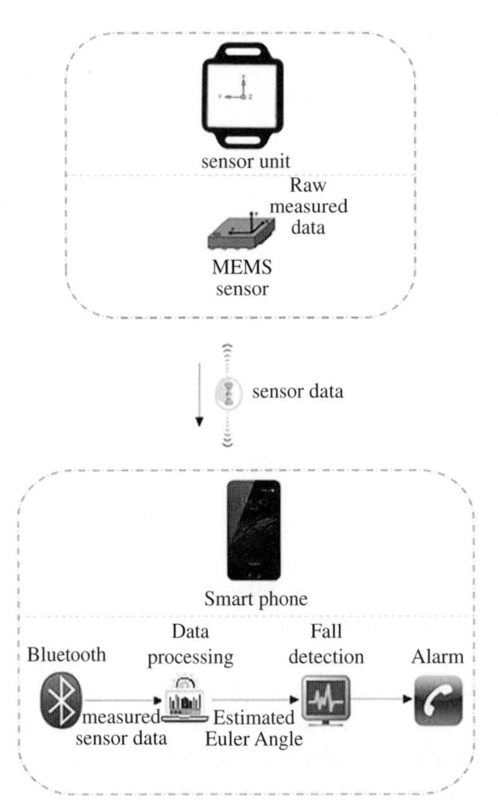


**Figure 8.3** Portable fall detection system using a MEMS sensor placed on the waist [69]. Source: Courtesy of Mao, A., Ma, X., He, Y., and Luo, J.

adults, depression, and dementia. A framework called DemaWare2 has been proposed for dementia ambient care using wearable and ambient sensors [71]. Ambient sensors include depth cameras, IP (Internet protocol) cameras, microphones, Withings Aura (sleep sensor), and Gear4 for sleep recordings. The wearable sensors included the DIT-2, UP24, and GoPro cameras. The care of older patients with dementia has been of great importance. The computer vision systems developed for these purposes enhance their cognitive ability [72] to better cope with their daily activities [73].

A unified approach has been developed in [74] which integrated technologies such as RFID, GPS, GSM communications, and a geographic information system (GIS). The proposed approach provides four monitoring systems to perform indoor residence monitoring, outdoor activity monitoring, remote monitoring, and emergency rescue. A user interface was designed which provided the carers or family members with real-time information about position of the missing older or dementia patients using mobile phones, notebooks, or other portable devices with wireless connection capability [74]. The healthcare platform was made of a database service, a web service server message controller server, and a health GIS [74].

Portable devices such as smartwatches can be used for monitoring dementia patients. Tracking the location of patients is crucial for monitoring their presence and movement trajectory. As an example, a smartwatch containing accelerometer, ambient light, and GPS has been proposed to be used by dementia patients [75]. The smartwatch continuously transfers the patient's activity information from the accelerometer sensor, location information from GPS sensor, and amount of exposure to light through TCP/IP communication using the popular code division multiple access (CDMA) communication system to the server [75]. All the data are then analysed to produce a profile of each patient including detection of activity features such as his/her walking.

A technology assisted monitoring system called technology integrated health management (TIHM), has been proposed for the continuous monitoring of dementia patients in their home environments [76, 77]. TIHM exploits IoT-enabled solutions for efficient environmental data collection and analysis using machine learning algorithms. Both short- and long-term data analysis incorporating various temporal granularity are integrated within the TIHM system. A high-level interpretation of the TIHM architecture is introduced in [76] which comprises four modules: (i) sensors deployed at home; (ii) TIHM back-end that consists of data integration, data analysis, and data storage; (iii) integrated view that includes a user interface for data visualisation, alerts, notifications, and actions; and (iv) clinical pathways for healthcare practitioners to communicate with patients and caregivers, and respond to their needs. The ADLs and abnormal patterns of dementia patients have been recognised and detected through the development of a Markov chain model and by exploiting the entropy rate [76]. Historical data have been used to apply an entropy-based method to identify 'deviation boundaries', for detecting any anomaly in the test data. Since the ADLs do not always follow similar patterns over a short period, they have demonstrated specific patterns in long-term or longitudinal observations. Therefore, using an information theory-based parameter (i.e. entropy) a degree of randomness in ADLs has been quantified. The Markov chain has been formed based on each patient's historical data to learn the regular ADLs and their sequence. Then, the complexity through randomness using entropy rate has been measured based on the consecutive state transitions from the trained Markov chain [76].

A number of passive sensors have been used in the TIHM architecture [76], when nine sensors were deployed in various locations of the residential home. The sensors include two PIR sensors, three motion sensors, two pressure sensors, one sensor for the main entrance door, and one sensor for central energy consumption. The PIR sensors are installed in the hallway and living room; the motion sensors are installed in the kitchen, bedroom, and on bathroom doors; and pressure sensors are deployed on the person's chair and bed. Two categories are constructed as high-active (being home for most periods) and low-active days from applying $k$-means clustering technique [76]. For each category, one separate Markov chain has been trained where the variation of ADLs is measured and analysed for each subject.

One challenging issue in the detection of anomaly in ADLs is the identification of the actual source. A detected anomaly can originate from a technical problem in data collection, such as sensor malfunctioning or connectivity problems, or a change in the environment, such as being away from home for a couple of days. An anomaly can also be related to a decline in a patient's cognitive state (e.g. being restless or agitated) which is directly related to the subject's cognitive decline status known as agitation, irritation, and aggression (AIA). In order to verify the detected anomaly, both environmental and physiological data should be collected simultaneously in real time. These are used to generate environmental and clinical notifications.

### 8.3.6 Monitoring Patients with Parkinson's Disease

PD is a neurodegenerative brain disease which affects nearly five million people worldwide [78]. Common symptoms of PD include resting tremor, gait disorder, bradykinesia, and postural instability. In many cases where the main problem is tremor, deep brain stimulation (DBS) is performed by implanting an electrode in the brain to electrically stimulate dopamine generator cells, so the problem can be solved and the patient can have a normal life.

Management of PD symptoms as the disease progresses is of particular interest to clinicians. The impairment in motor symptoms related to losses in the dopaminergic neurons of the basal ganglia leads to gait disturbances and imbalance for PD patients. This affects their ADLs, thus degrading their quality of life. The gait impairments result in both alterations in the walking pattern and an increase likelihood of falls. Healthcare practitioners use clinical scales in order to obtain information on health status of PD patients and refer them to the carers for managing PD symptoms.

Unified Parkinson Disease Rating Scale (UPDRS) is considered as an internationally accepted measure of motor symptoms used in clinical settings [79]. Traditional clinical settings for monitoring and treatment of PD are often limited. Advanced technologies such as wearable sensors have been deployed in many trials to automate the assessment of PD patients in free-living environments. The movement-related data collected from PD patients during their ADLs without interference are analysed to better understand behavioural patterns and aid in the clinical assessment and management of PD patients.

One particular motor symptom of PD patients is freezing of gait (FoG), which has been targeted in various wearable sensing studies because real-time identification of FoG is important when evaluating the efficacy of therapeutic approaches [80]. FoG is defined as 'an episodic inability (lasting seconds) to generate effective stepping in the absence of any known cause other than parkinsonism or high-level gait disorders. It is most commonly experienced during turning and step initiation but also when faced with spatial constraint, stress, and distraction. Focused attention and external stimuli (cues) can overcome the episode' [81].

Acceleration signals are often used to identify FoG. Smartphones have been used to detect FoG during ADLs, too, for example in [80]. The vertical acceleration signals sampled at 100 Hz are filtered and a sliding Hamming window is used to segment the data and apply fast Fourier transform (FFT) to calculate the power spectrum. Then, the freezing indices called FI and EI [82, 83] are calculated. FI is defined as the power in the 'freeze' band divided by the power in the 'locomotor' band (0.5–3 Hz). A threshold is chosen such that FI values above this limit are labelled as FoG.

In addition, step cadence (step frequency) has been calculated using the second harmonic in the power spectrum. The integrals over the 3–8 Hz interval, as the Freeze band, and 0.5–3 Hz interval as the Loco band, are computed. FI is the ratio between Freeze and Loco bands while EI is the sum of Freeze and Loco bands. Calculated EI, FI, and cadence are used as the inputs to a FoG detector working based on the Moore–Bächlin algorithm [83] to detect FoG where FI and EI exceed a user-specified threshold value. Figure 8.4 illustrates the overall procedure [82].

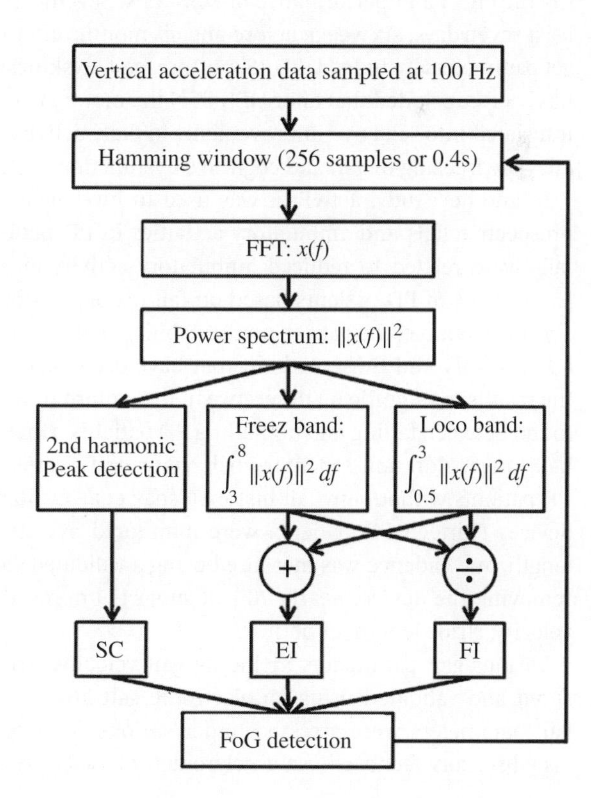


**Figure 8.4** Structure of FoG implementable in hardware [82]. Source: Courtesy of Moore, S.T., MacDougall, H.G., and Ondo, W.G.

Smart technologies have been exploited in various studies to monitor motor symptoms in PD patients in free-living environments. These wearable sensor-based mobility monitoring include PD and healthy control (usually gender matched) and target symptoms include tremor, UPDRS, ADL, gait, balance, posture, health youth survey (HYS), bradykinesia, dyskinesia, freezing of gait, akinesia, limb rigidity, motor fluctuation, and hypokinesia. The participants are monitored continuously for from one or more days to several weeks and even one year. Various health monitoring devices have been deployed to monitor symptoms. In a study, 17 PD and 17 matched health controls (in terms of age and gender) participated in the experiments [84]. The activPAL™, which includes a uniaxial accelerometer sensor, was used. The purpose of the study was the quantification of sedentary behaviour in advanced PD patients, which is useful for detecting the changes in PD symptoms. UPDRS III and HYS were utilised in these trials. By analysis of ADLs, the study has shown a significant difference in the distribution and pattern of accumulation of sedentary time across PD patients and controls [85]. The activPAL has been used for the recording and processing of acceleration signals.

Understanding the effect of DBS on PD patients with different backgrounds and symptoms over a period after the operation is indeed important and can lead to clinically significant conclusions on PD treatment. In a study activPAL was used for monitoring PD symptoms following deep brain stimulation of the subthalamic nucleus (DBS-STN) [86]. Several measures – including gait speed, disease severity, freezing of gait, number of steps, and variability of walking time – were obtained from the accelerometry system to quantify the effect and performance of DBS-STN. Seventeen PD patients were monitored for at least seven days, six weeks before and six months after implantation of DBS-STN. The target parameters included UPDRS, posture, gait, dyskinesia, and ADLs. Based on this study it has been concluded that after DBS-STN the motor symptom severity has been reduced and translated into selective improvements in daily activity. Recently, DBS has been applied to alleviate freezing of gait and cognitive dysfunction in PD [87].

In another study, activPAL was used to investigate a relationship between 12 months prospective falls and ambulatory activities in PD patients [88]. Based on the results, the falls were related to reduced ambulatory activity and consequently to disease severity. Evaluations of PD patients based on fall history has been conducted in other studies. As another example, the pattern and variability of physical activities between PD patients with a fall history and those without that have been evaluated using activPAL [89]. Based on the results, no significant difference in the pattern or trajectory of sedentary behaviour was found between falling and nonfalling PD patients. However, a PD patient with a fall history had longer duration sedentary behaviour and shorter standing time compared to those PD patients without any fall history. Espay et al. evaluated the efficiency of a gait training device. Thirteen PD patients were monitored over two weeks [90]. Gait velocity, stride length, and cadence was measured using a validated electronic gait analysis system. After removing the device, nearly 70% of subjects improved by at least 20% in either walking velocity, stride length, or both.

Various gait parameters including gait velocity, stride length, and cadence were measured and validated using an electronic gait analysis system (GAITRite walkway). The gait parameters were assessed under various conditions such as having no auditory or visual sensory feedback. An accelerometry based device, namely Mobi8, has been used to evaluate the mobility of PD patients in daily-life settings [91]. The PD patients' gait quality was assessed based on interstride variability using the data collected at home. The data were found to be similar to the data recorded in the clinic where the frequency-derived parameters could be estimated. In [92], StepWatch 3 Activity Monitor (SAM) has been used for the long-term monitoring of the changes in ambulatory activity in terms of the amount and intensity of daily activity. The device incorporates a microprocessor link unit which integrates acceleration, position, and timing information. Thirty-three PD patients were monitored for seven days at baseline and at a one-year follow-up procedure. Posture, balance, and ADLs were monitored using tri-axial accelerometers and gyroscopes [93]. Using audio-biofeedback system for training stability, improvements in posture and balance were observed among patients. A multiple learning instance algorithm was used in [94] for the detection of dyskinesia using a set of five tri-axial accelerometers. An innovative mobile health technology was developed by Global Kinetics Corporation (Melbourne, Australia) called Parkinson's KinetiGraph® system (Figure 8.5) for continuous monitoring of PD symptoms including tremor, bradykinesia, and dyskinesia [95]. The PKG® Watch produces a data-driven report known as PKG in addition to the assessment of daytime drowsiness and identification of impulsive behaviour. A sample report is shown in Figure 8.6.

Accelerometers are used in [96] in a large set of trials by employing 467 PD patients for 14 days where the intensity of daily physical activity has been evaluated. Based on the results, the PD patients spent more time on sedentary or low-intensity activities, although considerable variance across subjects was observed.

In [97], the Restricted Boltzmann Machines with a deep learning algorithm were applied to movement data recorded using accelerometer sensor. Thirty-four PD patients were monitored for at least six days in a home environment. The symptoms of interest included dyskinesia and HYS. The aim was to distinguish between PD states using deep learning approaches in naturalistic environments rather than in laboratory environments. An accelerometry-based approach using a PKG system has been designed to devise a fluctuation score [98]. It has been shown that the score is able to discriminate between fluctuating and nonfluctuating patients with high sensitivity and selectivity.

Difficulty in turning during gait is a major indicator of mobility disability, falls, and reduced quality of life in PD patients. Unfortunately, the assessment of mobility in the

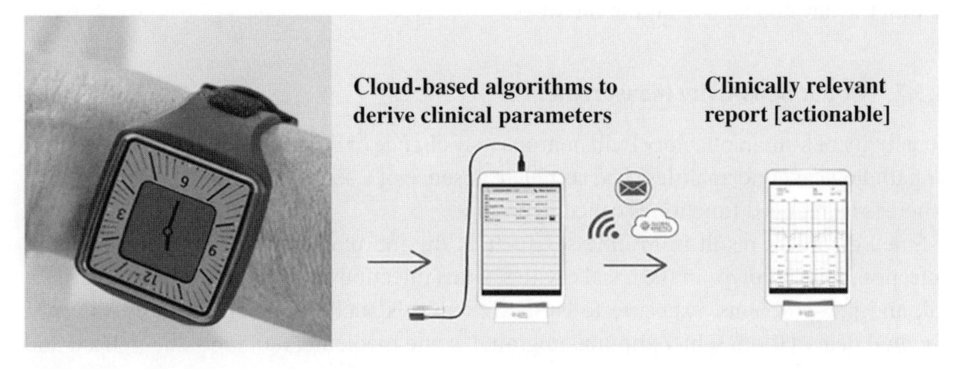


**Figure 8.5** The PKG system (https://medtechengine.com/article/global-kinetics-parkinsons-wearable). Source: Courtesy of Medtech Engine.

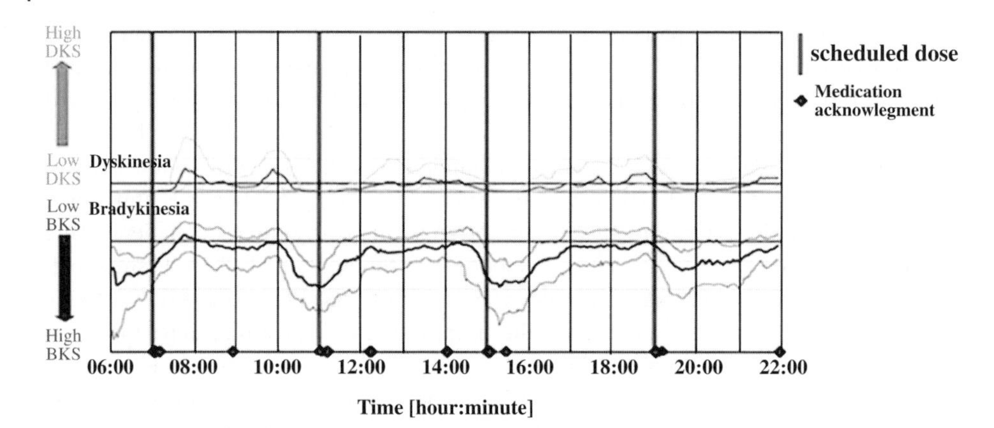


**Figure 8.6** Recorded data from the wrist-worn PKG watch are processed and the sample scores associated with dyskinesia and bradykinesia are shown in the produced graph. In the same graph, the medication reminder and acknowledgment are visually demonstrated. https://www.globalkineticscorporation.com/the-pkg-system.

clinic may not adequately reflect typical mobility function or its variability during daily life. It is hypothesised that the quality of turning mobility, rather than overall quantity of activity, would be impaired over time. Continuous monitoring and characterisations of turns in PD population using wearable sensors have been explored in several studies [99, 100].

The Opal wearable inertial sensors, which integrate tri-axial accelerometer, gyroscope, and magnetometer, have been used to monitor turns and overall physical activities of PD patients [100]. The results have provided sufficient evidence that the PD patients present an impaired quality of turning and higher variability throughout the day and across days compared to healthy subjects. On the other hand, no significant difference between the overall physical activity level of PD patients and healthy subjects has been realised. According to the results obtained from the study in [99], the turn characteristics have been successfully quantified using the activPAL system. It has been demonstrated that the PD patients normally take shorter turns with smaller angles and more steps, including more variability within the day and across days of the week.

### 8.3.7 Odour Sensitivity Measurement

Sensitivity of human olfactory bulb neurons may change or deteriorate due to physiological or pathological abnormalities. The complete absence of a sense of smell is called anosmia, whereas its reduced function is called hyposmia.

Smell disorders result from illnesses such as flu and blocking nose, upper respiratory infection, injury, polyps in the nasal cavities, sinus infections, hormonal disturbances, dental, and jaw problems, exposure to certain chemicals such as insecticides and solvents, cerebral degradation, schizophrenia, migraine, some medicine, radiation due to head and neck cancers, or presence of tumours around the olfactory pulp. Allergy and smoking may cause hyposmia too.

On the other hand, hyperemesis gravidarum and hyperosmia, as an increased olfactory sensitivity to smell, is usually caused by a lower threshold for odour. This may be genetic, hormonal, environmental, or the result of benzodiazepine withdrawal syndrome.

In recent years, psychophysical and electrophysiological tests have been developed to quantify olfactory function in the clinical setting. Additionally, modern structural and functional imaging procedures are applied to better define the underpinnings of functional losses, such as damage to or the lack of olfactory bulbs and tracts [101]. Nevertheless, olfactory tests vary in terms of sensitivity and practicality, ranging from brief tests of odour identification to sophisticated olfactometers yoked to electrophysiological recording equipment capable of quantifying odour-induced changes in electrical activity at the level of the olfactory epithelium (the electro-fractogram; EOG) and cortex (odour event-related potentials; OERPs). Electro-olfactography is a type of electrography effective in the study of olfaction. It measures and records the electrical potentials of olfactory epithelium in a way similar to electroencephalography (EEG) recordings.

Psychophysical tests are more practical and less costly than electrophysiological tests, making them much more popular, mainly because of technical issues with electrophysiological testing. For example, the EOG cannot be reliably measured in all patients, given epithelial sampling issues and the intolerance of some subjects to electrodes that are placed within their non-anaesthetised noses. Since the EOG is present in some anosmics and can be recorded even after death, it cannot be used by itself as a reliable indicator of general olfactory function. Unlike the auditory brainstem evoked potential, the OERP is presently incapable of localising anomalies within the olfactory pathways. OERP recording sessions can be quite long due to the need for relatively long interstimulus intervals to prevent adaptation [102].

The EOG sensor is placed in the olfactory bulb under endoscopic operation. Normally, this is not pleasant for patients. Therefore, for many patients, psychophysical examinations are more practical and less costly than electrophysiological tests, making them much more popular. The University of Pennsylvania Smell Identification Test (UPSIT) is a popular psychophysical test [103].

However, in combination with nasal endoscopy and air-dilution olfactometry, the EOG is a unique technique among others, used to provide comprehensive information about the processing of olfactory information in humans [104]. Figure 8.7 shows the EOG measures for a healthy (control) individual and a schizophrenic patient suffering from hyposmia.

Other methods such as multichannel near-infrared spectroscopy (NIRS) has been suggested to monitor the activity of frontal cortex as the result of hemodynamic responses subjected to olfactory stimulation [106]. There have been some attempts in measuring the olfactory response to odours from EEG signals. Very recently, a noninvasive olfactory test has been suggested by measuring the EEG signals from the electrodes placed at the nasal bridge which are expected to represent responses from the olfactory bulb [107]. In this work some localisation methods have been employed to localise the olfactory source signals before their evaluation.

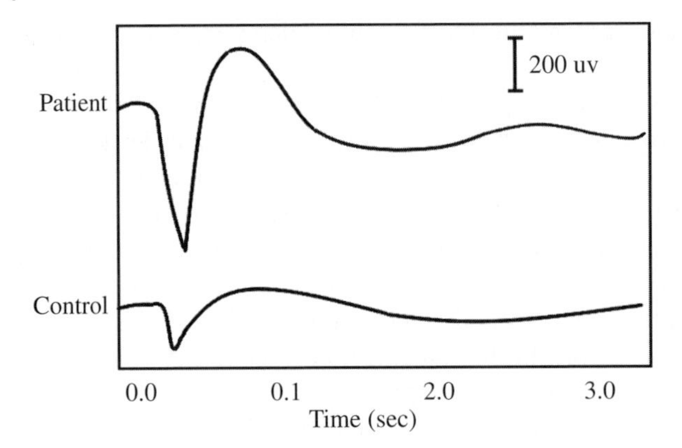


**Figure 8.7** EOG measures for a healthy (control) individual and a schizophrenic patient suffering from hyposmia. Here, olfaction is closely associated with the frontal and temporal brain regions most implicated in schizophrenia [105]. Source: Courtesy of Turetsky, B.I., Hahn, C.G., Borgmann-Winter, K., and Moberg, P.J.

## 8.4 Conclusions

Emerging new technologies in sensing and sensor networking encourages personalised healthcare. This is the result of wiring individuals with appropriate sensor networks. Implementation of these systems in their home environment is necessary to transform current healthcare systems to improve patients' quality of life, enable assistive technology, and predict and prevent possible life-threatening hazards. In this chapter various HMSs using wearable and ambient sensors have been described for monitoring patients inside ICUs, those with chronic conditions such as COPD, and individuals with neurological disease such as dementia and PD patients. Moreover, measuring olfactory sensitivity to odours, which changes significantly due to various diseases, and therefore, as a biomarker, is of interest to researchers. Future research needs to target improvement and validation of these systems using large clinical trials and further promote personalised healthcare.

## References

**1** Hayajneh, T., Almashaqbeh, G., Ullah, S., and Vasilakos, A.V. (2014). A survey of wireless technologies coexistence in WBAN: analysis and open research issues. *Wireless Networks* 20 (8): 2165–2199.

**2** Acampora, G., Cook, D.J., Rashidi, P., and Vasilakos, A.V. (2013). A survey on ambient intelligence in healthcare. *Proceedings of the IEEE* 101 (12): 2470–2494.

**3** Mshali, H., Lemlouma, T., Moloney, M., and Magoni, D. (2018). A survey on health monitoring systems for health smart homes. *International Journal of Industrial Ergonomics* 66: 26–56.

**4** Chen, M., Gonzalez, S., Vasilakos, A. et al. (2011). Body area networks: a survey. *Mobile Networks and Applications* 16 (2): 171–193.

5 Negra, R., Jemili, I., and Belghith, A. (2016). Wireless body area networks: applications and technologies. *Procedia Computer Science* 83: 1274–1281.

6 Vallejos de Schatz, C.H., Medeiros, H.P., Schneider, F.K., and Abatti, P.J. (2012). Wireless medical sensor networks: design requirements and enabling technologies. *Telemedicine and e-Health* 18 (5): 394–399.

7 Aleksandar, M., Chris, O., and Emil, J. (2006). Wireless sensor networks for personal health monitoring: issues and an implementation. *Computer Communications* 29 (13–14): 2521–2533.

8 National Institute for Health and Clinical Excellence (2007). *Acutely Ill Patients in Hospital: Recognition of and Response to Acute Illness in Adults in Hospital.* London: National Institute for Health and Clinical Excellence.

9 Royal College of Physicians (2017). *National Early Warning Score (NEWS) 2: Standardising the Assessment of Acute-Illness Severity in the NHS. Updated Report of a Working Party.* London: RCP.

10 Tarassenko, L., Clifton, D.A., Pinsky, M.R. et al. (2011). Centile-based early warning scores derived from statistical distributions of vital signs. *Resuscitation* 82 (8): 1013–1018.

11 Watkinson, P.J., Pimentel, M.A.F., Clifton, D.A., and Tarassenko, L. (2018). Manual centile-based early warning scores derived from statistical distributions of observational vital-sign data. *Resuscitation* 129: 55–60.

12 Moya, M., Koch, M., and Hostetler, L. (1993). One-class classifier networks for target recognition applications. In: *Proceedings of the World Congress on Neural Networks, International Neural Network Society*, 797–801. IEEE.

13 He, H. and Garcia, E. (2009). Learning from imbalanced data. *IEEE Transactions on Knowledge and Data Engineering* 21 (9): 1263–1284.

14 Lee, H.-J. and Cho, S. (2006). The novelty detection approach for different degrees of class imbalance. In: *Neural Information Processing, Lecture Notes in Computer Science*, vol. 4233 (eds. I. King, J. Wang, L.-W. Chan and D. Wang), 21–30. Berlin: Springer.

15 Tarassenko, L., Hann, A., Patterson, A. et al. (2005). Multi-parameter monitoring for early warning of patient deterioration. In: *Proceedings of the 3rd IEE International Seminar on Medical Applications of Signal Processing*, 71–76. IET.

16 Tarassenko, L., Hann, A., and Young, D. (2006). Integrated monitoring and analysis for early warning of patient deterioration. *British Journal of Anaesthesia* 97 (1): 64–68.

17 Hravnak, M., Edwards, L., Clontz, A. et al. (2008). Defining the incidence of cardiorespiratory instability in patients in step-down units using an electronic integrated monitoring system. *Archives of Internal Medicine* 168 (12): 1300–1308.

18 Tompkins, C. and Orwat, J. (2010). A randomized trial of telemonitoring heart failure patients. *Journal of Healthcare Management* 55 (5): 312–322.

19 Wade, M.J., Desai, A.S., Spettell, C.M. et al. (2011). Telemonitoring with case management for seniors with heart failure. *American Journal of Managed Care* 17 (3): 71–79.

20 Trabelsi, D., Mohammed, S., Chamroukhi, F. et al. (2013). An unsupervised approach for automatic activity recognition based on hidden Markov model regression. *IEEE Transactions on Automation Science and Engineering* 10 (3): 829–835.

**21** Noury, N. and Hadidi, T. (2012). Computer simulation of the activity of the elderly person living independently in a health smart home. *Computer Methods and Programs in Biomedicine* 108 (3): 1216–1228.

**22** Nguyen, N., Phung, D., Venkatesh, S., and Bui, H. (2005). Learning and detecting activities from movement trajectories using the hierarchical hidden Markov model. *IEEE Computer Society Conference on Computer Vision and Pattern Recognition* 2: 955–960.

**23** Park, H.-S., Oh, K., and Cho, S.-B. (2011). Bayesian network-based high-level context recognition for mobile context sharing in cyber-physical system. *International Journal of Distributed Sensor Networks* 7 (1) https://doi.org/10.1155/2011/650387.

**24** Hong, J., Yang, S., and Cho, S. (2010). ConaMSN: a context-aware messenger using dynamic Bayesian networks with wearable sensors. *Expert Systems with Applications* 37 (6): 4680–4686.

**25** Du, Y., Chen, F., Xu, W., and Li, Y. (2006). Recognizing interaction activities using dynamic Bayesian network. In: *18th International Conference on Pattern Recognition*, 618–621. IEEE.

**26** Khan, A.M., Lee, Y., Lee, S.Y., and Kim, T. (2010). A triaxial accelerometer-based physical-activity recognition via augmented-signal features and a hierarchical recognizer. *IEEE Transactions on Information Technology in Biomedicine* 14 (5): 1166–1172.

**27** Fleury, A., Vacher, M., and Noury, N. (2010). SVM-based multimodal classification of activities of daily living in health smart homes: sensors, algorithms, and first experimental results. *IEEE Transactions on Information Technology in Biomedicine* 14 (2): 274–283.

**28** Mannini, A., Intille, S., Rosenberger, M. et al. (2013). Activity recognition using a single accelerometer placed at the wrist or ankle. *Medicine and Science in Sports and Exercise* 45 (11): 2193–2203.

**29** Varkey, J., Pompili, D., and Walls, T. (2012). Human motion recognition using a wireless sensor-based wearable system. *Personal and Ubiquitous Computing* 16 (7): 897–910.

**30** Riboni, D. and Bettini, C. (2011). COSAR: hybrid reasoning for context-aware activity recognition. *Personal and Ubiquitous Computing* 15 (3): 271–289.

**31** Gao, L., Bourke, A., and Nelson, J. (2014). Evaluation of accelerometer based multi-sensor versus single-sensor activity recognition systems. *Medical Engineering & Physics* 36 (6): 779–785.

**32** Parkka, J., Cluitmans, L., and Ermes, M. (2010). Personalization algorithm for real-time activity recognition using PDA, wireless motion bands, and binary decision tree. *IEEE Transactions on Information Technology in Biomedicine* 14 (5): 1211–1215.

**33** Bao, L. and Intille, S.S. (2004). Activity recognition from user-annotated acceleration data. In: *Pervasive Computing* (eds. A. Ferscha and F. Mattern), 1–17. Springer.

**34** Brulin, D., Benezeth, Y., and Courtial, E. (2012). Posture recognition based on fuzzy logic for home monitoring of the elderly. *IEEE Transactions on Information Technology in Biomedicine* 16 (5): 974–982.

**35** Shimokawara, E., Kaneko, T., Yamaguchi, T. et al. (2013). Estimation of basic activities of daily living using ZigBee 3D accelerometer sensor network. In: *International Conference on Biometrics and Kansei Engineering*, 251–256. IEEE.

**36** Medjahed, H., Istrate, D., Boudy, J. et al. (2011). A pervasive multi- sensor data fusion for smart home healthcare monitoring. In: *IEEE International Conference on Fuzzy Systems*, 1466–1473. IEEE.

**37** Wongpatikaseree, K., Ikeda, M., and Buranarach, M. (2012). Activity recognition using context-aware infrastructure ontology in smart home domain. In: *International Conference on Knowledge, Information and Creativity Support Systems*, 50–57. IEEE.

**38** Magnusson, L. and Hanson, E. (2005). Supporting frail older people and their family carers at home using information and communication technology: cost analysis. *Journal of Advanced Nursing* 51 (6): 645–657.

**39** Tomita, R., Mann, W., Stanton, K. et al. (2007). Use of currently available smart home technology by frail elders process and outcomes. *Topics in Geriatric Rehabilitation* 23 (1): 24–34.

**40** Chandola, V., Banerjee, A., and Kumar, V. (2009). Anomaly detection: a survey. *ACM Computing Surveys* 41 (3): 15.

**41** Ivanov, S., Foley, C., Balasubramaniam, S., and Botvich, D. (2012). Virtual groups for patient WBAN monitoring in medical environments. *IEEE Transactions on Biomedical Engineering* 59 (11): 3238–3246.

**42** Alam, M., Reaz, M.B.I., and Ali, M.A.M. (2012). A review of smart homes: past, present, and future. *IEEE Transactions on Systems, Man, and Cybernetics, Part C: Applications and Reviews* 42: 1190–1203.

**43** Forkan, A.R.M., Khalil, I., Tari, Z. et al. (2015). A context-aware approach for long-term behavioural change detection and abnormality prediction in ambient assisted living. *Pattern Recognition* 48 (3): 628–641.

**44** Kartsakli, E., Antonopoulos, A., Lalos, A.S. et al. (2015). Reliable MAC design for ambient assisted living: moving the coordination to the cloud. *IEEE Communications Magazine* 53 (1): 78–86.

**45** Murray, C.J. and Lopez, A.D. (1997). Alternative projections of mortality and disability by cause 1990–2020: global burden of disease study. *Lancet* 349 (9064): 1498–1504.

**46** Pedone, C., Chiurco, D., Scarlata, S., and Incalzi, R.A. (2013). Efficacy of multiparametric telemonitoring on respiratory outcomes in elderly people with COPD: a randomized controlled trial. *BMC Health Services Research* 13 (1): 82–90.

**47** Antoniades, N., Rochford, P., Pretto, J. et al. (2012). Pilot study of remote telemonitoring in COPD. *Telemedicine and e-Health* 18 (8): 634–640.

**48** Velardo, C., Shah, S.A., Gibson, O. et al. (2017). Digital health system for personalised COPD long-term management. *BMC Medical Informatics and Decision Making* 17 (1).

**49** Farmer, A., Toms, C., Hardinge, M. et al. (2014). Self-management support using an internet-linked tablet computer (the EDGE platform)-based intervention in chronic obstructive pulmonary disease: protocol for the EDGE-COPD randomised controlled trial. *BMJ Open* 4 (1): e004437.

**50** Shah, S.A., Velardo, C., Farmer, A., and Tarassenko, L. (2017). Exacerbations in chronic obstructive pulmonary disease: identification and prediction using a digital health system. *Journal of Medical Internet Research* 19 (3): e69.

**51** Sparks, R.S., Celler, B., Okugami, C. et al. (2016). Telehealth monitoring of patients in the community. *Journal of Intelligent Systems* 25 (1): 37–53.

**52** Sparks, R.S. and Okugami, C. (2016). Tele-health monitoring of patient well-ness. *Journal of Intelligent Systems* 25 (4): 515–528.

**53** Celler, B.G. and Sparks, R.S. (2015). Home telemonitoring of vital signs: technical challenges and future directions. *IEEE Journal of Biomedical and Health Informatics* 19 (1): 82–91.

**54** Sparks, R., Okugami, C., and Dearin, J. (2015). Establishing the influence that polypharmacy has on patient well-being. *Journal of Pharmacy Practice & Health Care* 1: 100101.

**55** Yu, L., Chan, W.M., Zhao, Y., and Tsui, K.-L. (2018). Personalized health monitoring system of elderly wellness at the community level in Hong Kong. *IEEE Access* 6: 35558–35567.

**56** Kau, L. and Chen, C. (2014). A smart phone-based pocket fall accident detection system. In: *International Symposium on Bioelectronics and Bioinformatics*, 1–4. IEEE.

**57** Kau, L. and Chen, C. (2015). A smart phone-based pocket fall accident detection, positioning, and rescue system. *IEEE Journal of Biomedical and Health Informatics* 19 (1): 44–56.

**58** Liao, L., Patterson, D., Fox, D., and Kautz, H. (2007). Learning and inferring transportation routines. *Artificial Intelligence* 171 (5–6): 311–331.

**59** Franco, C., Demongeot, J., Villemazet, C., and Vuillerme, N. (2010). Behavioral telemonitoring of the elderly at home: detection of nyctohemeral rhythms drifts from location data. In: *International Conference on Advanced Information Networking and Applications Workshops*, 759–766. IEEE.

**60** Ariani, A., Redmond, S.J., Zhang, Z. et al. (2013). Design of an unobtrusive system for fall detection in multiple occupancy residences. In: *35th Annual International Conference of the IEEE Engineering in Medicine and Biology Society (EMBC)*, 4690–4693. IEEE.

**61** Taleb, T., Bottazzi, D., Guizani, M., and Nait-Charif, H. (2009). ANGELAH: a framework for assisting elders at home. *IEEE Journal on Selected Areas in Communications* 27 (4): 480–494.

**62** Hsu, C. and Chen, J. (2011). A novel sensor-assisted RFID-based indoor tracking system for the elderly living alone. *Sensor (Basel)* 11 (11): 10094–10113.

**63** Jih, W., Hsu, J., Wu, C. et al. (2006). A multi-agent service framework for context-aware elder care. In: *Workshop on Service-Oriented Computing and Agent-Based Engineering*, 61–75. Service-Oriented Computing and Agent-Based Engineering.

**64** Hori, T. and Nishida, Y. (2005). Ultrasonic sensors for the elderly and caregivers in a nursing home. In: *International Conference on Enterprise Information Systems*, 110–115. ACM.

**65** Kearns, W.D., Fozard, J.L., Becker, M., and Jasiewicz, J.M. (2012). Path tortuosity in everyday movements of elderly persons increases fall prediction beyond knowledge of fall history, medication use, and standardized gait and balance assessments. *Journal of the American Medical Directors Association* 13 (7): 665.e7–665.e13.

**66** Matjacic, Z., Bohinc, K., and Cikajlo, I. (2010). Development of an objective balance assessment method for purposes of telemonitoring and telerehabilitation in elderly population. *Disability and Rehabilitation* 32 (3): 259–266.

**67** Bian, Z., Hou, J., Chau, L., and Magnenat-Thalmann, N. (2015). Fall detection based on body part tracking using a depth camera. *IEEE Journal of Biomedical and Health Informatics* 19 (2): 430–439.

**68** Núñez-Marcos, A., Azkune, G., and Arganda-Carreras, I. (2017). Vision-based fall detection with convolutional neural networks. *Wireless Communications and Mobile Computing* 2017: 9474806.

**69** Mao, A., Ma, X., He, Y., and Luo, J. (2017). Highly portable, sensor-based system for human fall monitoring. *Sensors (Basel)* 17 (9): E2096.

**70** World Health Organization (2019) Dementia. https://www.who.int/news-room/fact-sheets/detail/dementia (accessed 25 November 2019).

**71** Stavropoulos, T.G., Meditskos, G., and Kompatsiaris, I. (2017). Demaware2: integrating sensors, multimedia and semantic analysis for the ambient care of dementia. *Pervasive and Mobile Computing* 34 (Suppl. C): 126–145.

**72** Adlam, T., Faulkner, R., Orpwood, R. et al. (2004). The installation and support of internationally distributed equipment for people with dementia. *IEEE Transactions on Information Technology in Biomedicine* 8 (3): 253–257.

**73** Mihailidis, A., Carmichael, B., and Boger, J. (2004). The use of computer vision in an intelligent environment to support aging in place, safety, and independence in the home. *IEEE Transactions on Information Technology in Biomedicine* 8 (3): 238–247.

**74** Lin, C.-C., Chiu, M.-J., Hsiao, C.-C. et al. (2006). Wireless health care service system for elderly with dementia. *IEEE Transactions on Information Technology in Biomedicine* 10 (4): 696–704.

**75** Shin, D.-M., Shin, D., and Shin, D. (2013). Smart watch and monitoring system for dementia patients. In: *Grid and Pervasive Computing*, vol. 7861 (eds. J.J.H. Park, H.R. Arabnia, C. Kim, et al.), 577–584. Berlin: Springer.

**76** Enshaeifar, S., Zoha, A., Markides, A. et al. (2018). Health management and pattern analysis of daily living activities of people with dementia using in-home sensors and machine learning techniques. *PLoS One* 13 (5): e0195605.

**77** Enshaeifar, S., Barnaghi, P., Skillman, S. et al. (2018). The Internet of things for dementia care. *IEEE Internet Computing* 22 (1): 8–17.

**78** Son, H., Park, W.S., and Kim, H. (2018). Mobility monitoring using smart technologies for Parkinson's disease in free-living environment. *Collegian* 25 (5): 549–560.

**79** Goetz, C.G., Tilley, B.C., Shaftman, S.R. et al. (2008). Movement disorder society-sponsored revision of the Unified Parkinson's Disease Rating Scale (MDS-UPDRS): scale presentation and clinimetric testing results. *Movement Disorders* 23 (15): 2129–2170.

**80** Capecci, M., Pepa, L., Verdini, F., and Ceravolo, M.G. (2016). A smartphone-based architecture to detect and quantify freezing of gait in Parkinson's disease. *Gait & Posture* 50: 28–33.

**81** Giladi, N. and Nieuwboer, A. (2008). Understanding and treating freezing of gait in parkinsonism, proposed working definition, and setting the stage. *Movement Disorders* 23: S423–S425.

**82** Moore, S.T., MacDougall, H.G., and Ondo, W.G. (2008). Ambulatory monitoring of freezing of gait in Parkinson's disease. *Journal of Neuroscience Methods* 167 (2): 340–348.

**83** Bächlin, M., Plotnik, M., Roggen, D. et al. (2010). Wearable assistant for Parkinson's disease patients with the freezing of gait symptom. *IEEE Transactions on Information Technology in Biomedicine* 14: 436–446.

**84** Chastin, S.F., Baker, K., Jones, D. et al. (2010). The pattern of habitual behaviour is different in advanced Parkinson's disease. *Movement Disorders* 25 (13): 2114–2120.

**85** Auvinet, B., Berrut, G., Touzard, C. et al. (2002). Reference data for normal subjects obtained with an accelerometric device. *Gait & Posture* 16 (2): 124–134.

**86** Rochester, L., Chastin, S.F., Lord, S. et al. (2010). Understanding the impact of deep brain stimulation on ambulatory activity in advanced Parkinson's disease. *Journal of Neurology* 259 (6): 1081–1086.

**87** Huang, C., Chu, H., Zhang, Y., and Wang, X. (2018). Deep brain stimulation to alleviate freezing of gait and cognitive dysfunction in Parkinson's disease: update on current research and future perspectives. *Frontiers in Neuroscience* 12: 29.

**88** Mactier, K., Lord, S., Godfrey, A. et al. (2015). The relationship between real world ambulatory activity and falls in incident Parkinson's disease: influence of classification scheme. *Parkinsonism and Related Disorders* 21 (3): 236–242.

**89** Hiorth, Y.H., Larsen, J.P., Lode, K. et al. (2016). Impact of falls on physical activity in people with Parkinson's disease. *Journal of Parkinson's Disease* 6 (1): 175–182.

**90** Espay, A.J., Baram, Y., Dwivedi, A.K. et al. (2011). At-home training with closed-loop augmented-reality cueing device for improving gait in patients with Parkinson disease. *Journal of Rehabilitation Research and Development* 47 (6): 573–581.

**91** Weiss, A., Sharifi, S., Plotnik, M. et al. (2011). Toward automated, at-home assessment of mobility among patients with Parkinson disease, using a body-worn accelerometer. *Neurorehabilitation and Neural Repair* 25 (9): 810–818.

**92** Cavanaugh, J.T., Ellis, T.D., Earhart, G.M. et al. (2012). Capturing ambulatory activity decline in Parkinson disease. *Journal of Neurology and Physical Therapy* 36 (2): 51–57.

**93** Mirelman, A., Herman, T., Nicolai, S. et al. (2011). Audio-biofeedback training for posture and balance in patients with Parkinson's disease. *Journal of Neuro Engineering and Rehabilitation* 8: 35.

**94** Das, S., Amoedo, B., De la Torre, F., and Hodgins, J. (2012). Detecting Parkinson's symptoms in uncontrolled home environments: a multiple instance learning approach. In: *Conference Proceedings of the IEEE Engineering in Medicine and Biology Society*, 3688–3691. IEEE.

**95** Griffiths, R.I., Kotschet, K., Arfon, S. et al. (2012). Automated assessment of bradykinesia and dyskinesia in Parkinson's disease. *Journal of Parkinson's Disease* 2 (1): 47–55.

**96** Dontje, M.L., de Greef, M.H., Speelman, A.D. et al. (2013). Quantifying daily physical activity and determinants in sedentary patients with Parkinson's disease. *Parkinsonism and Related Disorders* 19 (10): 878–882.

**97** Hammerla, N.Y., Fisher, J.M., Andras, P. et al. (2015). PD disease state assessment in naturalistic environments using deep learning. In: *29th AAAI Conference on Artificial Intelligence*, 1742–1748. Association for the Advancement of Artificial Intelligence.

**98** Horne, M.K., McGregor, S., and Bergquist, F. (2015). An objective fluctuation score for Parkinson's disease. *Public Library of Science* 10 (4): e0124522.

**99** El-Gohary, M., Pearson, S., McNames, J. et al. (2014). Continuous monitoring of turning in patients with movement disability. *Sensors (Basel)* 14 (1): 356–369.

**100** Mancini, M., El-Gohary, M., Pearson, S. et al. (2015). Continuous monitoring of turning in Parkinson's disease: rehabilitation potential. *NeuroRehabilitation* 37 (1): 3–10.

**101** Gottfried, J.A. (2015). Structural and functional imaging of the olfactory system. In: *Handbook of Olfaction and Gustation* (ed. R.L. Doty), 279–304. Hoboken, NJ: Wiley.

**102** Osman, A. and Silas, J. (2015). Electrophysiological measurement of olfactory function. In: *Handbook of Olfaction and Gustation* (ed. R.L. Doty), 261–278. Hoboken, NJ: Wiley.

**103** Doty, R.L. (2015). Olfactory dysfunction and its measurement in the clinic. *World Journal of Otorhinolaryngology: Head & Neck Surgery* 1 (1): 28–33.

**104** Knecht, M. and Hummel, T. (2004). Recording of the human electro-olfactogram. *Elsevier Journal of Physiology & Behavior* 83 (1): 13–19.

**105** Turetsky, B.I., Hahn, C.G., Borgmann-Winter, K., and Moberg, P., J. (2009). Scents and nonsense: olfactory dysfunction in schizophrenia. *Schizophrenia Bulletin* 35 (6): 1117–1131.

**106** Kobayashi, E., Karaki, M., Touge, T. et al. (2012). Olfactory assessment using near-infrared spectroscopy. In: *ICME International Conference on Complex Medical Engineering (CME)*, 517–520. IEEE.

**107** Iravania, B., Arshamiana, A., Ohlac, K. et al. (2019). Non-invasive recording from the human olfactory bulb. *BioRXiv* http://doi.org/10.1101/660050.

# 9

# Machine Learning for Sensor Networks

## 9.1 Introduction

'Machine learning' refers to a combination of learning from data and their clustering or classification. It is often combined with some signal processing techniques to extract the best discriminating features of the data prior to taking learning and classification steps. The emerging machine learning techniques, such as deep neural networks (DNNs), are also capable of feature learning.

Often, in body sensor networks (BSNs), multimodal data are used, and thus features from various measurement modalities should be combined or exploited. This leads to another area in machine learning called sensor (data) fusion.

Sensors are used almost everywhere. The newly emerging sensor technology is beginning to closely mimic the ultimate sensing machine, i.e. the human being. The sensor fusion technology leverages a microcontroller (which mimics the human brain) to fuse individual data collected by multiple sensors. This allows for a more accurate and reliable realisation of the data than one would acquire from an individual sensor.

Sensor fusion enables context awareness, which has significant potential for the Internet of Things (IoT). Advances in sensor fusion for remote emotive computing (emotion sensing and processing) could also lead to exciting new applications in the future, including smart healthcare. This approach motivates personalised healthcare providers to fine tune and customise systems which best suit individuals' needs.

However, new technology is shifting towards consensus or diffusion communication networks whereby sensors can communicate to each other and decide on their next action without communicating to any hub or master node. Unlike current systems, this is a decentralised approach in which the decision is made at each sensor. This also brings a new direction in machine learning called cooperative learning, which exploits the information from a number of sensors in a neighbourhood to decide on the action of each sensor within that neighbourhood.

There are mainly two types of machine learning algorithms for classification of data: supervised and unsupervised. In an unsupervised learning algorithm the classifier clusters the data into the groups farthest from each other. A popular example for these algorithms is $k$-means algorithm. On the other hand, for supervised classifiers, the target is known during the training phase and the classifier is trained to minimise a difference between the actual output and the target values. Traditional methods based on adaptive filters using

*Body Sensor Networking, Design and Algorithms*, First Edition. Saeid Sanei, Delaram Jarchi and Anthony G. Constantinides.

Companion Website: www.wiley.com/go/sanei/algorithm-design

mean squared error (MSE) or least square error (LSE) minimisations are still in use. Good examples of such classifiers are support vector machines (SVM), and multilayered perceptron (MLP). In both cases, effective estimation of data features often helps reduce the number of input samples, thus enhancing the overall algorithm speed. The more efficiently the features are estimated, the more accurate the result of the clustering or classification will be.

Many algorithms can be used for feature estimation. These algorithms are often capable of changing the dimensionality of the signals to make them more separable. Numerous statistical measures can be derived from the data with known distributions. Various order statistics such as mean, variance, skewness, and kurtosis are very popular statistical measures to describe the data in scalar forms. The vector forms may become necessary for multichannel data.

In addition to the above two main classes of learning systems, semi-supervised techniques benefit from partially known information during training or classification. This group of learning systems benefits from various constraints on the input, output, or classifier in order to best optimise the system for each particular application. One useful application of such systems is for the rejection of anomalies in the datasets during classifier training [1].

In the clustering methods there is no need to label the classes in advance and only the number of clusters may be identified and fed into the clustering algorithm by the user. Classification of data is similar to clustering, except the classifier is trained using a set of labelled data before being able to classify a new data. In practice, the objective of classification is to draw a boundary between two or more classes and to label them based on their measured features. In a multidimensional feature space this boundary takes the form of a separating hyperplane. The objective here is to find the best hyperplane, maximising the distance from all the classes (interclass members) while minimising proximities within members of each class (intraclass members).

Clustering and classification become ambiguous when deciding which features to use and how to extract or enhance those features. Principal component analysis (PCA) and independent component analysis (ICA) are two very common approaches to extract the dominant data features. Traditionally, a wide range of features (such as various order statistics) are measured or estimated from the raw data and used by the machine learning algorithms for clustering or classification.

In the context of human-related signals, classification of the data in feature space is often required. For example, the power of alpha and beta waves in electroencephalography (EEG) signals [2] may be classified to not only detect the brain abnormalities but also determine the stage of the disease. In sleep studies, one can look at the power of alpha and theta waves as well as appearance of spindles and *K*-complexes to classify the stages of sleep. As another example, in brain–computer interfacing (BCI) systems for left and right finger movement detection one needs to classify the time, frequency, and spatial features.

There have been several popular machine learning approaches developed in the past 50 years. Among them artificial neural networks (ANNs), linear discriminant analysis (LDA), hidden Markov model (HMM), Gaussian mixture model (GMM), decision tree, random forest, and SVMs are the most popular ones. *k*-mean clustering and fuzzy logic have also been widely used for clustering, pre-classification, and other artificial intelligent (AI) based applications [3]. These techniques are developed and well explained in the literature [4]. Detailed explanation of all these methods is beyond the objective of this chapter.

Instead, here we provide a summary of the most popular machine learning methods which have been widely used in sensor networking and health-related data analysis.

Unlike many mathematical problems in which some forms of explicit formula based on a number of inputs results in an output, in data classification there is no model or formula of this kind. Instead, the system should be trained (using labelled data) to be able to recognise the new (test) inputs.

In the case of data with considerable noise or artefacts, successful classifiers are those which can minimise the effect of outliers. Outliers are the data samples or features which turn up very distant from the class centres, and therefore they do not belong to any class. In many machine learning algorithms, outliers have destructive effect during the classifier training and increase the cross-validation error. Pre-processing (including smoothing, denoising, etc.) can significantly reduce the outliers or their impact. Right choice of classifiers and their associated feature detection systems can also reduce the influence of outliers.

For sensor network applications, machine learning is important since:

- Sensor networks usually monitor dynamic environments which rapidly change over time. It is therefore desirable to develop sensor networks that can adapt and operate efficiently in such environments.
- Sensor networks may be used to collect new information about unreachable and risky locations such as inside the body [5]. Owing to the unexpected behaviour patterns that may arise in such scenarios, the initial solutions provided by system designers may not operate as expected. Thus, robust machine learning algorithms, able to calibrate itself to newly acquired information, would be desirable.
- The sensors may be deployed in complicated environments where the researchers cannot build accurate mathematical models to describe the system behaviour. Meanwhile, some tasks in wireless sensor networks (WSNs) can be prescribed using simple mathematical models but may still need complex algorithms to solve them (e.g. the routing problem [6, 7]). In such circumstances, machine learning circumvents the problem and provides low-complexity estimates for the system model.
- Sensor networks often provide large amounts of data but may not be able to make clear sense of them. For example, in addition to ensuring communication connectivity and energy sustainability, WSN application often comes with minimum data coverage requirements which need to be fulfilled by limited sensor hardware resources [8]. Machine learning techniques can then be used to discover the data and recognise important correlations in the sensor data before proposing an improved sensor deployment for maximum data coverage.
- New applications and integrations of WSNs – such as in-patient status monitoring, cyber-physical systems (CPSs), machine-to-machine (M2M) communications, and IoT technologies – have been introduced with a motivation to support more intelligent decision making and autonomous control [9]. Here, machine learning is important to extract the different levels of abstractions necessary to perform the AI tasks with limited human intervention [10].

Many machine learning algorithms do not perform efficiently when

- The number of features or input samples is high.
- There is a limited time to perform the classification.

- There is a nonuniform weighting amongst features.
- There is a nonlinear map between inputs and outputs.
- Distribution of data is not known.
- There are significant number of outliers.
- The convergence of the algorithm is not convex (monotonic), so it may fall into a local minimum.

In addition, in using machine learning for sensor networks:

- Being a resource limited framework, WSN with a distributed management system drains a considerable percentage of its energy budget for conditioning, pre-processing, or processing the data samples. Thus, a trade-off between the algorithm's computational requirements and the learnt model's accuracy has to be taken into consideration. Specifically, the higher the required accuracy, the higher the computational requirements and the higher energy consumptions are.
- Generally speaking, learning by synthetic or simulated data requires a large dataset of samples to achieve the intended generalisation capabilities (i.e. fairly small error bounds), and there won't be a full control over the knowledge formulation process [11].

For the data with unknown distributions often nonparametric techniques are applied. In these techniques a subset of labelled data is used to estimate or model the data distributions through learning before application of any feature selection or classification algorithm.

To assess the performance quality of the classifiers a multifold cross-validation is often necessary. For this, only the training data are considered. In each fold a percentage of the labelled data is used to train the classifier and the rest used for testing. The cross-validation error is then calculated as the amount of misclassified data over a number of folds, say $k$-fold.

Lastly, but probably most importantly, is dimensionality reduction. Dimensionality reduction plays an important role in machine learning mainly for high-dimensionality datasets, where the number of available features is large. Reducing dimensionality often alleviates the effect of outliers and noise and therefore enhances the classifier performance. PCA is the most popular dimensionality reduction technique. PCA projects the data into a lower dimensional space using singular value decomposition (SVD), $\mathrm{X} = \mathrm{U}\Sigma\mathrm{V}^{\mathrm{T}}$ where U and V are singular vectors and $\Sigma$ represents the square matrix of singular values or eigenvalues. Input to SVD can have mixed-signs. Nonnegative matrix factorisation (NMF), as another approach, restricts factors to be nonnegative and can be used when the input data is nonnegative. Hence, it works based on setting a nonnegativity constraint on the extracted components, $\mathrm{X}_+ = \mathrm{W}_+\mathrm{H}_+$ where $(.)_+$ represents the nonnegativity of all elements of the input data and components. Binary matrix factorisation is another extension of NMF for the binary data by constraining components to be binary, $\mathrm{X}_{0-1} = \mathrm{W}_{0-1}\mathrm{H}_{0-1}$ where $(.)_{0-1}$ represents the binary elements. Sparsity constraints (by constraining factorised components' norm) can be added to the optimisation of PCA and NMF to enhance the interoperability and stability of components: SPCA (sparse principal component analysis) and SNMF (sparse non-negative matrix factorisation). Sparsity constraints are particularly important where the data are naturally sparse [12].

## 9.2 Clustering Approaches

### 9.2.1 *k*-means Clustering Algorithm

*k*-means algorithm [13] is an effective and, generally, simple clustering tool that has been widely used for many applications, such as discovering spike patterns in neural responses [14]. It is employed either for clustering only or as a method to find the target values (cluster centres) for a supervised classification in a later stage. This algorithm divides a set of features (such as points in Figure 9.1) into *k*-clusters.

The algorithm is initialised by setting '*C*' as the desired or expected number of clusters. Then, the centre for each cluster *k* is identified by selecting *k* representative data points. The next step is to assign the remaining data points to the closest cluster centre. Mathematically, this means that each data point needs to be compared with every existing cluster centre and the minimum distance between the point and the cluster centre found. Most often, this is performed in the form of error checking (which is discussed below). However, prior to this, the new cluster centres are calculated. This is essentially the remaining step in *k*-means clustering: once clusters have been established (i.e. each data point is assigned to its closest cluster centre), the geometric centre of each cluster is recalculated.

The Euclidean distance of each data point within a cluster to its centre can be calculated. This can be repeated for all other clusters, whose resulting sums can themselves be summed together. The final sum is known as the *sum of within-cluster sum-of-squares*. Consider the within-cluster variation (*sum of squares* for cluster *c*) error as $\varepsilon_c$:

$$\varepsilon_c = \sum_{i=1}^{n_c} \|x_i^c - \bar{x}_c\|_2^2 \qquad \forall c \tag{9.1}$$

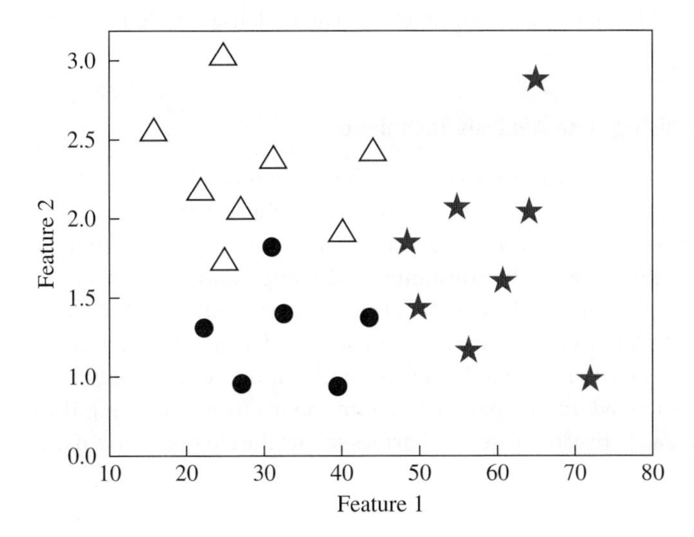


**Figure 9.1** A 2D feature space with three clusters, each with members of different shapes (circle, triangle, and asterisk).

where $\|.\|_2^2$ is the squared Euclidean distance between data point $i$ and its designated cluster centre $\bar{x}_c$, $n_c$ is the total number of data points (features) in cluster $c$, and $x_i^c$ is an individual data point in cluster $c$. The cluster centre (mean of data points in cluster $c$) can be defined as:

$$\bar{x}_c = \frac{1}{n_c} \sum_{i=1}^{n_c} x_i^c \tag{9.2}$$

and the total error is:

$$E_C = \sum_{c=1}^{C} \varepsilon_c \tag{9.3}$$

The overall $k$-means algorithm may be summarised as:

1. Initialisation
   a. Define the number of clusters ($C$).
   b. Designate a cluster centre (a vector quantity that is of the same dimensionality of the data) for each cluster, typically chosen from the available data points.
2. Assign each remaining data point to the closest cluster centre. That data point is now a member of that cluster.
3. Calculate the new cluster centre (the geometric average of all the members of a certain cluster).
4. Calculate the sum of within-cluster sum-of-squares. If this value has not significantly changed over a certain number of iterations, stop the iterations. Otherwise, go back to Step1.

Therefore, an optimum clustering procedure depends on an accurate estimation of the number of clusters. A common problem in $k$-means partitioning is that if the initial partitions are not chosen carefully enough the computation will run the chance of converging to a *local* minimum rather than the *global* minimum solution. The initialisation step is therefore very important.

### 9.2.2 Iterative Self-organising Data Analysis Technique

One way to estimate the number of clusters is to run the algorithm several times with different initialisations and by iteratively increasing the number of clusters from the lowest value. If the results converge to the same partitions, then it is likely that a global minimum has been reached. However, this is very time consuming and computationally expensive. Another solution is to dynamically change the number of partitions (i.e. number of clusters) as the iterations progress. The ISODATA (iterative self-organising data analysis technique algorithm) is an improvement on the original $k$-means algorithm that does exactly this. ISODATA involves a number of additional parameters into the algorithm allowing it to progressively check within- and between-cluster similarities so that the clusters can dynamically split and merge.

### 9.2.3 Gap Statistics

Another approach for solving this problem is gap statistics [15]. In this approach the number of clusters is iteratively estimated. The steps of this algorithm are:

1. For a varying number of clusters $k = 1, 2, \ldots, C$, compute the error $E_c$ using Eq. (9.3).
2. Generate a *B number* of reference datasets. Cluster each one with the $k$-means algorithm and compute the dispersion measures, $\tilde{E}_{kb}$, $b = 1, 2, \ldots, B$. The gap statistics are then estimated using:

$$G_k = \frac{1}{B} \sum_{b=1}^{B} \log(\tilde{E}_{kb}) - \log(E_k) \tag{9.4}$$

   where the dispersion measure $\tilde{E}_{kb}$ is the $E_k$ of the reference dataset $B$.
3. To account for the sample error in approximating an ensemble average with $B$ reference distributions, compute the standard deviation $S_k$ as:

$$S_k = \left[ \frac{1}{B} \sum_{b=1}^{B} (\log(\tilde{E}_{kb}) - \overline{E}_b)^2 \right]^{1/2} \tag{9.5}$$

   where:

$$\overline{E}_b = \frac{1}{B} \sum_{b=1}^{B} \log(\tilde{E}_{kb}) \tag{9.6}$$

4. By defining $\tilde{S}_k = S_k \left(1 + \frac{1}{B}\right)^{1/2}$, estimate the number of clusters as the smallest $k$ such that $G_k \geq G_{k+1} - \tilde{S}_{k+1}$,
5. With the number of clusters identified, utilise $k$-means algorithm to partition the feature space into $k$ subsets (clusters).

The above clustering method has several advantages over $k$-means since it can estimate the number of clusters within the feature space. It is also a multiclass clustering system and unlike SVM can provide the boundary between the clusters.

### 9.2.4 Density-based Clustering

The density-based clustering approach clusters the point features within surrounding noise based on their spatial distribution. It works by detecting areas where the points are concentrated and are separated by areas that are empty or sparse. Points that are not part of a cluster are labelled as noise. In this clustering method unsupervised machine learning clustering algorithms are used, which automatically detect patterns based purely on spatial location and the distance to a specified number of neighbours. These algorithms are considered unsupervised as they do not need any training on what it means to be a cluster [16].

Density-based spatial clustering of applications with noise (DBSCAN) is a popular density-based clustering algorithm with the aim of discovering clusters from approximate density distribution of the corresponding data points. DBSCAN does not need the number of clusters, and instead has two parameters to be set: an *epsilon* that indicates the closeness of the points in each and *minPts*, the minimum neighbourhood size a point should fall into to be considered a member of that cluster. The routine is initialised randomly. The neighbourhood of this point then retrieved and if it consists of an acceptable number of elements, a cluster is formed; otherwise, the element is considered as noise. Hence, DBSCAN may result in some samples which are not clustered.

Usually, DBSCAN parameters are not known in advance and there are several ways to select their values. One way is to calculate the distance of each point to its closest nearest neighbour and use the histogram of distances to select epsilon. Then, a histogram can be obtained of the average number of neighbours for each point using the epsilon. Some of the samples do not have enough any neighbouring point and are counted as noise samples. Implementation of the parameter selection is available at spark DBSCAN (https://github.com/alitouka/spark_dbscan).

DBSCAN can find arbitrary-shaped clusters and is robust to outliers. Nevertheless, it may not identify clusters of various densities or may fail if the data are very sparse. It is also sensitive to the selection of its parameters and the distance measure (usually Euclidean distance), which affects other clustering techniques too.

### 9.2.5 Affinity-based Clustering

This combines $k$-means clustering with a spectral-based clustering method where the algorithms cluster the points using eigenvectors of matrices derived from the data. It uses k eigenvectors simultaneously and identifies the best setting under which the algorithm can perform favourably [17].

### 9.2.6 Deep Clustering

Deep clustering is another modification to clustering which combines $k$-means with neural network (NN) to jointly learn the parameters of an NN and the cluster assignments of the resulting features. It iteratively groups the features using a $k$-means and uses the subsequent assignments as supervision to update the weights of the network. This organises an unsupervised training of convolutional neural networks (CNNs). This approach, demonstrated in Figure 9.2, is similar to the standard supervised training of a CNN and it integrates CNN in its structure [18]. The CNN is explained later in this chapter.

In addition to the above clustering approaches, there are other clustering methods such as power iteration clustering (PIC) [19], which use similar concepts. PIC finds a very low-dimensional embedding of a dataset using truncated power iteration on a normalised pair-wise similarity matrix of the data. It has been shown that it is fast for processing large datasets.

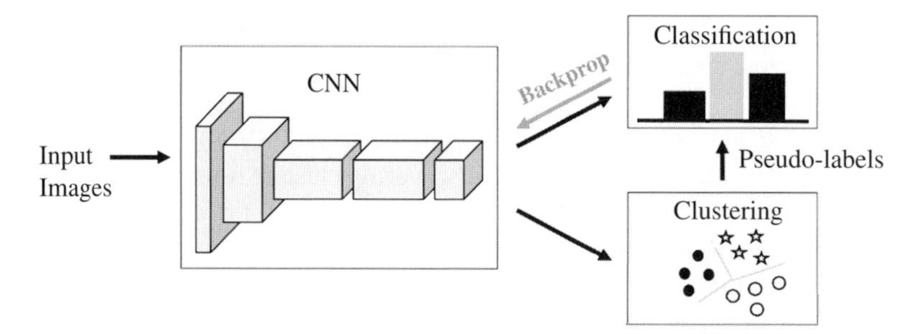


**Figure 9.2** Schematic diagram of deep clustering [17]. The deep features are iteratively clustered and the cluster assignments are used as pseudo-labels to learn the parameters of a CNN. Source: Courtesy of Ng, Y.A., Jordan, M.I., and Weiss, Y.

### 9.2.7 Semi-supervised Clustering

Conventional clustering methods are unsupervised, meaning that there is no available label (target) or anything known about the relationship between the observations in the dataset. In many situations, however, information about the clusters is available in addition to the feature values. For example, the cluster labels of some observations may be known, or certain observations may be known to belong to the same cluster. In other cases, one may wish to identify clusters that are associated with a particular outcome (or target).

The main purposes for so-called semi-supervised methods are the accurate determination of number of clusters and the minimisation of clustering error. Very similar to the concept in Section 9.2.6, there are semi-supervised clustering techniques where supervised and unsupervised learnings are combined to iteratively improve the performance of clustering methods in terms of number of clusters and clustering error. Among these algorithms, deep networks are often used for supervised clustering [17, 20, 21]. The algorithm in [20] simultaneously minimises the sum of supervised and unsupervised cost functions by back-propagation.

#### 9.2.7.1 Basic Semi-supervised Techniques

Two basic semi-supervised techniques are label spreading (LS) and label propagation (LP). LS is based on considering samples (isolates) as nodes in a graph in which their relations defined by edge weights, for example $w_{ij} = \exp(-\|x_i - x_j\|^2/2\sigma^2)$ if $i \neq j$ and $w_{ij} = 0$. The weight matrix is symmetrically normalised for better convergence. Each node receives information from its neighbouring points and its label is selected based on the class with most received information. A regularisation term was also introduced for a better label assignment based on having a smooth classifier (not changing between similar points). LP is based on an iterative technique considering a transition matrix to update labels which starts with a random initialisation. The transition matrix refers to the probability of moving from one node to another by propagating in high-density areas of the unlabelled data.

#### 9.2.7.2 Deep Semi-supervised Techniques

As described in a later section of this chapter, autoencoder (AE) is a DNN which has two main stages: an encoder that maps the input space to a lower dimension (latent space) and a decoder that reconstructs the input space back from the latent representation. Hence, the network is based on an unsupervised learning. AE can be stacked to form a deep stacked autoencoder (SAE) network that adds more power to the network and provides nonlinear latent representations per each hidden layer. Moreover, to improve generalisation and increasing possibility of extracting more interesting patterns in the data, various noise levels can be added to the input of each layer, denoted as stacked denoising autoencoder (SDAE). The optimisation for SDAE is based on minimising the reconstruction error between the original data and the predicted network output.

Although AE, SAE, and SDAE are based on unsupervised learning (using the reconstruction error), they have been used for semi-supervised learning by adding a layer to the learnt encoders and fine tuning the network based on the labelled data for the supervised task (supervised SDAE). The main disadvantage of this technique is to have two stages for supervised and unsupervised learning. Consequently, the first stage may extract information that are not useful for the final supervised task. Owing to this, the fine-tuning stage cannot extensively change a fully learnt network.

Ladder is an extension of SAE/SDAE considering a joint objective function of reconstruction error for unsupervised learning and likelihood of estimating the correct labels for supervised learning. It has two noiseless and noisy encoder paths and a denoising decoder.

### 9.2.8 Fuzzy Clustering

In this clustering method each element has a set of membership coefficients corresponding to the degree of being in a given cluster [22]. This is different from $k$-means, where each object belongs exactly to one cluster. For this reason, unlike $k$-means, which is known as hard or nonfuzzy clustering, this is called the soft clustering method.

In fuzzy clustering, points close to the centre of a cluster may be in the cluster to a higher degree than points in the edge of a cluster. The degree to which an element belongs to a given cluster is a numerical value varying from 0 to 1.

However, fuzzy $c$-means (FCM), the most widely used fuzzy clustering algorithms, is very similar to $k$-means. In FCM the centroid of a cluster is calculated as the mean of all points, weighted by their degree of belonging to the cluster. The aim of $c$-means is to cluster the points into $k$ cluster by minimising the objective function defined as:

$$\sum_{j=1}^{k} \sum_{x_i \in c_j} u_{ij}^{m} (x_i - \mu_j)^2 \tag{9.7}$$

where $\mu_j$ is the centre of the cluster $j$, $m$ the fuzzifier, often selected manually, and $u_{ij}^m$ is the degree to which an observation $x_i$ belongs to a cluster $c_j$, defined as:

$$u_{ij}^{m} = \frac{1}{\sum_{l=1}^{k} \left( \frac{|x_i - c_j|}{|x_i - c_l|} \right)^{\frac{2}{m-1}}} \tag{9.8}$$

The degree of belonging, $u_{ij}^m$, is linked inversely to the distance from $x$ to the cluster centre.

The parameter $m$ is a real number within $1.0 < m < \infty$ and defines the level of cluster fuzziness. For $m$ close to 1 the solution becomes very similar to hard clustering such as $k$-means, whereas a value of $m$ close to infinity leads to complete fuzzyness. The centroid of a cluster, $c_j$, is the mean of all points, weighted by their degree of belonging to the cluster:

$$c_j = \frac{\sum_{x \in c_j} u_{ij}^{m} x}{\sum_{x \in c_j} u_{ij}^{m}} \tag{9.9}$$

The algorithm of fuzzy clustering can be summarised as follow:

1. Set the number of clusters $k$.
2. Assign randomly to each point coefficients for being in the clusters.
3. Repeat until the maximum number of iterations is reached, or when the algorithm has converged for a predefined error:
   a. Compute the centroid for each cluster using Eq. (9.9).
   b. For each point, use Eq. (9.8) to compute its coefficients of being in (degree of belonging to) the clusters.

This concludes the well-established clustering approaches. Other clustering methods such as $k$-medoid are very similar to the above methods. Most of these methods are sensitive to outliers and the objective of new weighted or regularised techniques is to reduce this sensitivity.

## 9.3 Classification Algorithms

### 9.3.1 Decision Trees

Decision trees (DTs) are popular, simple, powerful, and generally nonlinear tools for classification and prediction. They represent *rules* which can be deciphered by human beings and utilised in knowledge-based systems.

The main steps in a DT algorithm are:

1. Select the *best* attribute(s) to split the remaining instances and make that attribute a decision node.
2. Repeating this process recursively for each child.
3. Stop when:
   a. All the instances have the same target attribute value.
   b. There are no more attributes.
   c. There are no more instances.

One simple DT example for a humidity sensor can be:

> If the time is 9:00 a.m. and there is no rain what would be the humidity level (out of low, medium, and severe)?

Depending on the number of attributes, a DT may look like Figure 9.3. DTs can perfectly be fitted to any training data, with zero bias and high variance. Different algorithms are used to determine the 'best' split at a node.

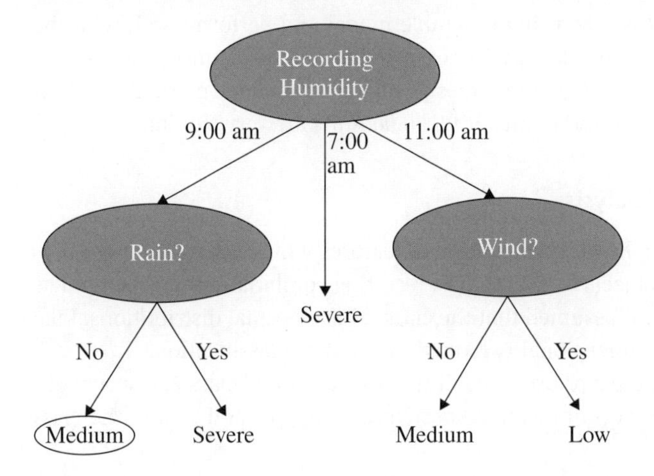


**Figure 9.3** An example of a DT to show the humidity level at 9 a.m. when there is no rain.

DTs are often used to predict data labels by iterating the input data through a learning tree [23]. During this process, the feature properties are compared relative to the decision conditions to reach a specific category. Among many applications, a DT provides a simple, but efficient method to identify link reliability in WSNs by identifying a few critical features such as loss rate, corruption rate, mean time to failure (MTTF) and mean time to restore (MTTR). However, DT works only with linearly separable data and the process of building optimal learning trees is NP (neural process) complete [24]. DT-based sensor activity recognition and detection is one practical example [25].

### 9.3.2 Random Forest

A random forest (or random forests) classifier creates a set of DTs from randomly selected subsets of training set. It then aggregates the votes from different DTs to decide the final class of the test data [26]. The term *random forest* comes from random decision forests initially proposed by Tin Kam Ho of Bell Labs in 1995. This method combines Breiman's 'bagging' idea and random feature selection. Bagging or *bootstrap aggregation* is a technique for reducing the variance of an estimated prediction function. Random forest classifier is in fact an extension to bagging which uses *de-correlated* trees.

The main advantages of random forest classifiers are that there is no need for pruning trees, accuracy, and variable importance are generated automatically, overfitting is not a problem, they are not very sensitive to outliers in training data, and setting their parameters is easy.

However, there are some limitations: the regression process cannot predict beyond the available range in the training data and in regression the extreme values are often not predicted accurately. This causes underestimation of highs and overestimation of lows.

Unlike for a traditional random forest, which is a bagging technique, in *boosting* as the name suggests, one learns from other which in turn boosts the learning. The bagging scheme trains a bunch of individual models in a parallel way. Each model is trained by a random subset of the data. On the other hand, boosting trains a bunch of individual models in a sequential way. Each individual model learns from the mistakes made by the previous model. *AdaBoost* is a boosting ensemble model and performs well with the DT. AdaBoost learns from the mistakes by increasing the weight of misclassified data points. The optimisation of AdaBoost is sometimes by means of adaptive methods, such as gradient decent, and therefore, AdaBoost and *gradient boosting* are very similar.

### 9.3.3 Linear Discriminant Analysis

LDA is a method used to find a linear combination of features which characterises or separates two or more classes of objects or events. The resulting combination may be used as a linear classifier. In an LDA it is assumed that the classes have normal distributions. Like PCA, an LDA is used for both dimensionality reduction and data classification.

In a two-class dataset, given the a priori probabilities for class 1 and class 2, respectively, as $p_1$ and $p_2$, class means and overall mean, respectively, as $\mu_1$, $\mu_2$, and $\mu$, and the class variances as $cov_1$ and $cov_2$.

$$\mu = p_1 \times \mu_1 + p_2 \times \mu_2 \tag{9.10}$$

Then, within-class and between-class scatters are used to formulate the necessary criteria for class separability. Within-class scatter is the expected covariance of each of the classes. The scatter measures for multiclass case are computed as:

$$S_w = \sum_{j=1}^{C} p_j \times cov_j \tag{9.11}$$

where $C$ refers to the number of classes and:

$$cov_j = (x_j - \mu_j)(x_j - \mu_j)^T \tag{9.12}$$

Slightly differently, the between-class scatter is estimated as:

$$S_b = \frac{1}{C}\sum_{j=1}^{C}(x_j - \mu_j)(x_j - \mu_j)^T \tag{9.13}$$

Then, the objective would be to find a discriminant plane, $\mathbf{w}$, to maximise the ratio between between-class and within-class scatters (variances):

$$J_{LDA} = \frac{\mathbf{w}S_b\mathbf{w}^T}{\mathbf{w}S_w\mathbf{w}^T} \tag{9.14}$$

In practice, class means and covariances are not known but they can be estimated from the training set. In the above equations, either the maximum likelihood estimate or the maximum a posteriori estimate may be used instead of the exact values.

### 9.3.4 Support Vector Machines

Amongst all supervised classifiers, SVM is very popular for many applications whilst outperforming them in many linear (separable) and nonlinear (nonseparable) cases. The SVM concept was introduced by Vapnik in 1979 [27] and explored in many publications such as [28–32]. To understand the SVM concept, consider a binary classification for the simple case of a 2D feature space of linearly separable training samples $S = \{(x_1, y_1), (x_2, y_2), \ldots, (x_m, y_m)\}$ (Figure 9.4), where $\mathbf{x} \in R^d$ is the input vector and $\mathbf{y} \in \{1, -1\}$ is the class label. A discriminating function is defined as:

$$f(\mathbf{x}) = sgn(\langle \mathbf{w}, \mathbf{x} \rangle + b) = \begin{cases} +1 & \mathbf{x} \in \textit{first class} \\ -1 & \mathbf{x} \in \textit{second class} \end{cases} \tag{9.15}$$

In this formulation $\mathbf{w}$ determines the orientation of a discriminant plane (or hyperplane). Clearly, there are an infinite number of possible planes that could correctly classify the training data. An optimal classifier finds the hyperplane for which the best generalising hyperplane is equidistant or farthest from each set of points. Optimal separation is achieved when there is no separation error and the distance between the closest vector and the hyperplane is maximal.

One way to find the separating hyperplane in a separable case is by constructing the so-called convex hulls of each dataset. The encompassed regions are the convex hulls for the datasets. By examining the hulls one can then determine the closest two points lying on the hulls of each class (note that these do not necessarily coincide with actual data points).

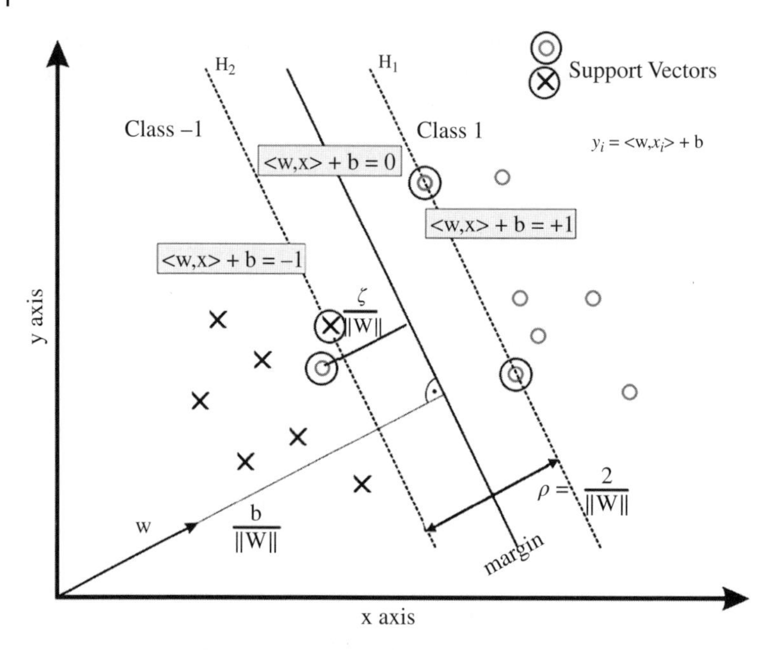


**Figure 9.4** The SVM separating hyperplane and support vectors for a separable data case.

By constructing a plane perpendicular and equivalent to these two points, an optimal hyperplane with robust classifier should result.

For an optimal separating hyperplane design, often few points, referred to as support vectors (SVs), are utilised (e.g. the three circled black data feature points in Figure 9.4).

SVM formulation starts with the simplest case: linear machines are trained on separable data (for general case analysis, nonlinear machines trained on nonseparable data result in a very similar quadratic programming problem). Then, the training data $\{\boldsymbol{x}_i, \boldsymbol{y}_i\}$, $\boldsymbol{i} = \mathbf{1}, \ldots, \boldsymbol{m}$, $\boldsymbol{y}_i \in \{\mathbf{1}, -\mathbf{1}\}, \mathbf{x}_i \in \boldsymbol{R}^d$ are labelled. Consider having a hyperplane which separates the positive from the negative examples. Points x lying on the hyperplane satisfy $\langle \mathbf{w}, \mathbf{x} \rangle + \boldsymbol{b} = \mathbf{0}$, where w is normal to the hyperplane, $\boldsymbol{b}/\|\mathbf{w}\|_2$ is the perpendicular distance from the hyperplane to the origin, and $\|\mathbf{w}\|_2$ is the Euclidean norm of $\mathbf{w}$. Define the 'margin' of a separating hyperplane illustrated in Figure 9.4, and for the linearly separable case, the algorithm simply looks for the separating hyperplane with largest margin. Here, the approach is to reduce the problem to a convex optimisation problem by minimising a quadratic function under linear inequality constraints. To find a plane farthest from both classes of data, the margin between the supporting canonical hyperplanes for each class is maximised. The support planes are pushed apart until they meet the closest data points, which are then deemed to be the support vectors (circled in Figure 9.4). Therefore, the SVM problem to find $\mathbf{w}$ is stated as:

$$\langle \mathbf{x}_i, \mathbf{w} \rangle + b \geq +1 \ \textit{for} \ y_i = +1 \tag{9.16a}$$

$$\langle \mathbf{x}_i, \mathbf{w} \rangle + b \geq -1 \ \textit{for} \ y_i = -1 \tag{9.16b}$$

which can be combined into one set of inequalities as $y_i(\langle \mathbf{x}_i, \mathbf{w} \rangle + b) - 1 \geq 0 \, \forall \, i$. The margin between these supporting planes ($H_1$ and $H_2$) can be shown to be $\gamma = 2/\|\mathbf{w}\|_2$. Therefore, to maximise this margin we therefore need to:

$$\begin{aligned}&\text{minimise } \langle \mathbf{w}, \mathbf{w}\rangle \\ &\text{subject to } y_i(\langle \mathbf{x}_i, \mathbf{w}\rangle + b) - 1 \geq 0 \qquad i = 1, \dots, m\end{aligned} \tag{9.17}$$

This constrained optimisation problem can be changed into an unconstrained problem by using Lagrange multipliers. This leads to minimisation of an unconstrained empirical risk function (Lagrange), which consequently results in a set of conditions called Kuhn–Tucker (KT) conditions. The new optimisation problem brings about the so-called primal form as:

$$L(\mathbf{w}, b, \boldsymbol{\alpha}) = \frac{1}{2}\langle \mathbf{w}, \mathbf{w}\rangle - \sum_{i=1}^{m} \alpha_i [y_i(\langle \mathbf{x}_i, \mathbf{w}\rangle + b) - 1] \tag{9.18}$$

where the $\alpha_i$, $i = 1, \dots, m$ are the Lagrange multipliers. The Lagrange primal has to be minimised with respect to $\mathbf{w}$, b and maximised with respect to $\alpha_i \geq 0$. Constructing the classical Lagrange dual form facilitates this solution. This is achieved by setting the derivatives of the primal to zero and re-substituting them back into the primal. Hence, the dual form is derived as:

$$L(\mathbf{w}, b, \boldsymbol{\alpha}) = \sum_{i=1}^{m} a_i - \frac{1}{2}\sum_{i=1}^{m}\sum_{j=1}^{m} y_i y_j \alpha_i \alpha_j \langle \mathbf{x}_i \cdot \mathbf{x}_j\rangle \tag{9.19}$$

and:

$$\mathbf{w} = \sum_{i=1}^{m} y_i \alpha_i \mathbf{x}_i \tag{9.20}$$

considering that $\sum_{i=1}^{m} y_i \alpha_i = 0$ and $\alpha_i \geq 0$. These equations can be solved using many different publicly available quadratic programming (QP) algorithms such as those proposed at www.support-vector.net and www.kernel-machines.org.

In nonseparable cases where the classes have overlaps in the feature space, the maximum margin classifier described above is no longer applicable. With the help of some complicated mathematical derivations, we are often able to define a nonlinear hyperplane to accurately separate the datasets. As we will see later, this causes an overfitting problem which reduces the robustness of classifier. The ideal solution where no points are misclassified and no points lie within the margin is no longer feasible. This implies that we need to relax the constraints to allow for a minimum of misclassifications. In this case, the points that subsequently fall on the wrong side of the margin are considered errors. However, they have less influence on the location of the hyperplane (according to a pre-set *slack* variable) and as such are considered to be support vectors. The classifier obtained in this way is called a *soft margin classifier*.

To optimise the soft margin classifier, we must allow violation of the margin constraints according to a pre-set *slack* variable $\xi_i$ in the original constraints, which then become:

$$\langle \mathbf{x}_i, \mathbf{w}\rangle + b \geq +1 - \xi_i \ \textit{for}\ y_i = +1 \tag{9.21a}$$

$$\langle \mathbf{x}_i, \mathbf{w}\rangle + b \geq -1 + \xi_i \ \textit{for}\ y_i = -1 \tag{9.21b}$$

and $\xi_i \geq 0 \, \forall \, i$

For an error to occur, the corresponding $\xi_i$ must exceed unity. Therefore, $\sum_{i=1}^{m} \xi_i$ is an upper bound on the number of training errors. Hence, a natural way to assign an extra cost for errors is to change the objective function to:

$$\text{minimise}\langle \mathbf{w}, \mathbf{w} \rangle + C\sum_{i=1}^{m} \xi_i$$
$$\text{subject to } y_i(\langle \mathbf{x}_i, \mathbf{w} \rangle + b) \geq 1 - \xi_i \text{ and } \xi_i \geq 0 \quad \text{for } i = 1, \ldots, m \tag{9.22}$$

The primal form will then be:

$$L(\mathbf{w}, b, \xi, \boldsymbol{\alpha}, \mathbf{r}) = \frac{1}{2}\langle \mathbf{w}, \mathbf{w} \rangle + C\sum_{i=1}^{m} \xi_i - \sum_{i=1}^{m} \alpha_i[y_i(\langle \mathbf{x}_i, \mathbf{w} \rangle + b) - 1 + \xi_i] - \sum_{i=1}^{m} r_i \xi_i \tag{9.23}$$

Hence, by differentiating this cost function with respect to $\mathbf{w}$, $\xi$, and $b$ we can achieve:

$$\mathbf{w} = \sum_{i=1}^{m} y_i \alpha_i \mathbf{x}_i \tag{9.24}$$

and:

$$\alpha_i + r_i = C \tag{9.25}$$

By substituting these into the primal form and again considering that $\sum_{i=1}^{m} y_i \alpha_i = 0$ and $\alpha_i \geq 0$, the dual form is derived as:

$$L(\mathbf{w}, b, \xi, \boldsymbol{\alpha}, \mathbf{r}) = \sum_{i=1}^{m} a_i - \frac{1}{2}\sum_{i=1}^{m}\sum_{j=1}^{m} y_i y_j \alpha_i \alpha_j \langle \mathbf{x}_i \cdot \mathbf{x}_j \rangle \tag{9.26}$$

This is similar to the maximal marginal classifier with the only difference in that here we have a new constraint of $\alpha_i + r_i = C$, where $r_i \geq 0$, hence $0 \leq \alpha_i \leq C$. This implies that the value $C$ sets an upper limit on the Lagrange optimisation variables $\alpha_i$. This is sometimes referred to as the box constraint. The value of $C$ offers a trade-off between accuracy of data fit and regularisation. A small value of $C$ (i.e. $C < 1$) significantly limits the influence of error points (or outliers), whereas if $C$ is chosen to be very large (or infinite) then the soft margin (as in Figure 9.5) approach becomes identical to the maximal margin classifier. Therefore, in the use of a soft margin classifier, the choice of $C$ strongly depends on the data. Appropriate selection of $C$ is important and itself is an area of research. One way to set $C$ is by gradually increasing $C$ from max $(\alpha_i)$ for $\forall i$ and find the value for which the error (outliers, cross validation, or number of misclassified points) is minimum. Eventually, $C$ can be found empirically [33].

There will be no change in formulation of the SVM for multidimensional cases, only the dimension of the hyperplane changes depending on the number of feature types.

In many nonseparable cases the use of a nonlinear function may help to make the datasets separable. As can be seen in Figure 9.6, the datasets are separable if a nonlinear hyperplane is used. *Kernel mapping* offers an alternative solution by nonlinearly projecting the data into a (usually) higher dimensional feature space to allow the separation of such cases.

The key success to Kernel mapping is that special types of mapping that obey Mercer's theorem, sometimes called reproducing kernel Hilbert spaces (RKHSs) [27], offer an

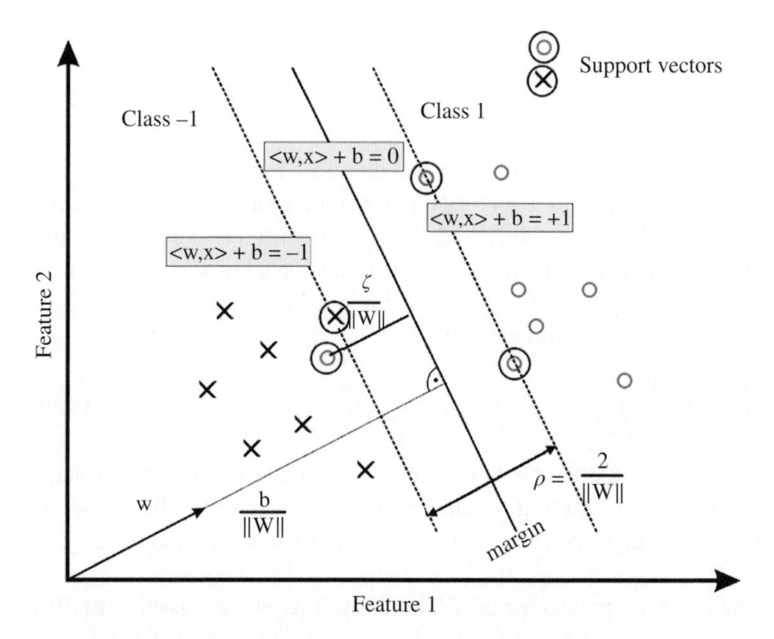


**Figure 9.5** Soft margin and the concept of slack parameter.

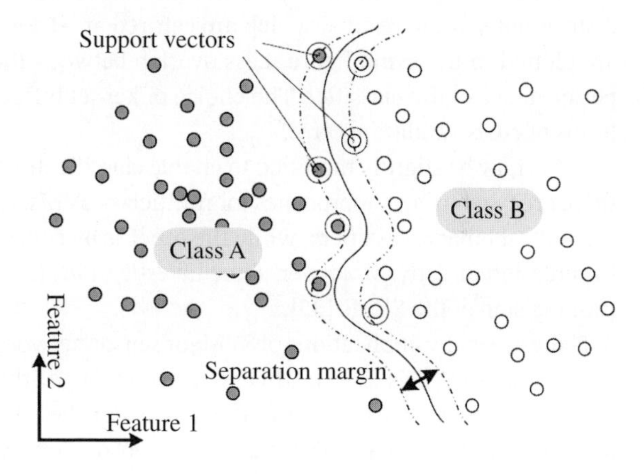


**Figure 9.6** Nonlinear discriminant hyperplane (separation margin) for SVM.

implicit mapping into feature space:

$$K(\mathbf{x}, \mathbf{z}) = \langle \varphi(\mathbf{x}), \varphi(\mathbf{z}) \rangle \tag{9.27}$$

This implies that there is no need to know the explicit mapping in advance, rather the inner-product itself is sufficient to provide the mapping. This simplifies the computational burden significantly and in combination with the inherent generality of SVMs largely alleviates the dimensionality problem. Moreover, this means that the input feature inner-product can simply be substituted with the appropriate Kernel function to obtain the mapping whilst having no effect on the Lagrange optimisation theory. Hence:

$$L(\mathbf{w}, b, \xi, \boldsymbol{\alpha}, \mathbf{r}) = \sum_{i=1}^{m} a_i - \frac{1}{2} \sum_{i=1}^{m} \sum_{j=1}^{m} y_i y_j \alpha_i \alpha_j K(\mathbf{x}_i \cdot \mathbf{x}_j) \tag{9.28}$$

The relevant classifier function then becomes:

$$f(\mathbf{x}) = sgn\left(\sum_{i=1}^{nSVs} y_i a_i K(\mathbf{x}_i \cdot \mathbf{x}_j) + b\right) \tag{9.29}$$

In this way all the benefits of the original linear SVM method are maintained. We can train a highly nonlinear classification function such as a polynomial or a radial basis function (RBF), or even a sigmoidal NN, using a robust and efficient algorithm that does not suffer from local minima. The use of Kernel functions transforms a simple linear classifier into a powerful and general nonlinear classifier [33]. RBF is the most popular nonlinear kernel used in SVM for nonseparable classes and defined as:

$$K(u, v) = exp\left(\frac{\|u - v\|_2^2}{2\sigma^2}\right) \tag{9.30}$$

where $\sigma^2$ is the variance. As mentioned previously, a very accurate nonlinear hyperplane, estimated for a particular training dataset, is unlikely to generalise well. This is mainly because the system may no longer be robust since a testing or new input can easily be misclassified.

Another issue related to the application of SVMs (as well as other classifiers) is the cross-validation problem. The classifier output distribution, without the hard limiter 'sign' in Eq. (9.24), for a number of inputs for each class may be measured. The probability distributions of the results (which are centred at −1 for class '−1' and at +1 for class '+1') are plotted in the same figure. Less overlap between the distributions represents a better performance of the classifier. The choice of kernel influences the classifier performance in terms of cross-validation error.

SVMs may be slightly modified to enable classification of multiclass data [34]. Currently, there are two types of approaches for multiclass SVMs. One is by constructing and combining several binary classifiers, while the other is by considering all data in one optimisation formulation directly. Moreover, some investigations have been undertaken to speed up the training step of the SVMs [35].

There are many applications of SVM for sensor networks. SVM has been utilised for target localisation in WSNs [36]. In this research, the algorithm partitions the sensor field using a fixed number of classes. The authors consider the problem of estimating the geographic locations of nodes in a WSN where most sensors are without an effective self-positioning functionality. Their SVM-based algorithm, called localised support vector machine (LSVM), localises the network merely based on connectivity information, addresses the border and coverage-hole problems, and performs in a distributed manner.

Yoo and Kim [37] introduced a semi-supervised online SVM, also called support vector regression (SVR), to alleviate the sensitivity of the classifier to noise and the variation in target localisation using multiple wireless sensors. This is achieved by combining the core concepts of manifold regularisation and the supervised online SVR.

Despite its excellent performance and very wide range applications, SVM doesn't perform very well for large-scale samples and imbalanced data classes. Natural data are often imbalanced and consist of multiple categories or classes. Learning discriminative models from such datasets is challenging due to lack of representative data and the bias of traditional classifiers towards the majority class. Many attempts have been made to alleviate this problem. Sampling methods such as synthetic minority oversampling technique (SMOTE) has been one of the popular methods in tackling this problem. Mathew et al. [38] tried to solve this problem by proposing a weighted kernel-based SMOTE that overcame the limitation of SMOTE for nonlinear problems by oversampling in the SVM feature space. Compared to other baseline methods on multiple benchmark imbalanced datasets, their algorithm along with a cost-sensitive SVM formulation was shown to improve the performance. In addition, a hierarchical framework with progressive class order has been developed for multiclass imbalanced problems.

Kang et al. [39] propose a weighted undersampling (WU) methodology for SVM based on space geometry distance. In their algorithm, the majority of samples were grouped into some subregions and different weights were assigned to them according to their Euclidean distance to the hyperplane. The samples in a subregion with higher weight have more chance to be sampled and used in each learning iteration. This retains the data distribution information of original datasets as much as possible.

### 9.3.5 *k*-nearest Neighbour

*k*-nearest neighbour (*k*NN) is another supervised learning algorithm which classifies a data sample (called a query point) based on the labels (i.e. the output values) of the near data samples. For example, missing readings of a sensor node can be predicted using the average measurements of neighbouring sensors within specific diameter limits. There are several functions to determine the nearest set of nodes. One simple method is to use the Euclidean distance between the signals from different sensors. *k*NN has low computational cost since the function is computed relative to the local points (i.e. *k*-nearest points, where $k$ is a small positive integer). This factor coupled with the correlated readings of neighbouring nodes makes *k*NN a suitable distributed learning algorithm for WSNs. It has been shown that the *k*NN algorithm may provide inaccurate results when analysing problems with high-dimensional spaces (more than 10–15 dimensions) as the distance to different data samples becomes invariant (i.e. the distances to the nearest and farthest neighbours are slightly similar) [40].

One important application of the *k*NN algorithm for WSN is in the query processing subsystem; see for example [41].

### 9.3.6 Gaussian Mixture Model

GMM is mainly used for estimating the observation probability given the features and follows the Bayes formula as:

$$p(object\ property \mid context) = p(O \mid v) = \frac{p(v \mid O)}{p(v)} p(O) \tag{9.31}$$

where $p(v)$ can easily be calculated as:

$$p(v) = p(v \mid O)p(O) + p(v \mid \overline{O})p(\overline{O}) \tag{9.32}$$

where $p(v \mid O)$ is the likelihood of the features given the observation (presence of an object) and $p(v \mid \overline{O})$ is the likelihood of the features when the object is absent. GMM is then used to estimate these likelihoods as:

$$p(v \mid O) = \sum_{i=1}^{M} w_i \cdot G(v; \mu_i, \Sigma_i) \tag{9.33}$$

which is the sum of weighted Gaussians with unknown means and variances. The parameters $w_i$, $\mu_i$, and $\Sigma_i$ have to be estimated using a kind of optimisation technique. Often expectation maximisation is used for this purpose [42].

### 9.3.7 Logistic Regression

Logistic regression is generally to estimate the probability of an observation subject to an attribute. It estimates the observation probability from the given features. A logistic function $F(v)$ is computed from the features as:

$$Logit = log\frac{p(O|v)}{p(\overline{O}|v)} = F(v) \tag{9.34}$$

where $F(v)$ depends on the sequence of attributes (i.e. previous experiences):

$$F(v) = a_0 + \sum_{i=1}^{D} a_i \cdot v(i) \tag{9.35}$$

and for testing:

$$p(O \mid v) = \frac{1}{1 + e^{-F(v)}} \tag{9.36}$$

To give an example, consider:

Observation $O$: suffering from back pain & $\overline{O}$: not suffering from back pain
Attribute $v$: age:

$$log\frac{p(O \mid v)}{p(\overline{O} \mid v)} = a_0 + a_1 \cdot (age - 20)$$

During the training stage $a_0$ (the log odds for a 20-year-old person) and $a_1$ (the log odds ratio when comparing two persons who differ by one year in age) are estimated. In the testing stage the probability of having back pain given the age is estimated as:

$$p(O \mid age) = \frac{1}{1 + e^{-a_0 + a_1(age-20)}}$$

### 9.3.8 Reinforcement Learning

In reinforcement learning an agent learns a good behaviour. This means that it modifies or acquires new behaviours and skills incrementally [43]. It exploits the local information as well as the restrictions imposed on the WSN application to maximise the influences of numerous tasks over a time period. Therefore, during the process of reinforcement learning, the machine learns by itself based on the penalties and rewards it receives in order to decide on its next step action. Thus, the learning process is sequential. With this algorithm, each WSN node discovers the minimal needed resources to carry out its routine tasks with benefits allocated by the Q-learning technique. This technique is mainly used to assess the value or quality of action taken by the learning system (embedded in each sensor for distributed cases) for the estimation of the award to be given [44] and consequently undertaking the next step.

### 9.3.9 Artificial Neural Networks

ANNs are other supervised nonlinear machine learning algorithms constructed by cascading chains of decision units (neurons followed by nonlinear functions and decision-makers) used to recognise nonlinear and complex functions [10]. In WSNs, using NNs in distributed manners is still not so pervasive due to the high computational requirements for learning the network weights and the high management overhead. However, in centralised solutions, NNs can learn multiple outputs and decision boundaries at once [45], which makes them suitable for solving several network challenges using the same model.

As seen in Figure 9.6, each ANN consists of input and output layers plus a number of hidden layers. Each layer contains a number of neurons and each neuron performs a nonlinear operation on its input. Hard limiter (sign function), sigmoid, exponential, and tangent hyperbolic (tanh) functions are the most popular ones.

The major challenges in using ANNs are their optimisation (i.e. estimation of the link weights) and selection of the number of layers and neurons. Backpropagation algorithms are often used in multilayer ANNs [46, 47]. This is very similar to the optimisation of adaptive filter coefficients, where the error between the achieved label (output) and the desired label is minimised in order to obtain the best set of link weights. The computational cost exponentially increases with the numbers of neurons and layers.

In an example of NN application for WSN node localisation, the input variables are the propagating angle and distance measurements of the received signals from anchor nodes [48]. Such measurements may include received signal strength indicator (RSSI), time of arrival (TOA), and time difference of arrival (TDOA), as illustrated in Figure 9.7. After supervised training, the NN generates an estimated node location as vector-valued coordinates in 3D space. The associated NN algorithms include self-organising map (or Kohonen maps) and learning vector quantisation (LVQ) (see [49] and references therein for an introduction to these methods). In addition to function estimation, one important application of NNs is in big data (high-dimensional and complex dataset) feature detection, classification, and dimensionality reduction [50].

Using backpropagation algorithm to train an ANN, the error between the output of each output neuron is compared with a target value (or the centre of a cluster) in order to find the link weights between various layers. For the two-layer network of Figure 9.7, assuming $z_k$ is the output of neuron $k$ and $t_k$ the target, the cost is defined as:

$$J(\mathbf{w}) = \frac{1}{2}\sum_{k=1}^{C}(t_k - z_k)^2 \tag{9.37}$$

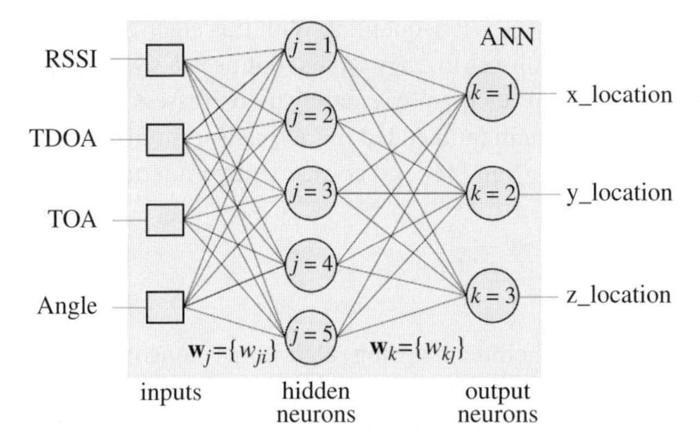


**Figure 9.7** A simple three-layer NN for node localisation in WSNs in 3D space.

where $C$ represents the number of outputs (classes). Estimation of the output, $z_k$, is straightforward; assuming the input vector to the ANN is $\mathbf{x}$, the output of neuron $k$ is:

$$z_k = f(\mathbf{w}_k^T f(\mathbf{w}_j^T \mathbf{x})) = f\left(\sum_{j=1}^{n_1} w_{kj} f\left(\sum_{i=1}^{n_2} w_{ji} x_i + w_{j0}\right) + w_{k0}\right) \tag{9.38}$$

In the above equation, $n_1$ and $n_2$ are the number of neurons in the first and second layers respectively, $w_{j0}$ and $w_{k0}$ are bias values (not shown in Figure 9.7), and $f$ ($y$), the neuron activation function is a continuous function which best approximates a hard limiter often called as rectified linear unit (ReLU). A popular example of such a function is the following exponential function:

$$f(y) = \frac{1}{1 + e^{-y}} \tag{9.39}$$

which looks like the curve in Figure 9.8. Many other functions can also be used, as stated previously. Different optimisation techniques such as MSE minimisation [51] can be used to minimise $J(\mathbf{w})$ and find the optimum $w_{ji}$ and $w_{kj}$ values.

#### 9.3.9.1 Deep Neural Networks

In parallel with the development of powerful computers and accessing to local and remote memory clusters as well as the cloud, DNNs have become widely popular. These classifiers often have a larger number of neurons and layers and a greater ability to learn, are more scalable, and have further processing capability on their layers. Unlike in the past, currently, large data can be processed using DNNs in real-time. Deep learning using DNNs allows computational models that are composed of multiple processing layers to learn data representations with multiple abstraction levels.

With the advancement of deep learning algorithms, developers, analysts, and decision-makers can explore and learn more about the data and their exposed relationships or hidden features. The new practices in developing data-driven application systems and decision making algorithms seek adaptation of deep learning algorithms and techniques in many application domains. In data-driven learning applications, the data govern the system behaviour.

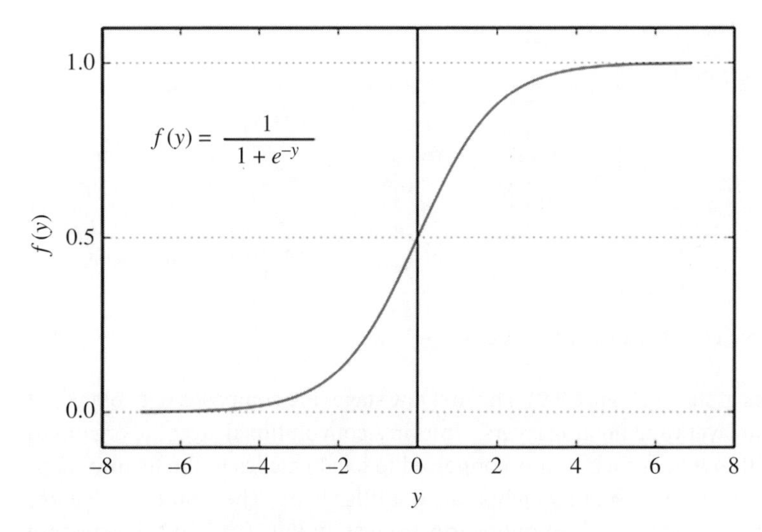


**Figure 9.8** An exponential activation function (ReLU).

The availability of deep structure as well as data-driven models paves the path for developing a new generation of deep networks called generative models. Unlike discriminative networks, which estimate a label for a test data, the generative networks assume they have the class label, and they wish to find the likelihood of particular features. They are often called generative adversarial networks (GANs) and include two nets, pitting one against the other (thus the 'adversarial'). GANs were introduced by Goodfellow et al. [52] and were called the most interesting idea in machine learning in the last 10 years be LeCun, Facebook's AI research director.

In GANs, the generative models are designed via an adversarial process, in which two models are trained simultaneously: a generative model G that captures the data distribution, and a discriminative model D that estimates the probability that a sample is taken from the training data rather than G. To train the G model the probability of D making a mistake is maximised. This framework resembles a minimax two-player game. For arbitrary functions G and D, a unique solution exists, with G recovering the training data distribution and D equal to 0.5 everywhere. In the case where G and D are defined by multilayer perceptrons, the entire system can be trained with a backpropagation algorithm [52].

One particular type of deep feedforward network is the convolutional neural network (CNN) [53, 54]. This network is easier to train and can generalise much better than networks with full connectivity between adjacent layers. It often performs better in terms of both speed and accuracy for many computer vision and video processing applications. Recently, CNNs have found many other applications, such as for the detection of interictal epileptiform discharges from intracranial EEG recordings [55, 56].

#### 9.3.9.2 Convolutional Neural Networks

CNNs are designed to process multiple array data such as colour images composed of three 2D arrays containing pixel intensities in the three colour channels. These networks have four main features that benefit from the properties of natural signals, namely local connections, shared weights, pooling, and the use of numerous layers. Typical CNN architecture is

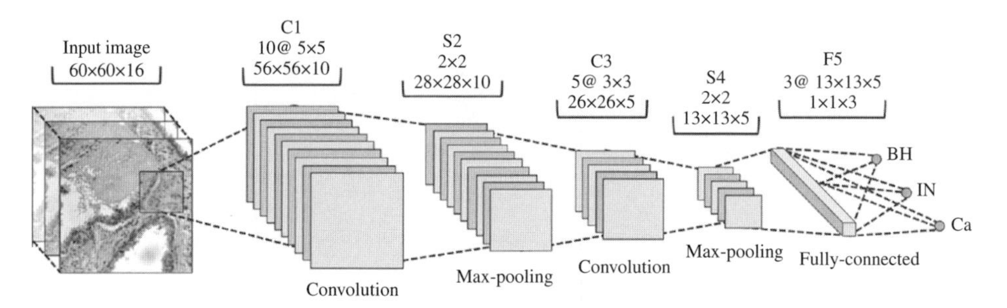


**Figure 9.9** An example of a CNN and its operations.

structured as a series of stages (Figure 9.9). The first few stages are composed of two types of layers: convolutional layers and pooling layers. Units in a convolutional layer are organised in feature maps, within which each unit is connected to local patches in the feature maps of the previous layer through a set of weights called a filter bank. The result of this local weighted sum is then passed through a nonlinearity such as an ReLU. All units in a feature map share the same filter bank. In a conventional CNN different feature maps in a layer use different filter banks. The reason for this architecture is twofold. First, in array data such as images, local groups of values are often highly correlated, forming distinctive local motifs that are easily detected. Second, the local statistics are invariant to location. In other words, if a motif can appear in one part of the image, it may appear anywhere in the image. Hence, we can have units at different locations sharing the same weights and detecting the same pattern in different parts of the array. From a mathematical point of view, the filtering operation performed by a feature map is a discrete convolution, hence the name.

Although the role of a convolutional layer is to detect local conjunctions of features from the previous layer, the role of the pooling layer is to merge semantically similar features into one. Since the relative positions of the features forming a motif can vary somewhat, its reliable detection can be performed by coarse-graining the position of each feature. A typical pooling unit computes the maximum of a local patch of units in one feature map (or in a few feature maps).

The neighbouring pooling units take input from patches that are shifted by more than one row or column, thereby reducing the representation dimension and creating an invariance to small shifts and distortions.

Stages of convolution, nonlinearity, and pooling are followed by more convolutional and fully connected layers. The backpropagation operation is then used for weight optimisation in a CNN as for the normal multilayer NN.

DNNs exploit the property that many natural signals follow compositional hierarchies. Based on this property, the higher-level features are obtained by composing lower-level ones. In images, local combinations of edges form motifs, motifs assemble into parts, and parts form objects. Similar hierarchies exist in speech and text from sounds to phones, phonemes, syllables, words, and sentences. The pooling allows representations to vary very little when elements in the previous layer vary in position and appearance.

The convolutional and pooling layers in CNNs are directly inspired by the classic notions of simple cells and complex cells in visual neuroscience [54], and the overall architecture is reminiscent of the LGN–V1–V2–V4–IT hierarchy (where LGN stands for lateral geniculate nucleus) in the visual cortex ventral pathway [57].

When the CNN models and monkeys are projected/shown the same picture, the activation of high-level units in the CNN explains half of the variance of random sets of 160 neurons in the monkeys' inferotemporal cortex [58]. CNNs have their roots in the neocognitron [59], the architecture of which is somewhat similar, but does not have an end-to-end supervised-learning algorithm such as backpropagation. Although CNNs were initially developed for image classification, an effective 1D CNN was developed by Antoniades et al. [55, 56] for the detection of interictal epileptiform discharges from intracranial EEG recordings.

#### 9.3.9.3 Recent DNN Approaches

One of the early introduced DNNs was LeNet proposed by LeCun in 1988 for hand digit recognition [60]. LeNet is a feed-forward NN constituted of five consecutive layers of convolutional and pooling, followed by two fully connected layers. A later DNN approach was AlexNet [61]. This approach, proposed by Krizhevsky et al., is considered the first deep CNN architecture which showed ground-breaking results for image classification and recognition tasks.

Learning mechanism of CNN was largely based on hit-and-trial, without a deep understanding of the exact reason behind the improvement before 2013. This lack of understanding limited the performance of deep CNNs on complex images. In 2013, Zeiler and Fergus [62] proposed a multilayer deconvolutional NN (DeconvNet), which became known as ZefNet to quantitatively visualise the network performance.

A deeper network, namely visual geometry group (VGG), was later proposed [63]. VGG suggested that the parallel placement of small size filters makes the receptive field as effective as that of large size filters. GoogleNet (also known as Inception-V1) is another CNN which won the 2014-ILSVRC (ImageNet Large-Scale Visual Recognition Challenge) competition [64]. The main objective of the GoogleNet architecture was to achieve high accuracy with a reduced computational cost. It introduced the new concept of inception module (block) in CNN, whereby multiscale convolutional transformations are incorporated using split, transform, and merge operations for feature extraction. Other deep networks such as ResNet [65], DenseNet [66], and many other DNN structures were later introduced for enhancing the accuracy of feature learning and classification.

In large NNs, model pruning seeks to induce sparsity in a DNN's various connection matrices, thereby reducing the number of nonzero-valued parameters in the model. Weights in an NN that are considered unimportant or rarely fire can be removed from the network with little or no consequence. Often, many neurons have a relatively small impact on the model performance, meaning we can achieve acceptable accuracy even when eliminating a large number of parameters. Reducing the number of parameters in a network becomes increasingly important as neural architectures and datasets becoming larger in order to obtain reasonable execution times of models.

Among NNs, recurrent neural networks (RNNs) and long short-term memory network (LSTM) rely on their previous states and therefore can learn state transitions for detection or recognition of particular/desired trends within the data. RNNs can be also used for prediction purposes. A shallow RNN, however, is not capable of using long-term dependencies between the data samples. LSTMs are a special kind of RNNs, which are capable of learning long-term dependencies. They were introduced by Hochreiter and Schmidhuber [67] and were refined and popularised by many people, such as Xu et al. [68]. Unlike RNN, which has only a single layer in each repeating module, the repeating module in an LSTM contains four interacting layers. A similar idea has been followed and explored for speech data generation. WaveNet DNN, as a probabilistic and autoregressive DNN approach, has been designed and applied for speech waveform generation [69]. This architecture exploits the long-range temporal dependencies needed for raw audio generation.

Despite heavy DNN computational cost, some DNN structures such as ENet (efficient neural network), have been proposed to enable real-time applications. ENet, meant for spatial classification and segmentation of images, uses a significantly smaller number of parameters.

### 9.3.10 Gaussian Processes

Gaussian processes (GPs) [70] are other algorithms for solving *regression* and *probabilistic classification* problems. These supervised classifiers, though not effective for large dimension data, use prediction to smooth the observation and are versatile enough to use different kernels. Therefore, they are suitable for recovering and classifying biomedical signals which often have a regular structure.

The GP approach performs inference using the noisy, potentially artefactual, data obtained from wearable sensors. Of fundamental importance is the GP notion as a distribution over functions, which is well suited to the analysis of the time series of patients' physiological data, in which the inference over functions may be performed. This approach contrasts with conventional probabilistic approaches which define distributions over individual data points.

As an example, a GP was used in an e-health platform for personalised healthcare to develop a patient-personalised system for the analysis and inference in the presence of data uncertainty [71]. This uncertainty is typically caused by sensor artefact and data incompleteness. The method was used for the clinical study and monitoring of 200 patients [71]. This provides the evidence that personalised e-health monitoring is feasible within an actual clinical environment, at scale, and that the method is capable of improving patient treatment outcome via personalised healthcare.

In another example a Bayesian Gaussian process logistic regression (GP-LR) with linear and nonlinear covariance functions has been used for the classification of Alzheimer's disease and mild cognitive impairment from resting-state fMRI [72]. These models can be interpreted as a Bayesian probabilistic system analogue to kernel SVM classifiers. However, GP-LR methods confer some benefits over kernel SVMs. Whilst SVMs only return a binary class label prediction, GP-LR, being a probabilistic model, provides a principled estimate of the probability of class membership. Class probability estimates are a measure of confidence the model has in its predictions. Such a confidence score can be very useful in the clinical setting.

### 9.3.11 Neural Processes

Garnelo et al. [73] introduced a class of neural latent variable models, namely neural processes (NPs), by combining GPs and NNs. Like GPs, NPs define distributions over functions, are capable of rapid adaptation to new observations, and can estimate the uncertainty in their predictions. Like NNs, NPs are computationally efficient during training and evaluation but also learn to adapt their prior probability distributions.

NPs perform regression by learning to map a context set of observed input–output pairs to a distribution over regression functions. Each function models the distribution of the output given an input, conditioned on the context. NPs have the benefit of fitting observed data efficiently with linear complexity in the number of context input–output pairs. They are able to learn a wide family of conditional distributions; they learn predictive distributions conditioned on context sets of arbitrary size.

Nonetheless, it has been shown that NPs suffer underfitting, giving inaccurate predictions at the inputs of the observed data they condition on [74]. This problem has been addressed by incorporating attention into NPs, allowing each input location to attend to the relevant context points for the prediction. It has been shown that this greatly improves the accuracy of predictions, speeds up the training process, and expands the range of functions that can be modelled [74].

### 9.3.12 Graph Convolutional Networks

Graph convolutional networks (GCNs) are powerful NNs which enable machine learning on graphs. Even small-size GCNs can produce useful feature representations of nodes in the network. In a GCN structure, the relative proximity of the network nodes is preserved in the 2D representation even without any training. There are two different inputs to a GCN: one is the input feature matrix and the other a matrix representation of the graph structure, such as the adjacency matrix of the graph. As an example, consider the classification of EEG signals for healthy and dementia subjects. Using a GCN, the brain connectivity estimates can be incorporated into the training process as the proximity matrix.

As an example, for a two-layer GCN the following relation between the input and output can be realised [75]:

$$z = f(\mathbf{X}, \mathbf{A}) = softmax(\hat{\mathbf{A}} ReLU(\hat{\mathbf{A}}\mathbf{X}\mathbf{W}^{(0)})\mathbf{W}^{(1)}) \tag{9.40}$$

where $\mathbf{A}$ is the adjacency matrix, $\hat{\mathbf{A}} = \tilde{\mathbf{D}}^{-1/2}\tilde{\mathbf{A}}\tilde{\mathbf{D}}^{-1/2}$, $\hat{\mathbf{A}} = \mathbf{A} + \mathbf{I}_N$, $\mathbf{I}_N$ is an $N \times N$ identity matrix, and $\tilde{D}_{ii} = \sum_j \tilde{A}_{ij}$.

### 9.3.13 Naïve Bayes Classifier

Naïve Bayes classification method is a supervised learning algorithm applying Bayes' theorem with the 'naïve' assumption of independence between every pair of features. Therefore, the class label of the data can be estimated using a naïve Bayes classifier through the following procedure.

Given a class variable $y$ and a dependent feature vector $x_1$ through $x_n$, the Bayes theorem states the following relationship:

$$p(y \mid x_1, \dots, x_n) = \frac{p(y)p(x_1, \dots, x_n \mid y)}{p(x_1, \dots, x_n)} \tag{9.41}$$

Using the naïve independence assumption:

$$p(y \mid x_1, \quad \dots, x_n) = \frac{p(y)\prod_{i=1}^{n} p(x_i \mid y)}{p(x_1, \dots, x_n)} \tag{9.42}$$

Since $p(x_1, \ldots, x_n)$ is constant for any given input, the following classification rule can be deduced:

$$p(y \mid x_1, \ldots, x_n) \propto p(y) \prod_{i=1}^{n} p(x_i \mid y) \tag{9.43}$$

The likelihood probability $p(x_i \mid y)$ is considered known either empirically or by assumption. Gaussian, multinomial, and Bernoulli are among the most popular distributions considered for these probabilities. This results in estimation of a class label as:

$$\hat{y} = arg \max_{y} p(y) \prod_{i=1}^{n} p(x_i \mid y) \tag{9.44}$$

Naïve Bayes classifiers are simple and work quite well in many real-world situations. They require a small selection of training data to estimate the necessary parameters [76].

Naïve Bayes learners and classifiers can be extremely fast compared to more computationally intensive methods. The decoupling of the class conditional feature distributions means that each distribution can be independently estimated as a 1D distribution. This in turn helps alleviate the problem of the curse of dimensionality.

### 9.3.14 Hidden Markov Model

HMMs are state-space machines meant to detect a correct data sequence which may happen within a longer data sequence. HMMs are presented as the state diagrams. Therefore, for each application, an HMM has a certain number of states and knows the probability of each possible event happening in each state. As an example, consider the detection of QRS in an electrocardiogram (ECG) signal depicted in Figure 9.10.

From this figure, an ECG is recognised when the symbols turn up in a correct sequence in the signal. Assuming the waveform in Figure 9.10 is the ECG of a healthy subject, the state diagram in Figure 9.11 classifies the subject as healthy as long as the above ECG sequence is detected in his/her ECG record.

In this diagram, state $Q_0$ is a wait state to detect the beginning of sequence starting from a bar (b). State $Q_1$ is a state which shows a bar has been detected. State $Q_2$ simply demonstrates that a bar and a small hump (p) have been already recognised. $Q_3$ represents the

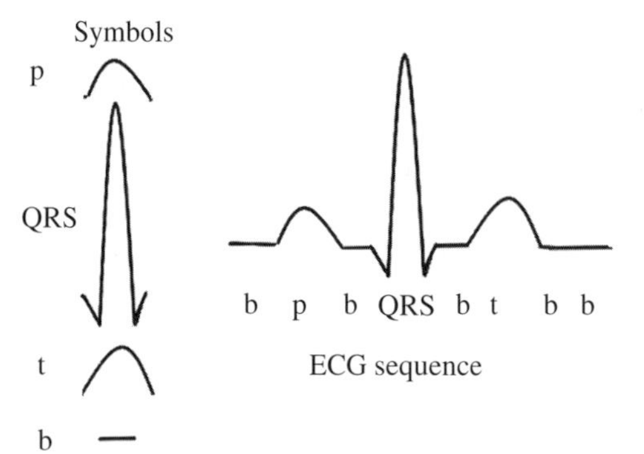


**Figure 9.10** A synthetic ECG segment of a healthy individual and its corresponding symbols.

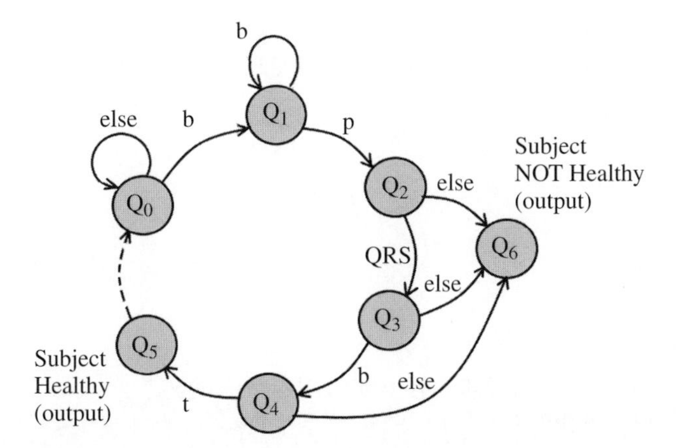


**Figure 9.11** An HMM for the detection of a healthy heart from an ECG sequence.

state of the system when a bar, a small hump, and a QRS wave are correctly sequenced. $Q_4$ and $Q_5$ similarly show the states after a bar and a large hump (t) have been respectively added to the data sequence. After detecting the correct sequence, the output is 'the subject is healthy'. After each state, if a wrong symbol or noise is detected, the system output is 'the subject is not healthy'.

In places where the objective is to detect a particular heart condition, separate HMM should be designed to recognise that condition. Beside their applications in biomedical signal analysis and disease recognition, HMMs have a wide range of applications in speech, video, bioscience (such as genomics), and face recognition.

In order to use an HMM for classification, the initial state ($\boldsymbol{\pi} = \pi_i$, where $i = 1, \ldots, N$, where $N$ is the number of states), the state probabilities ($\boldsymbol{A} = \{a_{ij}\}$, $i,j = 1, \ldots, N$), and the state transition probabilities ($\boldsymbol{B} = \{b_j(k)\}$, $j = 1, \ldots, N$, $k = 1, \ldots, T$, where $T$ is the observation sequence length) should be known. For training the HMMs (i.e. to find the model parameters), however, a set of observations and the number of states should be known.

There are three major problems (objectives) in using or designing HMMs:

- Given the observation sequence $O = o_1, o_2, \ldots, o_T$, and the model $\lambda = (\boldsymbol{A}, \boldsymbol{B}, \boldsymbol{\pi})$, what will be the probability of observation sequence $p(O|\lambda)$?
- Given the observation sequence and the model as above, how can we find the state sequence $\mathrm{Q} = Q_1, \ldots, Q_N$ which generates such an observation sequence?
- Given the observations and the number of states, how can the HMM model be built up, i.e. how can we find $\lambda = (\boldsymbol{A}, \boldsymbol{B}, \boldsymbol{\pi})$?

To solve the first problem, often, the iterative forward–backward algorithm is used, while the Viterbi algorithm is employed to solve the second problem. Finally, quite a few algorithms, such as the most popular one, Baum–Welch (Leonard E. Baum and Lloyd R. Welch, 1960s) [77, 78], can be used to solve the third problem which is the most important one.

To go through forward–backward algorithms, set $\lambda = (\boldsymbol{A}, \boldsymbol{B}, \boldsymbol{\pi})$ with random initial conditions (unless some information about the parameters is available already). Then, follow the iterative procedures below.

#### 9.3.14.1 Forward Algorithm

Let $\alpha_i(t) = p(o_1 = y_1, \ldots o_t = y_t, Q_t = i \mid \lambda)$ the probability of observing $y_1, y_2, \ldots, y_t$ and being in state $i$ at time $t$.

$$\alpha_i(1) = \pi_i b_i(y_1) \tag{9.45}$$

and iterate over $t$:

$$\alpha_i(t+1) = b_i(y_{t+1}) \sum_{j=1}^{N} \alpha_j(t) a_{ji} \tag{9.46}$$

#### 9.3.14.2 Backward Algorithm

Let $\beta_i(t) = p(o_{t+1} = y_{t+1}, \ldots, o_T = y_T, Q_t = i \mid \lambda)$ the probability of observing $y_{t+1}, y_2, \ldots, y_T$ partial sequence and being in state $i$ at time $t$.

$$\beta_i(T) = 1 \tag{9.47}$$

and iterate over $t$:

$$\beta_i(t) = \sum_{j=1}^{N} \beta_j(t+1) a_{ij} b_j(y_{t+1}) \tag{9.48}$$

#### 9.3.14.3 HMM Design

For HMM design we need to calculate the following auxiliary variables following the Bayes theorem:

$$\gamma_i(t) = p(Q_t = i \mid O, \lambda) = \frac{p(Q_t = i, O \mid \lambda)}{p(O \mid \lambda)} = \frac{\alpha_i(t)\beta_i(t)}{\sum_{j=1}^{N} \alpha_j(t)\beta_j(t)} \tag{9.49}$$

This is the probability of being in state $i$ at time $t$ for a given model $\lambda$ and the observed sequence $O$. Then, the probability of being in states $i$ and $j$ at times $t$ and $t+1$, respectively, for the given model $\lambda$ and the observed sequence $O$ is:

$$\xi_{ij}(t) = p(Q_t = i, Q_{t+1} = j \mid O, \lambda) = \frac{p(Q_t = i, Q_{t+1} = j, O \mid \lambda)}{p(O \mid \lambda)}$$
$$= \frac{\alpha_i(t) a_{ij} \beta_j(t+1) b_j(y_{t+1})}{\sum_{i=1}^{N} \sum_{j=1}^{N} \alpha_i(t) a_{ij} \beta_j(t+1) b_j(y_{t+1})} \tag{9.50}$$

The HMM parameters for the model $\lambda$ are then updated using:

$$\hat{\pi}_i = \gamma_i(1) \tag{9.51}$$

$$\hat{a}_{ij} = \frac{\sum_{t=1}^{T-1} \xi_{ij}(t)}{\sum_{t=1}^{T-1} \gamma_i(t)} \tag{9.52}$$

$$\hat{b}_i(v_k) = \frac{\sum_{\substack{t=1 \\ y_t = v_k}}^{T} \gamma_i(t)}{\sum_{t=1}^{T} \gamma_i(t)} \tag{9.53}$$

$p(O|\lambda)$ is then calculated utilising these parameters. After each iteration, this probability increases until the parameters $\boldsymbol{A}$, $\boldsymbol{B}$, and $\boldsymbol{\pi}$ reach their optimum values, where the algorithm terminates and there won't be any considerable change in the probability.

## 9.4 Common Spatial Patterns

The common spatial pattern (CSP) is a popular feature extraction and optimisation method for the classification of multichannel signals (such as EEG). CSP aims to estimate spatial filters which discriminate between two classes based on their variances. CSP ($\mathbf{w}$) minimises the Rayleigh quotient of the spatial covariance matrices to achieve the variance imbalance between two classes of data, $\mathbf{X}_1$ and $\mathbf{X}_2$. Before applying CSP, the signals are bandpass filtered and centred. The CSP goal is to find a spatial filter $\mathbf{w} \in \Re^c$ such that the variance of the projected samples of one class is maximised while the other's is minimised. The following maximisation criterion is used for CSP estimation [79, 80]:

$$\mathbf{w}^{(CSP)} = arg \max_{\mathbf{w}} \frac{tr(\mathbf{w}^T \mathbf{C}_1 \mathbf{w})}{tr(\mathbf{w}^T \mathbf{C}_2 \mathbf{w})} \tag{9.54}$$

where $\mathbf{C}_1$ and $\mathbf{C}_2$ are covariance matrices of the two clusters $\mathbf{X}_1$ and $\mathbf{X}_2$ and $tr$ refers to trace of a matrix. With $k$ as any real constant, this optimisation problem can be solved (though this is not the only way) by first observing that the function $J(\mathbf{w})$ remains unchanged even if the filter $\mathbf{w}$ is rescaled, i.e. $J(k\mathbf{w}) = J(\mathbf{w})$, Hence, extremising $J(\mathbf{w})$ is equivalent to extremising $\mathbf{w}^T \mathbf{C}_1 \mathbf{w}$ subject to the constraint $\mathbf{w}^T \mathbf{C}_2 \mathbf{w} = 1$, since it is always possible to find a rescaling of $\mathbf{w}$ such that $\mathbf{w}^T \mathbf{C}_2 \mathbf{w} = 1$. Employing Lagrange multipliers, this constrained problem changes to an unconstrained problem as [81]:

$$L(\lambda, \mathbf{w}) = \mathbf{w}^T \mathbf{C}_1 \mathbf{w} - \lambda(\mathbf{w}^T \mathbf{C}_2 \mathbf{w} - 1) \tag{9.55}$$

A simple way to derive optimum $\mathbf{w}$ is to take derivatives of $L$ and set to zero as follows:

$$\frac{\partial L}{\partial \mathbf{w}} = 0 \Rightarrow C_2^{-1} C_1 \mathbf{w} = \lambda \mathbf{w} \tag{9.56}$$

Based on the standard eigenvalue problem, spatial filters extremising Eq. (9.54) are then the eigenvectors of $\mathbf{M} = \mathbf{C}_2^{-1} \mathbf{C}_1$ corresponding to its largest and lowest eigenvalues. When using CSP, the extracted features are derived from logarithm of the signal variance after projection to filters $\mathbf{w}$.

Eigenvalue $\lambda$ measures the ratio of variances of the two classes. CSP is suitable for classification of both spectral [80] and spatial data since power of the latent signal is larger for the first cluster than for the second cluster.

Applying this approach to separation of event-related potentials (ERPs) from EEG signals enhances one of the classes against the rest. This allows a better discrimination between the two classes, thus can be separated easier.

In early 2000 in a two-class BCI setup, Ramoser et al. [82] proposed the application of CSP that learnt to maximise the variance of bandpass filtered EEG signals from one class while minimising their variance from the other class. Currently, CSP is widely used in BCI where evoked potentials or movement can cause alteration of signals.

The model uses the four most important CSP filters. The variance is then calculated from the CSP time series. Then, a log operation is applied and the weight vector obtained with the LDA is used to discriminate between left- and right-hand movement imaginations. Finally, the signals are classified and based on that an output signal is produced to control the cursor on computer screen. Patterns related to right-hand movement are shown in Figure 9.12a and those for left-hand movement in Figure 9.12b. These patterns are the strongest patterns.

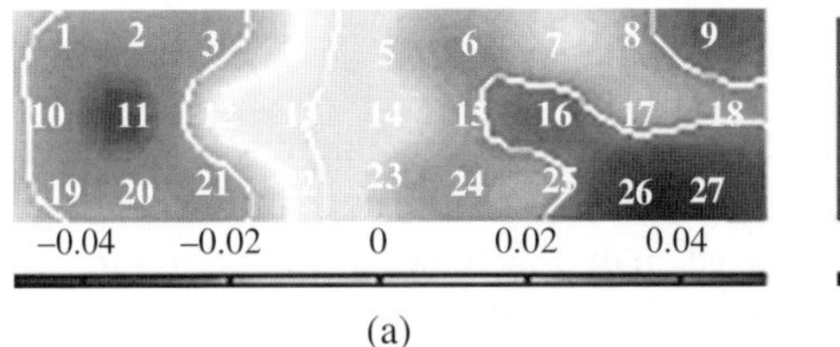


(a)

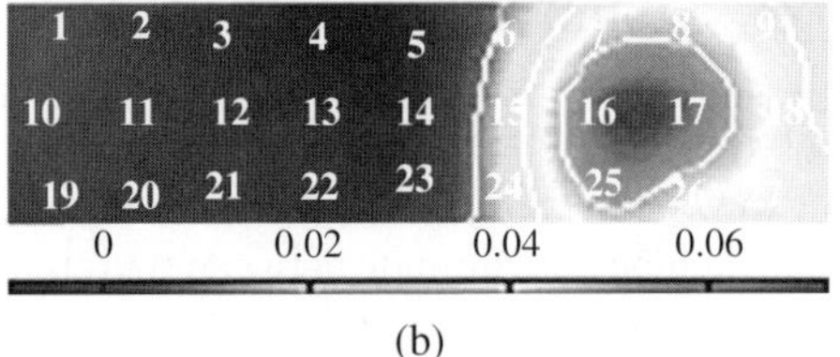


(b)

**Figure 9.12** CSPs related to right-hand movement (a) and left-hand movement (b). The EEG channels are indicated by numbers that correspond to three rows of channels (electrodes) within the central and centroparietal regions. (*See color plate section for color representation of this figure*)

Most of the existing CSP-based methods exploit covariance matrices on a subject-by-subject basis so that the inter-subject information is neglected. In that paper, CSP has been modified for subject-to-subject transfer, where a linear combination of covariance matrices of subjects under consideration has been exploited. Two methods have been developed to determine a composite covariance matrix that is a weighted sum of covariance matrices involving subjects, leading to composite CSP [83].

The accurate estimation of CSPs under weak assumptions (mainly noisy cases) requires derivation of an asymptotically optimal solution. This in sequel necessitates calculation of the loss in signal-to-noise-plus-interference ratio because of the finite sample effect in a closed form. This is often important in the detection/classification of ERPs since the EEG or magnetoencephalography (MEG) signals contains not only the spatiotemporal patterns bounded to the events but also ongoing brain activity as well as other artefacts such as eye blink and muscle movement artefacts. Therefore, to improve the estimation results CSPs are applied to the recorded multichannel signals.

In Eq. (9.54), the trace values are equivalent to $L_2$-norm. Owing to the sensitivity of $L_2$-norm to outliers [84], it is replaced by $L_1$-norm. This changes Eq. (9.54) to:

$$\mathbf{w}^{(CSP)} = arg\max_{\mathbf{w}} \frac{\|\mathbf{w}^T\mathbf{x}_1\|_1}{\|\mathbf{w}^T\mathbf{x}_2\|_1} = arg\max_{\mathbf{w}} \frac{\sum_{k=1}^{m}|\mathbf{w}^T\mathbf{x}_{1k}|}{\sum_{l=1}^{n}|\mathbf{w}^T\mathbf{x}_{2l}|} \tag{9.57}$$

where $m$ and $n$ are the total sampled points of the two classes and $\|\cdot\|_1$ refers to the $L_1$-norm. It has been shown that, using this approach, the effects of outliers are alleviated.

Most of the existing CSP-based methods exploit covariance matrices on a subject-by-subject basis so that inter-subject information is neglected. CSP and its variants have received much attention and have been one of the most efficient feature extraction methods for BCI. However, despite its straightforward mathematics, CSP overfits the data and is highly sensitive to noise.

To enhance CSP performance and address these shortcomings, recently it has been proposed to improve the CSP learning process using prior information in the form of regularisation terms or constraints. Eleven different regularisation methods for CSP have been categorised and compared by Lotte and Guan [81]. These methods included regularisation in the estimation of the EEG covariance matrix [85], composite CSP [83], regularised CSP with generic learning [86], regularised CSP with diagonal loading [80], and invariant CSP [87]. All the algorithms were applied to a number of trials from 17 subjects and the CSP with Tikhonov regularisation in which the optimisation of $J(\mathbf{w})$ was penalised by minimising

$||\mathbf{w}||^2$, and consequently minimising the influence of artefacts and outliers, was suggested as the best method.

To make CSP less sensitive to noise, the objective function in (9.57) may be regularised either during estimation of the covariance matrix for each class or during the minimisation process of the CSP cost function by imposing priors to the spatial filters $\mathbf{w}$ [81]. A straightforward approach to regularise the estimation of the covariance matrix for class $i$, i.e. $\widehat{\mathbf{C}}_i$, is denoted as:

$$\widehat{\mathbf{C}}_i = (1-\gamma)\mathbf{P} + \gamma\mathbf{I} \tag{9.58}$$

where $\mathbf{P}$ is estimated as:

$$\mathbf{P} = (1-\beta)\mathbf{C}_i + \beta\mathbf{G}_i \tag{9.59}$$

In these equations, $\mathbf{C}_i$ is the initial spatial covariance matrix for class $i$, $\mathbf{G}_i$, the so-called generic covariance matrix, is computed by averaging the covariance matrices of a number of trials for class $i$, though it may be defined based on neurophysiological priors only, and $\gamma$ and $\beta \in [0,1]$ are the regularising parameters. The generic matrix represents a given prior on how the covariance matrix for the mental state considered should be.

In regularising the CSP objective function on the other hand, a regularisation term is added to the CSP objective function in order to penalise the resulting spatial filters that do not satisfy a given prior. This results in a slightly different objective function as [81]:

$$J(\mathbf{w}) = \frac{\mathbf{w}^T\mathbf{C}_1\mathbf{w}}{\mathbf{w}^T\mathbf{C}_2\mathbf{w} + \alpha\mathbf{Q}(\mathbf{w})} \tag{9.60}$$

The penalty function $\mathbf{Q}(\mathbf{w})$ weighted by the penalty (or Lagrange) parameter $\alpha \geq 0$ indicates how much the spatial filter $\mathbf{w}$ satisfies a given prior. Similar to the first term in the denominator of (9.60), this term needs to be minimised in order to maximise the cost function $J(\mathbf{w})$. In the literature, different quadratic and nonquadratic penalty functions have been defined. Some exemplar methods can be seen in [81, 87, 88]. A quadratic constraint, such as $\mathbf{Q}(\mathbf{w}) = \mathbf{w}^T\mathbf{K}\mathbf{w}$, results in solving a new set of eigenvalue problems as [81]:

$$(\mathbf{C}_2 + \alpha\mathbf{K})^{-1}\mathbf{C}_1\mathbf{w} = \lambda\mathbf{w} \quad \text{for the 1st pattern} \tag{9.61a}$$

$$(\mathbf{C}_1 + \alpha\mathbf{K})^{-1}\mathbf{C}_2\mathbf{w} = \lambda\mathbf{w} \quad \text{for the 2nd pattern} \tag{9.61b}$$

The composite CSP algorithm proposed in [83] performs subject-to-subject transfer by regularising the covariance matrices using other subjects' data. In this approach $\alpha$ and $\gamma$ are zero and only $\beta$ is nonzero. The generic covariance matrix $\mathbf{G}_i$ is defined according to the covariance matrices of other subjects.

The approach based on regularised CSP with generic learning [86] uses both $\beta$ and $\gamma$ regularisation terms, which means it intends to shrink the covariance matrix towards both the identity matrix and a generic covariance matrix $\mathbf{G}_i$. Here, similar to composite CSP, $\mathbf{G}_i$ is computed from the covariance matrices of other subjects.

Diagonal loading (DL) is another form of covariance matrix regularisation used in BCI literature [80], which consists of shrinking the covariance matrix towards the identity matrix (sometimes called orthonormalisation). Hence, in this approach, only $\gamma$ is used ($\alpha = \beta = 0$) and its value can automatically be identified using Ledoit and Wolf's method [86] or by cross-validation [81].

In the approach based on Tikhonov regularisation [81], the penalty term is defined as $\mathbf{P}(\mathbf{w}) = ||\mathbf{w}||^2 = \mathbf{w}^T\mathbf{w} = \mathbf{w}^T\,\mathbf{I}\,\mathbf{w}$ which is obtained by using $\mathbf{K} = \mathbf{I}$ in (9.61a and b). Such regularisation is expected to constrain the solution to filters with a small norm, hence mitigating the influence of artefacts and outliers.

Finally, in the weighted Tikhonov regularisation method, it is considered that some EEG channels are more important than others (as expected). Therefore, signals from these channels are weighted more than those of others. This consequently causes different penalisations for different channels. To perform this regularisation method, define $\mathbf{P}(\mathbf{w}) = \mathbf{w}^T\mathbf{D}_\mathbf{w}\mathbf{w}$, where $\mathbf{D}_\mathbf{w}$ is a diagonal matrix whose entries are defined based on the measures of CSPs from other subjects [81].

The obtained results in implementing 11 different regularisation methods in [81] show that in places where the data are noisy the regularisation improves the results by approximately 3% on average. However, the best algorithm outperforms CSP by about 3–4% in mean classification accuracy and by almost 10% in median classification accuracy. The regularised methods are also more robust, i.e. they show lower variance across both classes and subjects. Among various approaches, the weighted Tikhonov regularised CSP proposed in [81] has been reported to have the best performance.

Regularisation of $\mathbf{w}$ can also be used to reduce the number of EEG channels without compromising the classification score. Farquhar et al. [87] converted CSP into a quadratically constrained quadratic optimisation problem with $l_1$-norm penalty and Arvaneh et al. [89] used $l_1/l_2$-norm constraint. Recently, in [90] a computationally expensive quasi $l_0$-norm-based principle was applied to achieve a sparse solution for $\mathbf{w}$. In [83] CSP was modified for subject-to-subject transfer, to exploit a linear combination of covariance matrices of subjects under study. In this approach a composite covariance matrix, i.e. a weighted sum of covariance matrices involving subjects, was used, leading to composite CSP.

## 9.5 Applications of Machine Learning in BSNs and WSNs

Numerous applications for most of the machine learning methods above have been briefly explained in the related sections. A few more examples are as follows.

### 9.5.1 Human Activity Detection

Physical and mental disabilities highly affect the human posture and movements during daily life. Measurements from a single sensor, the fusion of data from multiple body sensors [91], and the analysis of the data from distributed body sensors [92] have been practiced recently. Accelerometers with wireless transmission system have been the most popular sensors in many recent applications.

Most existing works proposed in [93–102] are based on off-line processing systems to perform gesture or activity recognition. There are some recent works focusing on real-time activity recognition. Tapia et al. [103] propose a real-time DT-based algorithm for recognition of physical activities (i.e. gestures). Three-axis accelerometer sensors transmit the information wirelessly to a laptop computer for processing.

A tree-based classifier is initially trained and then used to recognise gymnasium activities in real time. Krishnan et al. [104] propose an AdaBoost algorithm based on decision stumps for real-time classification of gestures (i.e. walking, sitting, and running) using three-axis accelerometer sensors. He et al. [105] present an HMM-based approach for real-time activity classification using acceleration data collected from a wearable WSN. The model is used to classify a number of gestures such as standing, sitting, and falling.

Wang et al. [91] propose a real-time, hierarchical model to recognise both simple gestures and complex activities using a wireless body sensor network (WBSN). In this model, they initially employed a fast, lightweight template matching algorithm to detect gestures at the sensor node level, and then used a discriminative pattern based real-time algorithm to recognise high-level activities at the portable device level.

Györbíró et al. [106] present a real-time mobile activity recognition system consisting of wireless body sensors, a smartphone, and a desktop workstation. A sensor node has an accelerometer, a magnetometer, and a gyroscope. They propose a recognition model based on feed-forward backpropagation NNs which are first trained at a desktop workstation, and then run at the smartphone to recognise six different gestures. Liu et al. [107] propose an efficient gesture recognition method based on a single accelerometer using dynamic time warping (DTW). They first defined a vocabulary of known gestures based on training, then used these predefined templates for recognising hand gestures.

Yang et al. [92] propose a distributed recognition framework to classify continuous human actions using a low-bandwidth wearable motion sensor network, called distributed sparsity classifier (DSC). The algorithm classifies human actions using a set of training motion sequences as prior examples. It is also capable of rejecting outlying actions that are not in the training categories. The classification is operated in a distributed fashion on individual sensor nodes and a computer base station. The distribution of multiple action classes is modelled as a mixture subspace model, one subspace for each action class. Given a new test sample, the method seeks the sparsest linear representation of the sample versus all the training examples. It has been shown that the dominant coefficients in the representation only correspond to the action class of the test sample, and hence its membership is encoded in the sparse representation.

Lombriser et al. [108] used a set of wireless accelerometers, light sensors, and a microphone to recognise the office daily activities such as drinking water, moving a computer, mouse writing on a whiteboard, opening a drawer, opening a cupboard, typing on a keyboard, and writing with a pen. They used *k*NN and DT classifiers and demonstrated a recognition rate of above 90%.

Ravi et al. [109] used DNN for recognition of activities of daily living. In their approach they use the spectrogram of an inertial signal, as a function of frequency and time. The convolution filters are then applied to the spectrogram vertically. The achieved temporal convolutions produce an output layer of the learnt network from small segment features, enabling real-time processing. This provides orientation invariance to the input signal. Finally, the last two layers are used to classify the features. They applied their algorithm on a number of standard datasets and report over 95% correct classification rates.

### 9.5.2 Scoring Sleep Stages

Prochazka et al. [110] show that the sleep stages can be classified and scored from multiple sensor polysomnography (PSG), including EEG, ECG, oximeter, and breathing information, data up to a very high accuracy. They effectively used multimodal data features in time and frequency for this purpose. They compared the results of applying DT, NNs, and SVMs to classify and score sleep stages. In this analysis time-frequency and time-scale features have been used [111].

### 9.5.3 Fault Detection

Machine learning techniques have also been widely used in WSNs for environmental and industrial monitoring, mainly for human welfare. In sensor networks, fault detection in both the system itself and the recorded data is important. In WSNs the collected data are prone to the faults which often happen in either the sensors, communication links, communication systems, or the energy supply, and are due to internal and external influences [112]. The authors in [113] focus on occurring faults due to low battery and calibration in WSNs. They used HMM to identify and classify various types of faults. Offset fault, gain fault, stuck-at fault, calibration problem, and low battery have been the major faults considered in a sensor network and an HMM has been trained for each fault separately. An average accuracy of 90% has been reported in their paper.

### 9.5.4 Gas Pipeline Leakage Detection

In [114] SVM, *k*NN, GMM, and naïve Bayes machine learning techniques have been used for large gas pipeline leakage detection and size estimation through WSNs. A number of pressure sensors collect the data from different positions across normal pipes and those with low, medium, and severe leakages. Various order statistics have been estimated and utilised as the features. It has been concluded that *k*NN and SVM outperform others for this application [114].

### 9.5.5 Measuring Pollution Level

In [115] the researchers present the results of application of CNN to detect the level of pollution using the data collected from 72 metal-oxide gas sensors. In addition, they include the air purifying fan speed information proportional to the degree of air pollution. The purpose of this work is to determine the optimal fan speed for air purification in a tunnel by analysing the degree of air pollution.

In their experiment, each sensor has been used as a feature to construct a vector composed of 72 features. The value of the feature is an integer value representing the air gas value and each feature vector has a fan speed as a class. Thirty classes of speed have been used. The accuracy of fan speed prediction has also been optimised. An average prediction accuracy of approximately 85% has been reported.

### 9.5.6 Fatigue-tracking and Classification System

The Neuro-Fuzzy Fatigue-Tracking and Classification System for Wheelchair Users was developed by Li et al. [116]. A neuro-fuzzy fatigue tracking and classification system has been proposed and deployed for the classification of fatigue severity of wheelchair users. In the proposed system, physiological, and kinetic data are collected, including surface electromyography (EMG), ECG, and acceleration signals. The relevant features are then extracted from the signals and integrated with a self-rating method to train the neuro-fuzzy classifier. The musculoskeletal disorders caused by underlying fatigue have been recognised using various degrees of fatigue measures.

These are only a very small number of machine learning applications. Many more applications are reported in the literature and some of them are addressed in the corresponding sections of this book.

### 9.5.7 Eye-blink Artefact Removal from EEG Signals

In [117] and [118] SVM has been combined with blind source separation (BSS) to separate and eliminate the eye-blink artefacts from multichannel EEG signals. BSS separates the multichannel signals into their constituent independent components and SVM recognises the components belonging to eye-blinks using some simple input features related to the morphology of eye-blink signatures within EEGs. This method is very simple, effective, and accurate.

### 9.5.8 Seizure Detection

One of the popular subjects in EEG analysis is detection of onset of seizure for epileptic patients. One way to do that is to characterise the morphology of the seizure EEG pattern mainly in terms of amplitude, frequency range, and the decline in frequency over the seizure onset duration. These features can then be classified into seizure and non-seizure segments. SVM has been used as the classifier for this purpose in [119].

### 9.5.9 BCI Applications

Brain-computer interfacing (BCI) is perhaps the most popular application of machine learning. Classification of the brain motor activity, mainly from the EEG signals, is important in understanding various body part movements and also detection of an intention to move the body part. In [120] the authors have shown that temporo-spatial EEG features can be accurately used by an SVM for detection of left and right hand movements.

## 9.6 Conclusions

In this chapter, the most popular machine learning methods and their applications for clustering or classification of signals and information collected by BSNs have been briefly introduced. Some machine learning algorithms have less sensitivity to noise and outliers. These artefacts often make the classifier applications less reliable. Nonlinear techniques and regression-based algorithms are becoming more popular due to their advantages in classification of nonlinearly separable data. New approaches in deep learning are capable of handling nonseparable and nonlinear cases of very large datasets.

The trained large deep networks may be partially reused for the classification of data of a different nature through *transfer learning* [121]. Following this approach part of the network parameters may be kept fixed and the remaining final layers trained using the new data.

In places where the training data size is limited, the generation and application of surrogate data may boost the quality/accuracy of the results. In such cases, the classifiers should deal with the data distribution and statistics rather than the data samples. This is mainly because the methods for the generation of surrogate data are able to generate new data with same distribution as the data distribution. The surrogate data may not look like the data themselves. In the field of deep networks, one option is to use variational AE, which uses data distribution for learning [122].

Distributed and cooperative learning may become more popular in the near future due to the availability of on-sensor processing and decision making as well as the nature of multisensor scenarios and the large data size. Often, signal processing methods boost machine learning techniques as they can extract and provide more meaningful data features prior to machine learning applications. On the other hand, the need for on-board classification and state recognition requires many algorithms to be developed in Python, Java, and machine languages suitable for compact wearable devices, tablets, and mobile phones.

Moreover, real-time applications require faster systems which are able to process, communicate, and make decisions within one sample interval. Many regression-based classifiers can cope with real-time applications. In the area of DNN, a deep reinforcement learning approach has been claimed to be able to perform real-time learning classification [123]. Nevertheless, these networks need considerable time for training.

## References


1 Akcay, S., Atapour-Abarghouei, A., and Breckon, T. P. (2018). GANomaly: semi-supervised anomaly detection via adversarial training. arXiv:1805.06725.
2 Sanei, S. (2013). *Adaptive Processing of Brain Signals*. Wiley.
3 Fu, L. and Medico, E. (2007). FLAME: a novel fuzzy clustering method for the analysis of DNA microarray data. *BMC Bioinformatics* 8 (3).
4 Vapnik, V. (1998). *Statistical Learning Theory*. Wiley Press.
5 Paradis, L. and Han, Q. (2007). A survey of fault management in wireless sensor networks. *Journal of Network and Systems Management* 15 (2): 171–190.
6 Krishnamachari, B., Estrin, D., and Wicker, S. (2002). The impact of data aggregation in wireless sensor networks. In: *22nd International Conference on Distributed Computing Systems Workshops*, 575–578.
7 Al-Karaki, J. and Kamal, A. (2004). Routing techniques in wireless sensor networks: a survey. *IEEE Wireless Communications* 11 (6): 6–28.
8 Romer, K. and Mattern, F. (2004). The design space of wireless sensor networks. *IEEE Wireless Communications* 11 (6): 54–61.
9 Wan, J., Chen, M., Xia, F. et al. (2013). From machine-to-machine communications towards cyber-physical systems. *Computer Science and Information Systems* 10: 1105–1128.
10 Bengio, Y. (2009). Learning deep architectures for AI. *Foundations and Trends in Machine Learning* 2 (1): 1–127.

**11** Hoffmann, A.G. (1990). General limitations on machine learning. In: *European Conference on Artificial Intelligence*, 345–347. ECAI.

**12** Kouchaki, S., Yang, Y., Walker, T.M. et al. (2019). Application of machine learning techniques to tuberculosis drug resistance analysis. *Bioinformatics* 35 (13): 2276–2282. https://doi.org/10.1093/bioinformatics/bty949.

**13** Hartigan, J. and Wong, M. (1979). A *k*-means clustering algorithm. *Applied Statistics* 28: 100–108.

**14** Fellous, J.-M., Tiesinga, P.H.E., Thomas, P.J., and Sejnowski, T.J. (2004). Discovering spike patterns in neural responses. *Journal of Neuroscience* 24 (12): 2989–3001.

**15** Hastie, T., Tibshirani, R., and Walter, G. (2000) Estimating the number of clusters in a dataset via the gap statistic. *Tech. Rep. 208*, Stanford University, Stanford, CA.

**16** Ester, M., Kriegel, H.-P., Sander, J., and Xu, X. (1996). A density-based algorithm for discovering clusters in large spatial databases with noise. In: *KDD-96 Proceedings*, 226–231. AAAI.

**17** Ng, Y.A., Jordan, M.I., and Weiss, Y. (2002). On spectral clustering: analysis and an algorithm. In: *Conference on Neural Information Processing Systems*. MIT Press.

**18** Ioffe, S., Szegedy, C. (2015). Batch normalization: accelerating deep network training by reducing internal covariate shift, arXiv:1502.03167.

**19** Lin, F. and Cohen, W.W. (2010). Power iteration clustering. In: *Proceedings of the 27th International Conference on Machine Learning, ICML 2010*, 655–666. Omnipress.

**20** Rasmus, A., Mikko Honkala, V., Berglund, M., and Raiko. T. (2015) Semi-supervised learning with ladder networks, arXiv:1507.02672 [cs.NE].

**21** Bair, E. (2013). Semi-supervised clustering methods. *Wiley Interdisciplinary Reviews: Computational Statistics* 5 (5): 349–361.

**22** Bezdek, J.C. (1981). *Pattern Recognition with Fuzzy Objective Function Algorithms*. Springer.

**23** Ayodele, T.O. (2010). Types of machine learning algorithms. In: *New Advances in Machine Learning* (ed. Y. Zhang). InTechOpen https://doi.org/10.5772/9385.

**24** Safavian, S.R. and Landgrebe, D. (1991). A survey of decision tree classifier methodology. *IEEE Transactions on Systems, Man and Cybernetics* 21 (3): 660–674.

**25** Lu, C.-H. and Fu, L.-C. (2009). Robust location-aware activity recognition using wireless sensor network in an attentive home. *IEEE Transactions on Automation Science and Engineering* 6 (4): 598–609.

**26** Breiman, L. (2001). Random forests. *Machine Learning* 45: 5–32.

**27** Vapnik, V. (1995). *The Nature of Statistical Learning Theory*. New York: Springer.

**28** Bennet, K.P. and Campbell, C. (2000). Support vector machines: hype or hallelujah? *ACM SIGKDD Explorations Newsletter* 2 (2): 1–13.

**29** Christianini, N. and Shawe-Taylor, J. (2000). *An Introduction to Support Vector Machines*. Cambridge University Press.

**30** DeCoste, D. and Scholkopf, B. (2001). Training invariant support vector machines. *Machine Learning* 46: 161–190.

**31** Burges, C. (1998). A tutorial on support vector machines for pattern recognition. *Data Mining and Knowledge Discovery* 2: 121–167.

**32** Gunn, S., (1998) Support vector machines for classification and regression. Technical Reports, Dept. of Electronics and Computer Science, Southampton University. http://

ce.sharif.ir/courses/85-86/2/ce725/resources/root/LECTURES/SVM.pdf (accessed 25 November 2019).

**33** Chapelle, O. and Vapnik, V. (2002). Choosing multiple parameters for support vector machines. *Machine Learning* 46: 131–159.

**34** Weston, J. and Watkins, C. (1999). Support vector machines for multi-class pattern recognition. In: *Proceedings of the 7th European Symposium on Artificial Neural Networks*, 219–224. ESANN.

**35** Platt, J. (1998) Sequential minimal optimisation: A fast algorithm for training support vector machines. *Technical Report, MSR-TR-98-14, Microsoft Research*, 1-21.

**36** Tran, D. and Nguyen, T. (2008). Localization in wireless sensor networks based on support vector machines. *IEEE Transactions on Parallel and Distributed Systems* 19 (7): 981–994.

**37** Yoo, J. and Kim, H.J. (2015). Target localization in wireless sensor networks using online semi-supervised support vector regression. *Sensors (Basel)* 15 (6): 12539–12559.

**38** Mathew, J., Pang, C.K., Luo, M., and Leong, W.H. (2018). Classification of imbalanced data by oversampling in kernel space of support vector machines. *IEEE Transactions on Neural Networks and Learning Systems* 29 (9): 4065–4076.

**39** Kang, Q., Shi, L., Zhou, M. et al. (2018). A distance-based weighted undersampling scheme for support vector machines and its application to imbalanced classification. *IEEE Transactions on Neural Networks and Learning Systems* 29 (9): 4152–4165.

**40** Winter, J., Xu, Y., and Lee, W.-C. (2005). Energy efficient processing of k nearest neighbor queries in location-aware sensor networks. In: *Proceedings of the 2nd International Conference on Mobile and Ubiquitous Systems: Networking and Services*, 281–292. IEEE.

**41** Jayaraman, P.P., Zaslavsky, A., and Delsing, J. (2010). Intelligent processing of *k*-nearest neighbors queries using mobile data collectors in a location aware 3D wireless sensor network. In: *Trends in Applied Intelligent Systems* (eds. N. García-Pedrajas, F. Herrera, C. Fyfe, et al.), 260–270. Springer.

**42** Dempster, A.P., Laird, N.M., and Rubin, D.B. (1977). Maximum likelihood from incomplete data via the EM algorithm. *Journal of the Royal Statistical Society, Series B* 39 (1): 1–38.

**43** van Hasselt, H., Guez, A., and Silver, D. (2016). Deep reinforcement learning with double Q-learning. In: *Proceedings of the Thirtieth AAAI Conference on Artificial Intelligence (AAAI-16)*, 2094–2100. Association for the Advancement of Artificial Intelligence (AAAI).

**44** Jin, C., Allen-Zhu, Z., Bubeck, S., and Jordan, M. I. (2018) Is Q-learning Provably Efficient?. *International Conference on Neural Information Processing Systems (NIPS)*, Montreal, Canada.

**45** Lippmann, R. (1987). An introduction to computing with neural nets. *IEEE ASSP Magazine* 4 (2): 4–22.

**46** Selfridge, O.G. (1958). Pandemonium: a paradigm for learning in mechanisation of thought processes. In: *Proceedings of the Symposium on Mechanisation of Thought Processes*, 513–526. HM Stationery Office.

**47** Rosenblatt, F. (1957) The Perceptron: A Perceiving and Recognizing Automaton. *Tech. Rep. 85-460-1 (Cornell Aeronautical Laboratory)*.

**48** Dargie, W. and Poellabauer, C. (2010). *Localization*, 249–266. Wiley.
**49** Kohonen, T. (2001). *Self-Organizing Maps*, Springer Series in Information Sciences, vol. 30. Springer.
**50** Hinton, G.E. and Salakhutdinov, R.R. (2006). Reducing the dimensionality of data with neural networks. *Science* 313 (5786): 504–507.
**51** Lehmann, E.L. and Casella, G. (1998). *Theory of Point Estimation*, 2e. New York: Springer.
**52** Goodfellow, I.J., Pouget-Abadie, J., Mirza, M. et al. (2014). Generative adversarial nets. *Advances in Neural Information Processing Systems* 3: 2672–2680.
**53** LeCun, Y., Boser, B., Denker, J.S. et al. (1990). Handwritten digit recognition with a back-propagation network. In: *Proceedings of the Advances in Neural Information Processing Systems*, 396–404. Morgan Kaufmann.
**54** Hubel, D.H. and Wiesel, T.N. (1962). Receptive fields, binocular interaction, and functional architecture in the cat's visual cortex. *Journal of Physiology* 160: 106–154.
**55** Antoniades, A., Spyrou, L., Martin-Lopez, D. et al. (2017). Detection of interictal discharges using convolutional neural networks from multichannel intracranial EEG. *IEEE Transactions on Neural Systems and Rehabilitation Engineering* 25 (12): 2285–2294.
**56** Antoniades, A., Spyrou, L., Martin-Lopez, D. et al. (2018). Deep neural architectures for mapping scalp to intracranial EEG. *International Journal of Neural Systems* 28 (8).
**57** Felleman, D.J. and Essen, D.C.V. (1991). Distributed hierarchical processing in the primate cerebral cortex. *Cerebral Cortex* 1: 1–47.
**58** Cadieu, C.F., Hong, H., Yamins, D.L.K. et al. (2014). Deep neural networks rival the representation of primate it cortex for core visual object recognition. *PLoS Computational Biology* 10 (12): e1003963.
**59** Fukushima, K. and Miyake, S. (1982). Neocognitron: a new algorithm for pattern recognition tolerant of deformations and shifts in position. *Pattern Recognition* 15: 455–469.
**60** LeCun, Y., Jackel, L.D., Bottou, L. et al. (1995). Learning algorithms for classification: a comparison on handwritten digit recognition. *Neural Networks: The Statistical Mechanics Perspective*: 261–276.
**61** Krizhevsky, A., Sutskever, I., and Hinton, G.E. (2017). ImageNet classification with deep convolutional neural networks. *Communications of the ACM* 60: 84–90.
**62** Zeiler, M.D. and Fergus, R. (2014). Visualizing and understanding convolutional networks. In: *ECCV 2014, Part I, LNCS 8689*, 818–833. Springer.
**63** Simonyan, K. and Zisserman, A. (2015). Very deep convolutional networks for large-scale image recognition. *ICLR* 75: 398–406.
**64** Szegedy, C., Liu, W., Jia, Y. et al. (2015). Going deeper with convolutions. In: *IEEE Conference on Computer Vision and Pattern Recognition (CVPR), Boston MA*, 1–9. IEEE.
**65** He, K., Zhang, X., Ren, S., and Sun, J. (2015). Deep residual learning for image recognition. In: *IEEE Conference on Computer Vision and Pattern Recognition*, 770–778. IEEE.
**66** Huang, G., Liu, Z., Maaten, L.v.d., and Weinberger, K.Q. (2017). Densely connected convolutional networks. In: *IEEE Conference on Computer Vision and Pattern Recognition (CVPR), Honolulu, HI*, 2261–2269. IEEE.

**67** Hochreiter, S. and Schmidhuber, J. (1997). Long short-term memory. *Neural Computation* 9 (8): 1735–1780.

**68** Xu, K., Ba, J.L., Kiros, R. et al. (2015). Show, attend and tell: neural image caption generation with visual attention. *Proceedings of the 32nd International Conference on Machine Learning* 32: 2048–2057.

**69** Oord, A. V. D. N, Dieleman, S., Zen, H., Simonyan, K., Vinyals, O., Graves, A., Kalchbrenner, N., Senior, A., Kavukcuoglu, K. (2016) WaveNet: A generative model for raw audio, arXiv:1609.03499 [cs.SD].

**70** Rasmussen, C. and Williams, C. (2006). *Gaussian Processes for Machine Learning*. Cambridge, MA: The MIT Press.

**71** Clifton, L., Clifton, D.A., Pimentel, A.M.F. et al. (2013). Gaussian processes for personalized e-health monitoring with wearable sensors. *IEEE Transactions on Biomedical Engineering* 60 (1): 193–197.

**72** Hurley, P., Serra, L., Bozzali, M. et al. (2015). Gaussian process classification of Alzheimer's disease and mild cognitive impairment from resting-state fMRI. *NeuroImage* 112: 232–243.

**73** Garnelo, M., Schwarz, J., Rosenbaum, D., Viola, F., Rezende, D. J., Eslami, S. M. A., and Te, Y. W. (2018) Neural processes, arXiv:1807.01622 [cs.LG].

**74** Kim, H., Mnih, A., Schwarz, J., Garnelo, M., Eslami, S. M. A., Rosenbaum, D., Vinyals, O., and Te, Y. W. (2019) Attentive neural processes, arXiv:1901.05761 [cs.LG].

**75** Nipf, T. N. and Wellin, M. (2017) Semi-supervised classification with graph convolutional networks. arXiv:1609.02907 [cs.LG].

**76** Zhang, H. (2004). The optimality of naïve Bayes. In: *Proceedings of the Seventeenth International Florida Artificial Intelligence Research Society Conference (FLAIRS 2004)* (eds. V. Barr and Z. Markov). AAAI Press.

**77** Rabiner, L. (2013) First Hand: The Hidden Markov Model. *IEEE Global History Network*. https://ethw.org/First-Hand:The_Hidden_Markov_Model (accessed 6 January 2020).

**78** Baum, L.E. and Petrie, T. (1966). Statistical inference for probabilistic functions of finite state Markov chains. *The Annals of Mathematical Statistics* 37 (6): 1554–1563.

**79** Koles, Z. (1991). The quantitative extraction and topographic mapping of the abnormal components in the clinical EEG. *Electroencephalography and Clinical Neurophysiology* 79 (6): 440–447.

**80** Blankertz, B., Tomioka, R., Lemm, S. et al. (2008). Optimizing spatial filters for robust EEG single-trial analysis. *IEEE Signal Processing Magazine* 25 (1): 41–56.

**81** Lotte, F. and Guan, C. (2011). Regularizing common spatial patterns to improve BCI designs: unified theory and new algorithms. *IEEE Transactions on Biomedical Engineering* 58 (2): 355–362.

**82** Ramoser, H., Muller-Gerking, J., and Pfurtscheller, G. (2000). Optimal spatial filtering of single trial EEG during imagined hand movement. *IEEE Transactions on Rehabilitation Engineering* 8 (4): 441–446.

**83** Kang, H., Nam, Y., and Choi, S. (2009). Composite common spatial pattern for subject-to-subject transfer. *IEEE Signal Processing Letters* 16 (8): 683–686.

**84** Wang, H., Tang, Q., and Zheng, W. (2012). L1-norm-based common spatial patterns. *IEEE Transactions on Biomedical Engineering* 59 (3): 653–662.

**85** Lu, H., Plataniotis, K., and Venetsanopoulos, A. (2009). Regularized common spatial patterns with generic learning for EEG signal classification. In: *Proceedings of the IEEE Engineering in Medicine and Biology Conference, EMBC, Minnesota, USA*, 6599–6602. IEEE.

**86** Ledoit, O. and Wolf, M. (2004). A well-conditioned estimator for large dimensional covariance matrices. *Journal of Multivariate Analysis* 88 (2): 365–411.

**87** Farquhar, J., Hill, N., Lal, T., and Schölkopf, B. (2006). Regularised CSP for sensor selection in BCI. In: *Proceedings of the 3rd International BCI Workshop*, 14–15. Verlag der Technischen Universität Graz.

**88** Yong, X., Ward, R., and Birch, G. (2008). Sparse spatial filter optimization for EEG channel reduction in brain-computer interface. In: *Proceedings of the IEEE International Conference on Acoustics, Speech, and Signal Processing, ICASSP, Las Vegas, USA*, 417–420.

**89** Arvaneh, M., Guan, C.T., Kai, A.K., and Chai, Q. (2011). Optimizing the channel selection and classification accuracy in EEG-based BCI. *IEEE Transactions on Biomedical Engineering* 58 (6): 1865–1873.

**90** Goksu, F., Ince, N.F., and Tewfik, A.H. (2011). Sparse common spatial patterns in brain computer interface applications. In: *Proceedings of the IEEE International Conferenceon Acoustics, Speech, and Signal Processing, ICASSP, Prague, Czech Republic*, 533–536. IEEE.

**91** Wang, L., Gu, T., Chen, H. et al. (2010). Real-time activity recognition in wireless body sensor networks: from simple gestures to complex activities. In: *Proceedings of the 6th IEEE International Conference on Embedded and Real-Time Computing Systems and Applications*, 43–52. IEEE.

**92** Yang, A.Y., Jafari, R., Shankar Sastry, S., and Bajcsy, R. (2009). Distributed recognition of human actions using wearable motion sensor networks. *Journal of Ambient Intelligence and Smart Environments* 1: 1–5.

**93** Kela, J., Korpipaa, P., Mantyjarvi, J. et al. (2006). Accelerometer-based gesture control for a design environment. *Personal and Ubiquitous Computing* 10 (5): 285–299.

**94** Bao, L. and Intille, S. (2004). Activity recognition from user-annotated acceleration data. In: *Proceedings of the International Conference on Pervasive, LNCS 3001*, 1–17. Springer.

**95** Gu, T., Wu, Z., Tao, X. et al. (2009). *epSICAR: An emerging patterns based approach to sequential, interleaved and concurrent activity recognitionProceedings of the 7th Annual IEEE International Conference on Pervasive Computing and Communications (Percom '09)*, 1–9. IEEE.

**96** Huynh, T., Blanke, U., and Schiele, B. (2007). Scalable recognition of daily activities with wearable sensors. *Lecture Notes in Computer Science* 4718: 50.

**97** Philipose, M., Fishkin, K.P., Perkowitz, M. et al. (2004). Inferring activities from interactions with objects. *IEEE Pervasive Computing* 3 (4): 50–57.

**98** Modayil, J., Bai, T., and Kautz, H. (2008). Improving the recognition of interleaved activities. In: *Proceedings of the International Conference on Ubicomp, Seoul, South Korea*, 40–43. ACM.

**99** Wu, T., Lian, C., and Hsu, J. (2007). Joint recognition of multiple concurrent activities using factorial conditional random fields. In: *AAAI Workshop PAIR*, 82–87. Association for the Advancement of Artificial Intelligence (AAAI).

**100** Logan, B., Healey, J., Philipose, M. et al. (2007). A long-term evaluation of sensing modalities for activity recognition. In: *Proceedings of the 9th International Conference on Ubiquitous Computing*, 483–500. Springer-Verlag.

**101** Ward, J.A., Lukowicz, P., Troster, G., and Starner, T.E. (2006). Activity recognition of assembly tasks using body-worn microphones and accelerometers. *IEEE Transactions on Pattern Analysis and Machine Intelligence* 28 (10): 1553–1567.

**102** Palmes, P., Pung, H., Gu, T. et al. (2010). Object relevance weight pattern mining for activity recognition and segmentation. *Pervasive and Mobile Computing* 6: 43–57.

**103** Tapia, E.M., Intille, S., and Larson, K. (2007). Real-time recognition of physical activities and their intensities using wireless accelerometers and a heart rate monitor. In: *Proceedings of the 11th IEEE International Conference on Wearable Computers*, 37–40. IEEE.

**104** Krishnan, N. C., Colbry, D., Juillard, C., and Panchanathan, S. (2008) Real time human activity recognition using tri-axial accelerometers. Sensors Signals and Information Processing Workshop. Sedona, AZ (11–14 May).

**105** He, J., Li, H., and Tan, J. (2007). Real-time daily activity classification with wireless sensor networks using hidden Markov model. In: *Proceedings of the 29th Annual International Conference of the IEEE Engineering in Medicine and Biology Society*, 3192–3195. IEEE.

**106** Györbíró, N., Fabian, A., and Homanyi, G. (2009). An activity recognition system for mobile phones. *Journal of Mobile Networks and Applications* 14 (1): 82–91.

**107** Liu, J., Wang, Z., Zhong, L. et al. (2009). uWave: Accelerometer-based personalized gesture recognition and its applications. *Pervasive and Mobile Computing* 5: 657–675.

**108** Lombriser, C., Bharatula, N.B., Roggen, D., and Tröster, G. (2007). On-body activity recognition in a dynamic sensor network. In: *Proceedings of the ICST 2nd int. conference on Body area networks (BodyNet'07)*, 1–7. Wearable Computing Lab.

**109** Ravì, D., Wong, C., Lo, B., and Yang, G.-Z. (2016). Deep learning for human activity recognition: a resource efficient implementation on low-power devices. In: *Proceedings of the IEEE EMBS 13th Annual International Body Sensor Networks (BSN) Conference*, 1–6. Hamlyn Centre, Imperial College, London.

**110** Prochazka, A., Kuchynka, J., Vysata, O. et al. (2018). Sleep scoring using polysomnography data features. *Signal, Image and Video Processing* 12 (6): 1–9.

**111** Kingsbury, N.G. (2001). Complex wavelets for shift invariant analysis and filtering of signals. *Applied and Computational Harmonic Analysis* 10 (3): 234–253.

**112** Warriach, E.U. and Tei, K. (2017). A comparative analysis of machine learning algorithms for faults detection in wireless sensor networks. *International Journal of Sensor Networks* 24 (1): 1–13.

**113** Warriach, E.U. and Aiello, M. (2012). A machine learning approach for identifying and classifying faults in wireless sensor networks. In: *IEEE 15th International Conference on Computational Science and Engineering*, 618–625. IEEE.

**114** Rashida, S., Akram, U., and Khan, S.A. (2015). WML: wireless sensor network based machine learning for leakage detection and size estimation. In: *Proceedings of the 6th International Conference on Emerging Ubiquitous Systems and Pervasive Networks (EUSPN 2015)*, vol. 63, 171–176. Elsevier.

**115** Lee, K.-S., Lee, S.-R., Kim, Y., and Lee, C.-G. (2017). Deep learning–based real-time query processing for wireless sensor network. *International Journal of Distributed Sensor Networks* 13 (5).

**116** Li, W., Hu, X., Gravina, R., and Fortino, G. (2017). A Neuro-fuzzy fatigue-tracking and classification system for wheelchair users. *IEEE Access* 5: 19420–19431.

**117** Shoker, L., Sanei, S., and Chambers, J. (2005). Artifact removal from electroencephalograms using a hybrid BSS-SVM algorithm. *IEEE Signal Processing Letters* 12 (10): 721–724.

**118** Shoker, L., Sanei, S., Wang, W., and Chambers, J. (2004). Removal of eye blinking artifact from EEG incorporating a new constrained BSS algorithm. *IEEE Journal of Medical and Biological Engineering and Computing* 43 (2): 290–295.

**119** Gonzalez, B., Sanei, S., and Chambers, J. (2003). Support vector machines for seizure detection. In: *Proceedings of the IEEE, ISSPIT2003*, 126–129. Germany: IEEE.

**120** Shoker, L., Sanei, S., and Sumich, A. (2005). Distinguishing between left and right finger movement from EEG using SVM. In: *Proceedings of the IEEE, EMBS*, 5420–5423. IEEE.

**121** Yosinski, J. Clune, J. Bengio, Y. Lipson, H. (2014) How transferable are features in deep neural networks?, arXiv:1411.1792.

**122** Kingma, D., and Welling, M. (2014) Auto-encoding variational Bayes, arXiv:1312.6114.

**123** François-Lavet, V., Islam, R., Pineau, J., Henderson, P., and Bellemare, M. G. (2018) An introduction to deep reinforcement learning, arXiv:1811.12560v2.

# 10

# Signal Processing for Sensor Networks

## 10.1 Introduction

In the introductory chapters of this book it has been emphasised and clarified that there are three major components of sensing, processing, and communications involved in the design of sensor networks. Dealing with conventional computational machines, electronic gadgets, and pervasive systems, the sensor data need to be digitised before applying any learning or processing to them. Analogue-to-digital convertors are therefore embedded within all the data acquisition cards as well as sensor motes.

Although many hardware and software platforms have been introduced (some are addressed in Chapter 14 of this book) for hosting the necessary processing algorithms, yet the fundamentals, advances, and the applications of effective algorithms for sensor networks have not been explored in detail. Such algorithms have roots in fundamental digital signal processing concepts for sensor network theories.

Regarding the techniques in machine learning, most of the existing algorithms rely on a central system for fusing and aggregating the collected information. The concepts have been further advanced by introducing the decentralised systems and the notion of cooperative networks very recently. This requires migration from fusion, which is necessary for centralised systems, to consensus and diffusion techniques, deployed in decentralised networks. In decentralised systems, the sensors are smarter and act as intelligent agents capable of processing, learning, and decision-making.

Data denoising, feature extraction, communication channel identification and modelling, localisation, and sensor anomaly detection are perhaps the most important aspects of developing and implementing signal processing algorithms for sensor networks. In a wireless sensor network (WSN), we need to consider the fact that the sensors may enter or leave the network dynamically, resulting in unpredictable changes in the network's size and topology. Sensors may disappear permanently, either because of damage to the nodes or drained batteries; temporarily, because of topology, traffic, and channel conditions; or because of duty cycling or even severe Doppler effect during motion. This makes it necessary for distributed signal processing algorithms to be robust to the changes in network topology or size. These algorithms and protocols must also be robust to poor time synchronisation across the network and to inaccurate knowledge of sensor locations.

*Body Sensor Networking, Design and Algorithms,* First Edition. Saeid Sanei, Delaram Jarchi and Anthony G. Constantinides.

Companion Website: www.wiley.com/go/sanei/algorithm-design

In this chapter, first the fundamental and most widely used concepts in single and multichannel signal processing are explored and then some advances in signal processing made particularly for sensor and body sensor networks are discussed.

## 10.2 Signal Processing Problems for Sensor Networks

Most of the biomedical and human body generated signals have an effective frequency bandwidth of less than a few hundred hertz. This allows for the online processing of these signals using conventional signal processing algorithms operating in time, frequency, space, multidimensional, or multiway domains.

The rhythms in physiological signals can be viewed and well distinguished during different states of the human physical and mental states. The patterns often significantly differ for patients compared to those of healthy individuals.

The (often multichannel) signals related to body electrical activities, mainly those from central nervous system (CNS), such as electroencephalograph (EEG), electrocardiograph (ECG), electromyograph (EMG), and those from electromagnetic signatures, such as magnetoencephalograph (MEG), are called instantaneous in time since there is no time lag in travelling the waveforms from their sources to different sensors. Other types of signals – such as speech, human joint sound due to movement, heart and lung sounds, and ultrasound – are subject to delays due to the relatively lower speed of sound compared to the electromagnetic and electric signal propagations. In such cases the mixtures of their multichannel recordings are called convolutive.

In sensor networks, however, we deal with communication signals too. These signals propagate in the radio frequency range which can start from megahertz to hundreds of gigahertz. Such signals are propagated through wireless media and, therefore, are subject to noise, attenuation, interference, delays, fading, eavesdropping, Doppler (mobility) effect, and many other artefacts. In order to recover the original signals from the received distorted signals, often proper modelling and a good estimation of the communication channel become necessary.

Likewise, in multichannel recordings, such as for EEGs, the signals are actually considered combinations of the underlying sources. To have access to those sources for their recognition, powerful source separation algorithms need to be designed and deployed. Separation of the desired sources from the multisensor data has been a popular research area.

Localisation of an object as a passive or active source has been a major research in WSNs. In body area network (BAN), localisation of brain sources has been a popular line of research [1]. Localisation of endoscopic capsules inside intestine has become an area for investigation, too. Another popular example is localisation of sound (or ultrasound) sources using the data from multiple microphones (or piezoelectric sensors).

Any a priori information about the nature of the signal sources or the propagation media can be helpful in designing a reliable algorithm for their estimation, separation, or localisation.

The data can be single- or multichannel. Although multichannel signals have more inclusive information about the data origin, they often require complicated techniques to enable the extraction of sufficient information about the underlying sources from their mixtures.

Processing of single-channel signals, on the other hand, require basic techniques as long as the desired and undesired signal components have different subspaces or at least have distinct spatial, spectral, or temporal signatures.

In places where the single-channel data are a mixture of various components such as noise or interferences, with overlapping subspaces, their processing can be very complex.

For online (or real-time) monitoring of the signals, the algorithms should be fast and perform within a sample interval. As an example, in intensive care units (ICUs) the signals (such as ECG, oximeters, and respiratory information) are observed and the vital signs tracked continuously. Likewise, real-time monitoring of these signals becomes essential. Another example is the assessment of patients during physical rehabilitation by gait monitoring. Tracking the EEG signals to identify seizure onset or for scoring of sleep EEG in sleep abnormality detection also requires online processing of the data.

Time series analysis for the detection or evaluation of chaos, transient events, synchronicity, rhythmicity, and anomaly is very popular for both single- and multichannel applications. Seizure prediction from EEG, tremor detection from gait accelerometer data, heart rate variability monitoring, and many biomedical events of a cyclostationary nature are the popular examples of time series analysis.

In order to classify the human biomarkers, often necessary and efficient sets of features should be estimated from the signals. Features may be computed from time-domain signals or after they are transformed into some other domains. Also, statistical measures are possible when their distributions can be tracked accurately.

In the following sections some fundamental concepts used in developing signal processing tools and algorithms for applications to body sensor networks are explained and the mathematical foundations discussed. The reader should also be aware of the required concepts and definitions developed in linear algebra, further details of which can be found in [2].

## 10.3 Fundamental Concepts in Signal Processing

There is no need to emphasise that the signals recorded from the human body are naturally analogue. These signals have to be converted into digital to enable their processing using computers and transmission through limited bandwidth communication channels. An analogue-to-digital conversion involves three major operations of sampling, quantization, and coding. Most of the analogue-to-digital convertors are able to process the signals recorded from the human body as the frequency bandwidth is low. Accordingly, low speed microcontrollers can be used, too.

### 10.3.1 Nonlinearity of the Medium

In WSNs two kinds of nonlinearity are involved; one is due to the nonhomogeneity in the medium between sources and sensors. This significantly affects the complexity of the processing system. As an example, the head, as a mixing medium, combines the EEG signals locally generated within the brain at the sensor positions. As a system, the head may be more or less susceptible to such sources in different situations. Generally, an EEG signal can be considered the output of a nonlinear system, which may be characterised deterministically. The changes in the brain metabolism as a result of biological and physiological phenomena in the human body can change the mixing process. Some of these changes are influenced by the activity of the brain itself. These effects make the system nonlinear. Analysis or modelling of such a system is very complicated and yet nobody has fully modelled the system to aid in the analysis of brain signals. When the objective is localisation of the sources inside the brain using EEGs, the nonlinearity problem dramatically deteriorates the accuracy of the localisation system.

The second kind of nonlinearity is due to the time, frequency, or space variabilities of the wave propagation environment of the sensors. As an example, some sound sensors may involve reverberation and some may not. As another example, the heart and lung sound auscultated by a stethoscope travels through various air and tissue layers with different acoustic properties. Blood flow inside the vessels affects the frequency of the striking ultrasound through the Doppler effect. The change in lung volume during breathing changes the lung's acoustic model significantly. All these examples refer to the complexity of developing accurate signal processing algorithms.

### 10.3.2 Nonstationarity

The signal statistics such as single- or multivariate mean, variance, and higher-order statistics often change in time. As a result, the signals are very likely to be nonstationary. Nonstationarity of the signals can be quantified by measuring some statistics of the signals at different time lags. The signals can be deemed stationary if there is no considerable variation in these statistics. A signal is called stationary in wide sense (wide sense stationary) if the distribution of the signal does not change over time.

The signal distribution also changes over time. Even with multivariate Gaussian assumption, the mean and covariance properties generally change from segment to segment of the data. When processing biosignals, it is assumed that they are stationary only within short intervals.

This refers to the signals being quasistationary.

As an example, for the case of brain signals, the Gaussian assumption holds during normal brain conditions, whereas during mental and physical activities this assumption is not valid. Some examples of nonstationarity of the EEG signals can be observed during the change in alertness and wakefulness (where there are weaker alpha and stronger beta oscillations), eye blinking, the transitions between various ictal states, and elicitation of the event-related potential (ERP) and evoked potential (EP) signals.

The change in signal distribution from one segment to another can be measured in terms of both the parameters of a Gaussian process and the deviation of the distribution from Gaussian. The non-Gaussianity of the signals can be checked by measuring or estimating some higher-order moments, such as skewness, kurtosis, negentropy, and Kulback–Leibler (KL) divergence.

Skewness is a measure of asymmetry in the signal unimodal distribution. A distribution is symmetric if it looks the same to the left and right of the centre point. For real-valued signals the skewness is defined as:

$$Skewness = \frac{E[(x(n) - \mu)^3]}{\sigma^3} \tag{10.1}$$

where $\mu$ and $\sigma$ are, respectively, the mean and standard deviation, and $E$ denotes statistical expectation. If the distribution is more to the right of the mean point the skewness is negative, and vice versa. For a symmetric distribution, such as Gaussian, the skewness is zero.

Kurtosis is a measure of how peaky (sharp), relative to a normal distribution, the signal is. That is, datasets with high kurtosis tend to have a distinct peak near the mean, decline rather rapidly, and have heavy tails. Datasets with low kurtosis tend to have a flat top near the mean rather than a sharp peak. A uniform distribution would be the extreme case. The kurtosis for a signal $x(n)$ is defined as:

$$kurt = \frac{m_4(x(n))}{m_2^2(x(n))} \tag{10.2}$$

where $m_i(x(n))$ is the $i$th central moment of the signal $x(n)$, i.e. $m_i(x(n)) = E[(x(n) - \mu)^i]$. The kurtosis for signals with normal distributions is three. Therefore, an excess or normalised kurtosis is often used and defined as:

$$Ex\,kurt = \frac{m_4(x(n))}{m_2^2(x(n))} - 3 \tag{10.3}$$

which is zero for Gaussian distributed signals. Often, the signals are considered ergodic, hence the statistical averages can be approximated by their time averages.

The negentropy of a signal $x(n)$ [3] is defined as:

$$J_{neg}(x(n)) = H(x_{Gauss}(n)) - H(x(n)) \tag{10.4}$$

where, $x_{Gauss}(n)$ is a Gaussian random signal with the same covariance as $x(n)$ and $H(.)$ is the differential entropy [4] defined as:

$$H(x(n)) = \int_{-\infty}^{\infty} p(x(n)) log \frac{1}{p(x(n))} dx(n) \tag{10.5}$$

and $p(x(n))$ is the signal distribution. Negentropy is always non-negative.

The $KL$ distance between two distributions, $p_1$ and $p_2$, is defined as:

$$KL = \int_{-\infty}^{\infty} p_1(z) log \frac{p_1(z)}{p_2(z)} dz \tag{10.6}$$

It is clear that the $KL$ distance is generally asymmetric. Therefore, by changing the position of $p_1$ and $p_2$ in this equation the $KL$ distance changes. The minimum of the $KL$ distance occurs when $p_1(z) = p_2(z)$.

The difference between any two data sequences can also be described by their Euclidean or Mahalanobis distances. Mahalanobis distance between two random sequences $\mathbf{x}$ and $\mathbf{y}$ of the same distribution is defined as:

$$d_M(x, y) = \sqrt{(\mathbf{x} - \mathbf{y})^T \Sigma^{-1} (\mathbf{x} - \mathbf{y})}$$

where $\boldsymbol{\Sigma}$ is the covariance matrix between $\mathbf{x}$ and $\mathbf{y}$. If $\boldsymbol{\Sigma} = \mathrm{I}$, i.e. an identity matrix, then Mahalanobis distance changes to an Euclidean distance.

### 10.3.3 Signal Segmentation

Owing to the nonstationarity of the data and inherent computational limitations, often it is necessary to divide the signals into the segments of similar statistical, physiological,

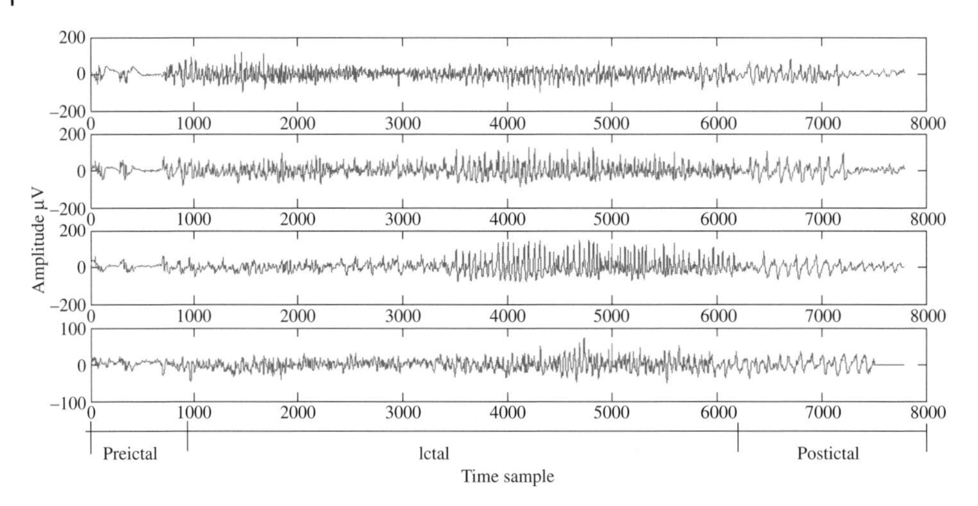


**Figure 10.1** Four channels of the EEG of a patient with tonic-clonic seizure signals including three segments of preictal, ictal, and postictal behaviour happening within approximately 40 seconds [1]. Source: Courtesy of John Wiley & Sons.

or visual characteristics a– to make them meaningful to clinicians [5, 6]. Within each segment, the signals are considered statistically stationary usually with similar time and frequency statistics. As an example, an EEG recorded from an epileptic patient may be divided into three segments of preictal, ictal, and postictal segments. Each may have a different duration. Figure 10.1 represents an EEG sequence including all the above segments.

In the segmentation of biological data, the time or frequency properties may be exploited. This eventually leads to a dissimilarity measurement denoted as $d(m)$ between the adjacent data frames, where $m$ is an integer value indexing the frame and the difference is calculated between the $m$ and ($m$-1)th (consecutive) signal frames. The boundary of the two different segments is then defined as the boundary between the $m$th and ($m$-1)th frames provided $d(m) > \eta_T$, where $\eta_T$ is an empirical threshold level. An efficient segmentation is possible by highlighting and effectively exploiting the diagnostic information within the signals with the help of expert clinicians. However, access to such experts is not always possible and therefore algorithmic (and often unsupervised) methods utilising the data statistics are required.

Rather than Euclidean and Mahalanobis distances defined in Section 10.3.2, many other different dissimilarity measures between two data segments may be defined based on the signal processing fundamentals. One criterion is based on the autocorrelations for segment $m$ defined as:

$$r_x(k, m) = E[x(n, m)x(n + k, m)] \tag{10.7}$$

The autocorrelation function of the $m$th length $N$ frame for an assumed time interval $n$, $n$+1,…, $n + (N - 1)$ can be approximated as:

$$\hat{r}_x(k, m) = \begin{cases} \frac{1}{N} \sum_{l=0}^{N-1-k} x(l + m + k)x(l + m), & k = 0, \ldots, N - 1 \\ 0 & k = N, N + 1, \ldots \end{cases} \tag{10.8}$$

Then, the first criterion is set to:

$$d_1(m) = \frac{\sum_{k=-\infty}^{\infty} (\hat{r}_x(k,m) - \hat{r}_x(k,m-1))^2}{\hat{r}_x(0,m)\hat{r}_x(0,m-1)} \tag{10.9}$$

A second criterion is based on higher-order statistics. The signals with more uniform distributions, such as normal brain rhythms, have a low kurtosis, whereas ECG or pulse oximeter signals often have peaky distributions. Kurtosis is defined as the fourth-order cumulant at zero time-lags and related to the second- and fourth-order moments as given in Eqs. (10.2) and (10.3). A second level discriminant $d_2(m)$ is then defined as:

$$d_2(m) = kurt_x(m) - kurt_x(m-1) \tag{10.10}$$

where $m$ refers to the $m$th frame of the signal $x(n)$. A third criterion is defined using the spectral error measure of the periodogram. A periodogram of the $m$th frame is obtained by Fourier transforming of the autocorrelation function of the signal:

$$S_x(\omega,m) = \sum_{k=-\infty}^{\infty} \hat{r}_x(k,m)e^{-j\omega k} \quad \omega \in [-\pi,\pi] \tag{10.11}$$

where $\hat{r}_x(.,m)$ is the autocorrelation function for the $m$th frame. A third criterion is then defined based on the normalised periodogram as:

$$d_3(m) = \frac{\int_{-\pi}^{\pi} (S_x(\omega,m) - S_x(\omega,m-1))^2 d\omega}{\int_{-\pi}^{\pi} S_x(\omega,m)d\omega \int_{-\pi}^{\pi} S_x(\omega,m-1)d\omega} \tag{10.12}$$

The test window sample autocorrelation for the measurement of both $d_1(m)$ and $d_3(m)$ may be updated through the following recursive equation over the windows of size $N$:

$$\hat{r}_x(k,m) = \hat{r}_x(k,m-1) + \frac{1}{N}(x(m-1+N)x(m-1+N-k) - x(m-1+k)x(m-1)) \tag{10.13}$$

and thereby computational complexity can be reduced in practice. A fourth criterion corresponds to the error energy in autoregressive (AR) based modelling of the signals. The prediction error in the AR model of the $m$th frame is simply defined as:

$$e(n,m) = x(n,m) - \sum_{k=1}^{p} a_k(m)x(n-k,m) \tag{10.14}$$

where $p$ is the prediction order and $a_k(m)$, $k = 1, 2, \ldots, p$ are the prediction coefficients. For certain $p$ the coefficients can be found directly (for example by using Durbin's method) in such a way as to minimise the error (residual) signal energy. In this approach it is assumed that the frames of length $N$ are overlapped by one sample. The prediction coefficients estimated for the ($m$-1)th frame are then used to predict the first sample in the $m$th frame, which we denote as $\hat{e}(1,m)$. For a small error, it is likely that the statistics of the $m$th frame are similar to those of the ($m$-1)th frame. On the other hand, a large error is likely to indicate a change. An indicator for the fourth criterion can then be the differencing (differentiating) of this prediction signal, which gives a peak at the segment boundary i.e.

$$d_4(m) = \max(\nabla_m \hat{e}(1,m)) \tag{10.15}$$

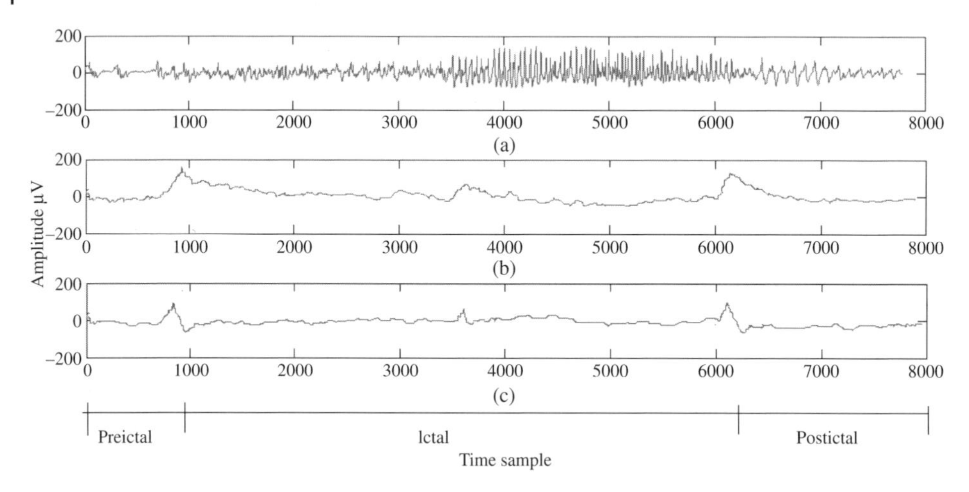


**Figure 10.2** (a) An EEG seizure signal including preictal, ictal, and postictal segments; (b) the error signal; (c) the approximate gradient of the signal, which exhibits a peak at each boundary between the segments. The number of prediction coefficients is $p = 12$.

where $\nabla_m(.)$ denotes the gradient with respect to $m$, approximated by a first order difference operation. Figure 10.2 shows the residual in Eq. (10.14) for $n$th channel and its gradient defined in Eq. (10.15). Finally, a fifth criterion, $d_5(m)$, may be defined by using the AR-based spectrum of the signals in the same way as short-time Fourier transform (STFT) for $d_3(m)$. A similar criterion may be defined when multichannel data such as EEGs are considered [7]. In such cases a multivariate autoregressive (MVAR) model is analysed. One interesting application can be seen in [8], where the signals are segmented mainly to exploit the stationarity of the signals for their blind source separation (BSS).

Given the concepts behind the above segmentation methods, in general, more efficient systems may be defined for the segmentation of the signals with particular behaviours [5, 9]. To do that, the features which best describe the behaviour of the signals have to be identified and used. Thus, the segmentation problem becomes a clustering or classification problem for which different classifiers can be used.

### 10.3.4 Signal Filtering

The concept of filtering (both analogue and digital) is very central in signal processing and the theory for the design of finite impulse response (FIR) and infinite impulse response (IIR) digital filters can be found in almost all signal processing textbooks. These filters can be lowpass, highpass, bandpass, or bandstop (or band reject), referring to the frequency bands the filters allow in. Lowpass filters are also used as anti-aliasing filters during sampling or down-sampling operations. Highpass filters are usually used in order to remove any DC offset or even slow trends due to slow body movements on the physiological signals. Narrow-band bandstop filters (also called notch filters) are often used for the rejection of 50 or 60 Hz grid frequency from most of the human body signals.

To design the digital filters, often two main approaches are followed; in the first approach, the equivalent analogue filter is first designed and then the formulation is transformed into digital domain such as by means of bilinear transformation or through impulse invariance methods. In the second approach, the filter is designed in the discrete domain directly.

The structure of most of these filters includes operational amplifiers mainly to alleviate the loading effects. In such cases the filters can be active instead of passive, to ensure that the desired signal components do not attenuate during the filtering process.

Later in this chapter we elaborate on another category of filters, namely adaptive filters, where the filter parameters can be adaptively estimated in order to shape the output signal as desired.

## 10.4 Mathematical Data Models

### 10.4.1 Linear Models

Although natural systems are generally nonlinear (i.e. the output is a nonlinear function of the input), in practice many of these systems are modelled as linear systems mainly for better modelling and ease of processing. This assumption is the foundation for the development and application of many signal processing methods. One of the most important and widely used models is a predictive one. This is mainly because the signal samples are correlated and predictable using their past samples. It also reflects that the system behaviour is deterministic.

#### 10.4.1.1 Prediction Method

The main objective of using prediction methods is to find a set of model parameters which best describe the signal generation system. Such models can then produce the desired signals by applying a random stationary noise to their inputs. In AR modelling of signals each sample of a single-channel measurement is defined to be linearly related to a few of its previous samples (this number is often called the memory of the predictor), i.e.

$$y(n) = -\sum_{k=1}^{p} a_k y(n-k) + x(n) \tag{10.16}$$

where $a_k$, $k = 1, 2, \ldots, p$, are the linear parameters, $n$ denotes the discrete sample time normalised to unity, and $x(n)$ is the noise input. In an autoregressive moving average (ARMA) linear predictive model each sample is obtained based on its previous input and output sample values, i.e.

$$y(n) = -\sum_{k=1}^{p} a_k y(n-k) + \sum_{k=0}^{q} b_k x(n-k) \tag{10.17}$$

where $b_k$, $k = 1, 2, \ldots, q$, are the additional linear parameters. The parameters $p$ and $q$ are the model orders. Among many options, the Akaike information criterion (AIC) can be used to determine the order of the appropriate model of a measurement signal by minimising the following equation [10] with respect to the model order.

$$AIC(i,j) = N \ln(\sigma_{ij}^2) + 2(i+j) \tag{10.18}$$

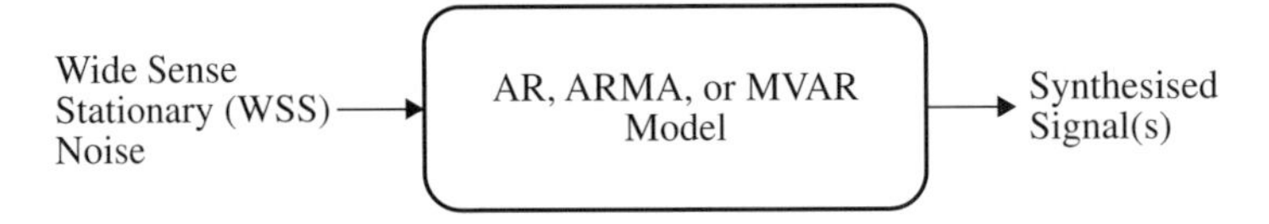


**Figure 10.3** A linear model for the generation of signals from the optimally estimated AR parameters.

where $i$ and $j$ represent, respectively, the assumed AR and MA model prediction orders, $N$ is the number of signal samples, and $\sigma_{ij}^2$ is the noise power of the ARMA model at the $i$th and $j$th stage.

For multichannel data a multivariate system is used. In an MVAR model a multichannel scheme is considered. Therefore, each signal sample is defined versus previous samples of both the signal itself and the signals from other channels, i.e. for channel $i$ we have:

$$y_i(n) = -\sum_{j=1}^{m}\sum_{k=1}^{p} a_{jk} y_j(n-k) + x_i(n) \tag{10.19}$$

where $m$ represents the number of channels and $x_i(n)$ represents the noise input to channel $i$. Similarly, the model parameters can be calculated iteratively in order to minimise the error between the actual and predicted values [11].

There are numerous applications for linear models. Different algorithms have been developed to find efficiently the model coefficients. In the maximum likelihood estimation (MLE) method [11–13] the likelihood function is maximised over the system parameters formulated from the assumed real, Gaussian distributed, and sufficiently long input signals (approximately 10–20 seconds for EEGs sampled at $f_s = 250$ Hz). Using AIC, the gradient of squared error is minimised using the Newton–Raphson approach applied to the resultant nonlinear equations [10, 11]. This is considered as an approximation to the MLE method. In the Durbin method [14] the Yule–Walker equations, which relate the model coefficients to the signals' autocorrelation, are iteratively solved. The approach and the results are equivalent to those using a least-squared-based scheme [15]. The MVAR coefficients are often calculated using the Levinson–Wiggins–Robinson (LWR) algorithm [16]. After the parameters are estimated the synthesis filter can be excited with wide sense stationary noise to generate the single- or multichannel signal samples. Figure 10.3 illustrates the simplified system.

#### 10.4.1.2 Prony's Method

Following Prony's method [17], an event can be considered as the impulse response (IR) of a linear IIR system. The original attempt in this area was to fit an exponentially damped sinusoidal model to the data [17]. This method was later modified to model sinusoidal signals [18]. Prony's method is used to calculate the linear prediction (LP) parameters. The angles of the poles in the z-plane of the constructed LP filter are then referred to the frequencies of the damped sinusoids of the exponential terms used for modelling the data. Consequently, both the amplitude of the exponentials and the initial phase can be obtained following the methods used for an AR model, as follows.

Based on the above approach we can consider the output of an AR system with zero excitation to be related to its IR as:

$$y(n) = \sum_{k=1}^{p} a_k y(n-k) = \sum_{j=1}^{p} w_j \sum_{k=1}^{p} a_k r_j^{n-k-1} \tag{10.20}$$

where y($n$) represents the exponential data samples, $p$ is the prediction order, $w_j = A_j e^{j\theta_j}$, $r_k = \exp((\alpha_k + j2\pi f_k)T_s)$, $T_s$ is the sampling period normalised to 1, $A_j$ is the amplitude of the exponential, $\alpha_k$ is the damping factor, $f_k$ is the discrete-time sinusoidal frequency in samples/sec, and $\theta_j$ is the initial phase in radians.

Therefore, the model coefficients are first calculated using one of the methods previously mentioned in this section, i.e. $\mathbf{a} = -\mathbf{Y}^{-1}\hat{\mathbf{y}}$, where:

$$\mathbf{a} = \begin{bmatrix} a_0 \\ a_1 \\ \cdot \\ \cdot \\ \cdot \\ a_p \end{bmatrix}, \ \mathbf{Y} = \begin{bmatrix} y(p)\dots y(1) \\ y(p-1)\dots y(2) \\ \cdot \\ \cdot \\ \cdot \\ y(2p-1)\dots y(p) \end{bmatrix}, \ \text{and } \hat{y} = \begin{bmatrix} y(p+1) \\ y(p+2) \\ \cdot \\ \cdot \\ \cdot \\ y(2p) \end{bmatrix} \tag{10.21}$$

and $a_0 = 1$. On the basis of Eq. (10.21), $y(n)$ is calculated as the weighted sum of its $p$ past values. $y(n)$ is then constructed and the parameters $f_k$ and $r_k$ are estimated. Hence, the damping factors are obtained as:

$$\alpha_k = \ln|r_k| \tag{10.22}$$

and the resonance frequencies as:

$$f_k = \frac{1}{2\pi} tan^{-1}\left(\frac{\text{Im}(r_k)}{\text{Re}(r_k)}\right) \tag{10.23}$$

where Re(.) and Im(.) denote, respectively, the real and imaginary parts of a complex quantity. The $w_k$ parameters are calculated using the fact that $y(n) = \sum_{k=1}^{p} w_k r_k^{n-1}$ or:

$$\begin{bmatrix} r_1^0 & r_2^0 & \dots & r_p^0 \\ r_1^1 & r_2^1 & & r_p^1 \\ \vdots & & \ddots & \vdots \\ r_1^{p-1} & r_2^{p-1} & \dots & r_p^{p-1} \end{bmatrix} \begin{bmatrix} w_1 \\ w_2 \\ \vdots \\ w_p \end{bmatrix} = \begin{bmatrix} y(1) \\ y(2) \\ \vdots \\ y(p) \end{bmatrix} \tag{10.24}$$

In vector form this can be illustrated as $\mathbf{Rw} = \mathbf{y}$, where $[\mathbf{R}]_{k,l} = r_l^k$, $k = 0, 1, \dots, p-1$, $l = 1, \dots, p$, denotes the matrix elements in the above equation. Therefore, $\mathbf{w} = \mathbf{R}^{-1}\mathbf{y}$, assuming $\mathbf{R}$ is a full-rank matrix, i.e. there are no repeated poles. Often, this is simply carried out by implementing the Cholesky decomposition algorithm [19]. Finally, using $w_k$, the amplitude and initial phases of the exponential terms are calculated as:

$$A_k = |w_k| \tag{10.25}$$

and:

$$\theta_k = tan^{-1}\left(\frac{\text{Im}(w_k)}{\text{Re}(w_k)}\right) \tag{10.26}$$

In the above solution, we considered that the number of data samples $N$ is equal to $N = 2p$, where $p$ is the prediction order. For the cases where $N > 2p$ a least-squares (LS) solution for $\mathbf{w}$ can be obtained as:

$$\mathbf{w} = (\boldsymbol{R}^H\boldsymbol{R})^{-1}\boldsymbol{R}^H\boldsymbol{y} \tag{10.27}$$

where $(.)^H$ denotes conjugate transpose. Another alternative for solving this problem is by using the Cholesky decomposition method. For real data, such as EEG signals, this equation changes to $\mathbf{w} = (\boldsymbol{R}^T\boldsymbol{R})^{-1}\boldsymbol{R}^T\boldsymbol{y}$, where $(.)^T$ represents the transpose operation. A similar result can be achieved using principal component analysis (PCA) [13].

In the cases where the data are contaminated with white noise, the performance of Prony's method is reasonable. However, for non-white noise the noise information is not easily separable from the data and therefore the method may not be sufficiently successful.

#### 10.4.1.3 Singular Spectrum Analysis

The domain of subspace signal analysis is extended to the analysis of one-dimensional signals by means of singular spectrum analysis (SSA). SSA can be effectively used for the prediction and forecasting of various data modalities [20].

In addition to decomposition and reducing the dimensionality of the data, as PCA does, SSA has become an emerging technique in time series and event prediction. Application of this method for financial forecasting has been well established. One example can be seen in [21]. Climatic time series and regional and global sea surface temperature prediction are some examples of using SSA for prediction [22].

SSA combined with a neural network (NN) has also been used for the prediction of daily rainfall [23]. Through experiments they verified that the role of SSA in decomposition of the signals before applying the NN improves the prediction outcome significantly.

The basic SSA involves the two stages of decomposition and reconstruction. In the first step the series is decomposed and in the second stage the desired series is reconstructed and used for further analysis. The main concept in studying the properties of SSA is separability, which characterises how well different components of a single-channel signal can be separated from each other. The absence of approximate separability is often observed in series with complex structures.

SSA is becoming an effective and powerful tool for time series analysis in meteorology, hydrology, geophysics, climatology, economics, biology, physics, medicine, and other sciences where short and long, one-dimensional and multidimensional, stationary and non-stationary, almost deterministic and noisy time series are to be analysed. A brief description of the SSA stages is given in the following subsections. For more information see [24].

***Decomposition***

This stage includes an embedding operation followed by singular value decomposition (SVD). Embedding operation maps a one-dimensional time series **f** into a $k \times l$ matrix with rows of length $l$ as:

$$\mathbf{X} = \{x_{ij}\} = [\mathbf{x}_1, \mathbf{x}_2, \ldots, \mathbf{x}_k] = \begin{bmatrix} f_0 & f_1 & f_2 & & f_{k-1} \\ f_1 & f_2 & f_3 & \cdots & f_k \\ f_2 & f_3 & f_4 & & f_{k+1} \\ & \vdots & & \ddots & \vdots \\ f_{l-1} & f_l & f_{l+1} & \cdots & f_{r-1} \end{bmatrix} \tag{10.28}$$

with vectors $\mathbf{x}_i = [f_{i-1}, f_i, \ldots, f_{i+l-2}]^T \in R^L$, where $k = r - l + 1$ is the window length $(1 \leq l \leq r)$, and subscript $T$ denotes the transpose of a vector. Vectors $\mathbf{x}_i$ are called *l-lagged vectors* (or, simply, *lagged vectors*). The window length $l$ should be sufficiently large. Note that the trajectory matrix **X** is a Hankel matrix, which means that all the elements along the diagonal $i + j = const.$ are equal.

In the SVD stage the SVD of the trajectory matrix is computed and represented as the sum of rank-one bi-orthogonal elementary matrices. Consider that the eigenvalues of the covariance matrix $\mathbf{C}_x = \mathbf{X}\mathbf{X}^T$ are denoted by $\lambda_1, \ldots, \lambda_l$ in decreasing order of magnitude $(\lambda_1 \geq \ldots \geq \lambda_l \geq 0)$ and the corresponding orthogonal eigenvectors by $\mathbf{u}_1, \ldots, \mathbf{u}_l$. Set $d = \max(i;$ such that $\lambda_i > 0) = rank(\mathbf{X})$. If we denote $\boldsymbol{v}_i = \mathbf{X}^T\boldsymbol{u}_i / \sqrt{\lambda_i}$ then the SVD of the trajectory matrix can be written as:

$$\mathbf{X} = \sum_{i=1}^{d} \mathbf{X}_i = \sum_{i=1}^{d} \sqrt{\lambda_i} \boldsymbol{u}_i \boldsymbol{v}_i^T \tag{10.29}$$

***Reconstruction***

During the reconstruction stage, the elementary matrices are first split into several groups (depending on the number of components in the time series – in this application EMG and ECG signals, the matrices within each group are added together). The size of the group or, in other words, the length of each subspace may be specified based on some a priori information. In the case of ECG removal from recorded EMGs this parameter and the desired signal components are jointly (and automatically) identified based on a constraint on the statistical properties of the signal to be extracted. Let $I = \{i_1, \ldots, i_p\}$ be the indices corresponding to the $p$ eigenvalues of the desired component. Then matrix $\hat{\mathbf{X}}_I$ corresponding to the group $I$ is defined as $\hat{\mathbf{X}}_I = \sum_{j=i_1}^{i_p} \mathbf{X}_j$. In splitting the set of indices $J = \{1, \ldots, d\}$ into disjoint subsets $I_1$ to $I_m$ we always have:

$$\mathbf{X} = \sum_{j=I_1}^{I_m} \hat{\mathbf{X}}_j \tag{10.30}$$

The procedure of choosing the sets $I_1, \ldots I_m$ is called eigentriple grouping. For a given group $I$ the contribution of the component $\mathbf{X}_I$ in expansion (10.29) is measured by contribution of the corresponding eigenvalues: $\sum_{i\in I}\lambda_i / \sum_{i=1}^{d} \lambda_i$. In the next step the obtained matrix is transformed to the form of a Hankel matrix which can be subsequently converted to a time series. If $z_{ij}$ stands for an element of a matrix $\mathbf{Z}$ then the $k$-th term of the resulting series is obtained by averaging $z_{ij}$ over all $i, j$ such that $i + j = k + 1$. This procedure is called diagonal averaging or the Hankelisation of matrix $\mathbf{Z}$.

***Prediction***

SSA-based forecasting is performed through the application of linear recurrent formulae (LRF). An infinite series is governed by some LRF if and only if it can be represented as a linear combination of products of exponential, polynomial, and harmonic series [24, 25].

We recall that the single most important parameter of embedding is the window length, $L$, an integer such that $2 < L < N$ ($N$ is the series length). This parameter should always be large enough to permit reasonable separability. It should not be greater than $N/2$ for optimum results. The vectors $X_i$, called the lagged vectors or $L$ lagged vectors (to emphasise their dimension), are the $K$ columns of the trajectory matrix $\mathbf{X}$.

If the rank of the trajectory matrix is smaller than the window length ($r < L$), the signal satisfies the LRF. For such signals, SSA could be applied as a forecasting algorithm. In usual practice, the reconstruction stage of SSA aims to smooth the original data by removing the eigentriples corresponding to noise. Then, a recurrent forecasting procedure is applied to the reconstructed signal [22].

Consider $\breve{\mathbf{q}} \in C^{L-1}$ a vector consisting of the last $(L - 1)$ elements of the eigenvector $\mathbf{q}$ and $v^2 = \eta_1^2 + \ldots + \eta_r^2$,where $\eta_i$ is the last component of the corresponding eigenvector. The vector $\boldsymbol{\delta} = (\delta_1, \ldots, \delta_{L-1})^T$ is then defined as:

$$\delta = \frac{1}{1 - v^2} \sum_{i=1}^{r} \eta_i \breve{\mathbf{q}}_i \tag{10.31}$$

where $i$ refers to the eigenvector index. The following equation is then used to forecast $h$ steps ahead using $\boldsymbol{\delta}$ and the reconstructed signal $\hat{x}(n)$ up to sample $n$:

$$\hat{x}(n) = \begin{cases} x(n) & n = 1, \ldots, N \\ \sum_{j=1}^{L-1} \delta_j \hat{x}(n-j) & n = N+1, \ldots, N+h \end{cases} \tag{10.32}$$

More SSA applications to time series prediction and forecasting can be seen in [20]. Recently, the SSA has been extended to nonlinear cases through nonlinear Laplacian spectral analysis (NLSA) [26]. This method has proven to be effective for the detection and prediction of nonstationary or intermittent time series.

### 10.4.2 Nonlinear Modelling

An approach similar to AR or MVAR modelling, in which the output samples are nonlinearly related to the previous samples, may be followed based on the methods developed for forecasting financial or stock exchange growth or decline.

In the generalised autoregressive conditional heteroskedasticity (GARCH) method [27] each sample relates to its previous samples through a nonlinear (or sum of nonlinear) function(s). This model was originally introduced for time-varying volatility (honoured with the Nobel Prize in Economic Sciences in 2003).

Generally, it is not possible to discern whether the nonlinearity is deterministic or stochastic in nature, nor can we distinguish between multiplicative and additive dependencies. The type of stochastic nonlinearity may be determined using a Hsieh test [28]. The additive and multiplicative dependencies can be discriminated by using this test. Nevertheless, the test itself is not used to obtain the model parameters.

Despite modelling the signals, the GARCH approach has many other applications. In some works [29] the concept of the GARCH modelling of covariance is combined with Kalman filtering to provide a more flexible model with respect to space and time for solving the inverse problem. There are several alternatives for the solution to the inverse problem. Many approaches fall into the category of constrained LS methods employing Tikhonov regularisation [30]. Among numerous possible choices the exponential GARCH (EGARCH) [31] has been used to estimate the variance parameter of the Kalman filter iteratively.

The application of nonlinear models in biomedical signal processing has been very limited. A better estimation of the media may pave the way for using such models. The problem of predicting the behaviour of nonlinear systems may be modelled using recurrent neural networks (RNNs), which are in fact a subset of adaptive systems [32].

### 10.4.3 Gaussian Mixture Model

In this very popular modelling approach the signals are characterised using their distribution parameters. The unimodal and multimodal distributions are modelled as a sum of some Gaussian functions with different means and variances which are weighted and delayed differently [33]. The overall distribution subject to a set of $K$ Gaussian components is defined as:

$$p(x|\theta_k) = \sum_{k=1}^{K} w_k p(x|\mu_k, \sigma_k) \tag{10.33}$$

The vector of unknown parameters $\theta_k = [w_k, \mu_k, \sigma_k]$ for $k = 1, 2, \ldots, K$. $w_k$ is equivalent to the probability (weighting) that the data sample is generated by the $k$th mixture component density subject to $\sum_{k=1}^{K} w_k = 1$, $\mu_k$, and $\sigma_k$ are mean and variances of the $k$th Gaussian distribution and $p(x\,|\,\mu_k, \sigma_k)$ is a Gaussian function of $x$ with parameters $\mu_k$, and $\sigma_k$. Expectation maximisation (EM) [34] is often used to estimate the above parameters by maximising the log-likelihood of the mixture of Gaussian (MOG) for an $N$-sample data defined as:

$$L(\theta_k) = \sum_{i=1}^{N} \log\left(\sum_{k=1}^{K} w_k p\left(x(i)|\mu_k, \sigma_k\right)\right) \tag{10.34}$$

The EM algorithm alternates between updating the posterior probabilities used for generating each data sample by the $k$th mixture component (in a so-called E-step) as:

$$h_i^k = \frac{w_k p(x_i|\mu_k, \sigma_k)}{\sum_{j=1}^{K} w_j p(x_i|\mu_k, \sigma_k)} \tag{10.35}$$

and weighted maximum likelihood updates of the parameters of each mixture component (in a so-called M-step) as:

$$w_k \leftarrow \frac{1}{N} \sum_{i=1}^{N} h_i^k \tag{10.36}$$

$$\mu_k \leftarrow \frac{\sum_{i=1}^{N} h_i^k x_i}{\sum_{j=1}^{N} h_j^k} \tag{10.37}$$

$$\sigma_k \leftarrow \sqrt{\frac{\sum_{i=1}^{N} h_i^k (x_i - \mu_k^*)^2}{\sum_{j=1}^{N} h_j^k}} \tag{10.38}$$

The EM algorithm (especially used for high-dimensional multivariate Gaussian mixtures) may converge to spurious solutions when there are singularities in the log-likelihood function due to small sample sizes, outliers, repeated data points or rank deficiencies leading to 'variance collapse'. Some solutions to these shortcomings have been provided by many researchers, such as those in [35–37]. Figure 10.4 demonstrates how an unknown multimodal distribution can be estimated using weighted sum of Gaussians with different means and variances.

Similar to prediction-based models, the model order can be estimated using AIC or by iteratively minimising the error for best model order. A mixture of exponential distributions can also be used instead of a mixture of Gaussians for data waveforms with sharp transients. It has been shown that these mixtures can be used for modelling a variety of physiological data such as EEG, electrooculography (EOG), EMG, and ECG [38]. In another work [39], an MOG model has been used for the segmentation of magnetic resonance brain images. As a variant of Gaussian mixture model (GMM), Bayesian Gaussian mixture models have been used for partial amplitude synchronisation detection in EEG signals [40]. This work introduces a method to detect subsets of synchronised channels that do not consider any baseline

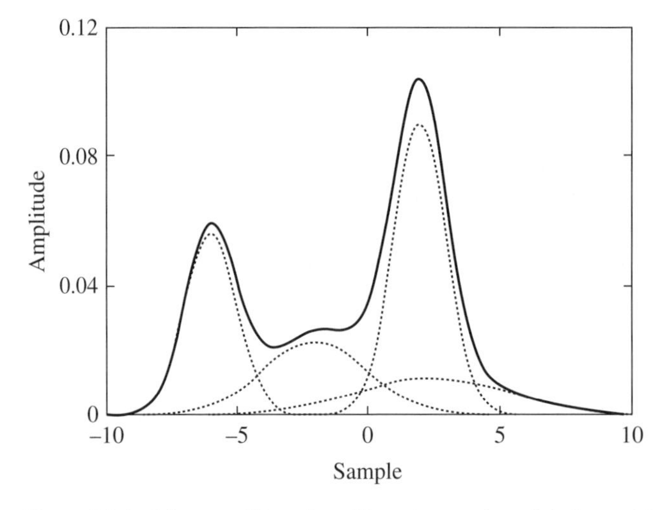


**Figure 10.4** Mixture of Gaussians (dotted curves) model of a multimodal unknown distribution (bold curve).

information. It is based on a Bayesian GMM applied at each location of a time-frequency map of the EEGs.

## 10.5 Transform Domain Signal Analysis

The field of signal processing was revolutionised in the early nineteenth century when Fourier transform was presented soon after Fourier series was introduced by Joseph Fourier in 1822 [41]. This transform represents the signal segments using their Fourier bases in frequency domain.

If the signals are statistically stationary, it is straightforward to characterise them in either time or frequency domain. The frequency domain representation of a finite-length signal can be found by using linear transforms such as the discrete Fourier transform (DFT), discrete cosine transform (DCT), or other semi-optimal transforms, which have kernels independent of the signal. One advantage of DCT over DFT is that the DCT basis functions are real-valued, while those of DFT are complex-valued. However, the results of these transforms can be degraded by spectral smearing due to the short-term time-domain windowing of the signals, before DFT application, and fixed transform kernels. An optimal transform such as Karhunen–Loève transform (KLT) requires complete signal statistical information, which may not be available in practice.

For a data segment of $N$ samples, $x(n)$, the DFT (also called discrete Fourier spectrum) and its inverse are respectively defined as:

$$X(k) = \sum_{n=0}^{N-1} x(n)e^{-j2\pi nk/N} \text{ for } k = 0, \dots, N-1 \tag{10.39}$$

$$x(n) = \sum_{k=0}^{N-1} X(k)e^{j2\pi nk/N} \text{ for } n = 0, \dots, N-1 \tag{10.40}$$

where $N$ is the number of DFT points in a so-called $N$-point DFT.

## 10.6 Time-frequency Domain Transforms

Natural signals particularly those from human body are nonstationary. Therefore, their variations may be best captured by time-frequency transforms. There are a number of well-known time-frequency transform methods among them STFT and wavelet transform (WT) are more popular. In some recent applications chirplet transform [42] and synchro-squeezed wavelet transform [43] have proved to best represent the underlying signal components for sleep analysis application [44], where the signal is nonstationary and includes bursts of transient components.

### 10.6.1 Short-time Fourier Transform

STFT is defined as the discrete-time Fourier transform evaluated over a sliding window. The STFT can be performed as:

$$X(n,\omega) = \sum_{\tau=-\infty}^{\infty} x(\tau)w(n-\tau)e^{-j\omega\tau} \tag{10.41}$$

where the discrete time index $n$ refers to the position of window $w(n)$. Analogous with the periodogram, a spectrogram is defined as:

$$S_x(n,\omega) = |X(n,\omega)|^2 \tag{10.42}$$

Based on the uncertainty principle, i.e. $\sigma_t^2\sigma_\omega^2 \geq \frac{1}{4}$, where $\sigma_t^2$ and $\sigma_\omega^2$ are, respectively, time and frequency domain variances, perfect resolution cannot be achieved simultaneously in both time and frequency domains. Windows are typically chosen to eliminate discontinuities at block edges and to retain positivity in the power spectrum estimate. The choice also impacts upon the spectral resolution of the resulting technique, which, in simple language, corresponds to the minimum frequency separation required to resolve two equal amplitude frequency components [45].

### 10.6.2 Wavelet Transform

WT is another alternative for time-frequency analysis. There is already a well-established literature detailing the WT, such as [45, 46]. Unlike the STFT, the time-frequency kernel for the WT-based method can better localise the signal components in time-frequency space. This efficiently exploits the dependency between time and frequency components. Therefore, the main objective of introducing WT by Morlet [45] was likely to have a coherence time proportional to the sampling period. To proceed, consider the context of a continuous time signal for which a continuous wavelet transform (CWT) is formulated.

#### 10.6.2.1 Continuous Wavelet Transform

The Morlet–Grossmann definition of the CWT for a one-dimensional signal $f(t)$ is:

$$W(a,b) = \frac{1}{\sqrt{a}}\int_{-\infty}^{\infty} f(t)\psi^*\left(\frac{t-b}{a}\right)dt \tag{10.43}$$

where $(.)^*$ denotes complex conjugate, $\psi(t)$ is the analysing wavelet, $a$ (>0) is the scale parameter (inversely proportional to frequency), and $b$ is the position parameter. The transform is linear and is invariant under translations and dilations, i.e.

$$\text{If } f(t) \to W(a,b) \text{ then } f(t-\tau) \to W(a,b-\tau) \tag{10.44}$$

and:

$$f(\sigma t) \to \frac{1}{\sqrt{\sigma}}W(\sigma a, \sigma b) \tag{10.45}$$

The last property makes the WT very suitable for analysing hierarchical structures. It is similar to a mathematical microscope with properties that do not depend on the magnification. Consider a function $W(a,b)$ which is the WT of a given function $f(t)$. The time-domain signal $f(t)$ can be recovered according to:

$$f(t) = \frac{1}{C_\varphi}\int_0^{\infty}\int_{-\infty}^{\infty}\frac{1}{\sqrt{a}}W(a,b)\varphi\left(\frac{t-b}{a}\right)\frac{dadb}{a^2} \tag{10.46}$$

where:

$$C_\varphi = \int_0^{\infty}\frac{\hat{\psi}^*(\nu)\hat{\varphi}(\nu)}{\nu}d\nu = \int_{-\infty}^{0}\frac{\hat{\psi}^*(\nu)\hat{\varphi}(\nu)}{\nu}d\nu \tag{10.47}$$

Although often it is considered that $\psi(t) = \varphi(t)$, other alternatives for $\varphi(t)$ may enhance certain features for some specific applications [46]. The reconstruction of $f(t)$ is subject to having $C_\varphi$ defined (admissibility condition). The case $\psi(t) = \varphi(t)$ implies $\hat{\psi}(0) = 0$, i.e. the mean of the wavelet function is zero.

#### 10.6.2.2 Examples of Continuous Wavelets

Different waveforms/wavelets/kernels have been defined for the CWTs. The most popular ones are given below.

Morlet's wavelet is a complex-valued waveform defined as:

$$\psi(t) = \frac{1}{\sqrt{2\pi}} e^{-\frac{t^2}{2} + j2\pi b_0 t} \tag{10.48}$$

This wavelet may be decomposed into its constituent real and imaginary components as:

$$\psi_r(t) = \frac{1}{\sqrt{2\pi}} e^{-\frac{t^2}{2}} \cos(2\pi b_0 t) \tag{10.49}$$

$$\psi_i(t) = \frac{1}{\sqrt{2\pi}} e^{-\frac{t^2}{2}} \sin(2\pi b_0 t) \tag{10.50}$$

where $b_0$ is a constant, and it is considered that $b_0 > 0$ to satisfy the admissibility condition. Figure 10.5 shows, respectively, the real and imaginary components.

As another wavelet, the Mexican hat defined by Murenzi [45], is given as:

$$\psi(t) = (1 - t^2)e^{-0.5t^2} \tag{10.51}$$

which is the second derivative of a Gaussian waveform.

#### 10.6.2.3 Discrete Time Wavelet Transform

A discrete approximation of the wavelet coefficients becomes necessary for processing the data using digital systems. The discrete wavelet transform (DWT) can be derived in accordance with the sampling theorem if we process a frequency band-limited signal.

The CWT may be discretised with some simple considerations on the modification of the wavelet pattern by dilation. Given that the wavelet function $\psi(t)$ is not generally band-limited, it is necessary to suppress the values outside the frequency components above half the sampling frequency to avoid aliasing (overlapping in frequency).

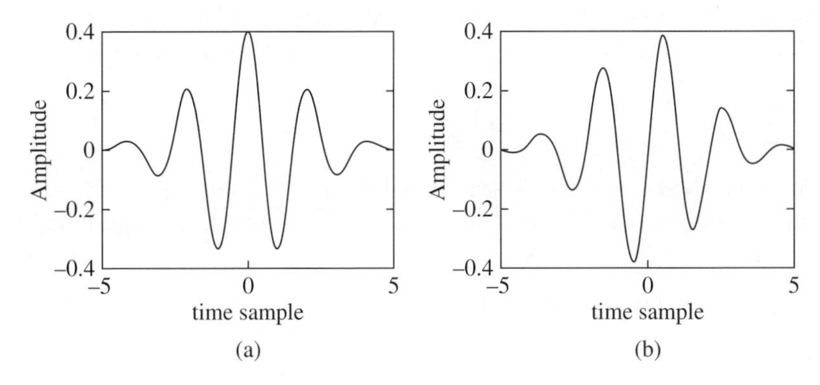


**Figure 10.5** Morlet's wavelet: real (a) and imaginary (b) parts.

A Fourier space may be used to compute the transform scale by scale. The number of elements for a scale can be reduced if the frequency bandwidth is also reduced. This requires a band-limited wavelet. The decomposition proposed by Littlewood Paley [47] provides a very nice illustration of the element reduction scale by scale. This decomposition is based on an iterative dichotomy of the frequency band. The associated wavelet is well localised in Fourier space, where it allows a reasonable analysis to be made, although not in the original space. The search for a discrete transform, which is well localised in both spaces, leads to multiresolution analysis.

### 10.6.3 Multiresolution Analysis

Multiresolution analysis refers the embedded data subsets generated by interpolations (or down-sampling and filtering) of the signal at different scales. A function $f(t)$ is projected at each step $j$ onto the subset $V_j$. This projection is defined by the scalar product $C_j(k)$ of $f(t)$ with the scaling function $\emptyset(t)$, which is dilated and translated:

$$C_j(k) = \langle f(t), 2^{-j}\emptyset(2^{-j}t - k)\rangle \tag{10.52}$$

where $\langle .,. \rangle$ denotes an inner product and $\emptyset(t)$ has the property:

$$\frac{1}{2}\emptyset\left(\frac{t}{2}\right) = \sum_{n=-\infty}^{\infty} h(n)\emptyset(t-n) \tag{10.53}$$

where the right side is convolution of $h$ and $\emptyset$. By taking the Fourier transform of both sides:

$$\Phi(2\omega) = H(\omega)\Phi(\omega) \tag{10.54}$$

where $H(\omega)$ and $\Phi(\omega)$ are the Fourier transforms of $h(t)$ and $\emptyset(t)$, respectively. For a discrete frequency space (i.e. using the DFT) the above equation permits computation of the wavelet coefficient $C_{j+1}(k)$ from $C_j(k)$ directly. If we start from $C_0(k)$ we compute all $C_j(k)$, with $j > 0$, without directly computing any other scalar product:

$$C_{j+1}(k) = \sum_n C_j(n)h(n-2k) \tag{10.55}$$

where k is the discrete frequency index. At each step, the number of scalar products is divided by two and consequently the signal is smoothed. Using this procedure, the first part of a filter bank is built up. In order to restore the original data, Mallat uses the properties of orthogonal wavelets [48], but the theory has been generalised to a large class of filters by introducing two other filters, $\tilde{h}$ and $\tilde{g}$, also called conjugate filters. The restoration is performed with:

$$C_j(k) = 2\sum_l [C_{j+1}(l)\tilde{h}(k+2l) + w_{j+1}(l)\tilde{g}(k+2l)] \tag{10.56}$$

where $w_{j+1}(.)$ are the wavelet coefficients at scale $j+1$. For an exact restoration, two conditions, namely anti-aliasing and exact restoration, must be satisfied for the conjugate filters:

*Anti-aliasing condition*:

$$H\left(\omega + \frac{1}{2}\right)\tilde{H}(\omega) + G\left(\omega + \frac{1}{2}\right)\tilde{G}(\omega) = 0 \tag{10.57}$$

*Exact restoration:*

$$H(\omega)\tilde{H}(\omega) + G(\omega)\tilde{G}(\omega) = 1 \tag{10.58}$$

In the decomposition stage the input is successively convolved with the two filters $H$ (low frequencies) and $G$ (high frequencies). Each resulting function is decimated by suppression of one sample out of two. In the reconstruction, we restore the sampling by inserting a zero between each two samples. Then, the signal is convolved with the conjugate filters $\widetilde{H}$ and $\widetilde{G}$, the resulting outputs added, and the result multiplied by 2. The iteration can continue until the smallest scale is achieved. Orthogonal wavelets correspond to the restricted case, where:

$$G(\omega) = e^{-2\pi\omega} H^*\left(\omega + \frac{1}{2}\right) \tag{10.59}$$

$$\widetilde{H}(\omega) = H^*(\omega) \tag{10.60}$$

$$\widetilde{G}(\omega) = G^*(\omega) \tag{10.61}$$

and:

$$|H(\omega)|^2 + \left|H\left(\omega + \frac{1}{2}\right)\right|^2 = 1 \quad \forall\omega \tag{10.62}$$

We can easily see that this set satisfies the two basic relations (10.57) and (10.58). Among various wavelets, Daubechies wavelets are the only compact solutions to satisfy the above conditions. For biorthogonal wavelets we have the relations:

$$G(\omega) = e^{-2\pi\omega} \widetilde{H}^*\left(\omega + \frac{1}{2}\right) \tag{10.63}$$

$$\widetilde{G}(\omega) = e^{2\pi\omega} H^*\left(\omega + \frac{1}{2}\right) \tag{10.64}$$

and:

$$H(\omega)\widetilde{H}(\omega) + H^*\left(\omega + \frac{1}{2}\right)\widetilde{H}^*\left(\omega + \frac{1}{2}\right) = 1 \tag{10.65}$$

In addition, the relations (10.57) and (10.58) need to be satisfied. A large class of compact wavelet functions can be used. Many sets of filters have been proposed, especially for coding [47]. It has been shown that the choice of these filters must be guided by the regularity of the scaling and the wavelet functions. The complexity is proportional to $N$.

### 10.6.4 Synchro-squeezing Wavelet Transform

Synchro-squeezing wavelet transform (SSWT) has been introduced as a post-processing technique to enhance the time-frequency spectrum obtained by applying WT [43]. Consider the WT of an input signal $f(t)$ using Eq. (10.43). Assuming that $f(t)$ is a pure harmonic signal ($f(t) = A\cos(\omega t)$), using Plancherel's theorem, the following equations are derived from (10.43) [43]:

$$\begin{aligned} W(a,b) &= \int_{-\infty}^{\infty} f(t) a^{-\frac{1}{2}} \overline{\psi}\left(\frac{t-b}{a}\right) dt \\ &= \frac{1}{2\pi} \int_{-\infty}^{\infty} \hat{f}(\varepsilon) a^{\frac{1}{2}} \overline{\hat{\psi}}(a\varepsilon) e^{ib\varepsilon} d\varepsilon \\ &= \frac{A}{4\pi} \int_{0}^{\infty} [\delta(\varepsilon - \omega) + \delta(\varepsilon + \omega)] a^{\frac{1}{2}} \overline{\hat{\psi}}(a\varepsilon) e^{ib\varepsilon} d\varepsilon \\ &= \frac{A}{4\pi} a^{\frac{1}{2}} \overline{\hat{\psi}}(a\omega) e^{ib\omega} \end{aligned} \tag{10.66}$$

One assumption in the above equation is that the selected mother wavelet is concentrated within the positive energy range, which means $\hat{\psi}(\varepsilon) = 0$, for $\varepsilon < 0$. If $\hat{\psi}(\varepsilon)$ is concentrated around $\varepsilon = \omega_0$ then $W(a, b)$ is concentrated around $a = \omega_0/\omega$. This is spread out over a region of horizontal line (e.g. $a = \omega_0/\omega$). In the situation where $\omega$ is almost but not exactly similar to the actual instantaneous frequency (IF) of the input signal then $W(a, b)$ has nonzero energy. By synchro-squeezing, the idea is to move away this energy from $\omega$. The proposed method in the SSWT aims to reassign the frequency locations which are closer to the actual IF to obtain an enhanced spectrum. To do that, first, the candidate IFs are calculated. For these IFs, $W(a, b) \neq 0$:

$$\omega(a, b) = -i(W(a, b))^{-1} \frac{\partial}{\partial b} W(a, b) \tag{10.67}$$

Considering the selected pure harmonic signal $f(t) = A\cos(\omega t)$, it is simple to observe that $\omega(a, b) = \omega$. The candidate IFs are exploited to recover the actual frequencies. Therefore, a reallocation technique has been used to map the time domain into time-frequency domain using $(b, a) \Rightarrow (b, \omega(a, b))$. Based on this, each value of $W(a, b)$ (computed at discrete values of $a_k$) is re-allocated into $T_f(\omega_l, b)$ as provided in the following equation:

$$T_f(\omega_l, b) = (\Delta\omega)^{-1} \sum_{a_k : |\omega(a_k, b) - \omega_l| \leq \frac{\Delta\omega}{2}} W(a_k, b) a_k^{-\frac{3}{2}} (\Delta a)_k \tag{10.68}$$

where $\omega_l$ is the nearest frequency to the original point $\omega(a, b)$, $\Delta\omega$ is the width of the frequency bins $\left[\omega_l - \frac{1}{2}\Delta\omega, \omega_l + \frac{1}{2}\Delta\omega\right]$, $\Delta\omega = \omega_l - \omega_{l-1}$, and $(\Delta a)_k = a_k - a_{k-1}$. $T_f(\omega_l, b)$ represents the synchrosqueezed transform at the centres $\omega_l$ of consecutive frequency bins. For each fixed time point $b$, the reassigned frequencies should be estimated for all scales using Eq. (10.67). For each desired IF of $\omega_l$, $T_f(\omega_l, b)$ is calculated using the summation of all $W(a_k, b)$ considering that the distance between the reassigned frequency $\omega(a_k, b)$ and $\omega_l$ must be within the specified frequency bin width $(\Delta\omega)$. It has been shown that the original signal can be reconstructed after the synchro-squeezing process [43].

## 10.7 Adaptive Filtering

Adaptive filters are mainly transversal filters whose tap coefficients are tuned directly or recursively in order to achieve a predefined (target) output. Figure 10.6 presents a schematic of an adaptive filter used for single-channel filtering.

The coefficients of an adaptive filter, $\mathbf{w}$, can be calculated using Weiner equations, which minimise the statistical average of the error between the output $y(n)$ and the target signal $d(n)$, i.e. $E[e^2(n)]$ as:

$$J = E[e^2(n)] = E[(d(n) - \mathbf{w}x(n))^2] \tag{10.69}$$

which results in a unique solution to the following system of equations called Wigner–Ville equations:

$$\mathbf{w} = \mathbf{R}^{-1}\mathbf{p} \tag{10.70}$$

where matrix $\mathbf{R}$ and vector $\mathbf{P}$ are, respectively, the autocorrelation of input $x(n)$ and its cross-correlation with target signal $d(n)$. The major problem with this solution is the need

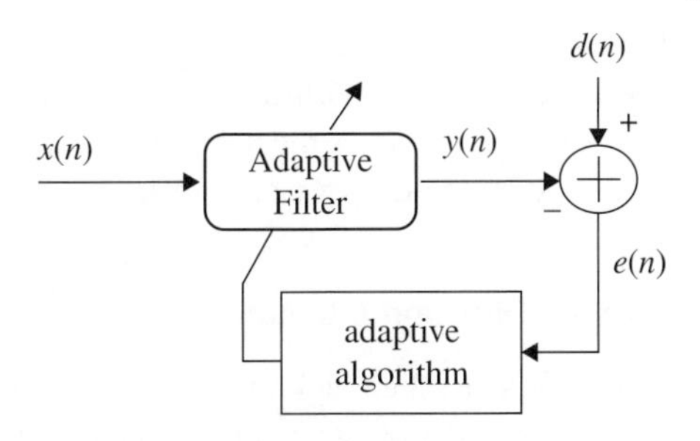


**Figure 10.6** Block diagram of an adaptive filter for single-channel filtering.

for inversing matrix **R**, which can be a large matrix for channels (systems) with a large number of tap coefficients, such as highly reverberant media. Recursive solutions are therefore recommended. The most popular solution is achieved by minimising the above cost function, $J$, through taking the gradient of error $J$ with respect to **w**, i.e. $\nabla_{\mathbf{w}}J$:

$$\mathbf{w}_k = \mathbf{w}_{k-1} - \mu\nabla_{\mathbf{w}}J \tag{10.71}$$

where $\mu$ is the stepsize parameter and often set empirically within $0 < \mu < 2/\lambda_{max}$, and $\lambda_{max}$ is the maximum eigenvalue of **R**. A stopping criterion such as a certain number of iterations or a reasonably low level of error is defined to end the iteration.

Many other approaches maximise the orthogonality between the input and error signal or between the error and target signal [49].

The notion of adaptive filters has been extended to multisensor networks through consensus or the diffusion of information. Adaptive networks in mobile communications consist of collections of nodes with learning and motion abilities that interact with each other locally in order to solve distributed processing and distributed inference problems in real time. The objective of such a network is to estimate some filter coefficients in a fully distributed manner and in real time, where each node is allowed to interact with its neighbours only.

Assuming the sensor nodes have limited functionalities, they are allowed to cooperate with their neighbours to optimise a common objective function.

An efficient way to collectively use the information recorded or sensed by a number of sensors within a neighbourhood is by cooperative adaptive filtering. The cooperative filters are similar to conventional adaptive filters, where the inputs are the aggregation of contributions (or cooperation) between the input nodes (sources). The cooperation pattern between the nodes themselves varies in time.

## 10.8 Cooperative Adaptive Filtering

In distributed adaptive networks the network nodes (agents) process the streaming data in a cooperative manner. These networks are ubiquitous and are present in various applications such as Internet-connected devices, sensor networks [50], wearable networks [51], intrabody molecular communication networks [52], biological and social networks [53], smart grids [54], game theory [55], and power control systems [56].

In multiagent distributed networks, the agents collaborate with each other in order to solve a global optimisation problem over the network. Several strategies are proposed in the literature to solve this problem. Three prominent approaches proposed for this purpose are incremental [57–61], consensus [62–65], and diffusion strategies [66–72]. In a general interconnected network, a diffusion strategy, which was introduced most recently, is preferred. This strategy alleviates the limitations of the incremental networks and also overcomes the instability problem of the consensus adaptive networks.

### 10.8.1 Diffusion Adaptation

Consider $N_k$ as the neighbours of node $k$, including $k$ itself. To model the problem, a connected network consisting of $N$ nodes is considered. A simple network of this type can be seen in Figure 10.7.

For each node a $1 \times M$ unknown vector $\mathbf{w}_k^o$ is estimated using the collected measurements. Each node $k$ of the network has access to a scalar measurement $d_k(i)$ and a $1 \times M$ regression vector $\mathbf{x}_k(i)$ at each time instant $i \geq 0$. The data at each node are assumed to be related to the unknown parameter vector $\mathbf{w}_k^o$ via a linear regression model:

$$d_k(i) = \mathbf{x}_k^T(i)\mathbf{w}_k^o(i) + n_k(i) \tag{10.72}$$

where $n_k(i)$ is the measurement noise at node $k$ and time instant $i$ and $T$ refers to the transposition of a vector. In the context of diffusion adaptation, this leads to an optimisation problem which minimises a cost function, such as $J_k(\mathbf{w})$ in Figure 10.7, proportional to the difference between an estimate and the corresponding objective, resulting in the following two-step solution including adaptation and combination [73]:

$$\begin{aligned} {}_k(i) &= \mathbf{w}_k(i-1) + {}_k(d_k(i) - \mathbf{x}_k^T(i)\mathbf{w}_k(i-1))\mathbf{x}_k(i) \\ \mathbf{w}_k(i) &= \sum_{l \in N_k} a_{lkl}(i) \end{aligned} \tag{10.73}$$

where $\mu_k \geq 0$ is the stepsize parameter used by node $k$, $\mathbf{w}_k$ denotes the estimate of $\mathbf{w}^o$, and $\{a_{lk}\}$ are nonnegative cooperation coefficients which satisfy the following conditions:

$$\sum_{l \in N_k} a_{lk} = 1, \text{ and } a_{lk} = 0 \ \textit{if} \ l \notin N_k \tag{10.74}$$

The adapt-then-combine (ATC) steps can be reordered to make the combination first and then the adaptation (i.e. combine-then-adapt, or CTA) without any significant change in

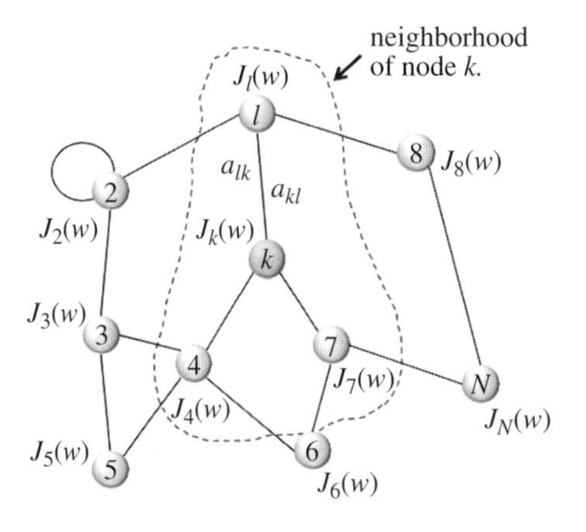


**Figure 10.7** A network of nodes with a cooperation neighbourhood around sensor $k$.

the outcome. In general applications, there is no clue about the values of $\{a_{lk}\}$ and often for an $N$ node neighbourhood each link weight is considered equal to $1/N$. In applications to multichannel EEG, however, the cooperation coefficients can be estimated through the measurement of brain connectivity. Least mean square (LMS) optimisation is often used to estimate the weights of the filter [73]. Other methods, such as Kalman filtering, are also employed [74].

For the assessment of cooperative filters mean-square deviation (MSD) is often used. MSD provides the squared error in the estimation of filter parameter vector which can be calculated in each iteration.

Diffusion adaptation filters have also been extended to multitask scenarios where the nodes are divided into groups or clusters each following a particular objective [75]. Both single- and multitask cooperative networks have numerous applications in sensor networks. For example, in [76] the authors try to solve the problem of positioning a source using angle-of-arrival measurements taken by a WSN where some of the nodes experience non-line-of-sight (NLOS) propagation conditions. In order to mitigate the resulting error, the EM and weighted least-squares (WLS) estimation of the source position algorithms are combined in order to identify and discard the nodes in NLOS. In addition, a distributed version of the algorithm based on a diffusion strategy that iteratively refines the position estimate while driving the network to a consensus has been presented. To enable the diffusion strategy, the nodes of the network should have the necessary computation ability to calculate the network parameters and perform the WLS operation in each iteration of the algorithm [76].

Although cooperative networking in general, and diffusion adaptation in particular, allow sharing of information among the nodes of a network, to facilitate distributed data processing at the nodes, different algorithms have been proposed. For example, a cooperative beamforming technique for BANs is proposed by Mijovic et al. [77]. They consider an indoor environment with multiple BANs that need to transmit data towards specific receivers, located in fixed positions. Nodes deployed on the same body may cooperate in order to form virtual antenna arrays (VAAs) and transmit the data towards one of the available receivers. The receivers are also equipped with multiple antennas, such that a virtual multiple input, multiple output (MIMO) channel is established. A cooperative beamforming technique is designed in both transmitter and receiver sides.

Considering the fact that wireless sensors have several constraints, such as short transmission range, poor processing capabilities, and limited available energy, Chiasserini et al. [78] designed an energy-efficient distributed processing system for sensor networks by exploiting the concept of collaborative signal processing. They considered a sensor network where each node corresponds to a processor each of them having a local memory and able to communicate with each other by sending and receiving messages. They showed that the processing operations can be distributed among sensors in such a way that the energy consumption of the overall network is reduced. This enhances the efficiency of energy usage. They applied their platform for performing a distributed fast Fourier transform (FFT) and presented their algorithm using a two-sensor scenario [78].

Distributed signal processing, as well as cooperative communications, is itself a hot topic due to the available large data sizes (big data) and the need for processing such data over multicore (multinode) processing systems.

## 10.9 Multichannel Signal Processing

The most important application of multichannel signal processing is to recover some underlying signal sources from their mixtures captured by a number of sensors simultaneously. Another application is localisation of the signal sources using a good model of the medium. Some practical examples are decomposition of communication signals received by multiple antennas in the receiver, separation of sound signals captured by multiple microphones, speaker localisation, and the recovery of original brain source activities from multichannel EEG or MEG signals.

Most physiological information, signals, and images are found as the mixtures of more than one component. Often, if the number of recording channels is large enough and the constituent sources are uncorrelated, independent, or disjoint in some linear domain, an accurate solution to their separation can be found. Although some a priori knowledge such as independency of the sources are exploited in solving the separation problem, the overall operation is often blind to the nature and characteristics of the constituent sources and their mixing system. In these cases, the approach is called blind source separation (BSS). Different approaches based on some fundamental mathematical concepts – such as adaptive filtering, PCA, matrix factorisation, independent component analysis (ICA), and tensor factorisation –have been very popular in the literature for solving the multichannel data decomposition and channel identification problems.

Most of the BSS approaches require the data to remain stationary within the segment under process. This is generally needed to enable the definition of some invariant statistics within each data segment. However, this may not be always true, particularly for many physiological signals. Nonstationarity of the sources, mainly due to the existing transient events; nonlinearity of the system, due to the involvement of a time-varying media or moving sources; and also other effects such as time-delays in reaching the source signals to the sensors and the way the sources are spread in space can be the main reasons for the degrading of BSS operations.

Assuming the mixing process is linear, there are still two ways the sources are combined at the sensors: instantaneous and convolutive. In instantaneous mixing systems there is no time delay involved in the propagation of the signals from their sources to their destinations (sensors). Good examples are electric and electromagnetic signals such as EEG, MEG, and ECG. The convolutive systems, however, involve time delays and the source signals arrive at the sensors at different times. This is simply because the propagation in the medium has limited speed. Communication, sound, and speech signals, as well as the internal body sounds (such as body joint sounds or heart and lung sounds), captured by multiple sensors, antennas, or stethoscopes are of this kind. Solutions to convolutive BSS are generally more complex than those of instantaneous BSS and the difficulty rises where the delay length (reverberation in the case of sound) increases.

Among different assumptions about the sources, independency is the most important one and used in the majority of BSS applications which are based on ICA. Considering the multichannel signal vector at time instant $n$ as $\mathbf{y}(n)$ and the constituent signal components as $y_i(n)$, the components are independent if:

$$p_{\mathbf{Y}}(\mathbf{y}(n)) = \prod_{i=1}^{m} p_y(y_i(n)) \qquad \forall n \tag{10.75}$$

where $p_{\mathbf{Y}}(\mathbf{y}(n))$ is the joint probability distribution, $p_y(y_i(n))$ are the marginal distributions, and $m$ is the number of independent components. The objective of the separation (or de-mixing) process is therefore maximising the independency among the separated sources. Nevertheless, a perfect separation of the signals takes into account the structure of the mixing process. In practice this process is unknown, but some assumptions, such as independency, may be made about the source statistics.

### 10.9.1 Instantaneous and Convolutive BSS Problems

Generally, the BSS algorithms do not make realistic assumptions about the environment in order to make the problem more tractable. The most simple and widely used model is the instantaneous BSS, where the source signals arrive at the sensors at the same time. This has been considered for the separation of biological signals, such as the EEG, where the signals have narrow bandwidths and the sampling frequency is normally low. The BSS model in this case can be easily formulated as:

$$\mathbf{x}(n) = \mathbf{H}\mathbf{s}(n) + \mathbf{v}(n) \tag{10.76}$$

For $n_e$ electrodes and $m$ sources, the dimensions of vector form of source signals $\mathbf{s}(n)$, observed signals $\mathbf{x}(n)$, and noise $\mathbf{v}(n)$ at discrete time $n$ are respectively $m \times 1$, $n_e \times 1$, and $n_e \times 1$. $\mathbf{H}$ is the mixing matrix of size $n_e \times m$. The separation is performed by means of a separating $m \times n_e$ matrix, $\mathbf{W}$, which uses only the information about $\mathbf{x}(n)$ to reconstruct the original source signals (or the independent components) as:

$$\mathbf{y}(n) = \mathbf{W}\mathbf{x}(n) \tag{10.77}$$

In the context of EEG signal processing, demonstrated in Figure 10.8, $n_e$ denotes the number of electrodes. The early approaches in instantaneous BSS include the work of Herault and Jutten [79] in 1986. In their approach, they considered non-Gaussian sources with a similar number of independent sources and mixtures. They proposed a solution based on recurrent artificial NN for the separation of the sources.

In acoustic applications, however, there are usually time lags between the arrival of the signals at the sensors. The signals also may arrive via multiple paths. This type of

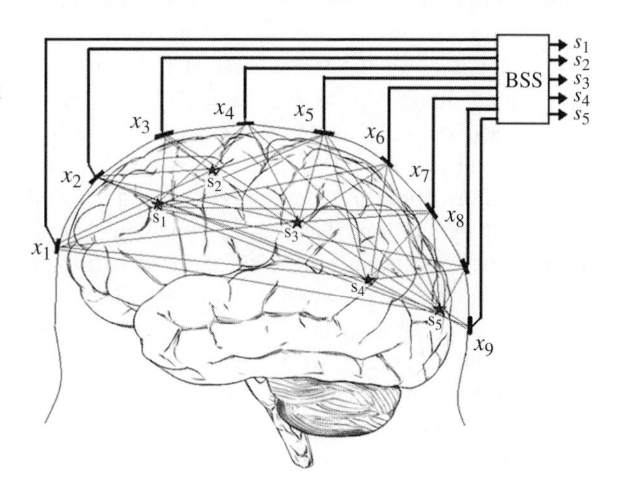


**Figure 10.8** Multichannel recording of brain signals (EEG) and separating them to their constituent sources.

mixing model is called a convolutive model, which can be echoic (with reverberation) or anechoic (involving no reverberation but time delays) [80]. In both cases the vector representations of mixing and separating processes are changed to $\mathbf{x}(n) = \mathbf{H}(n) * \mathbf{s}(n) + \mathbf{v}(n)$ and $\mathbf{y}(n) = \mathbf{W}(n) * \mathbf{x}(n)$, respectively, where * denotes convolution operation.

In an anechoic model, however, the expansion of the mixing process may be given as:

$$x_i(n) = \sum_{j=1}^{m} h_{ij} s_j(n - \delta_{ij}) + v_i(n),\ for\, i = 1, \ldots, n_e \tag{10.78}$$

where the attenuation, $h_{ij}$, and delay, $\delta_{ij}$, of source $j$ to reach sensor $i$ would be determined by the physical position of the source relative to the sensors. Then, the separating process is given as:

$$y_j(m) = \sum_{i=1}^{n_e} w_{ji} x_i(m - \delta_{ji}),\ for\, j = 1, \ldots, m \tag{10.79}$$

where the $w_{ji}$ are the elements of matrix $\mathbf{W}$. In an echoic mixing environment, it is expected that the signals from the same sources reach to the sensors through multiple paths and delays. Therefore, the expansion of mixing model is changed to:

$$x_i(n) = \sum_{j=1}^{m} \sum_{k=1}^{K} h_{ij}^k s_j(n - \delta_{ij}^k) + v_i(n),\ for\, i = 1, \ldots, N \tag{10.80}$$

where $K$ denotes the number of paths and $v_i(n)$ is the accumulated noise at sensor $i$. The separating system is formulated similarly to the anechoic one. Obviously, modelling such a system is more challenging, and for a known number of sources a more accurate result may be expected if the number of paths is known.

The aim of BSS using ICA is to estimate a separating matrix $\mathbf{W}$ such that $\mathbf{Y} = \mathbf{WX}$ best approximates the independent sources S, where $\mathbf{Y}$ and $\mathbf{X}$ are, respectively, matrices with columns $\mathbf{y}(n) = [y_1(n), y_2(n), \ldots, y_m(n)]^T$ and $\mathbf{x}(n) = [x_1(n), x_2(n), \ldots, x_{ne}(n)]^T$.

In any case, the separating matrix for the instantaneous case is expected to be equal to the inverse of the mixing matrix, i.e. $\mathbf{W} = \mathbf{H}^{-1}$. However, since all ICA algorithms are based upon restoring independence, the separation is subject to permutation and scaling ambiguities in the output independent components, i.e. $\mathbf{W} = \mathbf{PDH}^{-1}$, where $\mathbf{P}$ and $\mathbf{D}$ are, respectively, the permutation and scaling matrices.

In order to solve the instantaneous BSS problem Herault and Jutten used a simple but fundamental adaptive algorithm [79]. Linsker [81] proposed unsupervised learning rules based on information theory that maximised the average mutual information between the inputs and outputs of an ANN. Comon [82] performed minimisation of mutual information to make the outputs independent. The Infomax algorithm [83] was developed by Bell and Sejnowski, which in spirit, is similar to the Linsker method. Infomax uses an elegant stochastic gradient learning rule initially, which was proposed by Amari et al. [84]. Non-Gaussianity of the sources was first exploited by Hairline and Oja [85] in developing their fast independent component analysis (fICA) algorithm. fICA is actually a blind source extraction (BSE) algorithm, which extracts the sources one by one based on their kurtosis; the signals with transient peaks have high kurtosis. Later it was demonstrated that the Infomax algorithm and maximum likelihood estimation are, in fact, equivalent [86, 87].

On the other hand, convolutive source separation is a difficult problem as the received signals to different sensors are generally subject to delay, reverberation, and multipath fading. It is very difficult to estimate an accurate reverberation length. There are, however, many solutions to convolutive BSS. These solutions are not very accurate and are highly dependent on the reverberation length (level). The longer the reverberation period has more negative impact on the separation results. One of the fundamental solutions is to transform the signals into frequency domain where the convolution changes to multiplication and therefore the instantaneous solutions can be applied [86]. More accurate solutions to this problem can be achieved using tensor factorisation [88].

Moreover, any extra information about the source signals, such as sparsity, or the mixing system, such as the room dimensions, may be exploited in the formulation to regularise the separation process and achieve more accurate solutions.

### 10.9.2 Array Processing

An important problem in WSNs is localisation and tracking of the source objects. This requires localisation of the sensor nodes or the mobile units themselves and localisation of intruders or undesired objects [89]. Angle and range are the main factors in positioning the sources. The widely used range estimation models include received signal strength (RSS), time of arrival (TOA), and time difference of arrival (TDOA). These techniques are more useful for sensor nodes with direct communication links between the sensors within a neighbourhood. Estimation of the direction of incident signals or their angle of arrival (AOA) potentially locate the signal sources in a noncooperative, stealthy, and passive manner [90–92].

AOA estimation can be also used to enhance the communication efficiency as well as the network capacity, support location-aided routing, manage dynamic networking, and many location-based services [93–95]. Moreover, using smart antennas and through spatial filtering, the sender can focus the transmission energy towards the desired user while minimising the effect of interference. Thus, the receiver can form a desired beam towards the sender while simultaneously placing nulls in the directions of the other transmitters. This leads to an improved user capacity through reducing the power consumption, lowering the bit error rate (BER), and extending the range of coverage [96]. To further exploit the AOA capability, various medium access control (MAC) protocols have been developed.

An AOA problem is classified as narrow-band if the signal bandwidth is small compared to the inverse of a wave front transit time across the array aperture and also the array response is not a function of frequency over the signal bandwidth. The number of sources is assumed known (given or estimated from the signal detection algorithms [97]) and there are fewer sources than sensors; this is to guarantee the uniqueness of AOA estimation [98].

AOA estimation with a sensor array has received considerable attention from a variety of research communities like radar, sonar, radio astronomy, mobile communications, and wireless sensor networks. Thus, a range of useful and complementary algorithms have been developed, such as delay-and-sum beamforming; multiple signal classification (MUSIC) which exploits the eigen-structure of the data covariance matrix; estimation of signal parameters via rotational invariance technique (ESPRIT); minimum variance distortion less response (MVDR) or Capon's algorithm, which is often used as a beamformer;

maximum likelihood (ML) as a near optimal technique [99–101]; and many others [99–112]. In general, AOA estimation algorithms benefit from the statistical information and structure of the data covariance matrix computed from a batch of array snapshots or data samples.

Delay-and-sum beamforming, also known as digital beamforming (DBF), is one of the oldest and simplest array processing algorithms [103], and is a basis for other approaches. In this approach when a propagated signal is present at an array's aperture, if the sensor element outputs are delayed by appropriate amounts and added together, the signal will be reinforced with respect to noise or waves propagating in different directions. The delays that reinforce the signal are directly related to the length of time it takes for the signal to propagate between sensor elements. The AOA measurement process is similar to that of mechanically steering the array in different directions and measuring the output power. When a set of data samples is given, the DBF power spectrum is evaluated and the AOA estimates are obtained by detecting the peaks in the spectrum.

Popular communication channel modelling assumptions validated for cellular radio and other long-range wireless sensor networks are not appropriate for BAN. In BANs, short- and long-range fading principles are not justified and there are assumptions to be considered due to wave propagation through the body and around the body during the body movement. In general, there is neither any optimum nor any near-optimum model for BAN propagations which can be generalised to all the cases; instead, the emphasis is on the 'goodness' of the propagation model. A good choice of model for a BAN should satisfy some fundamental requirements, such as:

- *Generality*. It should apply to various scenarios related to different subjects, different sensor modalities, and different measurements.
- *Comprehensiveness*. It should accommodate the variations in both body and environment, as well as the sensor/recording parameters (distant, amplification, sampling rate, transmission rate, etc.).
- *Simplicity*. It can be formulated and implemented easily.

The communication and propagation models for BANs are discussed in the next chapter of this book.

## 10.10 Signal Processing Platforms for BANs

Establishing a useful signal processing platform for BANs has become very challenging as there are numerous types of sensing and communication methodologies for sensor networks. Various generations of motes, mobile devices, wristwatches, body-worn sensors, and the sensors used for occasional measurements have different data capturing parameters. Sampling frequency, communication rate, required power, geometry, mode (active or passive), etc., are the parameters used to select a suitable hardware or software platform. Therefore, it is not feasible to provide a platform which can accommodate all means of sensing, processing, and communication methods.

Some simple platforms, such as the signal processing in node environment (SPINE), are very basic software frameworks for the development of body sensor network (BSN) applications. SPINE provides developers of signal processing algorithms with application programming interfaces (APIs) and libraries of protocols, some utilities, and data processing functions for applying simple operations on the sensor signals [113]. SPINE1.x has been under development since 2008 and it has been adopted in several wireless BAN applications. It has been implemented for supporting different sensor architectures. Its programming model is mainly based on functions and signal features extraction. On the other hand, SPINE2 has been conceived and developed to achieve a high platform-independency through C-like programmable sensor architectures. SPINE2 offers programming abstraction based on a task-oriented paradigm so that distributed and collaborative applications can be programmed as a dynamically schedulable and reconfigurable set of tasks to be instantiated on the sensor nodes.

SPINE has been recently used as the wearable and mobile software layer of the BodyCloud open-source smart-Health platform. BodyCloud is a distributed software framework for the rapid prototyping of large-scale BSN applications. It is designed as a SaaS architecture to support the storage and management of sensor data streams and the processing and analysis of the stored data using software services hosted in the Cloud. In particular, BodyCloud endeavours to support several cross-disciplinary applications and specialised processing tasks. It enables large-scale data sharing and collaborations among users and applications in the Cloud, and delivers Cloud services via sensor-rich mobile devices. BodyCloud also offers decision support services to take further actions based on the analysed BSN data [113].

SPINE has been also integrated within Agent-based COoperating Smart Objects (ACOSO) which is an agent-oriented middleware for the development, management, and deployment of cooperating smart objects in any context of application which requires distributed computation, proactivity, knowledge management, and interaction among system operators, sensors, or actuators.

Seto et al. [114] present a mobile platform for body sensor networking based on a smartphone. The platform facilitates the local processing of data at both the sensor mote and the smartphone levels, reducing the overhead for data transmission to remote services. Their smartphone platform enables wearable signal processing and opportunistic sensing strategies, in which many of the onboard sensors and capabilities of modern smartphones may be collected and fused with body sensor data to provide environmental and social context.

To demonstrate the system performance, using simple signal processing operations, they implemented the system for monitoring the patients with congestive heart failure by using sensor motes (TelosB, SHIMMER) for measuring motion, heart, and breathing. The platform has also been used for obesity intervention and also GPS-based localisation. Such a platform, however, is not very useful in places where sophisticated algorithms are to be implemented.

## 10.11 Conclusions

The fundamentals in processing captured signals using single and multiple sensors together with some applications have been briefly described in this chapter. Pre-conditioning, mainly denoising, of the signals and extracting the clinically meaningful information are probably the most important requirements. Transforming the signals into a more meaningful and usable domain by means of signal transformation techniques is often a very useful approach for analysing and visualising the data. Decomposition of the information into its meaningful and constituent components is extremely useful in order to detect, recognise, and track the changes in the retrieved information through multiple communication channels. This is less structured and more difficult in the case of single-channel/ sensor data. There is also a wide field of array signal processing techniques used for the localisation, tracking, and channel estimation of WSNs in general.

There is also a great deal of signal processing research performed on the onboard (on-sensor) embedded systems for efficient sampling, filtering, denoising, and preconditioning of data. Moreover, the variety of signal processing algorithms has been created and implemented for communication systems, such as channel modelling, source recovery, and array processing. Finally, the art of signal processing has been extended to the network level of communication networks particularly to tackle problems like security and data packet routing.

## References


1 Sanei, S. (2013). *Adaptive Processing of Brain Signals*. Wiley.

2 Strang, G. (1998). *Linear Algebra and Its Applications*, 3e. Thomson Learning.

3 Hyvärinen, A., Kahunen, J., and Oja, E. (2001). *Independent Component Analysis*. Wiley.

4 Cover, T.M. and Thomas, J.A. (2001). *Elements of Information Theory*. Wiley.

5 Azami, H., Hassanpour, H., Escudero, J., and Sanei, S. (2014). An intelligent approach for variable size segmentation of non-stationary signals. *Elsevier Journal of Advanced Research* 6 (5): 687–698.

6 Azarbad, M., Azami, H., and Sanei, S. (2014). A time-frequency approach for EEG signal segmentation. *Journal of AI and Data Mining* 2 (1): 63–71.

7 Morf, M., Vieria, A., Lee, D., and Kailath, T. (1978). Recursive multichannel maximum entropy spectral estimation. *IEEE Transactions on Geoscience Electronics* 16 (2): 85–94.

8 Corsini, J., Shoker, L., Sanei, S., and Alarcon, G. (2006). Epileptic seizure predictability from scalp EEG incorporating constrained blind source separation. *IEEE Transactions on Biomedical Engineering* 53 (5): 790–799.

9 Azami, H., Sanei, S., Mohammadi, K., and Hassanpour, H. (2013). A hybrid evolutionary approach to segmentation of non-stationary signals. *Elsevier Journal of Digital Signal Processing* 23 (4): 1103–1114.

10 Akaike, H. (1974). A new look at statistical model order identification. *IEEE Transactions on Automatic Control* 19: 716–723.

11 Kay, S.M. (1988). *Modern Spectral Estimation: Theory and Application*. Prentice Hall.

12 Guegen, C. and Scharf, L. (1980). Exact maximum likelihood identification of ARMA models: a signal processing perspective. In: *Signal Processing Theory ApplicationsNorth-Holland Publishing Company* (eds. M. Kunt and F. de Coulon), 759–769.

13 Akay, M. (2001). *Biomedical Signal Processing*. Academic Press.

**14** Durbin, J. (1959). Efficient estimation of parameters in moving average models. *Biometrika* 46 (3/4): 306–316.

**15** Trench, W.F. (1964). An algorithm for the inversion of finite Toelpitz matrices. *Journal of the Society for Industrial and Applied Mathematics* 12 (3): 515–522.

**16** Ding, M., Chen, Y., and Bressler, S.L. (2006). Granger causality: basic theory and application to neuroscience. In: *Handbook of Time Series Analysis* (eds. B. Schelter, M. Winterhalder and J. Timmer), 451. Berlin: Wiley.

**17** De Prony, B.G.R. (1795). Essai experimental et analytique: sur les lois de la dilatabilite de fluids elastiques et sur celles de la force expansive de la vapeur de l'eau et de la vapeur de l'alkool, a differentes temperatures. *Journal of Engineering and Polytechnique* 1 (2): 24–76.

**18** Marple, S.L. (1987). *Digital Spectral Analysis with Applications.* Prentice-Hall.

**19** Lawson, C.L. and Hanson, R.J. (1974). *Solving Least Squares Problems.* Englewood Cliffs, NJ: Prentice Hall.

**20** Sanei, S. and Hassani, H. (2015). *Singular Spectrum Analysis of Biomedical Signals.* CRC Press.

**21** Hassani, H., Heravi, S., and Zhigljavsky, A. (2013). Forecasting UK industrial production with multivariate singular Spectrum analysis. *Journal of Forecasting* 32 (5): 395–408.

**22** Ghil, M., Allen, M.R., Dettinger, M.D. et al. (2002). Advanced spectral methods for climatic time series. *Reviews of Geophysics* 40 (1): 3.1–3.41.

**23** Chau, K.W. and Wu, C.L. (2010). A hybrid model coupled with singular spectrum analysis for daily rainfall prediction. *Journal of Hydroformatics* 12 (4): 458–473.

**24** Golyandina, N., Nekrutkin, V., and Zhigljavsky, A. (2001). *Analysis of Time Series Structure: SSA and Related Techniques.* New York/London: Chapman & Hall/CRC.

**25** Awichi, R.O. and Muller, W.G. (2013). Improving SSA predictions by inverse distance weighting. *REVSTAT – Statistical Journal* 11 (1): 105–119.

**26** Giannakis, D. and Majda, A.J. (2012). Nonlinear Laplacian spectral analysis for time series with intermittency and low-frequency variability. *Proceedings of the National Academy of Sciences* 109 (7): 2222–2227.

**27** Dacorogna, M., Muller, U., Olsen, R.B., and Pictet, O. (1998). Modelling short-term volatility with GARCH and HARCH models, nonlinear modelling of high frequency financial time series. In: *Econometrics* (eds. L.C. Dunis and B. Zhou). Wiley.

**28** Hsieh, D.A. (1989). Testing for nonlinear dependence in daily foreign exchange rates. *Journal of Business* 62 (3): 339–368.

**29** Galka, A., Yamashita, O., and Ozaki, T. (2004). GARCH modelling of covariance in dynamical estimation of inverse solutions. *Physics Letters A* 333 (3–4): 261–268.

**30** Tikhonov, A. (1992). *Ill-Posed Problems in Natural Sciences.* Coronet.

**31** Nelson, D.B. (1990). Stationarity and persistence in the GARCH(1,1) model. *Econometric Theory* 6 (3): 318–334.

**32** Schubert, M., Köppen-Seliger, B., and Frank, P.M. (1997). Recurrent neural networks for nonlinear system modelling in fault detection. *IFAC Proceedings Volumes* 30 (18): 701–706.

**33** Roweis, S. and Ghahramani, Z. (1999). A unifying review of linear Gaussian models. *Neural Computation* 11 (2): 305–345.

**34** Dempster, A.P., Laird, N.M., and Rubin, D.B. (1977). Maximum likelihood from incomplete data via the EM algorithm. *Journal of the Royal Statistical Society B* 39 (1): 1–38.

**35** Redner, R.A. and Walker, H.F. (1984). Mixture densities, maximum likelihood and the EM algorithm. *SIAM Review* 26 (2): 195–239.

**36** Ormoneit, D. and Tresp, V. (1998). Averaging, maximum penalized likelihood and Bayesian estimation for improving Gaussian mixture probability density estimates. *IEEE Transactions on Neural Networks* 9 (4): 639–650.

**37** Archambeau, C., Lee, J.A., and Verleysen, M. (2003). On convergence problems of the EM algorithm for finite Gaussian mixtures. In: *Proceedings of European Symposium on Artificial Neural Networks (ESANN 2003)*, 99–106. ESANN.

**38** Hesse, C.W., Holtackers, D., and Heskes, T. (2006). On the use of mixtures of gaussians and mixtures of generalized exponentials for modelling and classification of biomedical signals. In: *Belgian Day on Biomedical Engineering IEEE Benelux EMBS Symposium*. IEEE.

**39** Greenspan, H., Ruf, A., and Goldberger, J. (2006). Constrained gaussian mixture model framework for automatic segmentation of MR brain images. *IEEE Transactions on Medical Imaging* 25 (9): 1233–1245.

**40** Rio, M., Hutt, A., and Loria, B. G. (2010) Partial amplitude synchronization detection in brain signals using Bayesian Gaussian mixture models. *Cinquième Conférence Plénière Française de Neurosciences Computationnelles Neurocomp'10, Lyon, France.*

**41** Condon, E.U. (1937). Immersion of the Fourier transform in a continuous group of functional transformations. *Proceedings of the National Academy of Sciences of the United States of America* 23 (3): 158–164.

**42** Cui, J. and Wong, W. (2006). The adaptive chirplet transform and visual evoked potentials. *IEEE Transactions on Biomedical Engineering* 53 (7): 1378–1384.

**43** Daubechies, I., Lu, J., and Wu, H. (2011). Synchrosqueezed wavelet transforms: an empirical mode decomposition-like tool. *Applied and Computational Harmonic Analysis* 30 (2): 243–261.

**44** Jarchi, D., Sanei, S., and Prochazka, A. (2019). Detection of sleep apnea/hypopnea events using synchrosqueezed wavelet transform. Proceedings of the IEEE ICASSP 2019, Brighton, United Kingdom (12–17 May 2019). In: . IEEE.

**45** Murenzi, R., Combes, J.M., Grossman, A., and Tchmitchian, P. (eds.) (1988). *Wavelets*. Berlin: Springer.

**46** Chui, C.K. (1992). *An Introduction to Wavelets*. Academic Press.

**47** Serrano, E. and D'Attellis, C.E. (2006). Littlewood Paley spline wavelets. In: *Proceedings of the 6th WSEAS International Conference on Wavelet Analysis & Multirate Systems*, 6–10. ACM.

**48** Mallat, S.G. (1989). Multiresolution approximations and wavelet orthonormal bases of $L^2(R)$. *Transactions of the American Mathematical Society* 315 (1): 69–87.

**49** Haykin, S.O. (2013). *Adaptive Filter Theory*. Pearson.

**50** Akyildiz, I.F., Pompili, D., and Melodia, T. (2005). Underwater acoustic sensor networks: research challenges. *Journal of Ad hoc Networks* 3 (3): 257–279.

**51** Ashok, R.L. and Agrawal, D.P. (2003). Next-generation wearable networks. *Computer* 36 (11): 31–39.

**52** Atakan, B. and Akan, O.B. (2007). An information theoretical approach for molecular communication. In: *Proceedings of 2nd Conference Bio-Inspired Models of Network. Information and Computing System*, 33–40. Springer.

**53** Di Lorenzo, P. (2012) Bio-inspired dynamic radio access in cognitive networks based on social foraging swarms. PhD dissertation. Sapienza University of Rome.

**54** Latifi, M., Khalili, A., Rastegarnia, A. et al. (2017). A distributed algorithm for demand-side management: selling back to the grid. *Heliyon* 3 (11): 1–28.

**55** Latifi, M., Khalili, A., Rastegarnia, A., and Sanei, S. (2017). Fully distributed demand response using adaptive diffusion Stackelberg algorithm. *IEEE Transactions on Industrial Informatics* 13 (5): 2291–2301.

**56** de Azevedo, R., Cintuglu, M.H., Ma, T., and Mohammed, O.A. (2017). Multiagent-based optimal microgrid control using fully distributed diffusion strategy. *IEEE Transactions on Smart Grid* 8 (4): 1997–2008.

**57** Bertsekas, D.P. (1997). A new class of incremental gradient methods for least squares problems. *SIAM Journal on Optimization* 7 (4): 913–926.

**58** Tsitsiklis, J.N. and Athans, M. (1984). Convergence and asymptotic agreement in distributed decision problems. *IEEE Transactions on Automatic Control* 29 (1): 42–50.

**59** Lopes, C.G. and Sayed, A.H. (2007). Incremental adaptive strategies over distributed networks. *IEEE Transactions on Signal Processing* 55 (8): 4064–4077.

**60** Nedic, A. and Bertsekas, D.P. (2001). Incremental subgradient methods for nondifferentiable optimization. *SIAM Journal on Optimization* 12 (1): 109–138.

**61** Blatt, D., Hero, A.O., and Gauchman, H. (2007). A convergent incremental gradient method with a constant step size. *SIAM Journal on Optimization* 18 (1): 29–51.

**62** DeGroot, M.H. (1974). Reaching a consensus. *Journal of the American Statistical Association* 69 (345): 118–121.

**63** Xiao, L. and Boyd, S. (2004). Fast linear iterations for distributed averaging. *Systems & Control Letters* 53 (1): 65–78.

**64** Nedic, A. and Ozdaglar, A. (2009). Distributed subgradient methods for multi-agent optimization. *IEEE Transactions on Automatic Control* 54 (1): 48–61.

**65** Srivastava, K. and Nedic, A. (2011). Distributed asynchronous constrained stochastic optimization. *IEEE Journal of Selected Topics in Signal Processing* 5 (4): 772–790.

**66** Sayed, A.H. (2014). Adaptation, learning, and optimization over networks. *Foundations and Trends in Machine Learning* 7 (4–5): 311–801.

**67** Sayed, A.H., Tu, S.-Y., Chen, J. et al. (2013). Diffusion strategies for adaptation and learning over networks: an examination of distributed strategies and network behaviour. *IEEE Signal Processing Magazine* 30 (3): 155–171.

**68** Chen, J. and Sayed, A.H. (2012). Diffusion adaptation strategies for distributed optimization and learning over networks. *IEEE Transactions on Signal Processing* 60 (8): 4289–4305.

**69** Cattivelli, F.S. and Sayed, A.H. (2010). Diffusion LMS strategies for distributed estimation. *IEEE Transactions on Signal Processing* 58 (3): 1035–1048.

**70** Chen, J. and Sayed, A.H. (2013). Distributed pareto optimization via diffusion strategies. *IEEE Journal on Selected Topics in Signal Processing* 7 (2): 205–220.

**71** Lopes, C.G. and Sayed, A.H. (2008). Diffusion least-mean squares over adaptive networks: formulation and performance analysis. *IEEE Transactions on Signal Processing* 56 (7): 3122–3136.

**72** Chen, J. and Sayed, A.H. (2012). Distributed pareto-optimal solutions via diffusion adaptation. In: *Proceedings of IEEE Statistical Signal Processing Workshop*, 648–651. SSP.

**73** Sayed, A.H. (2014). Adaptive networks. *Proceedings of the IEEE* 102 (4): 460–497.

**74** Vahidpour, V., Rastegarnia, A., Khalili, A., and Sanei, S. (2018). Partial diffusion Kalman filtering for distributed state estimation in multi-agent networks. *IEEE Transactions on Neural Networks and Learning Systems* https://doi.org/10.1109/TNNLS.2019.2899052.

**75** Monajemi, S., Sanei, S., Ong, S.-H., and Sayed, A.H. (2015). Adaptive regularized diffusion adaptation over multitask networks. In: *Proceedings of IEEE Workshop on Machine Learning for Signal Processing, Boston*, 1–5. IEEE.

**76** Giménez-Febrer, P., Pagès-Zamora, A., Pereira, S.S., and López-Valcarce, R. (2015). Distributed AOA-based source positioning in NLOS with sensor networks. In: *2015 IEEE International Conference on Acoustics, Speech and Signal Processing (ICASSP)*. IEEE.

**77** Mijovica, S., Burattia, C., Zanellab, A., and Verdonea, R. (2014). A cooperative beamforming technique for body area networks. *Procedia Computer Science* 40: 181–189.

**78** Chiasserini, C.F. and Rao, R.R. (2002). On the concept of distributed digital signal processing in wireless sensor networks. In: *Proceedings of the IEEE Conf. on Military Communications (MILCOM) 2002*. IEEE.

**79** Jutten, C. and Herault, J. (1991). Blind separation of sources, part I: an adaptive algorithm based on neuromimetic architecture. *Signal Processing* 24 (1): 1–10.

**80** Yeredor, A. (2001). Blind source separation with pure delay mixtures. In: *Proceedings of the Independent Component Analysis and Blind Source Separation*. University of California in San Diego.

**81** Linsker, R. (1988). An application of the principle of maximum information preservation to linear systems. In: *NIPS'88: Proceedings of the 1st International Conference on Neural Information Processing Systems*, 186–194. MIT Press.

**82** Comon, P. (1994). Independent component analysis: a new concept. *Signal Processing* 36 (3): 287–314.

**83** Bell, A.J. and Sejnowski, T.J. (1995). An information-maximization approach to blind separation, and blind deconvolution. *Neural Computation* 7 (6): 1129–1159.

**84** Amari, S., Cichocki, A., and Yang, H.H. (1996). A new learning algorithm for blind signal separation. In: *NIPS'96: Proceedings of the International Conference on Neural Information Processing Systems*, 757–763. MIT Press.

**85** Hyvärinen, A. and Oja, E. (1997). A fast-fixed point algorithm for independent component analysis. *Neural Computation* 9 (7): 1483–1492.

**86** Parra, L. and Spence, C. (2000). Convolutive blind separation of non-stationary sources. *IEEE Transactions on Speech and Audio Processing* 8 (3): 320–327.

**87** Cardoso, J.-F. (1997). Infomax and maximum likelihood for blind source separation. *IEEE Signal Processing Letters* 4 (4): 112–114.

**88** Makkiabadi, B. and Sanei, S. (2012). A new time domain convolutive BSS of heart and lung sounds. In: *Proceedings of the IEEE International Conference on Acoustics, Speech and Signal Processing, ICASSP 2012*. IEEE.

**89** Li, M.H. and Lu, Y.L. (2008). Angle-of-arrival estimation for localization and communication in wireless networks. In: *Proceedings of 16th European Signal Processing Conference*. Lausanne, Switzerland (25–29 August 2008). EURASIP.

**90** Li, M.H. and Lu, Y.L. (2007). Optimal direction finding in unknown noise environments using antenna arrays in wireless sensor networks. In: *Proceedings of 7th International Conference on Intelligent Transportation Systems Telecommunications*. Sophia Antipolis, France (6–8 June 2007), 332–337. IEEE.

**91** Li, M., Lu, Y., and He, B. (2013). Collaborative signal and information processing for target detection with heterogeneous sensor networks. *International Journal of Sensor Networks and Data Communications* 1: 112.

**92** Li, M.H., Lu, Y.L., Chen, H.H. et al. (2009). Angle of arrival (AOA) estimation in wireless networks. In: *Wireless Networks: Research, Technology and Applications* (ed. J. Feng), 135–164. New York: Nova Science Publishers, Inc.

**93** Wang, B., Li, M.H., Lim, H.B. et al. (2009). Energy efficient information processing in wireless sensor networks. In: *Guide to Wireless Sensor Networks*, 1–26. London: Springer.

**94** Li, M.H., Lu, Y.L., and Wee, L. (2006). Target detection and identification with a heterogeneous sensor network by strategic resource allocation and coordination. In: *Proceedings of 6th International Conference on Intelligent Transportation Systems Telecommunications*. Chengdu, China (21–23 June 2006), 992–995. IEEE.

**95** Li, M.H., Wang, B., Lu, Y.L. et al. (2010). Smart antenna in intelligent transportation systems. In: *Wireless Technologies in Intelligent Transportation Systems* (eds. M.T. Zhou, Y. Zhang and L.T. Yang), 51–84. New York: Nova Science Publishers, Inc.

**96** Li, M.H., Ho, K.S., and Hayward, G. (2010). Accurate angle-of-arrival measurement using particle swarm optimization. *Wireless Sensor Network* 2 (5): 358–364.

**97** Li, M.H. and Lu, Y.L. (2007). A refined genetic algorithm for accurate and reliable DOA estimation with a sensor array. *Wireless Personal Communications* 43 (2): 533–547.

**98** Wax, M. and Ziskind, I. (1989). On unique localization of multiple sources by passive sensor arrays. *IEEE Transactions on Acoustics, Speech, and Signal Processing* 37 (7): 996–1000.

**99** Li, M.H. and Lu, Y.L. (2008). Maximum likelihood DOA estimation in unknown colored noise fields. *IEEE Transactions on Aerospace and Electronic Systems* 44 (3): 1079–1090.

**100** Li, M.H. and Lu, Y.L. (2009). Source bearing and steering-vector estimation using partially calibrated arrays. *IEEE Transactions on Aerospace and Electronic Systems* 45 (4): 1361–1372.

**101** Li, M., Lu, Y., and He, B. (2013). Array signal processing for maximum likelihood direction-of-arrival estimation. *Journal of Electrical and Electronics Systems* 3 (1): 117.

**102** Li, M.H. and Lu, Y.L. (2005). Null-steering beamspace transformation design for robust data reduction. In: *Proceedings of 13th European Signal Processing Conference*. Antalya, Turkey (4–8 September 2005). EURASIP.

**103** Li, M.H. and Lu, Y.L. (2006). Dimension reduction for array processing with robust interference cancellation. *IEEE Transactions on Aerospace and Electronic Systems* 42 (1): 103–112.

**104** Li, K., Lu, Y.L., and Li, M.H. (2005). Approximate formulas for lateral electromagnetic pulses from a horizontal electric dipole on the surface of one-dimensionally anisotropic medium. *IEEE Transactions on Antennas and Propagation* 53 (3): 933–937.

**105** He, B., Liang, Y., Feng, X. et al. (2012). AUV SLAM and experiments using a mechanical scanning forward-looking sonar. *Sensors* 12 (7): 9386–9410.

**106** Li, M.H., Ho, K.S., and Hayward, G. (2009). Beamspace transformation for data reduction using genetic algorithms. In: *Proceedings of IEEE International Ultrasonics Symposium*. Rome, Italy (20–23 September 2009). IEEE.

**107** Li, M.H., McGuire, M., Ho, K.S., and Hayward, G. (2010). Array element failure correction for robust ultrasound beamforming and imaging. In: *Proceedings of 2010 IEEE International Ultrasonics Symposium*. San Diego, California, (11–14 October 2010). IEEE.

**108** Li, M.H. and Lu, Y.L. (2007). Maximum likelihood processing for arrays with partially unknown sensor gains and phases. In: *Proceedings of 7th International Conference on Intelligent Transportation Systems Telecommunications*. Sophia Antipolis, France (6–8 June 2007), 185–190. IEEE.

**109** Lardner, T., Li, M.H., Gongzhang, R., and Gachagan, A. (2012). A new speckle noise suppression technique using cross-correlation of array sub-apertures in ultrasonic NDE of coarse grain materials. In: *Proceedings of Review of Progress in Quantitative Nondestructive Evaluation (QNDE)*. Nondestructive Evaluation Centre.

**110** Gongzhang, R., Li, M.H., Lardner, T. et al. (2012). Robust defect detection in ultrasonic nondestructive evaluation (NDE) of difficult materials. In: *Proceedings of 2012 IEEE International Ultrasonics Symposium* Dresden, Germany (7–10 October 2010). IEEE.

**111** Li, M.H., Hayward, G., and He, B. (2011). Adaptive array processing for ultrasonic non-destructive evaluation. In: *Proceedings of 2011 IEEE International Ultrasonics Symposium*. Orlando, Florida (18–21 October 2001). IEEE.

**112** Li, M.H. and Hayward, G. (2012). Ultrasound nondestructive evaluation (NDE) imaging with transducer arrays and adaptive processing. *Sensors* 12 (1): 42–54.

**113** Gravina, R., Guerrieri, A., Fortino, G. et al. (2008). Development of body sensor network applications using SPINE. In: *Proceedings of the IEEE International Conference on Systems, Man and Cybernetics (SMC 2008 Singapore Oct 2008)*, 2810–2815. IEEE.

**114** Seto, E., Martin, E., Yang, A. et al. (2010). Opportunistic strategies for lightweight signal processing for body sensor networks. In: *Proceedings of the 3rd International Conference on Pervasive Technologies Related to Assistive Environments*. ACM Samos, Greece (23–25 June 2010).

# 11

# Communication Systems for Body Area Networks

## 11.1 Introduction

The human body is a complex environment for the operation of wireless communications systems. The complex antenna–body electromagnetic interaction is perhaps the main reason for increasing the difficulty for researchers to accurately model the body area network (BAN) communication channel in different scenarios. This is further compounded by the impact of body movement and the propagation characteristics of the surroundings which all have significant impacts upon body-related communication channels. The successful design of complete BAN systems is therefore inextricably linked to a thorough understanding of these factors.

A sensor network for medical applications has to be wearable or implantable, reliable, secure, and interoperable (i.e. it should allow users to easily build a robust wireless BAN). Such a network may involve various technologies in sensor development to enable sophisticated physical, physiological, mental, or metabolic measurements from the human body. A simple representative example of such a sensor network is depicted in Figure 11.1.

A BAN is a self-contained wireless network in which the constituent nodes are usually worn on or in close proximity to the human body. BANs are used in an increasing number of areas such as the medical, military, sports, and recreational domains. For example, a BAN used for medical purposes may employ nodes with in-built sensors to monitor vital signs such as heart rate, body temperature, and gait with the aim of improving patient care.

The centralised BAN (or any) network topology consists of nodes and hubs. A node is an entity that contains a medium access control (MAC) sublayer and a physical layer, and optionally provides security services. A hub is an entity that possesses a node's functionality and coordinates the medium access and power management of the nodes. Nodes can be classified into different groups based on their functionality (personal devices, sensors, actuators), implementation (implant nodes, body surface nodes, external nodes) and role (coordinators, end nodes, relays) [1].

The main focus of this chapter is the on-body radio channel, for communications from one location on a given subject's body to another location on the body, which is envisaged as the most common BAN implementation. Off-body and body-to-body channels are also important possible cases. The body-to-body channel is important due to the anticipated prevalence of BANs, where this interfering channel, with multiple co-located BANs, can dominate the on-body radio channel.

*Body Sensor Networking, Design and Algorithms,* First Edition. Saeid Sanei, Delaram Jarchi and Anthony G. Constantinides.

Companion Website: www.wiley.com/go/sanei/algorithm-design

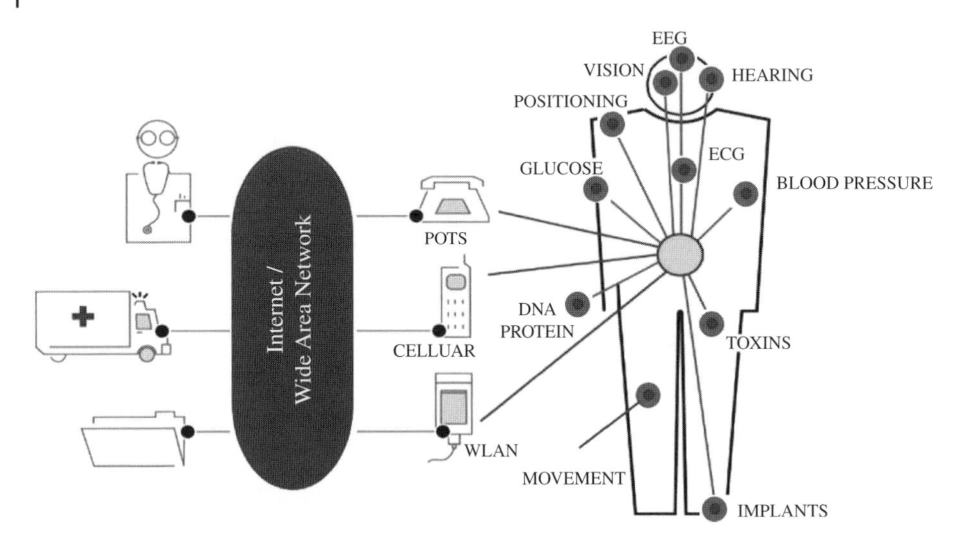


**Figure 11.1** A typical BAN network.

There are difficulties in channel modelling for BAN, which are particular to the BAN channel, emphasising the importance of BAN reliability and lifetime enhancing system design, such as relay-assisted communications, transmit power control, and link adaptation.

Short-range communications for sensor networks have been in use mainly for connecting sensors to central hubs (mobile phone, main stations, or central base stations within a building,) covering distances between a few centimetres to 10–100 m. In hospitals the vital information from patients is delivered to the visualisation or processing units through wired or wireless media.

To move on with modelling wireless BAN channels, we may consider the following possible scenarios related to node/sensor locations:

- On-body: for communications between sensors or devices mounted on two different parts of body surfaces.
- In-body: for communications between a sensor inside the body (which can be implanted or used for imaging, such as endoscopic capsule) and a device or sensor typically on the body's surface.
- Off-body: for communications between a sensor or device on the body surface to a device closely located up to few metres away from the body (or vice versa, i.e. from off the body to on the body).
- Body-to-body or interference between two networks: for radio communications, target or interfering, from one subject's body to another subject's body. This can happen between a BAN and a wireless sensor network (WSN) too.

A full BAN system – including sensor nodes, on-body, in-body, and off-body links, and a gateway (hub) – in a centralised setup is shown in Figure 11.2. The hub locations are typically near the torso, either at the hips or on the chest. Placing the hub in these locations allows a subject to comfortably wear a device that is expected to be larger than a sensor

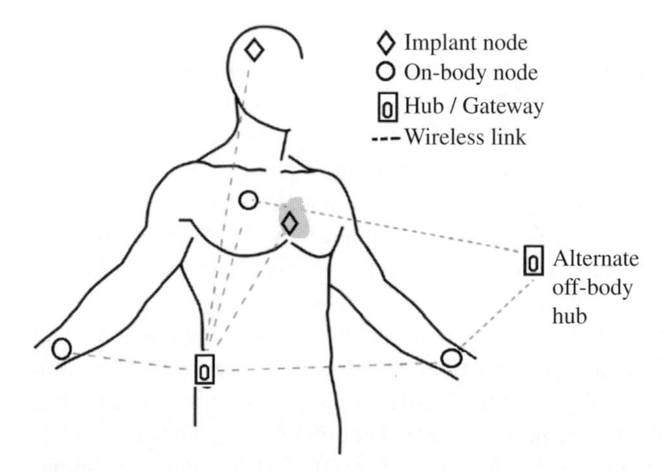


**Figure 11.2** BAN on a dummy subject, including sensors, in-body, on-body, and off-body links, and gateway (hub) [2]. Source: Courtesy of Smith, D.B., Miniutti, D., Lamahewa, T.A., and Hanlen, L.W.

node. These locations are also reasonably central on the human body and less affected by the body's movement. The four scenarios are explained later in this chapter in more detail, particularly with respect to challenges, operating environments, and their individual applications.

Wireless links between BAN nodes operating within the popular medical (ISM, or industrial, scientific, and medical) bands at 868/915 MHz, 2.45 and 5.8 GHz, and nearby frequencies are often formed using on-body surface waves [3–5] and the components reflected, diffracted, and scattered by other body parts [6]. Depending upon the exact geometry of the wireless link, some BAN channels may also utilise free-space propagation (e.g. waist to wrist) as well as multipath propagation through the environment whereby signal transmissions from an on-body node are returned toward the body from nearby surfaces [6–8]. Because the nodes designed for BAN applications are worn in close proximity to the human body and therefore are prone to antenna–body interaction effects, which include near-field coupling, radiation pattern distortion, and shifts in antenna impedance, which may degrade their efficiency and reduce signal reliability [3, 9–13]. To complicate matters further, due to the inherent mobility, variations in body shape caused by movement and physiological processes such as respiration and also the changes in local environment, the BAN channels are susceptible to the time-varying characteristics of the received signals [14–17].

Owing to these complex electromagnetic interactions and variable channel conditions, the design of the physical and MAC layer technologies to be used in BANs present many challenges for wireless systems designers. This task is made even more difficult when some inevitable constraints are imposed on the size and power consumption of BAN nodes. In the physical layer, the BAN nodes are required to be compact, lightweight, robust, unobtrusive to the user, and the necessary antennas mounted conformal to or in extremely close proximity to the body surface. MAC layer protocols must be resilient to the communication path being obscured or shadowed by the human body [18–22]. These stringent requirements must be met, while still maintaining high quality, reliability, and efficiency especially for niche and life critical applications in the medical and military domains [6, 23]. The standardisation of communications in BANs has largely been undertaken by the Institute of Electrical and Electronics Engineers (IEEE) 802.15 Task Group 6, also known as IEEE 802.15.6, which was set up in 2007. The IEEE 802.15.6 standard [24] appeared in February 2012 and defines the physical and MAC layers optimised for short-range transmissions in, on, or around the human body. Its purpose is to support low-complexity, low-cost, ultra-low-power, and highly reliable wireless communications for use in close proximity to or inside a human body (but not limited to humans) to serve a variety of applications including medical/healthcare and nonmedical uses. The broad range of possible application fields in which BANs could be used leads to an equally wide variety of system requirements that need to be met. The definition of a unique physical layer has yet to yield a feasible solution. Hence, the IEEE 802.15.6 has proposed three different ranges for various short-range communication modalities: a narrowband PHY centred at different frequencies, an ultra-wideband (UWB) PHY, and a human body communication (HBC) PHY. Figure 11.3 graphically represents the frequency bandwidth allocation chart of all available frequencies for BAN applications, with the specification of the related country or region where they could be used. As shown in Table 11.1, the standardisation group has identified a list of scenarios in which IEEE802.15.6 compliant devices will be expected to operate, along with their description and the frequency band of interest.

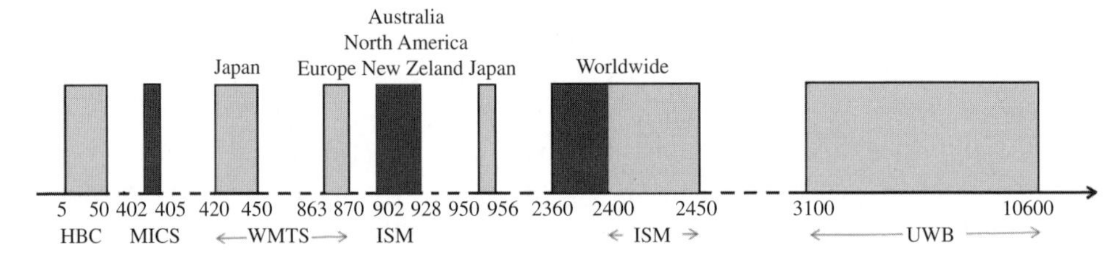


**Figure 11.3** The frequency ranges allocated to narrow and wideband BAN communication systems by different organisations, approved by the IEEE.

**Table 11.1** List of scenarios and their descriptions; LOS refers to line-of-sight and NLOS refers to non-line-of-sight.

| Scenario | Description | Frequency band | Channel model |
|---|---|---|---|
| S1 | Implant to implant | 402–405 MHz | CM1 |
| S2 | Implant to body surface | 402–405 MHz | CM2 |
| S3 | Implant to external | 402–405 MHz | CM2 |
| S4 | Body surface to body surface (LOS) | 13.5, 50, 400, 600, 900 MHz, 2.4, 3.1–10.6 GHz | CM3 |
| S5 | Body surface to body surface (NLOS) | 13.5, 50, 400, 600, 900 MHz, 2.4, 3.1–10.6 GHz | CM3 |
| S6 | Body surface to external (LOS) | 900 MHz 2.4, 3.1–10.6 GHz | CM4 |
| S7 | Body surface to external (NLOS) | 900 MHz 2.4, 3.1–10.6 GHz | CM4 |

In Table 11.1, the channel models are named as CM1 to CM4 for relatively different BAN communication scenarios. They cover narrow band channels such as the medical implant communication system (MICS) and ISM in Figure 11.3 and wideband channels such as UWB referred to in the same figure.

## 11.2 Short-range Communication Systems

Short-range communication systems are widely used in remote healthcare monitoring, assisted living, telemedicine, and recently in invasive diagnostic devices. Bluetooth, Wi-Fi, ZigBee, radio frequency identification device (RFID), and UWB are the most common short-range communication systems suitable for BANs. Among these networks, newly established protocols allow multinode networking for ZigBee systems. In the following sections, first of all, these popular short-range communications systems are explained and then some of their applications for BANs are detailed. The limitations, artefacts, and disadvantages are described next, and finally, some examples of communication channel models for these systems are discussed.

### 11.2.1 Bluetooth

It is named after Danish King Harald Bluetooth and operates at a transmission frequency mode of around 2.45 GHz. Bluetooth is an IEEE 802.15.1 standard commonly known as wireless personal area network (WPAN). Bluetooth technology is used for both voice and data communications. It uses spread spectrum frequency hopping (SSFH) technology to best tackle multipath using multiple carrier frequency. It utilises frequency hopping among 79 1 MHz channels at a nominal rate of 1600 hops/sec to avoid interference. SSFH provides more robustness against multipath fading as well as security breach. It also uses phase shift keying (PSK), frequency shift keying (FSK), or Gaussian frequency shift keying (GFSK) digital modulation providing a good trade-off between channel noise and bit error rate. The information about all these systems is widely available. As an example, a transmitted GFSK signal can be formulated as:

$$x(t) = \sqrt{\frac{2E}{T}} cos\left(j2\pi\left(f_c t + h\int_{-\infty}^{t} g(\tau)d\tau\right)\right) \tag{11.1}$$

where $E$ is the energy per symbol, $T$ is the symbol period, $f_c$ is the carrier frequency, $h$ is the modulation index, defined as $h = 2f_d T$, and g($t$) is the output of Gaussian lowpass filtered input signal. The frequency $f_d$ is the frequency deviation, the maximum frequency shift with respect to the carrier frequency when a '0' or '1' is being transmitted.

Bluetooth is good for linking a communication device, such as a mobile phone to a recording or broadcasting system, such as a car audio system, or transferring data between mobiles and computers. This modality, however, is not suitable for networking and connecting a number of computers (PC or laptops). Bluetooth is also used favourably for connecting a computer to its peripheral devices such as a mouse, printer, keyboard, and digital camera.

Bluetooth allows communication with approximately 1 Mbps speed. Different products have a different number of Bluetooth channels but most of the systems are equipped with only one channel.

Many modern electronic systems, mobile phones, notepads, and computers have Bluetooth installed in them. Bluetooth devices can also be connected to computers without Bluetooth using Bluetooth USB sticks.

### 11.2.2 Wi-Fi

Wi-Fi complies with an IEEE 802.11 standard for a wireless local area network (WLAN) [25]. Generally, it comes with four standards (802.11 a/b/g/n) that run in ISM band 2.4 and 5 GHz with a modest coverage of 100 m. Wi-Fi allows users to transfer data at broadband speed when connected to an access point (AP) or in ad hoc mode.

Wi-Fi covers a longer distance range of up to a few hundred metres, has a frequency band of between 2.5 to 5 GHz, and a communication rate of over 10 Mbps. Often, a limited number of channels are used (such as 10) and direct sequence spread spectrum (DSSS) and orthogonal frequency division multiplexing (OFDM) are usually used for their digital modulation. Both DSSS and OFDM allow diversity in the number of users and provide high fidelity secure communications. At the same time these modulation techniques are very efficient in battling multipath and Doppler frequency shift problems.

The Federal Communications Commission (FCC) has approved the allocation of 40 MHz of spectrum bandwidth for medical BAN low-power, wide-area radio links at the 2360–2400 MHz band. This will allow off-loading BAN communication from the already saturated standard Wi-Fi spectrum to a standard band [26].

In some new systems, the Wi-Fi devices can be used in data acquisition applications that allow a direct communication between sensors and smartphones/PCs even without an intermediate router. This short-range communication system is more suitable for large amounts of data transfers with high-speed wireless connectivity that allows videoconferencing, voice calls, and video streaming. It has the important advantage of being integrated in all smartphones, tablets, and laptops. Nevertheless, its main disadvantage is high energy consumption [27].

### 11.2.3 ZigBee

ZigBee is perhaps the most popular low-range communication system applicable to BAN. IEEE 802.15.4/ZigBee devices have been mainly developed for use in WSN applications, but due to characteristics such as low power and small form factor, they are also being widely applied to BAN design. The suitability of ZigBee is studied and evaluated in [28].

ZigBee is a wireless telecommunication designed for sensors and controls, and suitable for use in harsh or isolated conditions. One of the biggest advantages of the ZigBee network is its low power consumption. Figure 11.4 shows a typical ZigBee network topology which consists of three kinds of devices or nodes: coordinator, router, and end device. One coordinator exists in every ZigBee network. It starts the network and handles management functions as well as data routing functions. End devices are devices that are battery-powered due to their low power consumption. They are in standby mode most of the time and become active to collect and transmit data. The sensors, as end devices, are connected to the network through the routers. Routers disseminate the data across multihop ZigBee networks. In the cases where the network is point to point or point to multipoint the ZigBee network topology is formed without routers.

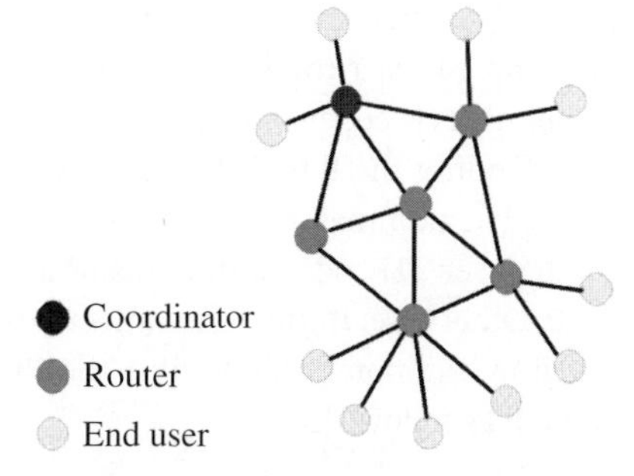


**Figure 11.4** A typical ZigBee network topology. (*See color plate section for color representation of this figure*)

ZigBee is aimed at radio frequency (RF) applications that require a low data rate, long battery lifespan, and secure networking. Through the standby mode, ZigBee enables devices to operate for several years. ZigBee uses three different frequency bands: 868 MHz, 915 MHz, and 2.4 GHz. Therefore, one substantial drawback of using the ZigBee network for wireless body area network (WBAN) applications is due to interference with WLAN transmission, especially at 2.4 GHz. Since ZigBee devices operate at a low data rate, they are not suitable for large-scale and real-time WBAN applications. However, they are very useful for personal use like assisted living, health monitoring, sports, and environment monitoring, within modest distances of 50–70 m [29].

Another important advantage of the system is its distributed networking capability through shortcut three routing [30].

### 11.2.4 Radio Frequency Identification Devices

RFIDs are often used in parallel or together with other sensor network techniques in order to detect the presence and location of an object.

Hospitals and home-based healthcare services use a combination of both active and passive RFID technology. A radio-frequency identification system uses tags or labels attached to the patient's body parts to be monitored. RFID readers send a signal to the tag and read its response. The RFID reader transmits an encoded radio signal to read the tag, and then the RFID tag receives the message and responds with its identification and other measured medical information [31]. Hospitals may use infrared monitors installed in medical facility rooms to collect data from the transmissions of RFID badges and tags worn by patients and employees, as well as from tags attached to the equipment, clinical tools, or hospital-related stationary or mobile devices.

### 11.2.5 Ultrawideband

UWB is a relatively new technology which uses a wider bandwidth but suffers from shorter-range communication. The most common UWB physical layer uses an impulse-radio-based signalling scheme in which each symbol of information is represented by a sequence of short-time duration pulses. The duration of an individual pulse is nominally considered the length of a chip. Chip duration is equal to 2.02429 ns. The modulation format implies a symbol duration of 1036.44 ns. A symbol period is composed of 32 bursts each defined by 16 chip times. For a symbol to be transmitted, a single time burst $T_s$ is picked up and used among the 32 available. In this burst, each chip time is occupied by a transmitted pulse. The transmission scheme foresees the use of time hopping (TH) codes to achieve multiple access. Pulse position modulation (PPM) is used for encoding the bits. PPM implies a time shift equivalent to multiples of the temporal burst. The nominal temporal burst during which a single user needs to transmit is defined by the user TH code value. This nominal time burst is used if the user wants to transmit bit 0. Otherwise, if the user wants to transmit bit 1, it needs to use another temporal burst given by the nominal one plus the PPM shift. The applicable temporal burst $T_u$ can be written as follows [32]:

$$T_u = c + 16T_s b \tag{11.2}$$

where $16T_s$ represents the PPM shift, $b$ is the transmitted bit, and $c$ is the nominal time burst. The reference pulse used by the UWB physical layer is a root raised cosine pulse with a roll-off factor of 0.6.

In a BAN setup [24] the information from ten 802.15.4a sensors around the body is propagated to a central (master) node using UWB. The master node then communicates with the outside world using a standard telecommunication infrastructure such as WLAN, Bluetooth, or 2G-3G cellular networks. To ease the communication, at the network layer a unique address format can be created for connecting nodes belonging to different underlying subnetworks. The packets originated by different subnets can be adapted to obtain a unified address format. To achieve this, the interworking modules must work by combining information of the different IEEE communication structures involved. This adaptation takes place at the network layer and makes the packet of the single subnetwork fitted for being transmitted through an overlying network, for example an IP network. Each subnetwork has its own physical layer and its own MAC protocol, and they can exchange packets by means of the interworking unit functionalities at the network layer.

UWB has other applications, such as for tissue imaging or for treating hyperthermia (for warming the internal body tissues such as a tumour). Very recently, an application of UWB in detection of bone fracture has been reported [33]. However, these applications are still being studied and in an experimental stage and haven't been commercialised or used for any human monitoring or treatment.

### 11.2.6 Other Short-range Communication Methods

IEEE 802.15.6; WBAN IEEE 802.15.6 is the latest addition to the WPAN standard that provides various medical and nonmedical applications and supports communications inside and around the human body [24]. It uses different frequency bands for data transmission with a data rate of 10 Mbps maximum. The first one is narrowband (NB) which operates within the range of 400, 800, 900 MHz, and 2.3, 2.4 GHz bands. The second one is the HBC which operates at a range of 50 MHz [34]. The UWB technology operates between 3.1 and 10.6 GHz, which supports a high bandwidth in short-range communication.

EnOcean is a system for transmitting data wirelessly that requires no power supply or maintenance, and instead uses energy harvesting technology to generate the small amount of energy needed from the environment (i.e. light or temperature difference). The frequency varies depending on geographical region. In Japan, this is a specified low power wireless system using 315 and 928 MHz frequencies. In the US, 315 and 902 MHz are employed, while the EU adopts 868 MHz.

The Wi-SUN standard has been attracting increased attention since its adoption by power companies in Japan for smart meter communication. Japan uses the 920 MHz band within a specified low power, which is characterised by slower communication speeds than Wi-Fi but features longer communication distance, lower power consumption, simpler connection, and greater performance at the presence of obstacles.

### 11.2.7 RF Modules Available in Market

A survey on RF modules, for both BANs and WSNs, available in the market has been provided in [35]. The main characteristics of these modules are frequency of operation and range, transmitter power, and receiver sensitivity in terms of dBm, electrical characteristics of the RF module including power supply levels for different modes and applications, and finally packaging and cost. The listed modules are NRF24L01, CC1101, RFM22B-S2\SMD Wireless Transceiver-434 MHz, TRM-433-LT, CC2500, and MC1322x. All these modules operate as transceivers (transmitter and receiver) and are used in both ends of communication systems.

Nevertheless, these days, the emerging technology in green energy and energy harvesting requires lower power consumption. This is particularly viable for BANs where the communication range and bandwidth can be better controlled.

## 11.3 Limitations, Interferences, Noise, and Artefacts

Sensor technology, pervasive computing, and intelligent information processing are widely studied and practised in BAN development. Each one of these technologies has its own limitations and challenges.

Although BANs have some similarities to WSN, the characteristics are somehow different because of their different application purposes and, more importantly, the effect of body. First of all, considering network deployment, WSNs can be deployed to inaccessible environments, such as forests, swamps, or mountains. Many redundant nodes are placed in the environments such as those mentioned above to solve the problem of node failures. Thus, it is possible for WSNs to have high node densities, whereas the body sensor network (BSN) nodes are deployed in, on, or around the human body, and so the total number of nodes is generally limited and not more than a few dozens. Each node ensures the accuracy of monitoring results by its robustness [36]. Secondly, considering attributes, the nodes in WSNs perform the same functions, and have the same properties. The size of nodes is not very critical. Once the node is deployed, it will probably no longer need to be moved. Owing to the diversity in BAN measurement modalities, different sensor types are often used [37]. Moreover, the nodes need to have high wearability and high biocompatibility in the case of in-body sensors [38]. Generally, the nodes move as the human body moves. Thirdly, considering energy supply, WSNs and BANs can be battery powered. The former, deployed outdoors, can also be powered by wind energy or solar energy, while the latter can also be powered by kinetic energy and heat [39, 40]. Finally, considering data transmission, the transfer rates of WSNs are almost the same, but those of BANs are different, as the data types and channel assignments are different among the nodes [37]. The rate is also different for different short-range communications used for BAN (e.g. narrowband or UWB). Additionally, BANs deployed in the human body are for monitoring human physiological data, which are subject to a user's personal safety and privacy protection issues. Therefore, the quality of service (QoS) and the real-time prosperity of data transmission must be taken into account [41, 42].

## 11.4 Channel Modelling

In majority of sensor network platforms, the communication between sensors in the network is the major concern for maintaining a reliable BAN data recording, transmission, or archival system. For healthcare applications, maintaining reliable low-power operation is particularly important. This requires accurately modelling the BAN radio-propagation channel. The model should include the BAN characteristics, including sensor locations, frequency range, body movement, and internal and external noise and interferences, which in many cases may be different from the longer-ranged WSNs, including cellular and sensor-network radio-propagation channels. A suitable and global channel modelling for all types of BANs is hard to achieve, if not impossible. However, there are scenarios where sensible and recognisable fixed, mobile line-of-sight (LOS) or multipath models can be envisaged.

Before we look at the possible approaches for channel modelling, it would be advisable to separate and explain various possible BAN channels. These channels are briefly referred to in Section 11.2.

### 11.4.1 BAN Propagation Scenarios

A comprehensive survey of radio propagation and channel modelling for wireless BANs is given in [43]. We have addressed both narrowband and UWB on-body channels, together with off-body and body-to-body scenarios.

#### 11.4.1.1 On-body Channel

The on-body channel is the most prevalent channel for wireless BANs and is the focus in this chapter. This channel operates in various environments and is dominated by slowly varying dynamics governed by human body movement and the variations in shadowing (obscuring) by body parts. Although there are some benefits in using this modality transmission, it imposes some difficulties, too. Both centralised, as often practised, and decentralised sensor networks face the same problems and share similar benefits. In terms of benefits, considering narrowband communications between the body-worn system and those in the environment we may refer to the following: (i) the channel shows reciprocity, that is the radio channel for communications from position A to position B on the body has the same channel profile as for communications B to A; (ii) the channel, for the majority of on-body BAN applications, is stationary for less than a second – typically 0.5 seconds, enabling relatively accurate channel prediction across multiple communications frames, which can help transmit power control and resource allocation; (iii) in a centralised BAN, although the direct sensor-to-hub link may be in outage, the slowly varying on-body channel, and possible postures of the human body, often requires another dual-hop link between the source and the hub through suitably located relays/transmission paths, which gives significant reliability benefits to radio communications; (iv) although the overall information transfer over the whole on-body BAN may be large, for typical applications, such as in healthcare, high data rates for particular links may not be required (often in orders of tens of kilobits per second); (v) finally, despite the fact that the on-body channel is up to some extent time-selective, the narrowband BAN communication is frequency nonselective, with no resolvable multipath, and one channel tap, such that the intersymbol interference (ISI) is ignored and does not need to be mitigated.

In terms of difficulties, when operating with small low-power transmissions, long sensor/actuator lifetime is desired, thus requiring small power demands on the battery of the RF communication system, as well as desired low electromagnetic radiation specific absorption rate (SAR) to the subject's body. This all leads to a desired transmit power of significantly less than 0 dBm (or 1 mW) though −10 dBm (or 0.1 mW) may often be desirable. Further, at typical carrier frequencies of several hundreds of MHz up to a few gigahertz, communication on the human body provides a difficult radio channel, where instantaneous path losses can become very significant and typical path losses for many on-body radio links are still (relatively) very large. Moreover, the variations in signal strength are not uniform, from one time interval to the next, such that generally the channel is not wide-sense stationary.

#### 11.4.1.2 In-body Channel

The in-body channel is usually applicable to medical cases and mostly operates at lower carrier frequencies than the on-body channel. The main frequency of operation is most likely to be within the MICS band, which operates from 402 to 405 MHz. The in-body channel is also predominantly from implants/devices, with miniature wireless transreceivers, to communication systems on the surface of, or just outside the body. The transmission from one transreceiver in the body directly to a system in another location inside the body is very unlikely and less common. The in-body channel, apart from transmissions at tens of MHz, suffers from significant attenuation for radio waves propagating through the body, and often depends on the RF propagation from the nearest body surface to the implant RF device [44].

Modelling the in-body propagation channel faces the same challenges as the on-body channel does. However, the output transmission power and the battery power consumption become more crucial for the in-body channels, as it is desirable for batteries inside the human body to have a lifetime of several years (frequent surgery is not desirable), as well as reducing radio-wave absorption inside the human body. The in-body communication channel includes various additional components (e.g. creeping waves). Some good description of in-body communications can be found in the literature, e.g. [44, 45].

#### 11.4.1.3 Off-body Channel

The off-body channel is the radio channel very similar to those of standard small cell and personal area network radio communications. However, the transmission from one part of the human body to a gateway/hub radio at a small distance from the human body is often affected by shadowing, similar to the on-body channel. It is also slowly time selective and a one-tap channel – but it can reasonably be expected that it is closer to wide-sense stationary than the on-body channel. Also, the path losses are often lower in this case, even though the channel length is likely to be more than that of the on-body links. In healthcare applications, suitable placement of the radio device(s) off the body may be particularly important to maximise the typical channel gains from the desired off-body transmission, or to enhance the on-body communications, where one or more relays is placed off the human body. Also, the off-body channel may have less energy-constrained relays than the on-body radio channel. Generally, all the nodes can be used as relays [46]. In an elegant work a dual-hop cooperative BAN with different schemes has been proposed [47]. Also, multihops or relay-based BANs with off-body channels have been examined and proved to be effective in reducing the packet error rate (PER) [48, 49]. All the other benefits for radio systems designed for the on-body channel also apply to the off-body channel, such as reciprocity, though the data rates may sometimes be larger than that for the on-body channels.

#### 11.4.1.4 Body-to-body (or Interference) Channel

In most wireless BANs, it is unlikely to have the network spread over multiple human bodies, apart from obvious exceptions for applications such as military or emergency services. But the body-to-body radio channel characteristics are still very important, as in many cases for BAN operation there are some mobility coupled with the anticipated large take-up of BANs without coordination between the BANs. Thus, understanding radio propagation from one BAN to the sensor, relay, or hub of another BAN becomes very important.

### 11.4.2 Recent Approaches to BAN Channel Modelling

Considering the BAN scenarios listed and explained in Section 11.4.1, there have been some research attempts for modelling the radio channels. Most of these approaches follow the conventional analytical methods for long-range RF communication systems. Nevertheless, the communication channel and the involved assumptions vary from a narrowband to a wide band system.

One approach applies static narrowband measurements and analytic modelling, e.g. [14, 45, 50–52]. Another approach involved narrowband measurements in anechoic chambers, e.g. [53, 54]. Electromagnetic modelling by computer simulation has also been performed. In the proposed model, finite-difference time-domain method has been used to model the radio interactions with the human body, which is particularly suited for inhomogeneous media such as the human body, e.g. [14, 25]. Also, a number of narrowband [55–58] and UWB [59–61] characterisations have been performed for both static and dynamic scenarios in indoor and outdoor environments. Particularly related to UWB, there has been specific interest in capturing second-order statistics, e.g. [60, 62, 63].

However, up to now, there has been little attempt on characterisation for a generic activity, where a particular measurement set contains a mixture of activities in some different environments. This has been attempted in [55], where first and second-order statistics have been measured and represented generic 'everyday' activities. In this work however, the subject variability in terms of gender and size hasn't been examined. Consideration of such variations in the BAN propagation channel may be seen in [64].

### 11.4.3 Propagation Models

In practice various short-range communication modalities, such as narrowband or UWB, are used for between-node BAN communications. A simple way of modelling the channels is to measure the received power during communication within a time interval and fit a distribution to the received power density for that interval (e.g. 12 hours).

There are three physical-layer propagation scenarios defined by the IEEE 802.15.6 BAN standard [24]:

- Narrowband communications: the use of narrowband in healthcare has been described extensively, e.g. [7, 65]. Of the three propagation methods, narrowband is better suited to a greater number of healthcare applications, due to its lower carrier frequencies that suffer less attenuation from the human body. Its smaller bandwidth (1 MHz or less) also implies that multipath is unlikely to cause significant ISI [66].
- UWB communications: frequency-modulated ultrawideband (FM-UWB) and impulse-radio ultrawideband (IR-UWB) are both supported by the standard, with IR-UWB being better suited for BANs (for IR-UWB, noncoherent receivers can be implemented very efficiently and promise low power consumption to meet stringent constraints on battery autonomy [67]). One particularly suitable application of UWB in BANs is in consumer electronics (although we note that UWB has been mooted for use in healthcare in some literature [68], although that is not common), as UWB offers higher throughput due to its larger bandwidth: each UWB channel has a bandwidth of 499 MHz in IEEE 802.15.6 [24]. A comparison between narrowband and UWB systems in [69] shows that UWB can support higher data rates and lower multipath fading. These make it more suitable for the design of BAN, although the narrowband interference may significantly degrade the performance of UWB.
- Human-body communications: the communication is performed via the galvanic coupling of signals over the human body surface, i.e. transmission is over the medium of human skin by an electrode, rather than by an antenna. This has been less popular compared with other propagation methods.

Table 11.2 summarises the specified carrier-frequency bands and channel bandwidths for these three propagation methods.

In [2] a number of experiments have been carried out for modelling the BAN channels in different scenarios mentioned above. This is basically done by looking at the distribution of channel effect on the input signal. More clearly, the error has been measured between the transmitted and received signals and the distribution is estimated by fitting the error distribution to one of the known distributions. Rayleigh, normal, log-normal, Rician, Nakagami-m, Weibull, and Gamma [7, 57, 70–73] are the most common ones which inclusively model noise, fading, and multipath. In some cases, the use of some other not so common (kappa-mu [$\kappa - \mu$]) distributions have been reported, too [74, 75]. Figure 11.5 shows the different estimated models for a narrow band short-range BAN on-body communication scenario using carrier frequency of 2.36 GHz and a bandwidth of 540 kHz. From this figure, the gamma and Weibull distributions provided the best fits. Gamma fading was a slightly better fit than Weibull according to a negative loglikelihood criterion of the parameter estimates. The very poor fits of the Rayleigh and normal distributions are also obvious in Figure 11.5.

**Table 11.2** The carrier-frequency bands and channel bandwidths for the three BAN propagation methods: narrowband, UWB, and human-body communications [24].

| Frequency band | Bandwidth | Frequency band | Bandwidth |
|---|---|---|---|
| Narrowband communications | | | |
| 402–405 MHz | 300 kHz | 420–450 MHz | 300 kHz |
| 863–870 MHz | 400 kHz | 902–928 MHz | 500 kHz |
| 950–956 MHz | 400 kHz | 2360–2400 MHz | 1 MHz |
| 2400–2483.5 MHz | 1 MHz | | |
| UWB communications | | | |
| 3.2–4.7 GHz | 499 MHz | 6.2–10.2 GHz | 499 MHz |
| Human-body communications | | | |
| 16 MHz | 4 MHz | 27 MHz | 4 MHz |

For all the methods, IEEE 802.15.6 compliant devices must operate in one of the associated bands.
Source: Courtesy of IEEE.

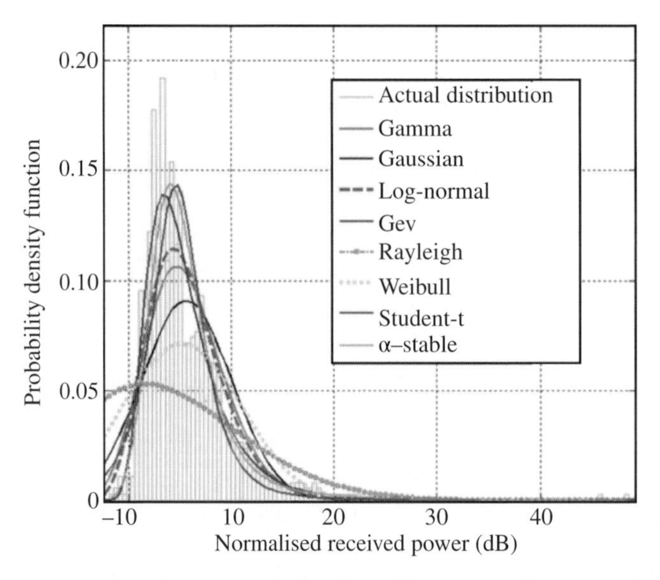


**Figure 11.5** The pdfs representing on-body channel-gain agglomerate from everyday activity of 10 subjects and 540 kHz bandwidth at 2360 MHz [70]. Source: Courtesy of El-Sallabi, H., Aldosari, A., and Abbasi, Q.H. (*See color plate section for color representation of this figure*)

In the case of narrowband channels, a simple exponential channel model [76] which provides a good compromise between simplicity and reality is often used. The taps are complex zero mean Gaussian random variables with variances that decay exponentially. The complex-valued taps are denoted as:

$$h_k = N\left(0, \frac{1}{2}\sigma_k^2\right) + jN\left(0, \frac{1}{2}\sigma_k^2\right) \quad for\ k = 0, 1, \ldots, k_{max} \tag{11.3}$$

where $k_{max}$ is the point where the potentially infinite number of taps is truncated and is related to the root mean square delay spread, $\tau_{rms}$, and the sampling rate (or the space between the taps) $T_s$, as:

$$k_{max} = \lceil 10\tau_{rms}/T_s \rceil \tag{11.4}$$

Here:

$$\sigma_k^2 = \sigma_0^2 e^{-kT_s/\tau_{rms}} \tag{11.5}$$

where $\sigma_0^2$ is the normalisation factor which ensures that the sum of average channel gain is one and defined as:

$$\sigma_0^2 = 1 - e^{-T_s/\tau_{rms}} \tag{11.6}$$

Also, it has been found that log-normal distribution best suits UWB BAN channels. More importantly, the addition of multiple log-normally distributed paths results in another log-normal distribution. For body-to-body communications gamma and Rician distributions have the best fits.

In a more realistic scenario such a distribution can be easily modelled using a sum of Gaussians. This allows a more flexible model fitting without much limitations.

On the other hand, in some cases where the communication through a direct link is not feasible a multihop communication using relays may be applicable [2]. The IEEE 802.15.6 standard thus includes the optional use of relays to help mitigate the deleterious effects of complicated BAN channels.

A communication channel won't be modelled perfectly unless the inherent temporal and statistical correlations due to multipath and the effect of motion are taken into account and incorporated within the model formulation. This is a more challenging problem in the area of channel modelling. For example, when compared to narrowband BAN, there is also more large-scale fading with UWB from larger path losses due to its higher carrier frequencies, and the accurate modelling of such scenario needs more research. Another influencing parameter is nonstationarity of the channel due to various body movements, change in the environment statistics and properties, obstacles in the venue, and also body peripheral variations. In a more adaptive design, the channel dynamics are more important than the static attenuation represented by the mean path loss for any individual link. Whilst a parameterised 'model' might give better fit by specifying the precise location of the sensor nodes, such a model is difficult to achieve for BAN, mainly because of practical issues such as subject-to-subject variabilities and mobility.

In order to deal with the channel variations due to movement, channel dynamics, and multipath correlations, second-order statistics are used for characterising BAN channels for both on-body and off-body communications. These statistics add further information to the channel distribution to reflect very important information about fading, multipath, Doppler shift, nonlinear power decay, and cross-channel interferences. The key second-order statistical characterisations for BAN radio propagation may be summarised as follows [2]:

- Delay spread and the power delay profile (PDP) of BAN channels can be used to estimate the number of channel taps, related to the number of paths in very common multipath scenarios, and hence the presence of ISI. This mostly occurs in the UWB BAN channels

[59, 60]. In general, hardly an accurate multipath UWB model can be achieved either theoretically or experimentally. On the other hand, although the multipath effect exists for the narrowband BAN channels, some researchers approximate it by a single-tap channel [66], which is for LOS single path models and therefore is not a good approximation of multipath environments.

- Average fading duration (i.e. the average time the received signal strength is below any given level) can be used to determine the amount of time for which successful packet transmission on a given Tx/Rx link may not be possible. This is therefore an important parameter for BAN communications. The level crossing rate (LCR), i.e. the average rate at which the signal strength crosses from above to below any given signal level (particularly at the mean path loss [77]), can be used to infer the rate of fading. The LCR can be used to determine the Doppler spread, which is approximately 1 Hz for slow varying and 4 Hz for fast varying (such as running) body movement. It has been determined that both average fade duration and LCR are highly dependent on channel dynamics, as they depend on the rate and amount of body movement [55, 77]. In many typical BAN channels, the average fading duration is 300 ms or more [77], significantly larger than the 250 ms latency requirement for many BAN applications [78].
- Autocorrelation of time-varying channel gain, which can be used to determine the coherence time [56], for any BAN link can determine the time during which a successful packet transmission is possible, as with average fade duration. If the coherence changes within the multipath environment, it can be estimated using the autocorrelation measure. In the simple cases of single-path stationary systems, the coherence time and Doppler spread are inversely related, i.e.

$$\textit{Coherence Time} \approx \frac{1}{\textit{Doppler Spread}} \tag{11.7}$$

So, if the transmitter, receiver, or the intermediate subjects move faster, the Doppler spread is larger and, accordingly, the coherence time is smaller. Thus, the autocorrelation drives the design of packet lengths and the placement of pilots for channel estimation, making it an important parameter for BAN communications. It is also important for power control based on channel prediction [79]. Longer coherence times of up to one second for the 'everyday' mixed activity for on-body narrowband BAN channel [55, 79] allow for successful transmit power control over the duration of multiple BAN superframes (even when a superframe is hundreds of milliseconds long). With continuous movement, the channel coherence time can drop to between 25 and 70 ms [56], indicating much smaller time for successful packet transmission. This is therefore a measure of system reliability for dynamic slot scheduling of advanced BAN systems [80].

- Cross-correlation is another influential factor [62, 81] as the BAN sensors may be densely placed on the body, and the quality of one gateway-to-sensor link could be used to determine the quality of the same gateway to another short proximity sensor link through the cross-correlation of their signal strengths. However, it has been found that with a medium density of 10 on-body sensors such spatial cross-correlation coefficients are 0.5 or lower. This may not be sufficient (for approximating the gateway quality of one sensor from that of its neighbouring sensor) given that spatial cross-correlation is generally considered significant for values of 0.7 or larger.

In [81] the spatial correlation of on-body radio propagation has been investigated using a K-weight-based spatial autocorrelation model and corresponding Z-score. It has been demonstrated that, due to rich multipath scattering in an indoor environment, both the spatial autocorrelation and the corresponding confidence level are higher compared to those in the chamber environment. The achievement has the implication that to achieve high repeatability in on-body channel measurements for less scattered environments the locations of transmitter and receiver need to be sufficiently accurate.

After all, the model fitting should be evaluated using some statistical measures. Such measures are often taken using simulated data and systems (including the propagation channel). For a parameterised model (taking into account the first- and second-order statistics) the goodness-of-fit criteria involve selection of right number of model parameters which result in the best fit. Simple criteria such as Akaike information criterion (AIC) [82], or more realistic criteria such as Cramér–Rao lower bound (CRB) may be used to evaluate the goodness-of-fit. Yet the evaluation criteria don't consider the variation in the status of the body (awake, sleep, etc.), co-interferences (of multiple bodies), or the changes in its peripheral situations.

### 11.4.4 Standards and Guidelines

The IEEE 802.15.6 standard has been under development since 2007 and, as of February 2012, has been approved by the IEEE [24, 83]. The standard communication systems are broadly defined, including narrowband communications, UWB, and also nonradio human-body communications [24]. Additionally, three communication channels are described in the 802.15.6 channel-modelling document [84], namely in-body, off-body, and on-body. These communication channels and the IEEE 802.15.6 standard are used as the basis for the insight into propagation characterisations in [85]. According to these characterisations, the models for different communication channels have been suggested as follows [85]:

- Small-scale fading in UWB communications is best modelled by a log-normal distribution.
- Small-scale fading in narrowband communications is best modelled by either a log-normal or a gamma distribution. The distribution called generalised gamma distribution includes both of them. Further, agglomerate, or combined, models that account for all typical on-body transceiver locations, in conjunction with a typical path loss for each transmitter/receiver location, can be found that are suitable to describe general BAN communications.
- For large-scale fading, traditional path-loss measures – characterising path loss in terms of distance and providing path-loss exponents, and also line-of-sight (LOS) and non-line-of-sight (NLOS) categorisations are not relevant descriptors of the BAN communications.
- Average fade duration and coherence time are of major interest because these are the best second-order measures to determine the available time for reliably sending packets over a given transmitter/receiver link.
- Using relays for implementations of IEEE 802.15.6 is important and can be effective in likely BAN propagation scenarios when compared to single-hop BANs.

Following these research attempts, some guidelines have been set to assess the IEEE 802.15.6 BAN standard proposal [24, 83]. These provide valuable insight into the expected application and needs of the IEEE 802.15.6 BAN standard and are mainly related to propagation and the physical layer communications. They include [2]:

- BANs should be scalable up to 256 nodes.
- A BAN link should support bit rates between 10 kbps and 10 Mbps.
- The PER for a 256-octet payload (i.e. $256 \times 8$ bits of data) should be measured for the 95% best-performing links.
- Maximum radiated transmitter power should be 0 dBm (or 1 mW), and all the devices should be able to transmit at −10 dBm (or 0.1 mW). This automatically meets the SAR guideline of the FCC of 1.6 W kg$^{-1}$ in 1 g of body tissue [85] (which equates to a max transmitter radiated power of 1.6 mW).
- The nodes should be addable to and removeable from the network in less than three seconds.
- Reliability, latency (delay), and jitter (variation of one-way transmission delay) should be supported for those BAN applications that need them. Latency in medical applications should be less than 125 ms. This can be up to 250 ms for nonmedical applications. Jitter should be less than 50 ms.
- Power-saving mechanisms (such as duty cycling) should be provided.
- The physical layer should support collocated operation of at least 10 randomly distributed BANs in a $6\,m \times 6\,m \times 6\,m$ volume.
- In-body and on-body BANs should coexist in and around the body.

## 11.5 BAN-WSN Communications

In most BAN applications the information captured or processed within a local network needs to be transmitted through wireless network to a distance not necessarily covered by any short-range communication system. In such scenarios two fundamental problems have to be solved; first of all, there should be a handshake between the short- and long-range systems and, secondly, the long-range system has to be robust enough to tackle communication channel issues, such as noise, multipath, and Doppler effect (for mobile systems).

In a multihop BAN or BAN-WSN communication, however, routing becomes an important issue and therefore another branch of research, namely routing, has been developed.

In a BAN complex communication environment the following problems and challenges must be considered in designing the routing algorithms [86, 87]:

- The topological structure dynamically changes: wireless transmission involves body surface transmission, body transmission, and free space transmission. There is a shadow effect caused by human motion, the distance and relative positions of the nodes change with the movement of limbs, and the topology is subject to time and various posture and health circumstances [88]. A reliable routing mechanism should therefore adapt to these changes in topology.

- Energy efficiency: due to the sensor positions (which may be inside the body) it is not enough to supply power only through micro-batteries. Now, although RF, electromagnetic (EM) or energy harvesting can be used for the power supply, the design should also be energy efficient and both, single node energy efficiency and the whole network energy balance, must be considered in the routing design.
- Node temperature: nodes generate heat when they are operational. This can harm the tissue [89, 90]. Therefore, the nodes temperature must be effectively controlled in the routing design.
- QoS requirements: nodes in a BAN measure different types of data, which must be archived and then processed differently to ensure the QoS requirements of different data types, such as emergency data, delay sensitive data (DSD), reliability sensitive data (RSD), and general data are fulfilled. QoS includes data quality, processing quality, delay in the data communication, and most importantly security of data.

The computer networking protocol design and implementation for BAN data communication are the subjects of the next section.

## 11.6 Routing in WBAN

Routing is one of the important problems in wireless communications. An effective routing algorithm should prevent delays and loss of data and deliver the data packets through their fastest paths from source to destination. On the other hand, the packets may have various priority levels and the BAN routing environment may vary as there are mobility and body motions involved. Based on such requirements, in [91–93] the routing approaches have been categorised into posture-based, temperature-based, cross-layer, cluster-based, and QoS-based routing classes.

### 11.6.1 Posture-based Routing

This is used to analyse the network topology of the human body in various dynamic postures to establish a fast and stable path. Experiments have verified that all kinds of human body movements have some regularity [94]. This can greatly improve the WBAN dynamics and flexibility. In a dynamic environment, the body movement affects the communication channel and consequently produces shadow effect. Nevertheless, if the new posture can be predicted the data transmission quality and speed can be significantly improved [95].

In an example of posture-based routing, called mobility handling routing protocol (MHRP) [96], the heart rate has been monitored. Seven nodes have been located on the left and right sides of the body. Two symmetrical sets of nodes have been considered. Each set has one sink node, two relay nodes, and one common acquisition node. A new type of fault-tolerant system is composed of two identical and independent sets of nodes. The design allows changing from one group to another when the current working set energy is scarce or the topology is interrupted so that the errors can be reduced and the reliability of communication improved. The routing configuration for MHRP structure is depicted in Figure 11.6.

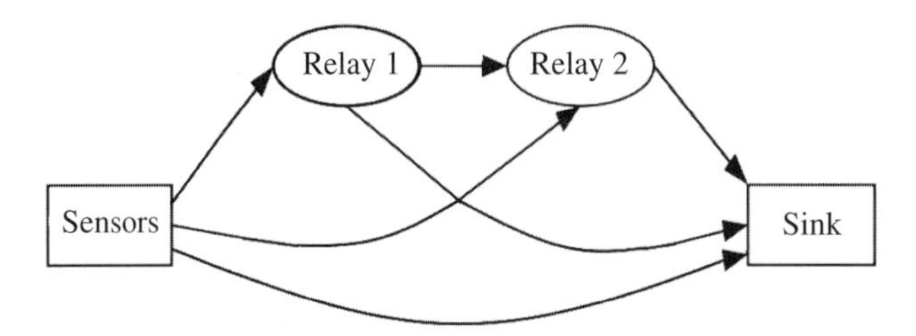


**Figure 11.6** Routing design for MHRP [96]. Source: Courtesy of Karmakar, K., Biswas, S., and Neogy, S.

For such a structure, the routing protocol selects the shortest path at the time of use to enable the fastest transmission of data from sensors to sink following the conventional routing algorithms.

In another example an opportunistic transmission link establishment algorithm, namely distributed network management cost minimization (NCMD) is proposed to minimise the cost of network management and optimise its QoS [97]. The algorithm provides a solution to the dynamic change in network topology, increase in the network management cost, and increase in the energy consumption by the sensors, all due to the limb movement, through the optimisation of a regularised energy-efficient prioritised opportunistic connectivity algorithm [97].

### 11.6.2 Temperature-based Routing

This method considers the node temperature to select the path in the routing process. The routing method avoids the nodes if their temperature rises [98]. Temperature-based routing has been very popular during the early stages of developing BANs. In recent years, however, a large number of studies have focused on energy, so the temperature-based routing method became of less interest. The routing process becomes complex where the body network or environment dynamics are considered too.

Examples of this type of routing are:

- Thermal aware routing algorithm (TARA) [99], where the temperature is taken as the only parameter for path selection and the neighbouring node with the lowest temperature is selected as the next hop. The so-called energy-efficient routing protocol (ER-ATTEMP) [100] not only avoids the nodes with high temperature but also chooses the path with the minimum number of hops as the best route.
- Trust and thermal aware routing protocol (TTRP) [101], which involves two parameters of trust and temperature. Different from other routing methods, in TTRP, additional relay nodes are added. These relays have the function of receiving and forwarding the data without participating in information collection and are provided with a higher energy to complete the function of forwarding information from other nodes. TTRP protocol includes three stages: trust estimation, routing discovery, and routing maintenance.
- Mobility-based temperature-aware routing (MTR)protocol [102] takes into account both temperature and mobility including change of posture. The protocol categorises the BAN nodes into static and dynamic nodes. Static nodes are those located in the human body

centre and so are not (or minimally) affected by human movement, while dynamic nodes are located in the moving positions such as on arms and legs.

### 11.6.3 Cross-layer Routing

This technique mainly integrates multiple protocol systems and exploits the advantages of each protocol stack to achieve a better network performance. It has been demonstrated that this method is more adaptable to dynamic BANs, and the collaboration between different layers can better serve different priority data, provide customised services for each type of data, and achieve a desirable network performance with a low latency, energy saving, and high reliability.

Some examples of the cross-layer approach are as follows:

- Priority-based cross-layer routing protocol (PCLRP) [103] has developed a protocol across MAC in the link and network layers. This allows prioritising the access to emergency, delay sensitive, and general data.
- In [104] a cross-layer design optimal (CLDO) scheme has been introduced to design and optimise the network parameters – such as reliability, energy efficiency, and network lifetime – through cooperation between physical, MAC, and network layers. Through a number of experiments and theoretical development, the goal has been to find the best transmission power, relay, and packet size to achieve the above objectives.
- Another cross-layer routing protocol, called cross-layer retransmitted strategy (CLRS), has been proposed by Wang and Guo [105]. Unlike other BAN routing methods, this protocol is meant to retransmit data packets that previously failed to transmit. In this work the collision and shadow effects are considered as the two main reasons for data transmission failure. In the receiver, the failure is detected based on the waveform characteristics and then a decision is made as to whether the data packet needs to be retransmitted.
- As often used in an effective routing strategy, a cross-layer optimisation based on link quality prediction has been proposed [106]. The method relies on the periodicity of human activities in order to design the routing algorithm. To do this, the instantaneous link quality is estimated through one-step prediction between the nodes as well as between the coordinator and the nodes separately. In order to allow more flexibility for adaptation to the mobile network environment, the routing protocol exploits the relay multihop transmission mode. In addition, instead of wasting energy in using a fixed amount of power, this study adopts an adaptive adjustment of the nodes transmission power. By comparing the predicted value of link quality with the standard value, for larger predicted values compared to the standard value, the transmission power is reduced to save energy and vice versa. However, for high-priority data, the power is increased to enhance the accuracy and ease of recovery of the received data.
- In another approach Zhang et al. [107] propose a cross-layer scheme in which the transmission power adaptively changes with respect to an autocorrelation coefficient, a joint transmission power control, dynamic slot scheduling, and a two-hop cooperative networking system. The network accounts for the trade-off between reliable transmission

and energy. This strategy consists of four stages: channel state prediction, adjustment of transmission power, rearrangement of slot sequence for direct transmission, and selection of relay nodes.

### 11.6.4 Cluster-based Routing

Cross-layer routing method inherits its concept from the low energy adaptive clustering hierarchy (LEACH) routing protocol, which is a classic routing scheme in WSN and proved to be suitable for WBANs [91]. When the number of nodes increases and the relative distance between the nodes increases, the clustering method can ensure network connectivity, balance the energy consumption of the network centre and edge, adapt to the dynamic topology structure, and improve the network robustness. Clustering routing protocol divides the network nodes into a number of clusters. Each cluster consists of several nodes and a cluster head. The cluster head is selected by the algorithm and is responsible for integrating and forwarding the information within the cluster, mainly to reduce the direct communication overhead [91].

A dual sink cluster-based (DSCB) node routing protocol proposed by Ullah et al. [108] uses dual sink nodes. This scheme aims at exploiting the network dynamics to analyse and resolve some of the shortcomings resulting from congestion, high data transmission failure rate, limited coverage, NLOS communication, etc. In a practical WBAN, the shadow effect caused by limb movement needs to be avoided. So, one needs to wait for it to disappear before it can continue to transmit, which may cause a relatively long transmission delay. Consequently, the resulting interruption delay may occur in the emergency data period and cause transmission failure, which can be life threatening. The method of double sink can improve the negative consequences of the shadow effect, balance the network load of a single sink node, and improve the practicability of the WBAN. DSCB protocol fixes the nodes in the network. The two sinks S1 and S2 are located on the front and back of the human body, respectively, defined as cluster head. Clusters 1 and 2 are composed of other nodes on the front and back of the human body, respectively. When the node has emergency data, it communicates directly with the corresponding cluster head node. For general data, a multi-hop mode is usually used. Through optimising a cost function the next hop node is selected.

An energy efficiency routing protocol, namely clustering based routing protocol for wireless body area network (CRPBA), is another cluster-based protocol [109], which refers to each sink node as a gateway node. The nodes in the network choose direct communication or cluster-based communication according to the distance between the gateway and node or data type. The two gateways are located at waist and neck. The protocol stipulates that some key data are communicated directly with the appropriate gateway node, to reduce transmission delay. The common data are transmitted to the cluster head by the cluster members, and then forwarded to the gateway by the cluster head, so the energy consumption is reduced and the success rate of data transmission is improved. With regards to the choice of cluster head, the protocol solves a suitable cost function to decide that the node with the lowest cost is the cluster head. The algorithm minimises the distance between gateway and source node, $d_{gateway,\ source\ node}$, divided by the residual energy of the node, $e_{residual}$, i.e.

$$J = \frac{d_{gateway,source\ node}}{e_{residual}} \tag{11.8}$$

### 11.6.5 QoS-based Routing

This is another routing approach which emphasises QoS. This algorithm has an important role in many applications particularly in resource constrained BANs. The targeted QoS include data priority, energy efficiency, link and data transmission reliability, low transmission delay, node temperature, and data security.

This approach may be categorised into the following routing methods:

Distributed lightweight QoS routing protocol (LRPD) [110] is a QoS-based routing approach which optimises the delay-based QoS. The data are first classified based on their priority. The data are divided into general packets (GPs) and delay sensitive packets (DP). DP has the highest priority. Then, the DP and GP data are processed by other system modules. As the result, the high-priority data do not need to wait for direct transmission, while the low-priority data have to wait until an appropriate transmission time, in order to ensure the nondelay transmission of DP data.

Another QoS-based routing, called hybrid data-centric routing protocol (HDPR) [111], adopts a modular technique. Similar to the LRPD methods, the data entered into the MAC are then classified into DSD, normal data (ND), critical data (CD), and RSD. The data priority starts from CD, as the highest, then DSD, RSD, and finally ND, as the lowest priority data. The data are classified based on their QoS perception. This protocol mainly considers the QoS indexes, such as path loss, link reliability, delay, and temperature to optimise the design. Moreover, the relay node used in this protocol only has the function of receiving and sending, which reduces the energy consumption of the acquisition node and increases the sensor network life time.

In addition to the above methods, there are many other approaches such as multiple-based routing which is a node-independent multipath routing method [112]. These methods allow for multiple transmission paths between source and destination without common nodes between them. The advantage of such a simple method ensures a stable communication.

Other routing algorithms, namely mobile-sink-based routing algorithms, allow the sink nodes to be mobile and move randomly. For such strategy, often high energy levels have to be produced in both the centre and edge nodes.

By looking at the variety of network optimisation parameters, it is highly expected that in future and for more complex BANs new inclusive routing protocols with more sophisticated algorithms be introduced and come to practice. These algorithms should cater for many particular applications which address human physical, biological, and mental states of the human body.

## 11.7 BAN-building Network Integration

There have been some attempts at developing smart sensor networks for human monitoring inside buildings. This can be elegantly performed by integrating the BAN within the building's sensor network.

Integration of the two networks aims at establishing smart and intelligent environments, namely human-aware smart buildings. The infrastructure effectively supports people while they enter, exit, or move inside the building. Through such a smart network many assistive services, including the following, can be provided:

- Human identification, which is meant to detect people inside the building. This service is the basis for more advanced services related to security and personalised people support.
- Human recognition benefits from the power of video face recognition, speech recognition, or even gait analysis in effectively recognising the people inside the building.
- Human localisation, which involves an information processing technique to identify and track the location of people inside the building. Using cameras are probably the most straightforward way to achieve this goal. However, this is not always allowed, mainly because of privacy issues.
- Information exchange, which enables data communication between the smart building and its residents. A good example is patient monitoring within hospitals or care homes. The network includes various human mount sensors and actuators through which the smart building monitors the vital parameters of people and decides on their healthcare needs.
- The information exchange can be very effective in providing the safety and security of people and predicting or preventing hazards such as fire, chemical spillage, and smoke.

The gateway between the two networks should use a suitable protocol to enable a secure handshaking, addressing, and communication between the two.

## 11.8 Cooperative BANs

The first step towards cooperative body area networks (CoBANs) has been taken by including some relay nodes in the network. As an example, in [47] the application of cooperative communications in UWB wireless BANs has been tested on a two-relay cooperative network such as the one in Figure 11.7. In this design, a group of on-body devices may collaborate together to communicate with other groups of on-body devices. A CoBAN has been introduced for a body-centric multipath channel. The important channel parameters – namely the path loss, power variation, power delay profile (PDP), and effective received power cross-correlation – are investigated and statistically analysed. Specifically, an intuitive measure is proposed to quantify the diversity gains in a single-hop cooperative network. This is defined as the number of independent multipaths that can be averaged over to detect the symbols.

In the proposed CoBAN channel model, the transmission is limited to the area around the human torso since many common physiological signals such as electrocardiography (ECG), electromyography (EMG), respiration and heart rates, heart sound, pulse oximetry,

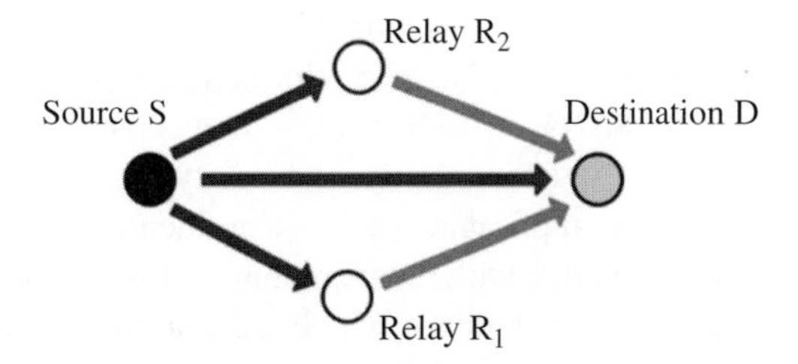


**Figure 11.7** Two-relay cooperative network used in [47]. Source: Courtesy of Chen, Y., Teo, J., Lai, J.C.Y, Gunawan, E., Low, K.-S., Soh, C.-B., and Rapajic, P.B.

and skin temperature can be detected there. Furthermore, directional antennas have been used to mitigate the effect of stray electromagnetic radiation so the antenna beam steering for collarbone pulse oximetry interference cancellation or reception enhancement could be achieved [47]. Unlike the typical methods where the setup frequency-domain measurements are carried out using vector network analyser (VNA), in this work a UWB pulse generator is used to send a short pulse around the body and the signal is picked up by a digital communication system analyser (sampling oscilloscope). This study would provide a full-bandwidth view of the UWB on-body channel that cannot be directly measured with a VNA. From the received data, subtractive deconvolution algorithms are used to extract the channel parameters relevant to CoBAN, including path loss, power variation, PDP, and effective received power cross-correlation.

Nevertheless, the focus is on the two-stage single-hop transmission model for the CoBAN. In the first stage, the source transmits to the destination and the relays. In the second stage, the relays transmit to the destination. The spatial diversity provided by the relays is dependent on the channel properties derived from the measurements. The cardinality of the overall 'diversity-paths' set offered by the CoBAN is then calculated. Since this value reflects the largest number of redundant copies of the transmitted signal, it can be used as a benchmark to assess the performance of various diversity-based cooperative BAN schemes.

The path loss is estimated by calculating the signal power within the time window that contains significant energy content. To study the power variation an exponential power distribution has been proposed to provide a wide variety of probability density functions. The time-domain CLEAN algorithm [113] is used to extract the path delay information and the signal strength. To model the effective received power cross-correlation for the proposed UWB BAN system, a RAKE receiver has been used to combine the information obtained from several multipath components to form a more refined and stronger version of the signal [114].

Further development in cooperative BAN networks may be performed by establishing a fully distributed system where each node has its own processing capability and therefore can be counted as an intelligent agent. The related problems to such networks may be solved through adaptive consensus or diffusion algorithms [115–117].

## 11.9 BAN Security

BAN transmission has to be made secure and accurate. Therefore, further efforts have to be made to comply with this requirement. Patient 'secure' data should be only derived from each patient's dedicated BAN system and not mixed up with other patients' data. Moreover, access to patient data should be limited to those directly involved in providing healthcare. Although security is a high priority in most networks, little study has been carried out in this area for BANs. As BANs are resource-constrained in terms of power, memory, communication rate, and computational capacity, the security solutions proposed for other networks may not be applicable to BANs. Confidentiality, authentication, integrity, and also reliability of data together with secure management are the major requirements for BAN security. The IEEE 802.15.6 standard, which is the latest standard for BAN, has tried to provide the necessary security protocols for BAN. However, the introduced conventions have several security problems [118]. More details on the IEEE security standards for BAN and the involved hierarchy [24] are discussed in a separate chapter of this book.

## 11.10 Conclusions

Current BAN communication systems involve both short-range communications, mainly for sensor to hub data transfer and long-term communications generally for making the data available to the long-distance user (doctor, caregiver, clinic, etc.). Multipath, fading, noise, Doppler effect, and channel interferences require a significantly complex model for BAN communication systems. As the consequence of body movement, body absorption of radio waves, blocking the LOS transmission due to shadowing, healthcare priorities, and constant changes in the environment (heat, humidity, etc.), effective routing strategies and protocols have to be designed and applied to ensure system reliability. In this chapter, the above problems have been discussed and some solutions proposed. Nevertheless, for real-time applications the environment nonlinearity as well as the nonstationarity in sensor recording must be carefully incorporated into the system design and channel modelling.

## References


1 Movassaghi, S., Abolhasan, M., Lipman, J. et al. (2014). Wireless body area networks: a survey. *IEEE Communications Surveys & Tutorials* 16 (3): 1658–1686.
2 Smith, D.B., Miniutti, D., Lamahewa, T.A., and Hanlen, L.W. (2013). Propagation models for body-area networks: a survey and new outlook. *IEEE Antennas and Propagation Magazine* 55 (5): 97–116.
3 Conway, G.A. and Scanlon, W.G. (2009). Antennas for over-body-surface communication at 2.45 GHz. *IEEE Transactions on Antennas and Propagation* 57 (4): 844–855.
4 Ryckaert, J., De Doncker, P., Meys, R. et al. (2004). Channel model for wireless communication around human body. *Electronics Letters* 40 (9): 543–544.
5 Tsouri, G.R., Sapio, A., and Wilczewski, J. (2011). An investigation into relaying of creeping waves for reliable low-power body sensor networking. *IEEE Transactions on Biomedical Circuits and Systems* 5 (4): 307–319.
6 Cotton, S.L., Conway, G.A., and Scanlon, W.G. (2009). A time-domain approach to the analysis and modeling of on-body propagation characteristics using synchronized measurements at 2.45 GHz. *IEEE Transactions on Antennas and Propagation* 57 (4): 943–955.

7 Fort, A., Desset, C., Wambacq, P., and Biesen, L. (2007). Indoor body-area channel model for narrowband communications. *IET Microwaves, Antennas and Propagation* 1 (6): 1197–1203.

8 Hu, Z., Nechayev, Y.I., Hall, P.S. et al. (2007). Measurements and statistical analysis of on-body channel fading at 2.45 GHz. *IEEE Antennas and Wireless Propagation Letters* 6: 612–615.

9 Conway, G.A., Scanlon, W.G., Orlenius, C., and Walker, C. (2008). In situ measurement of UHF wearable antenna radiation efficiency using a reverberation chamber. *IEEE Antennas and Wireless Propagation Letters* 7: 271–274.

10 Giddens, H., Paul, D.-L., Hilton, G.S., and McGeehan, J.P. (2012). Influence of body proximity on the efficiency of a wearable textile patch antenna. In: *6th European Conference on Antennas and Propagation (EUCAP)* Prague, Czech Republic (26–30 March 2012), 1353–1357. IEEE.

11 Salonen, P., Rahmat-Samii, Y., and Kivikoski, M. (2004). Wearable antennas in the vicinity of human body. In: *IEEE Antennas and Propagation Society International Symposium*, 467–470. IEEE.

12 Sanz-Izquierdo, B., Huang, F., and Batchelor, J.C. (2006). Covert dual-band wearable button antenna. *Electronics Letters* 42 (12): 3–4.

13 Wong, K.L. and Lin, C.I. (2005). Characteristics of a 2.4-GHz compact shorted patch antenna in close proximity to a lossy medium. *Microwave and Optical Technology Letters* 45 (6): 480–483.

14 Cotton, S. and Scanlon, W. (2009). Channel characterization for single- and multiple-antenna wearable systems used for indoor body-to-body communications. *IEEE Transactions on Antennas and Propagation* 57 (4): 980–990.

15 D'Errico, R. and Ouvry, L. (2009). Time-variant BAN channel characterization. In: *IEEE 20th International Symposium on Personal, Indoor and Mobile Radio Communications*, 3000–3004. IEEE.

16 Cotton, S.L., Scanlon, W.G., and Conway, G.A. (2009). Autocorrelation of signal fading in wireless body area networks. In: *2nd IET Seminar on Antennas and Propagation for Body-Centric Wireless Communications*, 1–5. IET.

17 Cotton, S.L., Scanlon, W.G., Conway, G.A., and Bentum, M.J. (2010). An analytical path-loss model for on-body radio propagation. 20th International URSI Symposium on Electromagnetic Theory, August 2010, Berlin, Germany.

18 Alomainy, A., Hao, Y., Owadally, A. et al. (2007). Statistical analysis and performance evaluation for on-body radio propagation with microstrip patch antennas. *IEEE Transactions on Antennas and Propagation* 55 (1): 245–248.

19 Van Torre, P., Vallozzi, L., Jacobs, L. et al. (2012). Characterization of measured indoor off-body MIMO channels with correlated fading, correlated shadowing and constant path loss. *IEEE Transactions on Wireless Communications* 11 (2): 712–721.

20 Wang, Y., Bonev, I.B., Nielsen, J.O. et al. (2009). Characterization of the indoor multi antenna body-to-body radio channel. *IEEE Transactions on Antennas and Propagation* 57 (4): 972–979.

21 Rosini, R. and D'Errico, R. (2012). Comparing on-body dynamic channels for two antenna designs. In: *Loughborough Antennas and Propagation Conference (LAPC)*, 1–4. IEEE.

**22** Cotton, S.L., McKernan, A., Ali, A.J., and Scanlon, W.G. (2011). An experimental study on the impact of human body shadowing in off-body communications channels at 2.45 GHz. In: *5th European Conference on Antennas and Propagation (EUCAP)*, 3133–3137. IEEE.

**23** Jovanov, E., Milenkovic, A., Otto, C., and De Groen, P.C. (2005). A wireless body area network of intelligent motion sensors for computer assisted physical rehabilitation. *Journal of NeuroEngineering and Rehabilitation* 2 (1) https://doi.org/10.1186/1743-0003-2-6.

**24** IEEE Std 802.15.6-2012 (2012) IEEE Standard for local and metropolitan area networks – Part 15.6: Wireless body area network, New York: IEEE

**25** Hall, P.S., Hao, Y., and Cotton, S.L. (2010). Progress in antennas and propagation for body area networks. In: *International Symposium on Signals, Systems and Electronics*, 1–7. IEEE.

**26** FCC (2012). FCC dedicates spectrum enabling medical body area networks to transform patient care, lower health care costs, and spur wireless medical innovation. Press release (24 May).

**27** Boulemtafes, A. and Badache, N. (2016). Design of wearable health monitoring systems: an overview of techniques and technologies. In: *mHealth Ecosystems and Social Networks in Healthcare* (eds. A.A. Lazakidou, S. Zimeras, D. Iliopoulou and D.-D. Koutsouris), 79–94. Switzerland: Springer International Publishing.

**28** Afonso, J.A., Taveira Gomes, D.M.F., and Rodrigues, R.M.C. (2014). An experimental study of ZigBee for body sensor networks. In: *Transactions on Engineering Technologies* (eds. G.I. Yang, S.L. Ao and L. Gelman), 467–481. Springer.

**29** Negra, R., Jemili, I., and Belghith, A. (2016) Wireless body area networks: applications and technologies. The 2nd International Workshop on Recent Advances on Machine-to-Machine Communications, Madrid.

**30** Kim, T., Kim, S.H., Yang, J. et al. (2014). Neighbor table based shortcut tree routing in ZigBee wireless networks. *IEEE Transactions on Parallel and Distributed Systems* 25 (3): 706–716.

**31** Angell, I. and Kietzmann, J. (2006). RFID and the end of cash? *Communications of the ACM* 49 (12): 90–96.

**32** Domenicali, D. and Di Benedetto, M. (2007). Performance analysis for a body area network composed of IEEE 802.15.4a devices. In: *4th Workshop on Positioning, Navigation and Communication 2007 (WPNC'07)*, 273–276. IEEE.

**33** Lee, D., Shaker, G., and Augustine, R. (2018). Preliminary study: Monitoring of healing stages of bone fracture utilizing UWB pulsed radar technique. 8th International Symposium on Antenna Technology and Applied Electromagnetics (ANTEM 2018). Waterloo, ON, Canada (19–22 August 2018).

**34** Jovanov, E. and Milenkovic, A. (2011). Body area networks for ubiquitous healthcare applications: opportunities and challenges. *Journal of Medical Systems* 35 (5): 1245–1254.

**35** Prasadh Narayanan, R., Veedu Sarath, T., and Veetil Vineeth, V. (2016). Survey on motes used in wireless sensor networks: performance & parametric analysis. *Wireless Sensor Network* (8): 51–60.

**36** Latrè, B., Braem, B., Blondia, G. et al. (2011). A survey on wireless body area networks. *Wireless Networks* 17 (1): 1–18.

**37** Chen, M., Gonzalez, S., Vasilakos, A. et al. (2010). Body area networks: a survey. *Mobile Networks and Applications* 16: 171–193.

**38** Garg, M.K., Kim, D.J., Turaga, D.S., and Prabhakaran, B. (2010). Multimodal analysis of body sensor network data streams for real-time healthcare. In: *Proceedings of the 11th ACM SIGMM International Conference on Multimedia Information Retrieval*, 469–478. ACM.

**39** Cooney, M.J., Svoboda, V., Lau, C. et al. (2008). Enzyme catalysed biofuel cells. *Energy & Environmental Science* 1: 320–337.

**40** Yoo, J., Yan, L., Lee, S. et al. (2010). A 5.2 mW self-configured wearable body sensor network controller and a 12 μW wirelessly powered sensor for a continuous health monitoring system. *IEEE Journal of Solid-State Circuits* 45: 178–188.

**41** Bui, F.M. and Hatzinakos, D. (2011). Quality of service regulation in secure body area networks: system modeling and adaptation methods. *EURASIP Journal on Wireless Communications and Networking* 2011: 56–69.

**42** Kumar, P. and Lee, H.J. (2012). Security issues in healthcare applications using wireless medical sensor networks: a survey. *Sensors (Basel)* 12 (1): 55–91.

**43** Cotton, S.L., D'Errico, R., and Oestges, C. (2014). A review of radio channel models for body centric communications. *Radio Science* 49: 371–388.

**44** Sayrafian-Pour, K., Yang, W.B., Hagedorn, J. et al. (2010). Channel models for medical implant communication. *International Journal of Wireless Information Networks* 17 (3–4): 105–112.

**45** Alomainy, A. and Hao, Y. (2009). Modeling and characterization of biotelemetric radio channel from ingested implants considering organ contents. *IEEE Transactions on Antennas and Propagation* 57 (4): 999–1005.

**46** Maman, M., Dehmas, F., D'Errico, R., and Ouvry, L. (2009). Evaluating a TDMA MAC for body area networks using a space-time dependent channel model. In: *IEEE 20th International Symposium on Personal, Indoor and Mobile Radio Communications*, 2101–2105. IEEE.

**47** Chen, Y., Teo, J., Lai, J.C.Y. et al. (2009). Cooperative communications in ultra-wideband wireless body area networks: channel modeling and system diversity analysis. *IEEE Journal on Selected Areas in Communications* 27 (1): 5–16.

**48** D'Errico, R., Rosini, R., and Maman, M. (2011). A performance evaluation of cooperative schemes for on-body area networks based on measured time-variant channels. In: *IEEE International Conference on Communications (ICC)*, 1–5. IEEE.

**49** Smith, D.B. and Miniutti, D. (2012). Cooperative body-area-communications: first and second order statistics with decode-and-forward. In: *IEEE Wireless Communications and Networking Conference (WCNC)*, 689–693. IEEE.

**50** Liu, L., Keshmiri, F., Craeye, C. et al. (2011). An analytical modeling of polarized time-variant on-body propagation channels with dynamic body scattering. *EURASIP Journal on Wireless Communications and Networking* 2011: 1–12.

**51** Oliveira, C., Pedrosa, L., and Rocha, R. (2008). Characterizing on-body wireless sensor networks. In: *Proceedings of New Technologies, Mobility and Security (NTMS)*, 1–6. IEEE.

**52** Zhen, B., Takizawa, K., Aoyagi, T., and Kohno, R. (2009). A body surface coordinator for implanted biosensor networks. In: *IEEE International Conference on Communications, ICC*, 1–5. IEEE.

**53** Katayama, N., Takizawa, K., Aoyagi, T. et al. (2008). Channel model on various frequency bands for wearable body area network. In: *ISABEL, International Symposium on Applied Sciences in Bio-Medical and Communication Technologies*, 1–5. IEEE.

**54** Gallo, M., Hall, P., Bai, Q. et al. (2011). Simulation and measurement of dynamic on-body communication channels. *IEEE Transactions on Antennas and Propagation* 59 (2): 623–630.

**55** Smith, D., Hanlen, L., Zhang, J. et al. (2011). First- and second-order statistical characterizations of the dynamic body area propagation channel of various bandwidths. *Annals of Telecommunications* 66 (3–4): 187–203.

**56** Smith, D., Hanlen, L., Zhang, J. et al. (2009). Characterization of the dynamic narrowband on-body to off-body area channel. In: *IEEE International Conference on Communications, ICC*, 1–6. IEEE.

**57** Scanlon, W.G. and Cotton, S.L. (2008). Understanding on-body fading channel at 2.45 GHz using measurements based on user state and environment. In: *Loughborough Antennas and Propagation Conference*. Loughborough, United Kingdom (17–18 March 2008), 10–13. IEEE.

**58** Nechayev, Y., Hall, P., Khan, I., and Constantinou, C. (2010). Wireless channels and antennas for body-area networks. In: *Proceedings of the 7th International Conference on Wireless On-demand Network Systems and Services (WONS)*, 137–144. IEEE.

**59** Fort, A., Desset, C., De Doncker, P. et al. (2006). An ultra-wideband body area propagation channel model – from statistics to implementation. *IEEE Transactions on Microwave Theory and Techniques* 54 (4): 1820–1826.

**60** Molisch, A.F., Cassioli, D., Chong, C.-C. et al. (2006). A comprehensive standardization model for ultrawideband propagation channels. *IEEE Transactions on Antennas and Propagation* 54 (11): 3151–3166.

**61** Fort, A., Ryckaert, J., Desset, C. et al. (2006). Ultra-wideband channel model for communication around the human body. *IEEE Journal on Selected Areas in Communications* 24 (4): 927–933.

**62** D'Errico, R. and Ouvry, L. (2010). Delay dispersion of the on-body dynamic channel. In: *Proceedings of the Fourth European Conference on Antennas and Propagation*, 1–5. IEEE.

**63** Takizawa, K., Aoyagi, T., Takada, J.-i. et al. (2008). Channel models for wireless body area networks. In: *Proceeding of the 30th Annual International Conference of the IEEE Engineering in Medicine and Biology Society (EMBS)*, 1549–1552. IEEE.

**64** Sani, A., Alomainy, A., and Hao, Y. (2009). Numerical characterization and link budget evaluation of wireless implants considering different digital human phantoms. *IEEE Transactions on Microwave Theory and Techniques* 57 (10): 2605–2613.

**65** Kim, M. and Takada, J.-I. (2009). Statistical model for 4.5 GHz narrowband on-body propagation channel with specific actions. *IEEE Antennas and Wireless Propagation Letters* 8: 1250–1125.

**66** Smith, D., Miniutti, D., Hanlen, L., Zhang, A., Lewis, D., Rodda, D., and Gilbert, B. (2009) *Power Delay Profiles for Dynamic Narrowband Body Area Network Channels.* ID: 802.15-090187-01-0006. IEEE submission.

**67** Luecken, H., Zasowski, T., Steiner, C. et al. (2008). Location-aware adaptation and precoding for low complexity IR-UWB receivers. In: *IEEE International Conference on Ultra-Wideband, 2008, ICUWB 2008*, vol. 3, 31–34. IEEE.

**68** Abbasi, Q., Sani, A., Alomainy, A., and Hao, Y. (2012). Numerical characterization and modeling of subject-specific ultrawideband body-centric radio channels and systems for healthcare applications. *IEEE Transactions on Information Technology in Biomedicine* 16 (2): 221–227.

**69** Kumar Rout, D. and Das, S. (2016). Narrowband interference mitigation in body surface to external communication in UWB body area networks using first-order Hermite pulse. *International Journal of Electronics* 103 (6): 985–1001.

**70** El-Sallabi, H., Aldosari, A., and Abbasi, Q.H. (2017). Modeling of fading figure for non-stationary indoor radio channels. In: *16th Mediterranean Microwave Symposium (MMS 2016).* Piscataway, NJ: IEEE.

**71** Pavlović, D.Č., Sekulović, N.M., Milovanović, G.V. et al. (2013). Statistics for ratios of Rayleigh, Rician, Nakagami-, and Weibull distributed random variables. *Mathematical Problems in Engineering* 2013: *252804.*

**72** Smith, D.B., Hanlen, L.W., Miniutti, D. et al. (2008). Statistical characterization of the dynamic narrowband body area channel. In: *ISABEL, International Symposium on Applied Sciences in Bio-Medical and Communication Technologies*, 1–5. IEEE.

**73** Cotton, S.L. and Scanlon, W.G. (2006). A statistical analysis of indoor multipath fading for a narrowband wireless body area network. In: *IEEE International Symposium on Personal, Indoor and Mobile Radio Communications, PIMRC*, 1–5. IEEE.

**74** Cotton, S.L., Scanlon, W.G., and Jim, G. (2008). The $\kappa$–$\mu$ distribution applied to the analysis of fading in body to body communication channels for fire and rescue personnel. *IEEE Antennas and Wireless Propagation Letters* 7: 66–69.

**75** Cotton, S.L. and Scanlon, W.G. (2007). Higher-order statistics for the $\kappa$–$\mu$ distribution. *Electronics Letters* 43 (22): 1215–1217.

**76** Halford, S., Halford, K., and Webster, M. (2000) *Evaluating the Performance of HRb Proposals in the Presence of Multipath: IEEE 802.11-00/282r2.* Intersil Corporation.

**77** Smith, D., Miniutti, D., Hanlen, L.W. et al. (2010). Dynamic narrowband body area communications: link-margin based performance analysis and second-order temporal statistics. In: *IEEE Wireless Communications and Networking Conference (WCNC)*, 1–6. IEEE.

**78** Lewis, D. (2008) *802.15.6 Call for Applications – Response Summary. ID: 802.15-08-0407-05.* IEEE submission. https://mentor.ieee.org/802.15/dcn/08/15-08-0407-06-0006-tg6-applications-summary.doc (accessed 6 January 2020).

**79** Smith, D., Lamahewa, T., Hanlen, L., and Miniutti, F. (2011). Simple prediction-based power control for the on-body area communications channel. In: *IEEE International Conference on Communications (ICC).* Kyoto, Japan (5–9 June 2011), 1–5. IEEE.

**80** Zhang, H., Safaei, F., and Tran, L.C. (2018). Channel autocorrelation-based dynamic slot scheduling for body area networks. *EURASIP Journal on Wireless Communications and Networking* 246 https://doi.org/10.1186/s13638-018-1261-8.

**81** Yang, X.D., Abbasi, Q., Alomainy, A., and Hao, Y. (2011). Spatial correlation analysis of on-body radio channels considering statistical significance. *IEEE Antennas and Wireless Propagation Letters* 10: 780–783.

**82** Akaike, H. (1974). A new look at the statistical model identification. *IEEE Transactions on Automatic Control* 19 (6): 716–723.

**83** Zhen, B., Patel, M., Lee, S., Won, E., and Astrin, A. (2008) TG6 Technical requirements document (TRD) ID: 802.15-080644. IEEE submission. https://mentor.ieee.org/802.15/dcn/08/15-08-0644-02-0006-tg6-technical-requirements-document.doc (accessed 6 January 2020).

**84** Yazdandoost, K., and Sayrafian-Pour, K. (2010) TG6 channel model ID: 802.15-08-0780-12-0006. IEEE submission. https://mentor.ieee.org/802.15/dcn/08/15-08-0780-12-0006-tg6-channel-model.pdf (accessed 6 January 2020).

**85** (1997) Federal Communications Commission (FCC) Guidelines. https://transition.fcc.gov/Bureaus/Compliance/Orders/1997/fcc97218.pdf (accessed 6 January 2020).

**86** Yessad, N., Omar, M., Tari, A., and Bouabdallah, A. (2017). QoS-based routing in wireless body area networks: a survey and taxonomy. *Computing* 100: 245–275.

**87** Amit, S. and Sudip, M. (2018). Dynamic connectivity establishment and cooperative scheduling for QoS-aware wireless body area networks. *IEEE Transactions on Mobile Computing* 17 (12): 2775–2788.

**88** Yuan, X., Li, C., Ye, Q. et al. (2018). Performance analysis of IEEE 802.15.6-based coexisting mobile WBANs with prioritized traffic and dynamic interference. *IEEE Transactions on Wireless Communications* 17: 5637–5652.

**89** Roy, M., Chowdhury, C., and Aslam, N. (2017). Designing an energy efficient WBAN routing protocol. In: *Proceedings of the International Conference on Communication Systems and Networks (COMSNETS)*, 298–305. IEEE.

**90** Jiang, W., Wang, Z.C., Feng, M., and Miao, T.T. (2017). A survey of thermal-aware routing protocols in wireless body area networks. In: *Proceedings of the IEEE International Conference on Computational Science and Engineering (CSE) and IEEE International Conference on Embedded and Ubiquitous Computing (EUC)*, 17–21. IEEE.

**91** Qu, Y., Zheng, G., Ma, H. et al. (2019). A survey of routing protocols in WBAN for healthcare applications. *Sensors (Basel)* 19 (7): 1–24.

**92** Bangash, J.I., Abdullah, A.H., Anisi, M.H., and Khan, A.W. (2014). A survey of routing protocols in wireless body sensor networks. *Sensors (Basel)* 14 (1): 1322–1357.

**93** Anand, J., Sethi, D. (2017) Comparative analysis of energy efficient routing in WBAN. In Proceedings of the International Conference on Computational Intelligence & Communication Technology, Ghaziabad, India.

**94** Hu, F.Y., Wang, L., Wang, S.S., and Guo, G. (2016). Hierarchical recognition algorithm of body posture based on wireless body area network. *Journal of Jilin University (Information Science Edition)* 34: 1–7.

**95** Kim, B.-S., Kim, K.H., and Kim, K.-I. (2017). A survey on mobility support in wireless body area networks. *Sensors (Basel)* 17 (4): 797.

**96** Karmakar, K., Biswas, S., and Neogy, S. (2017). MHRP: a novel mobility handling routing protocol in wireless body area network. In: *Proceedings of the 2017 International Conference on Wireless Communications, Signal Processing and Networking (WISPNET)*, 1939–1945. IEEE.

**97** Samanta, A. and Misra, S. (2018). Energy-efficient and distributed network management cost minimization in opportunistic wireless body area networks. *IEEE Transactions on Mobile Computing* 17: 376–389.

**98** Bhanumathi, V. and Sangeetha, C.P. (2017). A guide for the selection of routing protocols in WBAN for healthcare applications. *Human-centric Computing and Information Sciences* 7: 1–19.

**99** Tang, Q., Tummala, N., Gupta, S.K.S., and Schwiebert, L. (2005). TARA: thermal-aware routing algorithm for implanted sensor networks. In: *Distributed Computing in Sensor Systems (DCOSS)*. Marina del Rey, California (30 June–1 July 2005), 206–217. IEEE.

**100** Ahmad, A., Javaid, N., Qasim, U. et al. (2014). Re-attempt: a new energy-efficient routing protocol for wireless body area sensor networks. *International Journal of Distributed Sensor Networks* 10 (4) https://doi.org/10.1155/2014/464010.

**101** Bhangwar, A.R., Kumar, P., Ahmed, A., and Channa, M.I. (2017). Trust and thermal aware routing protocol (TTRP) for wireless body area networks. *Wireless Personal Communications* 97 (1): 349–364.

**102** Kim, B.S., Kang, S.Y., Lim, J.H. et al. (2017). A mobility-based temperature-aware routing protocol for wireless body sensor networks. In: *Proceedings of the International Conference on Information Networking (ICOIN)*, 63–66. IEEE.

**103** Ben Elhadj, H., Elias, J., Chaari, L., and Kamoun, L. (2016). A priority based cross layer routing protocol for healthcare applications. *Ad Hoc Networks* 42: 1–18.

**104** Chen, X., Xu, Y., and Liu, A. (2017). Cross layer design for optimizing transmission reliability, energy efficiency, and lifetime in body sensor networks. *Sensors (Basel)* 17 (4) https://doi.org/10.3390/s17040900.

**105** Wang, G.D. and Guo, K.Q. (2018). A cross-layer retransmission strategy for the body shadowing affect in IEEE 802.15.6-based WBAN. *Software Guide* 17: 200–202.

**106** Wang, L.L., Huang, C., and Wu, X.B. (2018). Cross-layer optimization for wireless body area networks based on prediction method. *Journal of Electronics and Information Technology* 40: 2006–2012.

**107** Zhang, H., Safaei, F., and Tran, L.C. (2018). Joint transmission power control and relay cooperation for WBAN systems. *Sensors (Basel)* 18 (12) https://doi.org/10.3390/s18124283.

**108** Ullah, Z., Ahmed, I., Razzaq, K. et al. (2017). DSCB: dual sink approach using clustering in body area network. *Peer-to-Peer Networking and Applications* 12: 357–370.

**109** Bahae, A., Abdelillah, J., and Haziti, M.E. (2018). An energy efficiency routing protocol for wireless body area networks. *Journal of Medical Engineering & Technology* 42: 290–297.

**110** Kuma, M.A. and Raj, C.V. (2017). On designing lightweight QoS routing protocol for delay-sensitive wireless body area networks. In: *Proceedings of the International Conference on Advances in Computing, Communications and Informatics (ICACCI)*, 740–744. IEEE.

**111** Vetale, S. and Vidhate, A.V. (2017). Hybrid data-centric routing protocol of wireless body area network. In: *Proceedings of the 5th IEEE International Conference on Advances in Computing, Communication and Control (ICAC3)*, 1–7. IEEE.

**112** Peng, Y. and Zhang, S. (2018). A power optimization routing algorithm for wireless body area network. *Journal of Electronic Science and Technology* 31: 38–41.

**113** Chen, Y., Gunawan, E., Low, K.S. et al. (2007). Time of arrival data fusion method for two-dimensional ultrawideband breast cancer detection. *IEEE Transactions on Antennas and Propagation* 55 (10): 2852–2865.

**114** Fort, A., Desset, C., Wambacq, P., and Van Biesen, L. (2006). Body area UWB RAKE receiver communication. In: *Proceedings of the IEEE ICC*, 4682–4687. IEEE.

**115** Monajemi, S., Jarchi, D., Ong, S.H., and Sanei, S. (2017). Cooperative particle filtering for detection and tracking of ERP subcomponents from multichannel EEG. *Entropy* 19 (5) https://doi.org/10.3390/e19050199.

**116** Sanei, S., Monajemi, S., Rastegarnia, A. et al. (2018). Multitask cooperative networks and their diverse applications. In: *Learning Approaches in Signal Processing* (eds. W.-C. Siu, L.-P. Chau, L. Wang and T. Tan), 543–578. Pan Stanford Publishing.

**117** Khalili, A., Rastegarnia, A., and Sanei, S. (2017). Performance analysis of incremental LMS over flat fading channels. *IEEE Transactions on Control of Network Systems* 4 (3): 489–498.

**118** Toorani, M. (2015). On vulnerabilities of the security association in the IEEE 802.15.6 standard. *Financial Cryptography and Data Security* 8976: 245–260.

# 12

# Energy Harvesting Enabled Body Sensor Networks

## 12.1 Introduction

Wireless and embedded systems are commonly powered using batteries and, over time, there has been advances in creating low-power and energy-efficient systems to enable running for a longer time while consuming less energy. For applications where the system is expected to operate for longer, energy becomes a severe bottleneck and much research effort needs to be spent on the efficient use of battery energy while creating new technologies. Recently, another alternative, namely harvesting energy from the environment, has been explored to supplement or even replace batteries.

Sensor networks typically are required to run for a long time, often several years, and are only powered by batteries. This makes energy awareness a major issue when designing these networks. Finite battery capacity is therefore a limitation in the battery-powered networks. This consequently limits the lifetime of the wireless sensor network (WSN) applications or additional cost and complexity to regularly change the batteries. On the other hand, depleted batteries constitute environmental pollution and hazards.

While the technology in battery design is constantly improving and the power requirement of electronics is also dropping, they are not keeping pace with the increasing energy demands of many WSN applications. Therefore, there has been considerable interest in the development of systems capable of extracting sufficient electrical energy from existing environmental sources.

The energy required by the sensors is consumed mostly for wireless communications, and it depends on the communication range. Some energy is also required for recording (or sensing) and processing the sensor data either by the sensors themselves or by the processing machines such as mobile phones or PCs. The data processing power is highly dependent on the working frequency, sampling frequency, and the computation cost for each particular algorithm. In addition, different communication systems require different energy consumptions. Increasing the data size, by adding overheads before transmission, for error protection, or security purposes further increases the energy demand. Table 12.1 shows the energy requirement for three different commercial nodes.

*Body Sensor Networking, Design and Algorithms*, First Edition. Saeid Sanei, Delaram Jarchi and Anthony G. Constantinides.

Companion Website: www.wiley.com/go/sanei/algorithm-design

**Table 12.1** Power consumption for Crossbow MICAz [1], Intel IMote2 [2], and Jennic JN5139 [3] commercial sensor network nodes (clock speed between 13 and 104 MHz).

| | Crossbow MICAz | Intel IMote2 | Jennic JN5139 |
|---|---|---|---|
| Radio standard | IEEE 802.15.4/ZigBee | IEEE 802.15.4 | IEEE 802.15.4/ZigBee |
| Typical range | 100 m (outdoor), 30 m (indoor) | 30 m | 1 km |
| Data rate (kbps) | 250 kbps | 250 kbps | 250 kbps |
| Sleep mode (deep sleep) | 15 $\mu$A | 390 $\mu$A | 2.8 $\mu$A (1.6 $\mu$A) |
| Processor only | 8 mA active mode | 31–53 mA* | 2.7 + 0.325 mA/MHz |
| RX | 19.7 mA | 44 mA | 34 mA |
| TX | 17.4 mA (+0dbm) | 44 mA | 34 mA (+3 dBm) |
| Supply voltage (minimum) | 2.7 V | 3.2 V | 2.7 V |
| Average | 2.8 mW | 12 mW | 3 mW |

Source: Courtesy of Xbow.

## 12.2 Energy Conservation

The efficient use of available energy in a network is always one of the fundamental metrics. This is even more critical for a body sensor network (BSN) or WSN where the nodes have limited power. Thus, in order to reduce energy consumption, a practical and efficient approach is to reduce the transmission power. On the other hand, the need for sensor power directly depends on the on-sensor processing capability, which can be very different from relay sensors or those directly connected (mostly through wire) to a separate processor.

Considering the wireless channel and energy consumption models in WSN, one observation is that, instead of using a long, energy-inefficient edge, the nodes should choose a multihop path composed of short edges that connect the two endpoints of a long distance (edge) for communication [4]. This observation is a fundamental idea in topology control to reduce energy consumption. Therefore, the topology control algorithms in a battery-powered WSN aim to choose short links between the nodes while preserving network connectivity. However, in an energy harvesting wireless sensor network (EH-WSN), with renewable energy the nodes need to manage the required energy smartly. This also depends on how they choose, communicate, or deal with their neighbouring sensors. In a smart EH-WSN each node should choose its neighbours based on not only its own energy consumption but also the energy levels available to its neighbours.

## 12.3 Network Capacity

Network capacity is the amount of traffic or the maximum data speed that a communication network can tolerate. In a wireless network, the communications share the same medium. This implies the existence of undesired noise, interference, attenuation, and multipath during communication, which negatively affect the network traffic capacity. The effects of various artefacts can be reduced by increasing the transmitted power. This in return requires more energy. An ideal situation is that the transmission range is reduced to minimise the overlapping area. A practical approach to decrease interference is to set their transmission power to the desired value, such that the transmission ranges are limited. However, if the transmission power is reduced too much, the network may be disconnected. So, when designing a topology control algorithm, it is necessary to keep a balance between the connectivity and the network performance. There is more on the topology control concept in a later section of this chapter.

## 12.4 Energy Harvesting

Energy harvesting is an approach to capture, harvest, or search an ambient energy and convert it into electrical energy which is directly used or stored-then-used for sensing or actuation. The captured energy by sensors is generally very small as compared to large-scale energy harvesting using renewable energy sources, such as solar and wind farms. Unlike the large-scale power stations which are fixed at a given location, the small-scale energy sources are portable and readily available for use. Energy harvested from the ambient environment are used to power small autonomous sensors that are deployed in remote locations for sensing or even to endure long-term experience to hostile environments. The operations of these small independent sensors are often restricted by a reliance on battery energy. Hence, the driving force behind the search for energy-harvesting practice is the desire to power WSNs and moveable devices for extended operation with the supplement of the energy storage elements if not completely avoiding the storage elements such as batteries. EH-WSNs can provide a solution to the energy problem by harvesting energy that already exists in the surrounding environment. If the harvested energy source is large and constantly (or periodically) available, a sensor node can be powered perpetually. In this way, the energy is essentially infinite; however, this is not always available.

Despite solar and wind energy, which have become very popular, one interesting example is vibration, which can power up WSNs. Vibration energy harvesting is the process by which otherwise wasted vibration (such as from a piece of industrial machinery) is harvested and converted to useful electrical energy to continuously power wireless sensor nodes. As an example, for system health monitoring application, a WSN can be used to monitor vibrations of a rotating machine and by analysing the vibrations in frequency domain to determine when the machine is going to fail and schedule the maintenance in time before the machine malfunctions. The fact that the WSN is being powered by the same vibration it's measuring allows the system to continuously monitor the machine's operation without the risk of the node running out of power.

Energy harvesting introduces a change to the fundamental principles based on which the necessary protocols for WSNs are designed. Instead of focusing on energy-efficient protocols that aim to maximise a sensor's lifetime, the main design objective in EH-WSNs is to maximise the network performance given the rate of available energy to be harvested from the environment. In other words, the surplus of harvested energy can be used to improve network performance. In general, energy harvesting provides numerous benefits to the end user. Some of the major benefits of energy harvesting suitable for WSNs are stated and elaborated in the following list.

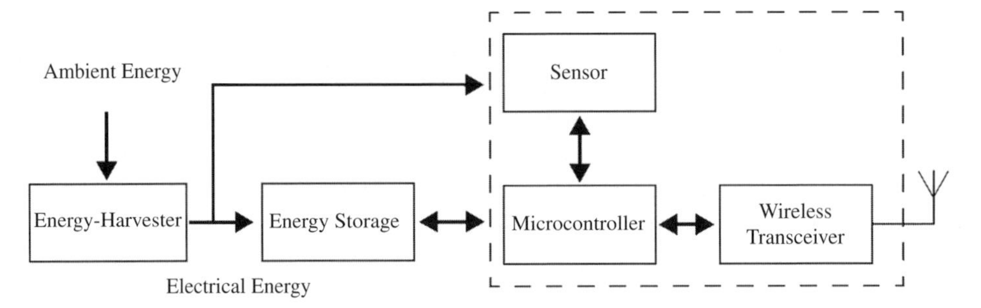


**Figure 12.1** A typical node in an energy harvesting sensor network [5]. Source: Courtesy of Seah, W.K.G., Tan, Y.K., and Chan, A.T.S.

Energy-harvesting technology provides numerous benefits, such as:

- Reducing the dependency on battery power. This is mainly because the nodes eliminate the use of battery power. The harvested ambient energy may be sufficient to eliminate the need for batteries completely.
- Reducing installation and maintenance costs by avoiding service visits to replace the batteries and avoiding other problems such as battery wearing if the batteries are rechargeable.
- Providing long-term solutions to design and deploy reliable nodes with energy-harvesting devices which can function in the cases where the ambient energy becomes unavailable to make it perfectly suited for long-term applications.

A typical node in EH-WSN can be viewed in Figure 12.1. Often, there are actuators which are triggered by the microcontroller which is responsible for processing and control, and normally consumes a large amount of energy. The microcontroller itself consumes energy depending on the processing load, information volume, sampling frequency, and speed.

## 12.5 Challenges in Energy Harvesting

Potential sources of energy harvesting are all around us. Examples include light, radio signals propagating through the air, wind, different kinds of vibration, and movement. With new materials and ways of transforming energy, efficiency of harvesting has risen to the level that can be used for powering WSNs. Nevertheless, harvesting sources are not powerful enough yet to allow the continuous operation of sensor nodes powered by energy harvesting. Therefore, effective design and optimisation of the system is necessary.

The first stage of design is finding a way of converting ambient energy to electrical energy. Harvesting light and vibration energy through use of small-scale solar cells for light harvesting and macro fibre composites for capturing vibration energy have become popular in recent applications. In order to harvest maximum energy from the source, the load on the circuit producing energy needs to be matched through appropriate circuitry and the load is always application-specific, making it hard to implement.

Another energy-harvesting example can be extracting energy from the ocean tides, which is technically more challenging as the tide peaks need to be effectively harvested to empower the energy source.

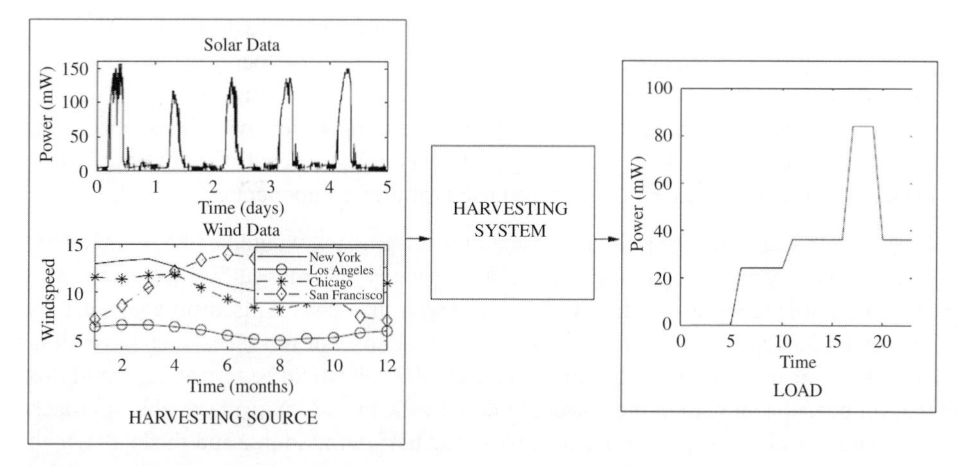


**Figure 12.2** Environment (solar and wind) energy harvesting [6]. Source: Courtesy of Kansal, A., Hsu, J., Zahedi, S., and Srivastava, M.B.

The second stage is to manage the energy stored in the most efficient way. The system needs to be in charge of selecting when and how the energy is distributed throughout the system being powered by it. Furthermore, this block needs to minimise the leakage from the storage and have the smallest possible quiescent current.

The third stage is the application running on the node. This can be very different depending on the sensor technology and functionality. For battery-powered systems the main challenge lies in prolonging the battery life, but with energy harvesting the designer is aware that the energy storage will be replenished. Therefore, other approaches, such as using the power when harvesting and keeping dormant when no energy is harvested, can be utilised. Changing the duty cycle of operation depending on the energy level may be another option.

The environment energy varies with time depending on the environmental conditions which are often outside the designer's control. For instance, Figure 12.2 shows two possible power output variations over time: a solar cell output on a diurnal scale and wind speeds at four arbitrarily chosen locations [7] over a year. In a distributed system, multiple such harvesting sources may be present at multiple nodes at different locations [6].

In a battery-powered device, the typical power management design goals are to minimise the energy consumption and to maximise the lifetime of the device while meeting the required performance constraints. In an energy-harvesting node, one mode of usage is to treat the harvested energy as a supplement to the battery energy and, again, a possible power management objective is to maximise its lifetime. However, in the case of harvesting nodes, another usage mode is possible: using the harvested energy at an appropriate rate such that the system continues to operate perennially. This mode may be called energy neutral operation. A harvesting node is said to achieve energy neutral operation if a desired performance level can be supported for ever (subject to hardware failure). In this mode, power management design considerations are very different from those of maximising lifetime. Two design considerations are apparent:

- *Energy neutral operation.* How to operate such that the consumed energy is always less than the harvested energy? The system may have multiple distributed components each

harvesting its own energy. In this case, the performance not only depends on the spatiotemporal profile of the available energy but also on how this energy is used to deliver network-wide performance guarantees.
- *Maximum performance*. While ensuring energy neutral operation, what is the maximum performance level that can be supported in a given harvesting environment? Again, this depends on the harvested energy at multiple distributed components.

For a simple scenario with a single sensor, a naive approach would be to develop a harvesting technology whose minimum energy output at any instant is sufficient to supply the maximum power required by the load. This, however, has several disadvantages, such as high cost, and may not even be feasible in many situations. For instance, when harvesting solar energy, the minimum energy output for any solar cell would be zero at night and this can never be made more than the power required by the load. A more reasonable approach is to add a power management system between the harvesting source and the load, which attempts to satisfy the energy consumption profile from the available generation profile (Figure 12.2).

In Figure 12.2 'HARVESTING SYSTEM' refers to the system designed specifically to support a variable load from a variable energy-harvesting source when the instantaneous power supply levels from the harvesting source are not exactly matched to the consumption levels of the load. In a harvesting network, this may also involve collaboration among the power management systems of the constituent nodes to support distributed loads from the available energy. On the other hand, 'LOAD' refers to the energy consuming activity being supported. A load, such as a sensor node, may consist of multiple subsystems, and energy consumption may be variable for its different modes of operation. For instance, the activity may involve sampling a sensor signal, transmitting the sensed value, and receiving an acknowledgment.

There are two ways in which the load requirements may be reliably fed from a variable supply. One is to use an energy storage unit in the harvesting system such as a battery or an ultra-capacitor. Another is to modify the load consumption profile according to the availability. In practice, neither of these approaches alone may be sufficient since the load cannot be arbitrarily modified, and energy storage technologies have nonideal behaviour that causes energy loss.

A fourth stage can be added to the system in Figure 12.3 to enable scheduling the sensors, systems, or appliances which use the energy. This can happen if a group of nearby sensors can make a wired network to enable energy transfer. Following this concept, which

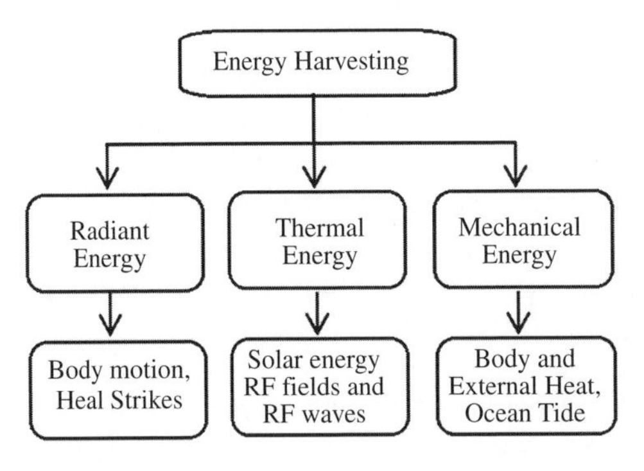


**Figure 12.3** Some methods of energy harvesting [8]. Source: Courtesy of Vijayaraghavan, K., and Rajamani, R.

stems from cooperative networking theory [9–11], the overall energy can be reduced and intelligently distributed among the sensors in a neighbourhood.

Much research and development work has been carried out on harnessing large-scale energy from various renewable energy sources, such as solar, wind, and water/hydro [12]. Little attention has been paid to small-scale energy-harvesting methods and strategies in the past as there hasn't been any need for it. Owing to the recent increase in small-scale sensor demands, there is quite a significant amount of research recorded in the literature that discusses scavenging or harvesting small-scale environmental energy for low-powered mobile electronic devices, especially wireless and wearable devices and sensors.

## 12.6 Types of Energy Harvesting

Figure 12.3 shows numerous types of ambient energy forms appropriate for energy harvesting along with examples of the energy sources. The energy types are thermal energy, radiant energy, and mechanical energy [8]. Substantial research has been devoted to the extraction of energy from kinetic motion. There are other vibration-based energy-harvesting techniques being reported, for instance piezoelectric generators, wearable electronic textiles, and electromagnetic vibration-based micro generator devices for intelligent sensor systems. In the thermal energy-harvesting research domain, consider the system design aspects for thermal energy scavenging via thermoelectric conversion that exploits the natural temperature difference between the ground and the air. Similarly, thermal energy harvesting through thermoelectric power generation from body heat to power wireless sensor nodes has been researched. Research on small-scale wind energy harvesting has also been performed by several groups of researchers [13].

Figure 12.4 further expands Figure 12.3 to show the common energy sources suitable for energy harvesting and their extraction techniques. Although for BSN very few of these

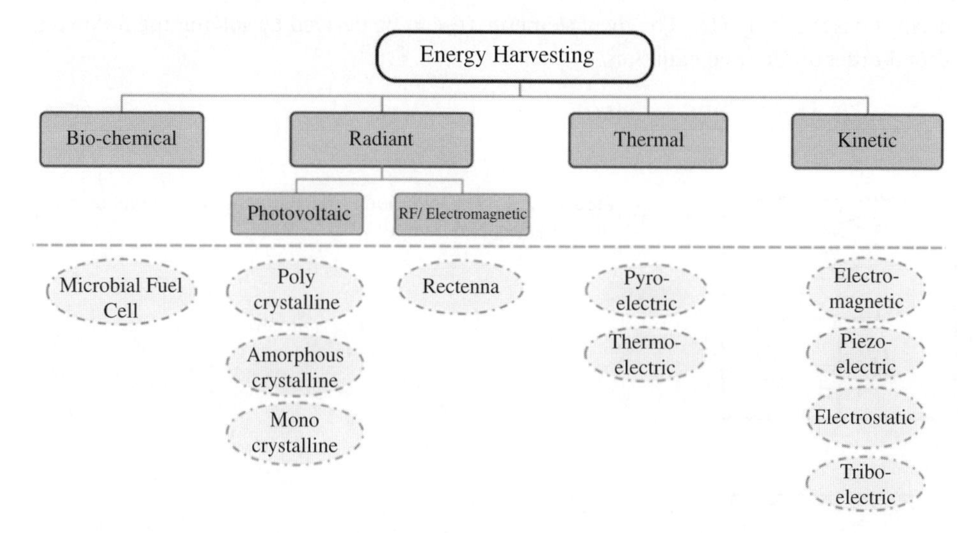


**Figure 12.4** Energy sources (rectangular blocks) and their extraction techniques (oval blocks) [14]. Source: Courtesy of Anwar Bhatti, N., Alizai, M.H., Syed, A.A. and Mottola, L.

energy sources are applicable, here we briefly look at the nature of these sources, their corresponding extraction techniques, along with some of their general applications. Despite the mechanism in energy harvesting, the efficiency in harvesting one type of energy is different from those for others.

### 12.6.1 Harvesting Energy from Kinetic Sources

Kinetic energy is generated by motion. It is one of the most fundamental forces of nature and is described as the work required to accelerate a body of a given mass from rest to a certain speed. The body gains the energy during its acceleration and maintains this amount of energy unless its speed changes. The same amount of work is performed by the body when decelerating to a state of rest. Beyond the computing domain, leveraging kinetic energy to power various devices is an established practice. One example is that of self-winding watches, where the mainspring is wound automatically as a result of the natural motion of one's arm.

Kinetic energy may take numerous forms. In the following, popular forms of kinetic energy together with the corresponding most commonly employed extraction techniques are discussed. These, however, should not be understood as mutually exclusive categories. A given form of kinetic energy may, for example, easily transform into a different one. As a result, extraction techniques employed for one form of kinetic energy are sometimes applicable when kinetic energy manifests in different ways.

One of the most common kinetic energy types is produced by vibration. Vibrations from manufacturing machines, mechanical stress, and sound waves are popular sources of kinetic energy. The vibration mechanical energy may be described using a simple mass-spring model. Assume there is a mass of $m$ connected to a spring of factor $k$. There is also a damper with a damping parameter of $b$, as illustrated in Figure 12.5. The entire device moves relative to the inertial frame with the position of the case at time instant $t$ described by $u(t)$. The displacement $x(t)$ can be derived by solving the following second-order differential equation:

$$m\ddot{x}(t) + b\dot{x}(t) + kx(t) = -m\ddot{u}(t) \tag{12.1}$$

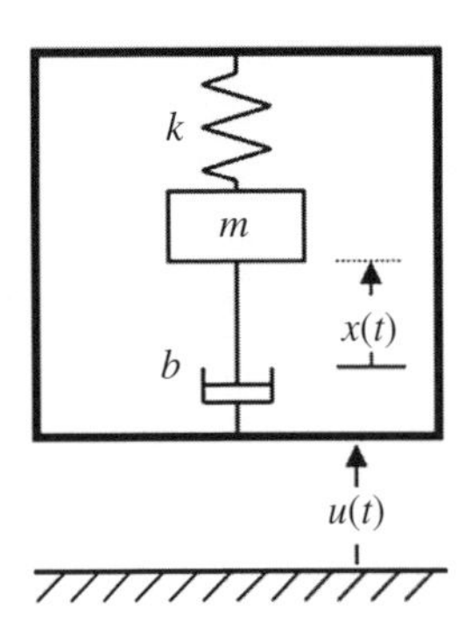


**Figure 12.5** A translational inertial generator model using mass, spring, and damper.

Assuming that the device vibration relative to the inertial frame is defined as $u(t) = \sin(\omega t)$ then the steady state solution for $x(t)$ is:

$$x(t) = \frac{\omega^2}{\sqrt{\left(\frac{k}{m} - \omega^2\right)^2 + \left(\frac{b\omega}{m}\right)^2}} \sin(\omega t + \psi) \tag{12.2}$$

In that case the total power dissipated in the damping element is [15]:

$$P_d = \frac{m\zeta_T G^2 \left(\frac{\omega}{\omega_n}\right)^3 \omega^3}{\left[1 - \left(\frac{\omega}{\omega_n}\right)^2\right]^2 + \left[2\zeta_T \left(\frac{\omega}{\omega_n}\right)\right]^2} \tag{12.3}$$

where the natural frequency of the spring is $\omega_n = \sqrt{k/m}$ and the damping factor is $\zeta_T = b/(2m\omega_n)$. For a given amplitude of acceleration, $A$, the amplitude of the displacement decreases as $G = A/\omega^2$. The peak power happens when $\omega = \omega_n$, resulting in:

$$P_d = \frac{mA^2}{4\omega_n \zeta_T} \tag{12.4}$$

This shows the relationship between the total power and the system parameters. Any parasitic element in the system can, however, affect or reduce this power.

*Piezoelectric* [4], *electromagnetic*, or *electrostatic* effects [16, 17] are the most important basis of kinetic energy measurement. As shown in Figure 12.6, these solutions share the fundamental mechanism of converting vibrations into electric energy. Two involved subsystems are the mass-spring system and a mechanical-to-electrical converter. The mass-spring system transforms vibrations into motion between two elements relative to a single axis. The mechanical-to-electrical converter transforms the relative motion into electric energy by exploiting any of the three aforementioned effects.

Solutions exploiting piezoelectric effect are based on a property of some crystals that generate an electric potential when they are twisted, distorted, or compressed [18]. When a piezoelectric material is under some external force, it causes a deformation of the internal

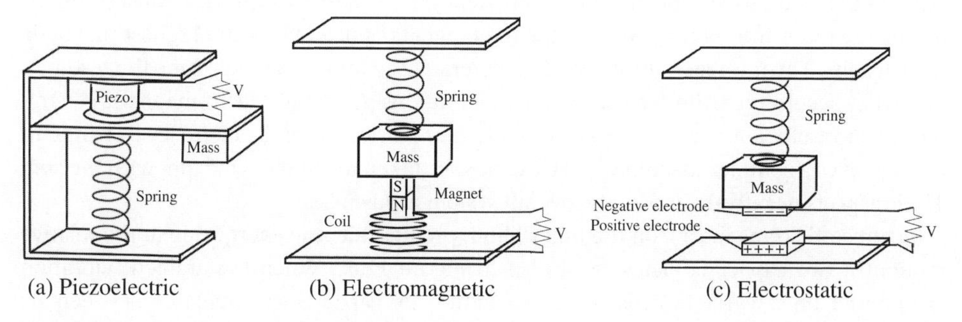


**Figure 12.6** Simplified models of different vibration energy harvesters [14]. Source: Courtesy of Anwar Bhatti, N., Alizai, M.H., Syed, A.A., and Mottola, L.

molecular structure that shifts positive and negative charge centres. This produces a macroscopic polarisation of the material directly proportional to the applied force. The resulting potential difference across the material generates an alternating current (AC) which is then converted into a direct current (DC).

Although piezoelectric materials are not necessarily used for harvesting energy from vibrations [19, 20], they are most often employed with a cantilever-like structure, as illustrated in Figure 12.6a. The cantilever acts as the mass-spring system. When the beam bends due to vibrations, it creates stress on the piezoelectric film, generating AC. The cantilever's resonant frequency is a key factor in determining the efficiency and can be adjusted by changing the mass at the end of the beam and the material. Owing to their mode of operation, these systems capture not only the periodic ambient vibrations but also sudden or sporadic motion. The overall efficiency of a piezo element clamped to a substrate and cyclically compressed at its resonant frequency is [21]:

$$\eta = \frac{\dfrac{k^2}{2(1-k^2)}}{\dfrac{1}{Q} + \dfrac{k^2}{2(1-k^2)}} \tag{12.5}$$

where $k$ is the electromechanical coupling coefficient, and $Q$ is the quality factor of the resonator. As $Q$ becomes larger, the efficiency tends towards unity but for typically achievable $Q$ factors the efficiency increases significantly for higher values of $k$.

The electromagnetic effect is ruled by Faraday's and Lenz's laws, stating that a change in the magnetic conditions of a coil's surrounding generates an electromotive force. This causes a voltage to be induced in the coil. To make a change in the magnetic conditions around a coil, a magnet acts as the mass in a mass-spring system that produces movement parallel to the coils axis [22, 23]. This concept can be seen in Figure 12.6b. This is not the only way to exploit the electromagnetic effect. For example, DeBruin et al. [24] have developed a current sensor that leverages the changes in the magnetic field induced by an AC current line. This, in turn, induces an AC signal on the secondary coil, producing sufficient energy to power the sensor device.

In the case of mass-spring systems, besides the magnet's mass, its material and the coils characteristics are used to determine the efficiency of the harvesting device. Most solutions use neodymium iron boron (NdFeB) for the magnet, as it provides the highest magnetic field density. The number of turns and the material used for the coil help tune the resonant frequency according to the expected ambient vibrations. Although electromagnetic generators are more efficient than piezoelectric ones, their fabrication at the micro level is difficult as it involves a complex assembly and care must be taken to align the magnet with the coil. These aspects negatively impact the overall system robustness.

Electrostatic transducers, on the other hand, produce electric energy due to the relative motion of two capacitor plates, as shown in Figure 12.6c. When the ambient vibration is imposed on a variable capacitance structure, its capacitance oscillates between its maximum and minimum values. An increase in the capacitance decreases the voltage and vice versa. If the voltage is constrained, the charges start flowing towards a storage device, converting the vibration energy into electrical energy. Consequently, electrostatic energy harvesters are modelled as current sources [25]. The main benefit of electrostatic transducers over piezoelectric and electromagnetic ones is the small form factor (i.e. they can be highly reduced in size), as they can be easily fused into a micro-fabrication process [15]. However, they require an initial charge for the capacitor.

Body movement and possible tremor are among those kinetic sources which can be harvested for BSNs. There are many other sources of kinetic energy with wide applications in WSNs such as wind and water flows used by wind and water turbines.

To extract kinetic energy from human motion, the same techniques employed for vibrations – such as those based on piezoelectric, electromagnetic, and electrostatic effects – are applied. For example, Paradiso and Feldmeier [20] have designed a system able to harvest energy from the push of a button through a piezoelectric material. The device then transmits a digital identifier wirelessly which can be used to control other electronic equipment.

In addition, it is possible to harvest electric energy from movements such as footfalls or finger motion through the triboelectric effect. This is very natural phenomena, which occurs when different materials contact by friction and become electrically charged. Rubbing glass with fur or passing a combing the hairs can, for example, yield triboelectricity. In another energy-harvesting approach using the triboelectric effect Hou et al. [26] showed that a triboelectric energy harvester embedded within a shoe is able to power 30 light-emitting diodes. These devices are typically composed of two films of different polymers laid in a 'sandwich' structure, as in Figure 12.7a. The charge is generated by rubbing two polymer films together and then captured by two electrodes.

To capture triboelectric energy generated by the touch of human skin, a single-electrode device is used, as in Figure 12.7b. When a positively charged surface, such as a human finger, comes in contact with the polymer, it induces a negative charge on the polymer. The overall circuit remains neutral as long as the two surfaces remain in contact, producing zero voltage on the load, as indicated on the left-hand side of Figure 12.7b. However, as the positively charged surface moves away, as shown on the right-hand side of Figure 12.7b, the polymer induces a positive charge on the electrode to compensate for the overall charge. This makes the free electrons flow from the polymer towards the electrode and then to the ground, producing output voltage on the load. Once the polymer returns to the original state, the voltage drops back to zero [14].

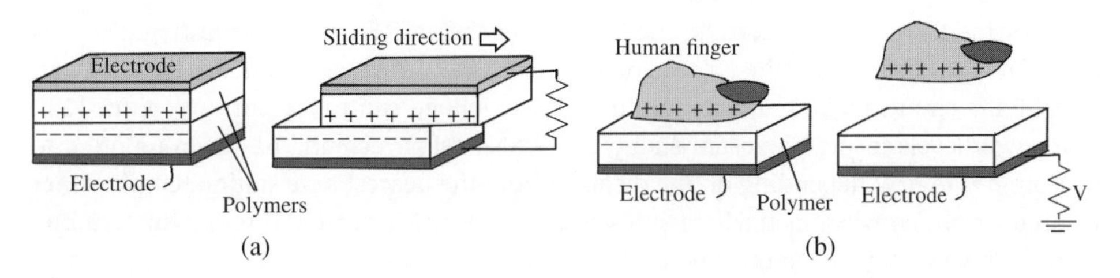


**Figure 12.7** Two modes of energy harvesting through the triboelectric effect [14]: (a) double electrode triboelectric and (b) single electrode triboelectric harvesters. Source: Courtesy of Anwar Bhatti, N., Alizai, M.H., Syed, A.A., and Mottola, L.

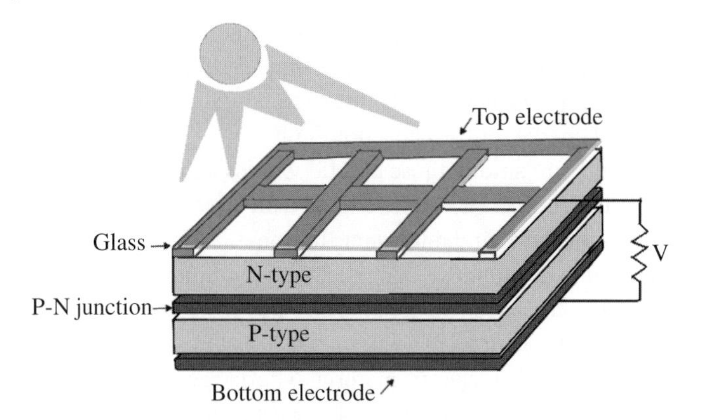


**Figure 12.8** Simplified model of a photovoltaic cell [14]. Source: Courtesy of Anwar Bhatti, N., Alizai, M.H., Syed, A.A., and Mottola, L.

### 12.6.2 Energy Sources from Radiant Sources

Radiant energy is the energy carried by light or electromagnetic radiations. Light, mainly solar light, is the most common form of radiant energy. Radio frequency (RF) and ultrasound are very popular radiant energy sources too.

One common way to extract energy from light is by means of the photoelectric effect for below the infrared spectrum, with solar (visible) light as the primary example. Photovoltaic cells are one of the most mature energy-harvesting devices. From Figure 12.8, a photovoltaic cell, also known as a solar cell, consists of a minimum of two layers of semi-conducting material, mostly silicon. The N-type layer is doped with impurities to increase the concentrations of electrons. Likewise, freely moving positive charges called holes are introduced by doping the silicon of the P-type layer. Each two layers make a P–N junction. When light hits the N-type layer, due to the photoelectric effect, the material absorbs the photons and thus releases the free electrons. The electrons travel through the P–N junction towards the P-type layer to fill the holes in the latter. Among the freed electrons, some do not find a hole in the P-type layer and move back to the N-type layer. This occurs through an external circuit. The produced current is directly proportional to the intensity of light. Thus, photovoltaic cells are generally modelled as current sources [27].

Solar cells are often characterised by the relation between their efficiency against the actual load and operating temperature.

As for the RF energy harvesting, the key element of an RF energy-harvesting device is a special type of antenna called a rectenna. A rectenna comprises a standard antenna and a rectifying circuit making a rectenna look like a voltage-controlled current source [28]. The power levels in the harvested energy for WSNs using rectenna has been reported to be from nW to mW depending on the distance from the nearest base station. Unlike other techniques, the conversion of RF transmissions into electrical current through the rectenna does not involve any mechanical process.

### 12.6.3 Energy Harvesting from Thermal Sources

Thermal energy refers to the internal energy of an object which follows the thermodynamic equilibrium conditions. Whenever these conditions cease to exist, the resulting matter or

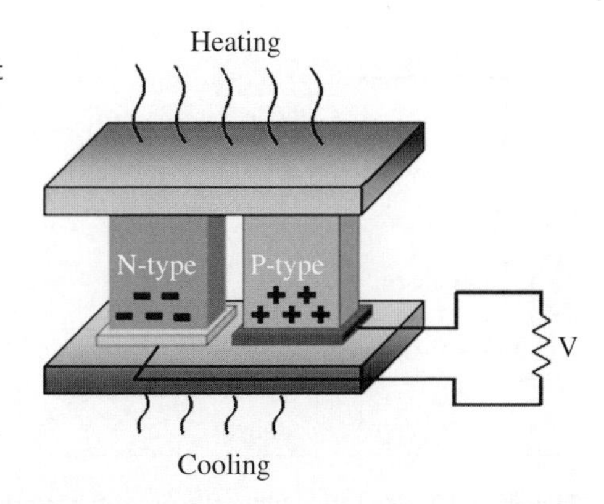


**Figure 12.9** Simplified illustration of thermoelectric effect leading to electrical current [14]. Source: Courtesy of Anwar Bhatti, N., Alizai, M.H., Syed, A.A. and Mottola, L.

energy flows become usable to harvest electric energy through thermoelectric or pyroelectric techniques.

Deriving energy from a thermal source requires a thermal gradient. The efficiency of conversion from a thermal source follows the Carnot efficiency as:

$$\eta \leq \frac{T_h - T_c}{T_h} \tag{12.6}$$

where $T_h$ is the absolute temperature on the 'hot' side of the device and $T_c$ is the absolute temperature on the 'cold' side. Therefore, the greater the temperature difference, the greater the efficiency of the energy conversion.

Thermoelectric techniques are based on Seebeck's effect, which is conceptually the same as the photovoltaic techniques described in Section 12.6.2. As an example, in Figure 12.9, as the temperature difference between opposite segments of the materials increases, the charges are driven towards the cold end. This creates a voltage difference proportional to the temperature difference across the base electrodes. Thus, the thermoelectric harvesters can be modelled as voltage sources [27]. Silicon wafers or aluminium oxides are typically used as the substrate material due to their high thermal conductivity.

In contrast, pyroelectric energy harvesters use materials with the ability to generate a temporary voltage as their temperature is made continuously varying, much like piezoelectric materials generate a potential when they are distorted. Specifically, temperature changes cause the atoms to reposition themselves in a crystalline structure, changing the polarisation of the material. This induces a voltage difference across the crystal, which gradually disappears due to leakage currents if the temperature stays constant.

Since the thermal sources have no moving parts, they are more robust to environmental effects compared with other micro-level harvesters. The devices may thus achieve operational lifetimes of several years without maintenance. Because of their mode of operation, it is also relatively simple to achieve small form factors. Both thermoelectric- and pyroelectric-based harvesters are becoming widely available in the market. The application of these harvesters is suitable for BAN applications as the body heat can be utilised.

### 12.6.4 Energy Harvesting from Biochemical and Chemical Sources

Energy may be harvested from biological or chemical processes. Among biochemical extraction techniques, microbial fuel cells (MFC) use biological waste to generate electrical

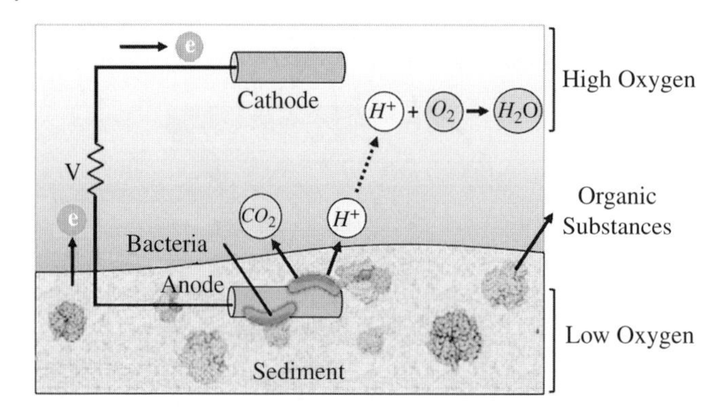


**Figure 12.10** Microbial fuel cell concept: bacteria remove electrons from organic material; the electrons flow through an anode–cathode, battery-like, structure where the two compartments are located in mediums with different $O_2$ concentration [14]. Source: Courtesy of Anwar Bhatti, N., Alizai, M.H., Syed, A.A., and Mottola, L. (*See color plate section for color representation of this figure*)

energy. As schematically shown in Figure 12.10, bacteria in water metabolise biological waste by breaking it down in an oxidisation process. This results in the creation of free electrons along with $CO_2$ and $H^+$ ions. If the environment has oxygen deficiency then an anode picks up the free electrons and transports them to a corresponding cathode where, in an environment abundant in oxygen, the reduction process completes, yielding a water molecule.

On the other hand, methane is a widely available abundant energy source used for power generation in thermal power plants via combustion, but direct conversion to electricity in fuel cells is challenging. Using recent technology, a microbial fuel cell has been demonstrated to efficiently convert methane directly to current by careful selection of a consortium of microorganisms [29].

The efficiency of an MFC is therefore dependent on the difference in oxygen concentration across the anode and the cathode tubes. This is why, for example, MFCs deployed in marine environments have the anode partly embedded in soil and the cathode placed close to the surface, at a higher oxygen concentration, as in Figure 12.10. MFCs are thus modelled as a voltage source in series with a resistance. They are robust and require little maintenance as long as the renewable resource, such as biological waste, is available. Recent results also indicate the potential to harvest energy from voltage differences across the xylem of a tree bark and the soil [30] following similar principles.

For miniaturised devices, such as biomedical implants, enzymatic biofuel cells are also considered [31]. Enzymes are proteins generated by living organisms to catalyse chemical reactions. Unlike MFCs, enzymatic biofuel cells do not use living organisms to trigger the oxidation process but only some of the enzymes produced by specific microorganisms, which are carefully extracted and purified for use. This results in higher energy efficiency compared with the MFCs, at the expense of higher production costs [32]. Moreover, the power densities of enzymatic biofuel cells vary significantly compared to MFCs, as they are typically several times smaller [33].

Extraction techniques based on chemical processes typically exploit and result from corrosion phenomena. The two main factors responsible for the corrosion of steel are

**Table 12.2** Properties of energy-harvesting sources. Taken from [34] and [35].

| Property | Energy harvesting sources: Biochemical | Visible light | RF waves | Thermal | Vibration | Air/Water flow | Human/Animal motion |
|---|---|---|---|---|---|---|---|
| Power density | Low (~300 $\mu$W) | Low (180 $\mu$W) to high (240 mW) | Low (~100 $\mu$W) | Low (218 $\mu$W to 250 $\mu$W) | Low (70 $\mu$W) to medium (12 mW) | Medium (7.7 mW to 18 mW) | Medium (~1 mW) |
| Conversion efficiency | N/A | ~0.1% to 15% | ~33% | ~0.1% to 10% | N/A | N/A | ~7.5% to 11% |
| Form factor | Large | Small | Medium | Small | Small | Large | Small |
| Robustness | High | High | High | High | Low | Low | Low |
| Cost | Medium | Low | Medium | Low | Low | High | Low |

Source: Courtesy of Vullers, R.J. M., van Schaijk, R., Doms, I., Van Hoof, C., and Mertens, R.

carbonation and oxidation under the presence of water. The water may be in touch with the iron or seeps through cement pores or other surrounding materials. These elements, when reacting with iron (Fe), form new compounds like hydrated iron oxides ($Fe_2O_3nH_2O$) or iron oxide-hydroxide, FeO(OH), also known as rust, while releasing electrons. The portion of steel that releases electrons acts as an anode, whereas the portion of metal that accepts the electrons acts as a cathode, with water acting as an electrolyte. The resulting flow of electrons is harvested in the form of electrical current.

Table 12.2 summarises the practical properties of various energy-harvesting sources and approaches. Although the majority of these sources are used in WSN, a number of them are applicable to BSNs too. Energy from body motion, light, and heat are among the very obvious ones.

## 12.7 Topology Control

A popular method to increase energy efficiency in WSNs is by employing topology control algorithms. There are some works, though limited, in topology control. The important one is by Tan et al., who presented a game-theory-based distributed energy harvesting aware (EHA) algorithm [36], which models the behaviours of sensor nodes as a game. This work analyses the energy consumption rate and energy-harvesting rate of each node at different time instants. In this game, the high harvesting power nodes cooperate with the low harvesting power nodes to maintain the connectivity of the network. The algorithm first constructs a preliminary topology based on the directed local spanning sub-graph (DLSS) algorithm [37]. Then, each node $k$ tries to find a neighbour that covers the farthest neighbour of node $k$ by adjusting the transmission power step by step. The idea is that a node may make a sacrifice by increasing its transmission power if it can help reduce energy consumption at another node with lower residual energy.

These algorithms focus only on the transmission power while constructing a static topology without taking into account the residual energy of the nodes, so, they may not be effective when the nodes have different energy levels and when the number of active nodes varies with time in EH-WSN. Since the number of operational nodes in EH-WSNs is varying, there is no possibility of having a centralised solution. Therefore, Wang et al. [38] suggest two localised energy-based topology control algorithms, namely EBTC-1 and EBTC-2, to alleviate the above shortcomings.

The basic idea of EBTC-x (x = 1, 2) is that the topology control in EH-WSN is not just about selecting links with low costs but also includes selecting neighbours according to various energy levels of the nodes. It is therefore aimed to design greedy algorithms to maximise the remaining nodes' energy and select neighbours with high residual energy. Consequently, since the nodes have 'high energy neighbours', their neighbours can receive and transmit more information, resulting in a more sustainable network. Both variants consist of two phases: topology construction and topology maintenance. The key idea in the construction phase is that the nodes select their neighbours according to the distances to the neighbours and the remaining energy of its neighbours. The nodes first collect their neighbour information, including the remaining energy and the distances between nodes. Then, each node selects neighbours according to a metric based on local information. Finally, the nodes adjust their transmission power to the lowest value that is needed to reach the farthest neighbours. In this case, the distance is no longer the only factor in selecting the neighbours. Topology maintenance is required in EH-WSN as a mechanism to update the topology whenever the nodes leave or rejoin the network, taking care of the nodes' energy in the heterogeneous network and keeping all active nodes in the topology.

There are two major differences between EBTC-1 and EBTC-2: first of all, they use different strategies to trigger nodes to initialize the neighbour information collection process and, second, the nodes select neighbours based on different criteria.

EBTC-1 is for converge cast applications of WSNs (where unlike broadcast the data from different nodes are sent and combined in a sink node) and EBTC-2 is for a generic scenario where all the nodes are required to be strictly connected. In some cases, to ensure fault tolerance, the network may be required to be k-connected. While typical topology control algorithms select a particular number of neighbours, the distinguishing feature of both these algorithms is that they select neighbours based on energy-levels and render the global topology strongly connected. It has been shown by simulation that EBTC-1 and EBTC-2 reduce the transmission power and let nodes have neighbours with high remaining energy [38]. It has been claimed that using these algorithms the remaining energy per neighbour is increased by approximately 33%. In addition, in terms of energy consumption and fault tolerance, the algorithms typically achieve significantly less energy compared to K-Neigh [38]. In the K-Neigh algorithm the $k$ neighbours of each node are selected based on distance. Using the algorithm, it is possible for a node to estimate the distance to another node by going through the following steps [38]:

1. Every node broadcasts its ID at maximum transmit power.
2. Upon receiving the broadcast messages from other nodes, every node stores the neighbour information.

3. Every node computes its $k$-closest neighbours, and broadcasts this information at maximum transmit power.
4. By exchanging neighbour information, the nodes are able to have lists of symmetric neighbours.

At the end of protocol execution, the nodes set the transmit power to the minimum value needed to reach the farthest node in the neighbour lists. In this approach, however, the connectivity of the $k$-neighbour graph should be guaranteed.

## 12.8 Typical Energy Harvesters for BSNs

As mentioned above, motion, thermal, and light are the most popular energy sources in a BSN. The early and most intuitive method of scavenging energy from footsteps utilised piezoelectric materials embedded in areas of footwear that experience the most strain. The method has been used for a long time to light up baby shoes when they walk around. In such cosmetic methods, the energy is intermittently generated but not harvested. Shenck and Paradiso demonstrated a method of harnessing foot-strike energy using piezoelectric by inserting a piezoelectric transducer (PZT) foil into the heel area of a shoe [39]. For their energy-harvesting device an average power of 8.4 mW was reported at a foot strike frequency of 0.9 Hz. The benefit of such a design is its unobtrusiveness with respect to the user's ergonomic experience with the product. However, the magnitude of energy that can be harvested is limited by the use of piezoelectric devices which rely on mechanical strain enforced by the heel inside the shoe to create a potential difference. Moro et al. also designed a shoe for energy harvesting with a piezoelectric vibrating cantilever device [40], where the generated power was limited by the low frequency of foot strike.

Akay et al. designed an energy-harvesting system installed in the boot for satellite positioning [41]. Their system converts able-bodied bipedal locomotion to usable electricity by way of unobtrusive hardware installed in footwear. This generates airflow from the compressive forces applied by feet inside shoes during walking [42]. The structure of this energy-harvesting system is displayed in Figure 12.11. Air bulbs are positioned in the sole,

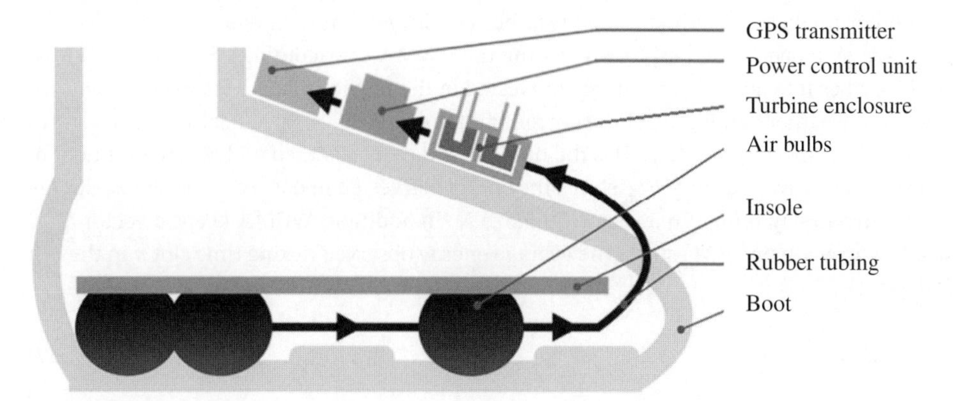


**Figure 12.11** The structure of boot-installed energy harvesting system [42]. Source: Courtesy of Akay, H., Xu, R., Han, D.C.X., Teo, T.H., and Kim, S.-G.

optimised relative to the length of the foot. Outflow from the bulbs is directed towards a two-stage turbine enclosure that drives two DC motors in series to generate electrical power. This energy is stored in a supercapacitor and used to charge a battery that powers a global positioning system (GPS) receiver. The key design parameters considered for the air bulbs are their effective stiffness and time period of regaining original shape after being compressed.

## 12.9 Predicting Availability of Energy

The effective use of energy-harvesting technologies needs to deal with the variable behaviour of the energy sources, which governs the amount and rate of harvested energy over time. In the case of predictable, noncontrollable power sources, such as the solar one, energy prediction methods can be used to forecast the source availability and estimate the expected energy intake [40–42]. Such a predictor can alleviate the problem of harvested power being neither constant nor continuous, allowing the system to take critical decisions about the utilisation of available energy. For prediction of solar and wind harvesters some methods have been proposed [43, 44].

In general, the idea is to develop models that can predict the measurements of a sensor given a subset of others and then use these prediction models to reduce the number of sensors or select a subset of sensors for taking the measurements in any particular time interval.

Kansal et al. [45] proposed a solar energy prediction model based on an exponentially weighted moving-average (EWMA) filter [46]. This method is based on the assumption that the energy available at a given time of the day is similar to that available at the same time of previous days. The time series is segmented into $N$ time slots of fixed length (usually 30 minutes each). The amount of energy available in previous days is maintained as a weighted average where the contribution of older data is exponentially decreasing. The EWMA model uses a smoothing factor $0 \leq \alpha \leq 1$ to predict that in time slot $n$ the amount of energy in day $d$ $\nu_n^{(d)} = \alpha \cdot x_n + (1 - \alpha) \cdot \nu_n^{(d-1)}$ is available for harvesting, where $x_n$ is the amount of energy harvested by the end of $n$th slot.

Another prediction method, called weather-conditioned moving average (WCMA), was proposed by Piorno et al. [47] for alleviating the EWMA shortcomings. Similar to EWMA, WCMA takes into account the energy harvested in the previous days. However, it also considers the weather conditions of the current and previous days. Specifically, WCMA stores a matrix E of size $D \times N$, where $D$ is the number of days considered and $N$ is the number of time slots per day. The entry $E_{d,n}$ stores the energy harvested in day $d$ at time slot $n$. Energy in the current day is kept in a vector $C$ of size $N$. In addition, WCMA keeps a vector $M$ of size $N$ whose $n$th entry $M_n$ stores the average energy observed during time slot $n$ in the last $D$ days:

$$M_n = \frac{1}{D} \cdot \sum_{i=1}^{D} E_{d-i,n} \tag{12.7}$$

At the end of each day $M_n$ is updated with the energy just observed, overwriting the date of the previous day. The amount of energy $P_{n+1}$ predicted by WCMA for the next time slot $n + 1$

of the current day is estimated as $P_{n+1} = \alpha \cdot C_n + (1-\alpha) \cdot M_{n+1} \cdot \rho_n$, where $C_n$ is the amount of energy observed during time slot $n$ of the current day, $M_{n+1}$ is the average harvested energy during time slot $n+1$ over the last $D$ days, and $\rho_n$ is a weighting factor providing an indication of the changing weather conditions during time slot $n$ of the current day with respect to the previous $D$ days.

Another algorithm, developed in Eidgenössische Technische Hochschule (ETH), aka the ETH algorithm, combines information about the energy harvested during the current time interval with the energy availability obtained in the past [48]. Similar to EWMA, the contribution of older data is exponentially decreasing with the past-to-now time interval. These algorithms have been further modified to enhance the prediction performances with minor differences in the outcome.

## 12.10 Reliability of Energy Storage

Energy-harvesting-enabled sensor networks not only have ultra-low power constraints but also should cope with the time-varying state of available energy. This means the energy stored at the energy buffer is constantly increasing and decreasing in a random manner. In addition to this, the ambient energy has a slow temporal dynamic which modulates the harvesting power in several orders of magnitude. The self-powered WSN approach aims to extend the sensor node life by means of energy harvesting. However, the low power density that these energy sources provide compared to the required energy for the communication process creates the necessity of temporal storage. Unfortunately, the random nature of the power sources implies that the energy storage unit might not be able to guarantee the communication at all times, thus giving a certain loss probability, which is a function of the statistics of the harvesting process and the energy storage capacity [49, 50]. A typical solution to reduce this loss probability is an over-dimensioning of the energy storage capacity, which implies a very large overhead in both volume and weight.

As an example, the available power of solar energy ranges from 10 μW cm$^{-2}$ to as high as 100 μW cm$^{-2}$ a day [51]. Often it is challenging for the network designer to dimension critical parameters such as the link capacity [52], the energy buffer capacity of each sensor node, or to design transmission policies [53]. Among other critical parameters, the energy buffer which is usually composed of supercapacitors or chargeable batteries is an expensive and larger volume subsystem. In particular, a typical supercapacitor requires an approximated volume of 2 cm in order to store 1 J of energy [9–11, 45, 54–56]. Thus, the design of the energy buffer might result either in an over-dimensioning of its maximum capacity which leads to significant downscaling impairments and an increase in cost or, on the contrary, an under-dimensioning, which would lead to unnecessary interruptions along the normal operation of the network. In order to capture the random patterns of the energy-harvesting process, the systems presented in [52] and [55] model the energy source as an uncorrelated process in the communications timescale, thus the model can handle specific communication patterns. As a result, these models show that very low buffer capacity (i.e. just a few tens of times the energy of a single data packet) is enough to maintain the communication. On the other hand, those suggested in [45, 51, 56] are aimed for solar energy harvesting and account for daily temporal variations. In this case, these models cannot provide detailed information regarding the communication energy harvesting random patterns, but point out to energy buffer of thousands to even millions of times larger than the energy of a single data packet.

Cid-Fuentes et al. [49] suggested a scalable energy model for estimation of the loss probability and applied their model in order to introduce some guidelines for battery dimensioning. They also provided a general-purpose framework for dimensioning the energy buffer [50]. They presented a dynamics-decoupled multisource-capable energy model to handle fast random patterns of communications and energy harvesting, while capturing slow variations of the ambient energy in both time and space. In their model they assume that the energy field is given by adding two separated dynamics:

$$P_H(\mathbf{r}, t) = P_S(\mathbf{r}, t) + p(\mathbf{r}, t) \tag{12.8}$$

where $p(\mathbf{r}, t)$ is a dimension-less spatiotemporally uncorrelated fast dynamic random process and $P_S(\mathbf{r}, t)$, in power units, stands for a spatiotemporally slow varying random process.

The model can more accurately evaluate the sensor node performance in terms of the energy storage capacity and estimation of the expected energy of the neighbouring nodes. The model based on energy-Erlang provides a link between the energy model, the environmental harvested power, and the energy buffer.

The above proposed source-versatile, dynamics-decoupled framework more accurately models the node energy-harvesting process and shows how the energy buffer helps to counteract the impact of harvested energy temporal evolution upon the energy outage probability, thus showing a strong compromise between performance and size.

## 12.11 Conclusions

Energy harvesting is now an essential integral part of sensor networks as the sensors may not have direct or indirect connections to the conventional energy suppliers. Energy from the environment and the environmental changes and processes have attracted much attention, particularly in WSNs, which are mostly used for wireless communications. For BSNs often smaller sensors are used in order to alleviate the discomfort of wearing intrusive devices. Therefore, more natural energy sources, suitable for various environments and activities are promoted for energy harvesting. Motion, vibration, thermal, and radiation (mostly visible light) are among the most desirable ones. Research has to be carried out to facilitate energy transfer among the nodes (sensors) as well as their efficiency in terms of power, size, and price. The major challenges in energy harvesting are the provision of the required energy to the sensor network and ensuring that sufficiently reliable energy storage is in place for continuous communication between the nodes in a sensor network. More advanced research may include a full topology control of the network to ensure the limited harvested energy is fairly and evenly distributed among the sensors. Moreover, in a decentralised distributed system with smart sensors, it is favourable to enable prediction of the power availability and limitations for each sensor by its neighbouring sensors.

## References

1 Crossbow Products. (2008). MiCAz: Wireless measurement system. http://www.openautomation.net/uploadsproductos/micaz_datasheet.pdf (accessed 13 January 2020).
2 Crossbow Products. (2008). Imote2: high-performance wireless sensor network node. http://wsn.cse.wustl.edu/images/e/e3/Imote2_Datasheet.pdf (accessed 13 January 2020).
3 Jennic Ltd. 2008. JN5139 Wireless Microcontroller (IEEE 802.15.4 and ZigBee). https://fccid.io/TYOJN5139M0/User-Manual/Module-data-sheet-773710 (accessed 13 January 2020).
4 Mathna, C., Donnell, T.O., Martinez, R.V. et al. (2008). Energy scavenging for long – term deployable wireless sensor networks. *Talanta* 75 (3): 613–623.
5 Seah, W.K.G., Tan, Y.K., and Chan, A.T.S. (2013). Research in energy harvesting wireless sensor networks and the challenges ahead. *Autonomous Sensor Networks*: 73–93.
6 Kansal, A., Hsu, J., Zahedi, S., and Srivastava, M.B. (2007). Power management in energy harvesting sensor networks. *ACM Transactions on Embedded Computing Systems (TECS) – Special Section LCTES'05* 6 (4): 1–38.
7 Wind Data. (2001). Wind and radiation data. http://www.esrl.noaa.gov/psd/data/gridded/reanalysis/ (accessed 13 January 2020).
8 Vijayaraghavan, K. and Rajamani, R. (2007). Active control based energy harvesting for battery-less wireless traffic sensors: theory and experiments. In: *Proceedings of American Control Conference*, New York, USA (11–13 July 2007), 4579–4584. IEEE.
9 Sanei, S., Monajemi, S., Rastegarnia, A. et al. (2018). Multitask cooperative networks and their diverse applications. In: *Learning Approaches in Signal Processing* (eds. W.-C. Siu, L.-P. Chau, L. Wang and T. Tan). Jenny Stanford Publishing.
10 Vahidpour, V., Rastegarnia, A., Khalili, A., and Sanei, S. (2019). Partial diffusion Kalman filtering for distributed state estimation in multi-agent networks. *IEEE Transactions on Neural Networks and Learning Systems* https://doi.org/10.1109/TNNLS.2019.2899052.
11 Sayed, A.H. (2014). Adaptation, learning, and optimization over networks. *Foundations and Trends in Machine Learning* 7 (4–5): 311–801.
12 Stojcev, M.K., Kosanovic, M.R., and Golubovic, L.R. (2009). Power management and energy harvesting techniques for wireless sensor nodes. In: *2009 9th International Conference on Telecommunication in Modern Satellite, Cable, and Broadcasting Services*, 65–72. IEEE.
13 Paulo, J. and Gaspar, P.D. (2010). Review and future trend of energy harvesting methods for portable medical devices. In: *Proceedings of the World Congress on Engineering*, London, UK (30 June–2 July 2010), 909–914. WCE.
14 Anwar Bhatti, N., Alizai, M.H., Syed, A.A., and Mottola, L. (2016). Energy harvesting and wireless transfer in sensor network applications: concepts and experiences. *ACM Transactions on Sensor Networks (TOSN)* 12 (3): 1–40.
15 Roundy, S., Wright, P.K., and Rabaey, J. (2003). A study of low level vibrations as a power source for wireless sensor nodes. *Computer Communications* 26 (11): 1131–1144.

**16** Carlos, A. and de Queiroz, M. (2013). Electrostatic generators for vibrational energy harvesting. In: *IEEE Fourth Latin American Symposium on Circuits and Systems (LASCAS)*, Cusco, Peru (27 February–March 2013), 1–4. IEEE.

**17** Chye, W.C., Dahari, Z., Sidek, O., and Miskam, M.A. (2010). Electromagnetic micro power generator: a comprehensive survey. In: *IEEE Symposium on Industrial Electronics Applications (ISIEA)*, Penang, Malaysia (3–6 October 2010), 376–382. IEEE.

**18** Poulin, G., Sarraute, E., and Costa, F. (2004). Generation of electrical energy for portable devices: comparative study of an electromagnetic and a piezoelectric system. *Sensors and Actuators A: Physical* 116 (3): 461–471.

**19** Zhu, D., Beeby, S.P., Tudor, M.J. et al. (2013). Novel miniature airflow energy harvester for wireless sensing applications in buildings. *IEEE Sensors Journal* 13 (2): 691–700.

**20** Paradiso, J.A. and Feldmeier, M. (2001). A compact, wireless, self-powered pushbutton controller. In: *Proceedings of the 3rd International Conference on Ubiquitous Computing*, 299–304. Springer.

**21** Beeby, S.P., Tudor, M.J., and White, N.M. (2006). Review paper: energy harvesting vibration sources for microsystems applications. *Measurement Science and Technology* 17 (12): 175–195.

**22** Kulah, H. and Najafi, K. (2004). An electromagnetic micro power generator for low-frequency environmental vibrations. In: *17th IEEE International Conference on Micro Electro Mechanical Systems*, Maastricht, Netherlands (25–29 January 2004), 237–240. IEEE.

**23** Mizunoand, M. and Chetwynd, D.G. (2003). Investigation of a resonance microgenerator. *Journal of Micromechanics and Microengineering* 13 (2): 209–216.

**24** DeBruin, S., Campbell, B., and Dutta, P. (2013). Monjolo: An energy-harvesting energy meter architecture. In: *Proceedings of the 11th ACM Conference on Embedded Networked Sensor Systems. SenSys'13.* ACM https://www.cs.virginia.edu/~bjc8c/papers/debruin13monjolo.pdf, accessed 14 December 2019.

**25** Torres, E.O. and Rincon-Mora, G.A. (2006). Electrostatic energy harvester and Li-ion charger circuit for micro-scale applications. In: *Proceedings of the 49th IEEE International Midwest Symposium on Circuits and Systems (MWSCAS)*, 65–69. IEEE.

**26** Hou, T.-C., Yang, Y., Zhang, H. et al. (2013). Triboelectric nanogenerator built inside shoe insole for harvesting walking energy. *Nano Energy* 2 (5): 856–862.

**27** Kang, K. (2012) Multi-source energy harvesting for wireless sensor nodes. Royal Institute of Technology (KTH). https://pdfs.semanticscholar.org/939c/cef633c4c076ffc7b6ab44e6c6ac4c74684a.pdf, accessed 14 December 2019.

**28** Nimo, A., Beckedahl, T., Ostertag, T., and Reindl, L. (2015). Analysis of passive RF-DC power rectification and harvesting wireless RF energy for micro-watt sensors. *AIMS Energy* 3 (2): 184–200.

**29** Zhiyong, J.R. (2017). Microbial fuel cells: running on gas. *Nature Energy* 2 (6): 17093.

**30** Love, C.J., Zhang, S., and Mershin, A. (2008). Source of sustained voltage difference between the xylem of a potted *Ficus benjamina* tree and its soil. *PLoS One* 3 (8): e2963.

**31** MacVittie, K., Halamek, J., Halamkova, L. et al. (2013). From "cyborg" lobsters to a pacemaker powered by implantable biofuel cells. *Energy and Environmental Science* 6: 81–86.

**32** Armstrong, R.E. (2010). *Bio-Inspired Innovation and National Security*. Smashbooks.

**33** Barton, S.C., Gallaway, J., and Atanassov, P. (2004). Enzymatic biofuel cells for implantable and microscale devices. *Chemical Reviews* 104 (10): 4867–4886.

**34** Vullers, R.J.M., van Schaijk, R., Doms, I. et al. (2009). Micropower energy harvesting. *Solid-State Electronics* 53 (7): 684–693.

**35** Sudevalayam, S. and Kulkarni, P. (2011). Energy harvesting sensor nodes: survey and implications. *IEEE Communications Surveys & Tutorials* 13 (3): 443–461.

**36** Tan, Q., An, W., Han, Y. et al. (2015). Energy harvesting aware topology control with power adaptation in wireless sensor networks. *Ad Hoc Networks* 27: 44–56.

**37** Li, N. and Hou, J.C. (2005). Localized topology control algorithms for heterogeneous wireless networks. *IEEE/ACM Transactions on Networking (TON)* 13 (6): 1313–1324.

**38** Wang, X., Rao, V.S., Prasad, R.V., and Niemegeers, I. (2016). Choose wisely: topology control in energy-harvesting wireless sensor networks. In: *2016 13th IEEE Annual Consumer Communications & Networking Conference (CCNC)*, 1054–1059. IEEE.

**39** Shenck, N.S. and Paradiso, J.A. (2001). Energy scavenging with shoe-mounted piezoelectrics. *IEEE Micro* 21 (3): 30–42.

**40** Moro, L. and Benasciutti, D. (2010). Harvested power and sensitivity analysis of vibrating shoe-mounted piezoelectric cantilevers. *Smart Materials and Structures* 19 (11).

**41** Akay, H., Xu, R., Seto, K., and Kim, S.-G (2017) Energy harvesting footwear. U.S. Patent No. 62/449208, filed 23 October 2017 and issued 26 July 2018.

**42** Akay, H., Xu, R., Han, D.C.X. et al. (2018). Energy harvesting combat boot for satellite positioning. *Micromechanics* 9 (5): 1–11.

**43** Kosunalp, S. (2017). An energy prediction algorithm for wind-powered wireless sensor networks with energy harvesting. *Energy* 139: 1275–1280.

**44** Kim, S.-G., Jung, J.-Y., and Sim, M.K. (2019). A two-step approach to solar power generation prediction based on weather data using machine learning. *Sustainability* 11 (1–16): 1501.

**45** Kansal, A., Hsu, J., Zahedi, S., and Srivastava, M.B. (2007). Power management in energy harvesting sensor networks. *ACM Transactions in Embedded Computing Systems* 6 (4): 32–38.

**46** Cox, D.R. (1961). Prediction by exponentially weighted moving averages and related methods. *Journal of the Royal Statistical Society. Series B (Methodological)* 23 (2): 414–422.

**47** Recas Piorno, J., Bergonzini, C., Atienza, D., and Simunic Rosing, T. (2009). Prediction and management in energy harvested wireless sensor nodes. In: *1st International Conference on Wireless Communication, Vehicular Technology, Information Theory and Aerospace & Electronic Systems Technology*. Aalborg, Denmark (19 May 2009), 6–10. IEEE.

**48** Moser, C., Thiele, L., Brunelli, D., and Benini, L. (2007). Adaptive power management in energy harvesting systems. In: *Proceedings of 2007 Design, Automation & Test in Europe Conference & Exhibition (DATE)*, 1–6. IEEE.

**49** Cid-Fuentes, R.G., Cabellos-Aparicio, A., and Alarcon, E. (2012). Energy harvesting enabled wireless sensor networks: energy model and battery dimensioning. In: *BodyNets '12 Proceedings of the 7th International Conference on Body Area Networks*, 131–134. ACM.

**50** Cid-Fuentes, R.G., Cabellos-Aparicio, A., and Alarcon, E. (2014). Energy buffer dimensioning through energy-Erlangs in spatio-temporal-correlated energy-harvesting-enabled wireless sensor networks. *IEEE Journal on Emerging and Selected Topics in Circuits and Systems* 4 (3): 301–312.

**51** Gorlatova, M., Wallwater, A., and Zussman, G. (2011). Networking low-power energy harvesting devices: measurements and algorithms. In: *Proceedings of the IEEE INFOCOM*, 1602–1610. IEEE.

**52** Rajesh, R., Sharma, V., and Viswanath, P. (2011). Information capacity of energy harvesting sensor nodes. In: *Proceedings of the IEEE International Symposium on Information Theory*, Saint Petersburg, Russia (31 July–5 August 2011), 2363–2367. IEEE.

**53** Ozel, O., Tutuncuoglu, K., Yang, J. et al. (2011). Transmission with energy harvesting nodes in fading wireless channels: optimal policies. *IEEE Journal on Selected Areas in Communications* 29 (8): 1732–1743.

**54** Pech, D., Brunet, M., Durou, H. et al. (2010). Ultrahigh-power micrometre-sized supercapacitors based on onion-like carbon. *Nature Nano* 5 (9): 651–654.

**55** Jornet, J. and Akyildiz, I. (2012). Joint energy harvesting and communication analysis for perpetual wireless nanosensor networks in the terahertz band. *IEEE Transactions on Nanotechnology* 11 (3): 570–580.

**56** Susu, A., Acquaviva, A., Atienza, D., and De Micheli, G. (2008). Stochastic modeling and analysis for environmentally powered wireless sensor nodes. In: *Proceedings of the 6th International Symposium on Modeling and Optimization in Mobile, Ad Hoc, and Wireless Networks and Workshops*. Berlin, Germany (1–3 April 2008), 125–134. IEEE.

# 13

# Quality of Service, Security, and Privacy for Wearable Sensor Data

## 13.1 Introduction

A great deal of clinical data, including radiological images, physiological signals, impression reports by physicians, laboratory reports, and above all patients' identities, are spread around the local and public networks, large core memories, and the cloud. The major concerns in recording, analysis, and disseminating the patient data are quality of service (QoS), network and data security, and, most importantly, protecting patient privacy. The confidentiality, authenticity, integrity, and freshness of patient data are certainly the major body area network (BAN) requirements. The communication systems, techniques, and platforms necessitate having to establish security systems or implement one of the existing data and network security platforms.

Data confidentiality requires that the transmitted information remain strictly private and can only be accessed by authorised persons, for example the doctor who takes care of the patient. It is usually achieved by encrypting the information before sending it using a secret key. Data authenticity is needed to ensure that the information is sent by the right sender for the intended purpose to a recipient through a secure system environment. For this, often a message authentication code is calculated using a shared secret key. Data integrity makes sure that the received data have not been tampered with. However, this can be inspected by verifying the media access control (MAC) address within the computer network layer. Data freshness is required as the received data should be recent and not be a replayed old message for any reason such as disruption or impersonation. Commonly, a counter controlled technique is employed where the counter is increased every time a message is sent.

In this chapter there is no intention to go through all of the details of the various security methods and approaches. Instead, we focus on some practical applications for wireless BAN which aim at keeping patients' information safe and secure while maintaining the QoS.

There have been variety of approaches to wireless sensor network (WSN) security. Recent advances in wireless communications, embedded systems, and integrated circuit technologies have enabled the wireless BANs with their diverse applications to become a promising networking paradigm.

The patient's personal and clinical data collected or transmitted in BANs are very sensitive, crucially personal, and not only related to their clinical states but also to their entire personal and family status and circumstances. Generally and based on clinical ethics, biomedical data are highly confidential and should be handled, transmitted, and

*Body Sensor Networking, Design and Algorithms,* First Edition. Saeid Sanei, Delaram Jarchi and Anthony G. Constantinides.

Companion Website: www.wiley.com/go/sanei/algorithm-design

stored with care to prevent information misused, alteration, or leaks to unauthorised users. Therefore, authentication, data confidentiality, integrity, nonrepudiation, and privacy preservation should be guaranteed during all communications while maintaining a high QoS. The QoS requires minimum delay in transferring the patient's data to ensure that their situation will not be negatively affected because of such delays. To comply with these requirements, IEEE 802.15.6 has been proposed to provide an international standard for reliable wireless communication for wireless BANs which supports data rates ranging from 75.9 kb $s^{-1}$ to 15.6 Mb $s^{-1}$ [1]. Adopted international standards also recommend four elliptic curve-based security schemes to achieve those goals. Nevertheless, in some recent works it has been shown that these security protocols are vulnerable to some attacks [2, 3]. Therefore, in the past few years, several anonymous authentication schemes for wireless body area networks (WBANs) have been proposed to enhance the information security for protecting patients' identities and through the encryption of such sensitive information.

## 13.2 Threats to a BAN

As for various wired or wireless communication systems and scenarios, the local area network (LAN) including BANs are prone to attacks, eavesdroppers, impersonation, and all other popular breaches of security. For a wearable device, the received information includes private and sensitive information coming from wearable devices and sensors [4, 5]. This underscores the importance of establishing secure a mutual authentication between the mobile terminal, wearable devices, and sensors. Such a scheme allows one to establish the necessary session key for subsequent secure communications [6]. Similar to a wireless communication environment [7], a wearable communication network is suspected for several well-known attacks, such as man-in-the-middle, replay, impersonations, stolen wearable device or mobile terminal, and ephemeral secret leakage attacks [8].

Assuming that the BSNs and other LANs are prone to similar types of threats, the most common types of network and system attacks are listed below.

### 13.2.1 Denial-of-service

Denial-of-service (DoS) attacks are probably the most common security problem which affects many unprotected systems including BANs. DoS attack is a type of cyberattack during which a malicious actor attempts to make a computer or other device unavailable to its intended users, thereby interrupting the device's normal functioning. These attacks typically function by overwhelming or flooding a targeted machine with requests until normal traffic is unable to be processed, resulting in DoS to new users. A DoS attack is usually launched by a single computer/user. However, a distributed denial-of-service (DDoS) attack is a type of DoS attack that comes from many distributed sources. A popular example of this kind is a botnet DDoS attack. DoS attacks typically fall in two categories: (i) buffer overflow attacks in which a memory buffer overflow can cause a machine to consume all available hard disk space, memory, or CPU time (this form of exploitation often results in sluggish behaviour, system crashes, or other deleterious server behaviours, resulting in DoS); (ii) flood attacks, in which by saturating a targeted server with a large number of packets, a malicious actor is able to oversaturate the server capacity resulting in DoS. For successful DoS attacks, the malicious actor must have more available bandwidth capacity than the target.

### 13.2.2 Man-in-the-middle Attack

This occurs when a hacker inserts itself between the communications of a client and a server (by stealing personal information, such as login credentials, account details, or credit card numbers) and sniffs the information. Common types of man-in-the-middle attacks include session hijacking, IP spoofing, and replay. The aim of this attack is either to eavesdrop or to impersonate one of the parties, making it appear as if a normal exchange of information is underway. This is common in places where a login is necessary to start the communication.

### 13.2.3 Phishing and Spear Phishing Attacks

A phishing attack is the practice of sending emails that appear to be from trusted sources with the goal of gaining personal information or influencing users to do something. It could involve an attachment to an email that loads malware onto the user computer. It could also be a link to an illegitimate website that can trick the user into downloading malware or handing over their personal information. On the other hand, spear phishing is a very targeted type of phishing activity. Attackers take the time to conduct research into targets and create messages that are personal and relevant. Because of this, spear phishing can be very hard to identify and even harder to defend against. One of the simplest ways that a hacker can conduct a spear phishing attack is email spoofing, which is when the information in the 'From' section of the email is falsified, making it appear as if it is coming from someone known to the legitimate user. Another technique that scammers use to add credibility to their story is website cloning: they copy legitimate websites to fool the user into entering personally identifiable information or login credentials.

### 13.2.4 Drive-by Attack

A drive-by download attack is a common method of spreading malware. Hackers look for insecure websites and plant a malicious script into HTTP (hypertext transfer protocol) or PHP (personal home page – hypertext processor) code on one of the pages. This script might install malware directly onto the computer of someone who visits the site, or it might redirect the victim to a site controlled by the hackers. Drive-by downloads can happen when visiting a website or viewing an email message or a pop-up window. Unlike many other types of cyber security attacks, a drive-by doesn't rely on a user to do anything to actively enable the attack, i.e. there is no need to click a download button or open a malicious email attachment to become infected. A drive-by download can take advantage of an app, operating system, or web browser that contains security flaws due to unsuccessful, or lack of, updates.

To protect a system or network from drive-by attacks, the browsers and operating systems need to be kept up to date and avoid websites that might contain malicious code. The users should stick to the sites they normally use, although even these sites can be hacked. Also, not too many unnecessary programs and apps should be keep on the system as the more plug-ins are active, the more vulnerabilities can be exploited by drive-by attacks.

### 13.2.5 Password Attack

Entering passwords are the most commonly used mechanism to authenticate users to an information system. Hence, obtaining passwords is a common and effective attack approach. Access to a person's password can be obtained by looking around the person's desk, 'sniffing' the connection to the network to acquire unencrypted passwords, using social engineering, gaining access to a password database, or outright guessing. The last approach can be done in either a random or a systematic manner.

### 13.2.6 SQL Injection Attack

SQL (structured query language) injection has become a common issue with database-driven websites. It occurs when a malefactor executes a SQL query to the database via the input data from the client to the server. SQL commands are inserted into data-plane input (for example instead of the login or password) in order to run predefined SQL commands. A successful SQL injection can read sensitive data from the database, modify (insert, update, or delete) database data, execute administration operations (such as shutdown) on the database, recover the content of a given file, and, in some cases, issue commands to the operating system.

### 13.2.7 Cross-site Scripting Attack

Cross-site scripting attack (XSS) attacks use third-party web resources to run scripts in the victim's web browser or scriptable application. Specifically, the attacker injects a payload with malicious JavaScript into a website's database. When the victim requests a page from the website, the website transmits the page, with the attacker's payload as part of the HTML body, to the victim's browser, executing the malicious script. For example, it might send the victim's cookie to the attacker's server, and the attacker can extract it and use it for session hijacking. The most dangerous consequences occur when XSS is used to exploit additional vulnerabilities. These vulnerabilities can enable an attacker to not only steal cookies but also log key strokes, capture screenshots, discover and collect network information, and remotely access and control the victim's machine.

### 13.2.8 Eavesdropping

Eavesdropping attacks occur through the interception of network traffic. By eavesdropping, an attacker can obtain passwords, credit card numbers, and other confidential information that a user might be sending over the network. Eavesdropping can be passive or active:

- *Passive eavesdropping*. A hacker detects the information by listening to the message transmission in the network.
- *Active eavesdropping*. A hacker actively grabs the information by disguising themselves as a friendly unit and by sending queries to the transmitters. This is called probing, scanning, or tampering.

Detecting passive eavesdropping attacks is often more important than spotting active ones, since active attacks require the attacker to gain knowledge of the friendly units by conducting passive eavesdropping before. Data encryption is the best countermeasure for eavesdropping.

### 13.2.9 Birthday Attack

Birthday attacks are made against hash algorithms that are used to verify the integrity of a message, software, or digital signature. A message processed by a hash function produces a message digest (MD) of fixed length, independent of the length of the input message. This MD uniquely characterises the message. The birthday attack stems from the probability of finding two random messages that generate the same MD when processed by a hash function. If an attacker calculates a similar MD for their message as that of the user, they can safely replace the user's message with their own message, and the receiver will not be able to detect the replacement even if it compares the MDs.

### 13.2.10 Malware Attack

Malicious software is an unwanted software installed in the system without the user's consent. It can attach itself to a legitimate code and propagate and lurk in useful applications or replicate itself across the Internet. There are a variety of factors and viruses affecting the system by performing this type of attack. Typical malware viruses and programs are macro viruses infecting applications such as Microsoft Word or Excel. Other file infectors attach themselves to executable codes such as.exe files. System or boot-recorded infectors (viruses) attach to the master boot record on hard disks and load into the memory when the system starts. Polymorphic viruses conceal themselves through varying cycles of encryption and decryption. Stealth viruses take over system functions to conceal themselves by compromising malware detection software so that the software will report an infected area as being uninfected. The popular Trojan virus hides in a useful program and usually has a malicious function and, unlike other viruses, does not self-replicate. A logic bomb is a type of malicious software appended to an application triggered by a specific occurrence, such as a logical condition or a specific date and time. Worms differ from viruses in that they do not attach to a host file but are self-contained programs that propagate across networks and computers through email attachments or such like. Droppers install viruses in computers. Finally, ransomware is a type of malware that blocks access to the victim's data and threatens to publish or delete it unless a ransom is paid.

Besides the above common network and system attacks, for BANs there should be a registration procedure built in the BAN security with authentication protocols to protect the wearable device or the mobile terminal against theft.

The protocols designed for BAN by different research groups aim to protect the system and the network against all these attacks. As an example, in the protocol designed by Des et al. [9] the BAN is protected against many attacks including stolen mobile terminal, replay, man-in-the-middle, stolen wearable device, impersonation, unanimity and untraceability, password change, and ephemeral secret leakage attacks. The network model, including a typical authentication model of wearable devices, and the threat model (widely used

Dolev-Yao threat model [10]) have been set to test the system. In their scheme the endpoint entities (e.g. wearable devices and mobile terminals) have been considered as untrustworthy nodes.

Following the work in [9], the group designed a remote user authentication scheme through which a user (a doctor) and a controller node can mutually authenticate each other and establish a session key for their future secure communication [11]. Apart from that, the pairwise key establishment between a controller node and its implantable medical devices is also provided in the proposed scheme for secure communication between them. In terms of computational cost, their scheme is comparable with the existing related schemes while it provides better security and more functionality features.

Liu and Chung [12] proposed a user authentication scheme through bilinear pairing and a trusted authority for authenticating the user. In addition, they established a secure communication between a node (sensor) and the user. In some other proposals the authors tested the vulnerability of Liu and Chung's system [13] and proved that it was not sensitive enough to some attacks. Then, they proposed a secure three-factor authentication and key agreement approach suitable for BANs.

## 13.3 Data Security and Most Common Encryption Methods

The IEEE 802.15.6 security hierarchy is illustrated in Figure 13.1. There are three security levels that the BAN sensors and hubs can comply with: level 0 for unsecured communication, level 1 for authenticated but not encrypted messages, and level 2 for the authenticated and encrypted cases. In a unicast communication, a pre-shared or a new master key (MK) is activated. A pairwise temporal key (PTK) is then generated which is utilised only once per session. In a multicast communication, on the other hand, a group temporal key (GTK) is generated that is shared with the corresponding group [1].

In a data security process, to keep the identities and messages secure, the information is encrypted just before propagating it into a network and decrypted in the receiver. The encryption key is somehow known by both transmitter and receiver. There are a variety of encryption/decryption algorithms, and to better protect the message against unlocking a more sophisticated encryption algorithm has to be used.

Although encryption of the information is good enough for many BAN communication scenarios, in many other cases using both encryption and authentication together becomes necessary to validate both the user's identity as well as the data.

Data encryption standard (DES), triple DES, Rivest–Shamir–Adleman (RSA), advanced encryption standard (AES), and Twofish are probably the most popular encryption

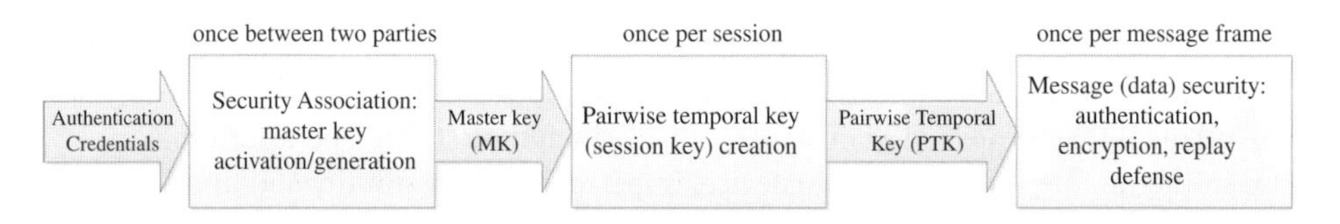


**Figure 13.1** IEEE 802.15.6 security hierarchy [1]. Source: Courtesy of IEEE.

methods. It is also common for different suppliers, users, and applications to have their own (but often very close to one of these five) algorithms.

### 13.3.1 Data Encryption Standard (DES)

This well-established encryption technique uses a symmetric-key block cipher published by the National Institute of Standards and Technology (NIST) [14]. DES implements a Feistel cipher and uses 16 round Feistel structure. The block size is 64-bit. Despite having the key length of 64-bit, DES has an effective key length of 56 bits, since 8 of the 64 bits in the key are not used by the encryption algorithm and are used as check bits only. The general structure of DES is depicted in Figure 13.2.

### 13.3.2 Triple DES

It is a type of computerised cryptography, introduced by IBM, where block cipher algorithms are applied three times to each data block. The key size is increased in triple DES to ensure additional security through encryption capabilities. Each block contains 64 bits of data [15].

### 13.3.3 Rivest–Shamir–Adleman (RSA)

This is an asymmetric cryptography algorithm which works on two different keys, namely a public key and a private key. As the name describes, the public key is given to everyone

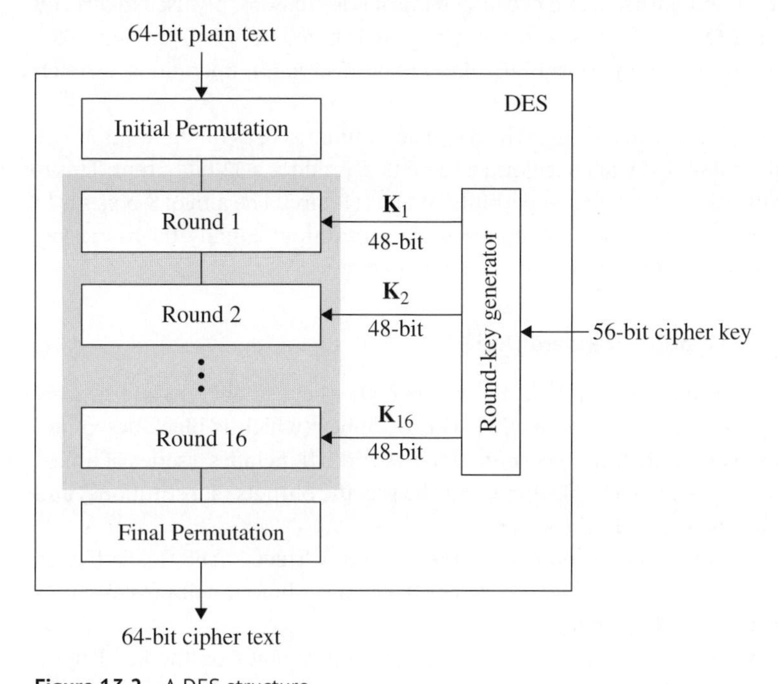


**Figure 13.2** A DES structure.

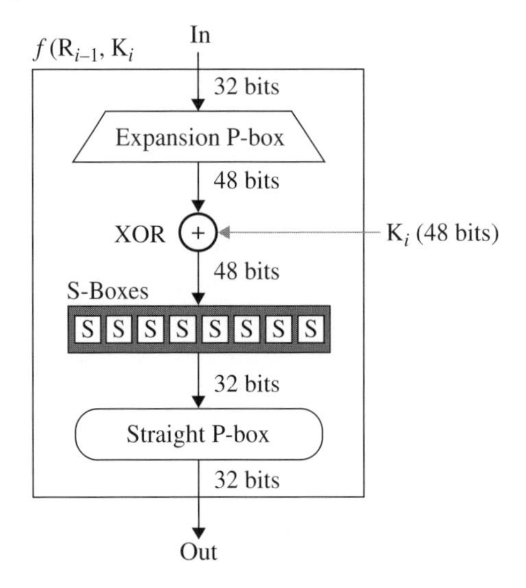


**Figure 13.3** A DES round function.

and the private key is kept private. Since this is asymmetric, nobody else except the browser can decrypt the data even if a third party has the browser's public key.

The RSA user generates a public key using two large prime numbers along with an auxiliary value. The prime numbers must be kept secret. Anyone can use the public key to encrypt a message, but with currently published methods, and if the public key is large enough, only someone with knowledge of the prime numbers can decode the message feasibly [16]. RSA decryption is as difficult as the factorisation problem and is an open question.

RSA is a relatively slow algorithm, and because of this it is less commonly used to directly encrypt the user's data. More often, RSA passes encrypted shared keys for symmetric key cryptography which in turn can perform bulk encryption-decryption operations at much higher speed.

The Round function is the heart of DES which can be summarised, as in Figure 13.3. The DES function applies a 48-bit key to the rightmost 32 bits to produce a 32-bit output. This is then extend to 48 bits using an expansion permutation box (P-box). The aim of S-boxes is to use the input as an address for a table look-up to generate the output. Finally, the round-key generator creates sixteen 48-bit keys out of a 56-bit cipher key.

### 13.3.4 Advanced Encryption Standard (AES)

The AES is a fast algorithm which uses three cryptographic keys to encrypt and decrypt electronic data. AES is an iterative rather than Feistel cipher (which is block-based and symmetric). It is based on 'substitution–permutation network'. It includes a series of linked operations, some of which involve replacing inputs by specific outputs (substitutions) and others shuffle bits around (permutations) [15].

AES performs all its computations on bytes rather than bits. Hence, AES treats the 128 bits of a plaintext block as 16 bytes. These 16 bytes are arranged in four columns and four rows so they can be processed as a matrix.

Unlike DES, the number of rounds in AES is variable and depends on the key length. AES uses 10 rounds for 128-bit keys, 12 rounds for 192-bit keys, and 14 rounds for 256-bit

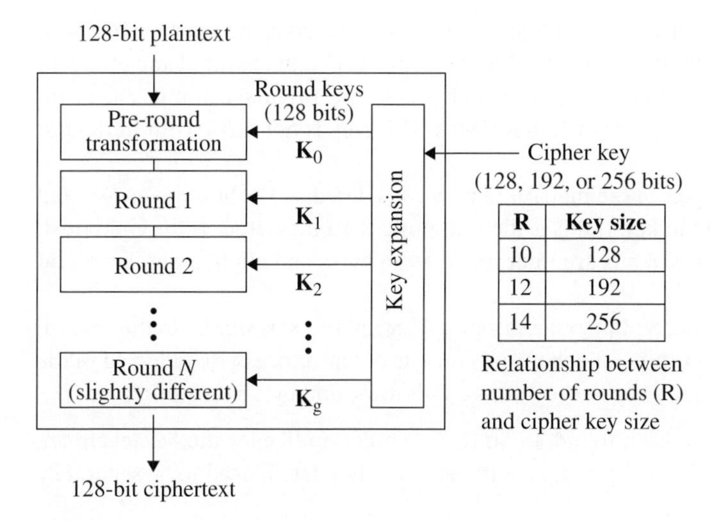


**Figure 13.4** An AES structure.

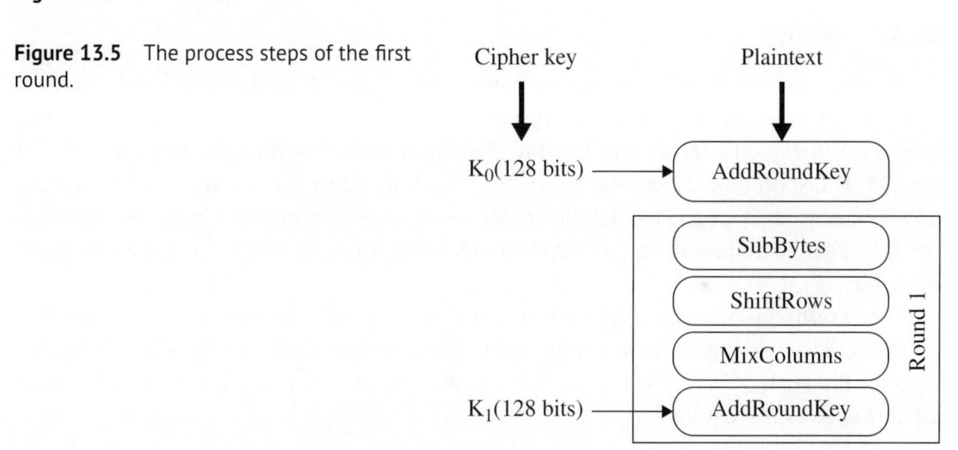


**Figure 13.5** The process steps of the first round.

keys. Each of these rounds uses a different 128-bit round key, which is calculated from the original AES key. Figure 13.4 presents a block diagram of the AES structure.

Each round comprises four subprocesses. The first-round process is depicted in Figure 13.5.

The blocks in Figure 13.5 are defined as follows:

- *Byte Substitution (SubBytes)*: The 16 input bytes are substituted by looking up a fixed table (S-box) given in the design. The result is in a square matrix form including four rows and columns.
- *Shiftrows*: Each of the four rows of the matrix is shifted to the left. Any entry that 'fall off' is re-inserted onto the right side of the row. The shift operation is carried out as follows: first row is not shifted, second row is shifted one (byte) position to the left, third row is shifted two positions to the left, and the fourth row is shifted three positions to the left. The result is a new matrix consisting of the same 16 bytes but shifted with respect to each other.

- *MixColumns*: Each column of the four bytes is now transformed using a particular mathematical function. This function takes the four bytes of one column as input and produces four completely new bytes as the outputs, which replace the original column. The result is another new matrix consisting of 16 new bytes. This step is not carried out in the last round.
- *Addroundkey*: The 16 bytes of the matrix are now considered as 128 bits and are XORed to the 128 bits of the round key. If this is the last round then the output is the ciphertext. Otherwise, the resulting 128 bits are interpreted as 16 bytes and another similar round begins.
- *Decryption process*: The decryption process for an AES ciphertext is similar to the encryption process in the reverse order. Each round consists of four processes conducted in the reverse order, i.e. add round key, mix columns, shift rows, and byte substitution.

Since the subprocesses in each round are in reverse order, unlike for the Feistel cipher, the encryption and decryption algorithms, although closely related, need to be separately implemented.

### 13.3.5 Twofish

Twofish is a symmetric key block cipher with a block size of 128 bits and key sizes up to 256 bits. A single key is used for encryption and decryption. Twofish accepts a key of any length up to 256 bits. (NIST requires the algorithm to work with 128-, 192-, and 256-bit keys.) It is fast on both 32-bit and 8-bit CPUs (such as smart cards and embedded chips), and in hardware. Owing to its high flexibility, it can be used in network applications where the keys change frequently and in applications where there is little or no RAM and ROM available [17].

There are many other encryption techniques with some similarities to the above five methods. These include but are not limited to block ciphers such as AES SHARK, Square, Anubis, Blowfish, KHAZAD, Manta, Hierocrypt, Kalyna, and Camellia, as well as stream ciphers such as MUGI.

## 13.4 Quality of Service (QoS)

Monitoring BAN applications is real time and life critical and requires a strict guarantee of QoS in terms of timeliness, reliability, security, power requirement, noise, and many other issues. Recently, there have been a number of proposals describing diverse approaches or frameworks to achieve QoS in BANs (i.e. for different layers or tiers and different protocols). The QoS also reflects the ability to set different priorities to different applications, users, or data flows.

These priorities directly depend on the BAN major characteristics. These characteristics include resource-constrained (mainly related to embedded computing and sensors, dynamic traffic patterns, network dynamics, data redundancy, heterogeneous traffic, criticality, as the patient data are life-related), dynamic pathloss, and environmental and user's context (which correspond to type of measurement, location, frequency range, number of channels, error rate that can be tolerated, etc.). Owing to all these characteristics, eventually there should be a measure of quality of information as the baseline for QoS quantification bearing in mind that there is no QoS support solution that can satisfy every application's requirements.

### 13.4.1 Quantification of QoS

As mentioned above, QoS is very application-specific and an inclusive measure for BAN QoS is difficult to achieve. However, the major requirements and the related metrics are discussed below [18].

#### 13.4.1.1 Data Quality Metrics

The quality of data and the metrics for measuring them concern:

- *Accuracy*: which refers to how clear and accurate the data are measured, processed, transmitted, or presented.
- *Reliability and believability*: to ensure that the data collected by the sensors are trustable, meaningful, and can be analysed.
- *Consistency*: to ensure that the information collected over time remains consistent and, therefore, the overall system is stable.
- *Ubiquitous access*: the patient's information should be accessible to those authorised (doctors, caregivers, etc.) whenever necessary.
- *Access security*: which is discussed in other sections of this chapter and is a measure for QoS. It also includes data freshness, ensuring that the available data are the most recent ones.
- *Accessibility of data*: it is important to ensure that the data are accessible, particularly at critical times, with no, or only a limited, delay.
- *Interpretability*: those involved in a patient's care have to be able to find and understand the patient's data.

#### 13.4.1.2 Network Quality Related Metrics

These metrics are related to the network function. The most popular metrics to measure the network quality are:

*Delay*: queueing, processing, measurement, communication, and propagation delays for wireless BAN should be small in order to ensure freshness, timeliness, and correctness of the patient information.

*Delay jitter*: it is important to measure and minimise the variation in delay, namely jitter, to ensure an in-order delivery of the data packets.

*Throughput*: there is need to have a good match between the channel capacity for different segments of the communication channel to enable accommodating a high rate of information transmission.

*Packet error rate*: most new systems can adapt to having low packet error rate, which is a measure of network quality when the system is prone to noise, multipath, and traffic.

*Energy efficiency*: energy suppliers and harvesters such as batteries, solar cells, and motion-based harvesters have to be evaluated and remain at a consistent level throughout the measurement and the BAN functioning.

*Interoperability*: the BAN system should be compatible with heterogeneous devices and applications without much change in the networking structure. Therefore, there won't be any need for changing, coding, or decoding the data to make them suitable for other networking systems.

In addition to the above, the devices within a BAN should be comfortable and easy to use, wearable, and safe.

A BAN, depending on the structure of its sensors, follows the hierarchy of computer networking layering system, though most of the design efforts are made for building and connecting the sensors. Most data quality measures such as accuracy, precision, completeness, consistency, and noise level are derived in relation to a physical layer and are dependent on the devices within this layer, and the technologies and standards of the BAN. Unlike other BAN parts, sensors are very different from one application to another and therefore a general-purpose BAN should be flexible enough to cater for such variabilities. The QoS is then evaluated based on the overall performance.

Nevertheless, for the network quality, the network standards and systems (Bluetooth, ZigBee, etc.) have their own QoS requirements. Most importantly, they use different frequency bands. Therefore, an acceptable QoS is often achieved by adhering to these standards and protocols.

On the other hand, data link layer, involving MAC protocols, has a crucial role in data framing, error detection, and recovery. As such, the MAC protocols become important. In particular, due to the clinical environment, through body or near-body communication channels, most of the QoS for BANs has to be arranged and conducted within the link layer rather than in the upper layers.

The last networking layer effective in QoS is the network layer. This is the layer where all the addressing actions and processing take place and therefore IP generation and identification followed by routing the packets from source to destination through several routers take place in this layer. Hence, developing an efficient and QoS-aware routing, including table look up by routers (address calculation) and forwarding in BANs, is a nontrivial task.

Some studies have shown that from a QoS perspective (requiring timeliness, reliability, and energy efficiency) multihop communications are preferable to single-hop ones. This is mainly to avoid shadowing caused by obstacles. Some examples of multihop routing can be found in [19, 20].

In [21] the authors focus on QoS-based routing protocols in BANs. They have classified the existing solutions into two main approaches: (i) multisink approach-based architecture and (ii) single-sink approach-based architecture. The latter category may be further divided into two subcategories to illustrate the different communication modes of the ensemble of solutions. The protocols are presented in a way to highlight many different approaches to QoS routing in WBANs. In order to assess the performance of QoS routing process in WBANs, a Markov chain has been used to evaluate the delivery process in terms of success/failure transmission probabilities under the probability of route rupture, followed by a comparison and analytical discussion of each category. They also studied the adaptability of the surveyed protocols related to the healthcare sector. The single-sink approach receives success in many aspects. However, the reliability and the transmission delay are deemed the most important requirement in WBANs scenarios. Therefore, this approach does not always provide the best results. In this context, the trend is to switch from a single-sink to a multisink approach with the main aim of further improving the reliability and transmission delay. In a general communication system multihop and single-hop transmission may be combined. The multihop approach may not be the optimal solution to guarantee some critical levels of QoS and may limit the required potential improvements. However, by combining these approaches the performance can be better tested for various QoS levels.

Finally, employing transmission control protocol (TCP) rather than user datagram protocol (UDP) in the transport layer allows a three-way hand-shaking in one hand and end-to-end acknowledgement (ACK) communication. This avoids both packet loss and transmission error up to a great extent. This, however, follows the standard transport layer design and is hardly changed by the BAN designers.

## 13.5 System Security

To guarantee secure communication in wireless BANs, Poon et al. [22] used physiological values (such as electrocardiogram, iris, and fingerprint) to design an authentication scheme. By measuring and comparing these values, a means of authentication can be achieved. To improve the performance, several improved schemes [23–25] based on physiological values have been proposed. However, the physiological value-based schemes [22–25] suffer from the DoS attack because there are differences between any two measured physiological signals recorded from the same person that may not be suitable enough for practical applications.

The received signal strength (RSS) in WBANs varies according to mobility and the channel state and variation. Therefore, channel-based authentication schemes [26–28] are also considered to implement mutual authentication in WBANs. Zeng et al. [26] used temporal RSS variation lists to deal with the identity-based attack. However, their scheme focuses on identification and cannot provide anonymity. Cai et al. [27] proposed a device pairing scheme using differential RSS involving at least two receiver antennas.

Several proximity-based authentication schemes [29–31] have been proposed for BAN applications. Varshavsky et al. [29] extended the Diffie–Hellman key exchange with the verification of device colocation and proposed an authentication method for WBANs. By monitoring the radio environment and generating a signature (including its RSS), the device could detect a similarity. With pairing devices transmitting and the trusted body-worn personal devices receiving, Kalamandeen et al. [30] proposed an authentication method by monitoring the transmissions. Later, Mathur et al. [31] developed a colocation-based pairing scheme by exploiting environmental signals. The main weakness of proximity-based authentication methods is that the devices have to be within half of the wavelength distance from each other, which can be restrictive for medical sensors deployed in a BAN. Shi et al. [32, 33] established that an off-body attacker has distinct RSS variation behaviour with an on-body sensor. Hence, they proposed a channel-based and a proximity-based authentication scheme for WBANs.

Unlike the physiological value-based schemes [22–25], the channel-based schemes [26–28], and the proximity-based schemes [29–31], the cryptography-based schemes require fewer restrictions (such as channel, distance, and location) from the applications' environments. Consequently, cryptography-based authentication schemes, as discussed in Section 13.3, have attracted increasing attention recently. These approaches may be implemented using the traditional public key cryptography (TPKC) [16, 34]. For example, Li et al. [35, 36] proposed a key management method and used the TPKC to implement mutual authentication. In the TPKC, the complex modular exponentiation operation is needed. However, the client device in WBANs has very limited computing capability and battery capacity. Therefore, the authentication schemes based on the TPKC are not suitable for WBAN applications. To avoid the modular exponentiation operation, several authentication schemes [37–39] based on the elliptic curve cryptography (ECC) have been proposed.

The concept of ECC was first introduced by Miller [40] and Koblitz [41] separately. Compared to TPKC, ECC is able to provide the same security with a much smaller key size [42]. Therefore, ECC is more suitable for environments with limited computing capabilities and battery capacity. However, a public key infrastructure is needed for the practical implementation of the ECC. With public key cryptography (PKC), every user has a certificate, generated by the certificate authority, to bind their identity and public key. The management of certificates becomes more tedious as the number of users grows. Therefore, the authentication schemes [37–39] based on ECC are not suitable for WBANs.

The idea of identity-based public key cryptography (ID-based PKC) was proposed by Shamir [43]. In the ID-based PKC, the key generation centre (KGC) uses its secret key to calculate the user's secret key according to their identity, and this identity plays the role of public key. Therefore, the ID-based PKC could address the problem of certificates management in the TCP. Yang and Chang [44] proposed an effective authentication scheme based on the ID-PKC. However, Yoon and Yoo [45] demonstrated that the scheme had serious security vulnerability by proposing an impersonation attack. Based on the ID-PKC, He et al. [46] used ECC to design a new provably secure authentication scheme. Unfortunately, Wang and Ma [47] found that the scheme [46] could not resist the reflection attack. They also pointed out that He et al.'s proposed method could not provide mutual authentication. Later, Islam and Biswas [48] used ECC to construct another authentication to solve security vulnerability in Yoon and Yoo's scheme. Unfortunately, Truong et al. [49] pointed out that this scheme could not resist the DoS attack. Although the above ID-based authentication schemes [44–49] perform better than previous schemes, they are not suitable for WBAN applications because they are designed for the client–server environment. To ensure secure communication in WBANs, Liu et al. [50] used the bilinear pairing defined on the elliptic curve to design a new certificateless signature scheme. Then, they presented a preliminary version authentication scheme for WBANs using their signature scheme. However, the scheme provides nontraceability because the user's identity is a constant value and the adversary could trace the client by observing the constant value. To enhance security, they also presented a security enhanced authentication method using their signature scheme and demonstrated that it could withstand various attacks.

However, in most of the practical applications, there is a privileged insider of the system, who is responsible for ensuring the device's normal functions. This insider can access the database of the system and modify the entries if necessary. Besides, a powerful adversary can actually penetrate into the system and modify the database. In other words, it is realistic and reasonable to assume that there is such an adversary who has the ability to modify the database of the system.

In the WBAN environment, the client and the application provider communicate wirelessly. Therefore, the authentication scheme for WBANs is susceptible to many attacks. To guarantee secure communication in WBANs, the authentication scheme should be able to withstand various attacks. In [51] a new anonymous authentication scheme for WBANs with provable security is proposed. The authors showed that the IEEE 802.15.6 is not secure against impersonation attacks. Then, they demonstrated that their own scheme not only overcomes the weakness in previous schemes but also reduces the computation burden for the clients. Finally, they argued that their system is secure and can meet the security requirements of BANs particularly for clinical applications.

## 13.6 Privacy

It is believed that, by establishing a secure system, network security as well as patient privacy are preserved. Although most of the techniques in security (such as network security and encryption) focus on data transmission, recently much attention has been paid towards data mining and archiving. Most importantly, cloud computing has become central to e-health. This requires an extra measure in securing patient privacy. The use of the cloud is becoming even more important nowadays due to the possibility of running very computationally costly algorithms, such as deep neural networks, over the cloud.

Vora et al. [52] have reviewed the privacy methods and looked at the privacy issues over the cloud as a low-cost data archiving and retrieval resource for e-health. To ensure that when two parties communicate a patient's data the traceability and identification of patients are achieved and the patients get access using anonymous credentials without revealing their identity or authentication credentials, an authentication scheme for e-health users using anonymous, adaptive authentication services has been proposed [52]. A level of confidentiality of the data has been achieved by adding a layer of anonymity. This layer follows the simple principle that the presence of a health record implies the retrieval of healthcare services for a particular condition, which violates patient privacy instead of maintaining patient confidentiality. Anonymous tickets work as the additional scheme of anonymity by offering consumption of the services, which allows the system not to depend on 'trust'. The 'trust' concept demands a baseline of confidence for assuring the user's anonymity. The access control layer blocks any unauthorised access to patient data while complying with the tasks of providing the necessary measures to ensure security and privacy of patient data.

## 13.7 Conclusions

Extra measures need to be undertaken for patient data as well as other human-body-related information. In addition, to enable efficient healthcare and timely medication, the QoS for BANs has to be high enough to secure, preserve, and speed up the flow of information whenever and wherever necessary. This becomes even more critical where more patients are connected to the WSN through their wearable networks. Therefore, the design of new protocols and conventions may become necessary after the expansion of personalised medicine.

## References


1 IEEE Std 802.15.6-2012 (2012). *IEEE standard for local and metropolitan area networks – Part 15.6: Wireless body area networks.* New York: IEEE.
2 Toorani, M. (2015) *On vulnerabilities of the security association in the IEEE 802.15. 6 Standard.* arXiv preprint arXiv:1501.02601.
3 Toorani, M. (2015). Cryptanalysis of two PAKE protocols for body area networks and smart environments. *International Journal of Network Security* 17 (5): 629–636.
4 He, D., Kumar, N., Wang, H. et al. (2018). A provably-secure cross-domain handshake scheme with symptoms matching for mobile healthcare social network. *IEEE Transactions on Dependable and Secure Computing* 15 (4): 633–645.
5 Meng, W., Li, W., Xiang, Y., and Choo, K.-K.R. (2017). A Bayesian inference-based detection mechanism to defend medical smartphone networks against insider attacks. *Journal of Network and Computer Applications* 78: 162–169.
6 Choo, K.-K.R. (2009). *Secure Key Establishment (Advances in Information Security 41).* New York: Springer.
7 Fang, H., Xu, L., Li, J., and Choo, K.-K.R. (2017). An adaptive trust-Stackelberg game model for security and energy efficiency in dynamic cognitive radio networks. *Computer Communications* 105: 124–132.
8 Lindstrom, J. (2007). Security challenges for wearable computing – a case study. In: *Proceedings of 4th International Forum on Applied Wearable Computing*, 1–8. IEEE.
9 Das, A.K., Wazid, M., Kumar, N. et al. (2018). Design of secure and lightweight authentication protocol for wearable devices environment. *IEEE Journal of Biomedical and Health Informatics* 22 (4): 1310–1322.
10 Dolev, D. and Yao, A. (1983). On the security of public key protocols. *IEEE Transactions on Information Theory* 29 (2): 198–208.
11 Wazid, M., Das, A.K., Kumar, N. et al. (2018). A novel authentication and key agreement scheme for implantable medical devices deployment. *IEEE Journal of Biomedical and Health Informatics* 22 (4): 1299–1309.
12 Liu, C.H. and Chung, Y.F. (2017). Secure user authentication scheme for wireless healthcare sensor networks. *Computers & Electronic Engineering* 59: 250–261.
13 Challa, S., Das, A.K., Odelu, V. et al. (2018). An efficient EEC based provably secure three-factor user authentication and key agreement protocol for wireless healthcare sensor networks. *Computers & Electronic Engineering* 69: 534–554.
14 Federal Information Processing Standards Publication, *FIPS Pub 46-3*, reaffirmed 25 October 1992. US Department of Commerce, National Institute of Standards and Technology.
15 Daemen, J. and Rijmen, V. (2002). *The Design of Rijndael: AES – The Advanced Encryption Standard.* Springer.
16 Rivest, R., Shamir, A., and Adleman, L. (1978). A method for obtaining digital signatures and public key cryptosystems. *Communications of the ACM* 21 (2): 120–126.
17 Schneier, B., Kelsey, J., Whiting, D., Wagner, D., Hall, C., and Ferguson, N. (1998) Twofish: A 128-Bit Block Cipher. in First Advanced Encryption Standard (AES) Conference. National Institute of Standards and Technology, 104.

18 Razzaque, M.A., Hira, M.T., and Dira, M. (2017). QoS in body area networks: a survey. *ACM Transactions on Sensor Networks* 13 (3) https://doi.org/10.1145/3085580.

19 Bangash, J.I., Khan, A.W., and Abdullah, A.H. (2015). Data-centric routing for intra wireless body sensor networks. *Journal of Medical Systems* 39 (9): 1–13.

20 Ababneh, N., Timmons, N., and Morrison, J. (2015). A cross-layer QoS-aware optimization protocol for guaranteed data streaming over wireless body area networks. *Telecommunication Systems* 58 (2): 179–191.

21 Yessad, N., Omar, M., Tari, A., and Bouabdallah, A. (2018). QoS-based routing in wireless body area networks: a survey and taxonomy. *Computing* 100 (3): 245–275.

22 Poon, C., Zhang, Y., and Bao, S. (2006). A novel biometrics method to secure wireless body area sensor networks for telemedicine and m-health. *IEEE Communications Magazine* 44 (4): 73–81.

23 Singh, K. and Muthukkumarasamy, V. (2007). Authenticated key establishment protocols for a home health care system. In: *Proceedings of 3rd International Conference on Intelligent Sensor, Sensor Network and Information (ISSNIP'07)*, 353–358. ACM.

24 Venkatasubramanian, K. and Gupta, S. (2010). Physiological value-based efficient usable security solutions for body sensor networks. *ACM Transactions on Sensor Networks* 6 (4): 1–36.

25 Venkatasubramanian, K., Banerjee, A., and Gupta, S. (2010). PSKA: usable and secure key agreement scheme for body area networks. *IEEE Transactions on Information Technology in Biomedicine* 14 (1): 60–68.

26 Zeng, K., Govindan, K., and Mohapatra, P. (2010). Non-cryptographic authentication and identification in wireless networks [Security and Privacy in Emerging Wireless Networks]. *IEEE Wireless Communications* 17 (5): 56–62.

27 Cai, L., Zeng, K., Chen, H., and Mohapatra, P. (2011). Good neighbor: ad hoc pairing of nearby wireless devices by multiple antennas. In: *Proceedings of the Network and Distributed System Security Symposium*. San Diego, California (6–9 February 2011), 1–15. The Internet Society.

28 Shi, L., Yuan, J., Yu, S., and Li, M. (2013). ASK-BAN: authenticated secret key extraction utilizing channel characteristics for body area networks. In: *Proceedings of the 6th ACM Conference on Security Privacy Wireless Mobile Networks*, 155–166. ACM.

29 Varshavsky, A., Scannell, A., LaMarca, A., and De Lara, E. (2007). Amigo: proximity-based authentication of mobile devices. In: *Proceedings of 9th International Conference on Ubiquitous Computating*, 253–270. Berlin: Springer-Verlag.

30 Kalamandeen, A., Scannell, A., De Lara, E. et al. (2010). Ensemble: cooperative proximity-based authentication. In: *Proceedings of the 8th International Conference on Mobile Systems, Applications, Services*, 331–344. New York: ACM.

31 Mathur, S., Miller, R., Varshavsky, A. et al. (2011). Proximate: proximity-based secure pairing using ambient wireless signals. In: *Proceedings of the 9th International Conference on Mobile Systems, Applications, and Services*, 211–224. ACM.

32 Shi, L., Li, M., Yu, S., and Yuan, J. (2012). BANA: body area network authentication exploiting channel characteristics. In: *Proceedings of the 5th ACM Conference Security and Privacy in Wireless Mobile Networks*, 1–12. Tucson, AZ: ACM.

**33** Shi, L., Li, M., Yu, S., and Yuan, J. (2013). BANA: body area network authentication exploiting channel characteristics. *IEEE Journal on Selected Areas in Communications* 31 (9): 1803–1816.

**34** Elgamal, T. (1985). A public key cryptosystem and a signature protocol based on discrete logarithms. *IEEE Transactions on Information Theory* 31 (4): 469–472.

**35** Li, M., Yu, S., Lou, W., and Ren, K. (2010). Group device pairing based secure sensor association and key management for body area networks. In: *Proceedings IEEE INFOCOM*, 1–9.

**36** Li, M., Yu, S., Guttman, J. et al. (2013). Secure ad hoc trust initialization and key management in wireless body area networks. *ACM Transactions on Sensor Networks* 9 (2): 1–35.

**37** Jiang, C., Li, B., and Xu, H. (2007). An efficient scheme for user authentication in wireless sensor networks. In: *Proceedings of the 21st International Conference on Advanced Information Networking and Applications Workshops*, 438–442. IEEE.

**38** Guo, P., Wang, J., Li, B., and Lee, S. (2014). A variable threshold-value authentication architecture for wireless mesh networks. *Journal of Internet Technology* 15 (6): 929–936.

**39** Shen, J., Tan, H., Wang, J. et al. (2015). A novel routing protocol providing good transmission reliability in underwater sensor networks. *Journal of Internet Technology* 16 (1): 171–178.

**40** Miller, V. (1985). Use of elliptic curves in cryptography. In: *Proceedings of Advance in Cryptology (CRYPTO'85)*, 417–426. Springer.

**41** Koblitz, N. (1987). Elliptic curve cryptosystem. *Mathematics of Computation* 48: 203–209.

**42** Hankerson, D., Menezes, A., and Vanstone, S. (2004). *Guide to Elliptic Curve Cryptography*. Berlin: Springer-Verlag.

**43** Shamir, A. (1984). Identity based cryptosystems and signature schemes. In: *Proceedings of Advance in Cryptology (CRYPTO'84)*, 47–53. Berlin: Springer-Verlag.

**44** Yang, J. and Chang, C. (2009). An ID-based remote mutual authentication with key agreement scheme for mobile devices on elliptic curve crypto system. *Computers & Security* 28 (3–4): 138–143.

**45** Yoon, E. and Yoo, K. (2009). Robust ID-based remote mutual authentication with key agreement protocol for mobile devices on ECC. In: *Proceedings of the International Conference on Computational Science and Engineering*, 633–640. Vancouver, Canada: IEEE.

**46** He, D., Chen, J., and Hu, J. (2012). An ID-based client authentication with key agreement protocol for mobile client-server environment on ECC with provable security. *Information Fusion* 13 (3): 223–230.

**47** Wang, D. and Ma, C. (2013). Cryptanalysis of a remote user authentication scheme for mobile client-server environment with provable security based on ECC. *Information Fusion* 41 (4): 498–503.

**48** Islam, S. and Biswas, G. (2011). A more efficient and secure ID-based remote mutual authentication with key agreement scheme for mobile devices on elliptic curve cryptosystem. *Journal of Systems and Software* 84 (11): 1892–1898.

**49** Truong, T., Tran, M., and Duong, A. (2012). Improvement of the more efficient and secure ID-based remote mutual authentication with key agreement scheme for mobile devices on ECC. In: *Proceedings of 26th International Conference on Advanced Information Networking and Applications Workshops*. Fukuoka, Japan (26–29 March 2012), 698–703. IEEE.

**50** Liu, J., Zhang, Z., Chen, X., and Kwak, K. (2014). Certificateless remote anonymous authentication schemes for wireless body area networks. *IEEE Transactions on Parallel and Distributed Systems* 25 (2): 332–342.

**51** He, D., Zeadally, S., Kumar, N., and Lee, J.-H. (2017). Anonymous authentication for wireless body area networks with provable security. *IEEE Systems Journal* 11 (4): 2590–2601.

**52** Vora, J., DevMurari, P., Tanwar, S. et al. (2018). Blind signatures based secured E-healthcare system. In: *International Conference on Computer, Information and Telecommunication Systems (CITS)*. Colmar, France (11–13 July 2018), 1–5. IEEE.

# 14

# Existing Projects and Platforms

## 14.1 Introduction

At the heart of body area network (BAN) architecture is the sensor platform equipped with a short- or even long-range communication capability. An on-board processing system is an extra but highly demanding and useful facility which is an emerging technology for smart sensors. The availability of smart sensors (remote system) further mobilises the research and technology for distributed and decentralised adaptive networks.

Several research groups and commercial vendors develop their own prototypes of wireless body sensor networks (WBSNs), also called wireless body area networks (WBANs) [1]. Some have more interest in building a system architecture and the associated service platform and others in developing the communication system focusing on the networking protocols, including routing algorithms. Here, a number of WBAN platforms are explained to give an overview of research and development as well as the market for developing body sensors, sensor networking, and communication strategies. Despite similarities, what makes them different are the short-range communication type, number of hops between the sensors and the sink, and of course the particular applications they focus on. These applications require particular body sensors as well as data acquisition and processing systems. On the other hand, the locations of sensors and sinks may be considered fixed or ad hoc and the overall system may include a diverse range of sensors. All these impose more complexity on the system design. In all these projects there are always the problem of energy constraint which has to be addressed and suitable solutions introduced.

The network developers should have solutions to the security and quality of service (QoS) problems, as discussed in Chapter 13. All these solutions somehow comply with IEEE 802.15.6 standard meant for WBAN [2]. However, the network topologies for BAN can be different. The most common ones are peer-to-peer, star, mesh, and clustered. Peer-to-peer topology doesn't rely on any coordinator station, though a pure peer-to-peer BAN topology hardly exists. Star topology is the most common BAN topology in which the coordinator acts as the star centre. Cluster topology is an extension of mesh topology where a number of meshes may interact through gateways. Similar to a mesh structure, a cluster topology allows more flexible communication infrastructures to interact with each

*Body Sensor Networking, Design and Algorithms,* First Edition. Saeid Sanei, Delaram Jarchi and Anthony G. Constantinides.

Companion Website: www.wiley.com/go/sanei/algorithm-design

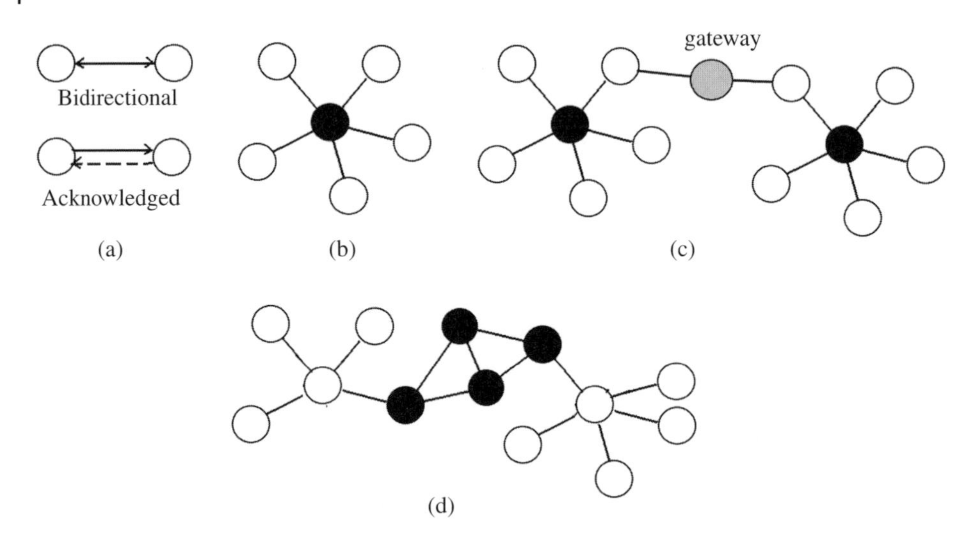


**Figure 14.1** Popular network topologies: (a) peer-to-peer; (b) star topology where the black circle is the central hub; (c) mesh topology; (d) clustered topology. The links without direction are assumed bidirectional.

other. Collaborative BANs and BAN–WSN interactive networks are two complex network infrastructures which may involve mesh or cluster topologies. Schematic diagrams of different popular topologies are presented in Figure 14.1.

Otto et al. [3] and Jovanov et al. [4] have proposed a system architecture which handles the communications both within and between the WBANs. The system also incorporates a medical server in a multitier telemedicine system. The communication between the sensors and the sink (central body-mounted or near-body hub) is single-hop, slotted, and uses ZigBee or Bluetooth. The slots are synchronised using beacons periodically sent by the sink. They use off-the-shelf wireless sensors to design a prototype WBAN such as the Tmote sky platform from Moteiv (www.moteiv.com), which is now named Sentilla (www.sentilla.com). The Tmote sky platform is also deployed in the CodeBlue project [5, 6], where WBANs are used in rapid disaster response scenarios. In this project, a wearable computer attached to the patient's wrist including a Tmote sky mote forms an ad hoc wireless network with a portable computer. A wireless two-lead electrocardiogram (ECG), a wireless pulse oximeter sensor, and a wireless electromyogram (EMG) have been developed, too. Ayushman [7] is a sensor-network-based medical monitoring infrastructure that can collect, query, and analyse patient health information in real-time.

A wireless system including ECG, gait monitoring, and environment monitoring sensors has been developed using the existing components with a Mica2 wireless transceiver. Further, the necessary software for consulting the data at a remote client has been developed. The Human++ project by IMEC-NL [8] was initiated with the aim of achieving highly miniaturised and autonomous sensor systems. These systems enable people to carry their personal BANs. In another project, an ambulatory EEG/ECG system with a transmitter working on 2.4 GHz has been developed. This system can run for approximately three months using two AA batteries. In order to obtain a longer autonomy, the project also investigates energy scavenging with thermoelectric generators (TEG). In 2006, a wireless pulse oximeter fully powered by the patient's body heat was introduced. Further, the project investigates other short-range wireless technologies such as ultrawideband (UWB) to make an ultra-low power transmitter. The European MobiHealth project [9] provides a complete end-to-end m-health platform for ambulant patient monitoring, deployed over UMTS (Universal Mobile Telecommunications Service) and GPRS (General Packet Radio Service) networks. The MobiHealth patient/user is equipped with different sensors that constantly monitor vital signals (e.g. blood pressure, heart rate, and ECG). The communication between sensors and the personalised device is Bluetooth or ZigBee based and is single-hop. The most important considerations in this design are security, reliability of communication resources, and the necessary QoS level.

The French project BANET (www.banet.fr) was established to provide a framework, models, and technologies for designing optimised wireless communication systems. Such systems facilitate the widest range of WBAN-based applications in consumer electronics, medical, and sport domains. Their focus has been on the study of WBAN propagation channel, MAC protocols, and the coexistence of WBANs and other wireless networks. The German BASUMA (Body Area System for Ubiquitous Multimedia Applications) project [10] aimed at developing a full platform for WBANs. For the communication module, a UWB frontend has been implemented and a MAC protocol based on IEEE 802.15.3 applied. This protocol also uses timeframes primarily divided into congestion free periods (with time slots) and those with possible traffics through carrier sense multiple access–collision avoidance (CSMA/CA) local area networking (LAN) protocol. A flexible and efficient WBAN solution suitable for a wide range of applications was developed in [11]. The main focus in this solution was on posture and activity recognition applications by means of practical implementation and on-the-field testing. The sensors are WiMoCA-nodes which are represented by tri-axial integrated micro-electro-mechanical systems (MEMS) accelerometers. The Flemish IBBT IM3-project (Interactive Mobile Medical Monitoring) focuses on the research and implementation of a wearable system for health monitoring (http://projects.ibbt.be/im3). Patient data are collected using a WBAN and analysed at the medical hub worn by the patient. If an event (e.g. heart rhythm problems) is detected, a signal is sent to a healthcare practitioner who can view and analyse the patient data remotely for diagnostic purposes.

## 14.2 Existing Wearable Devices

Variety of activity trackers and smartwatches with the promise of healthier living have hit store shelves over the last couple of years. These devices (e.g. Apple's smartwatch, Microsoft's band, Jawbone's wristband, or Fitbit's fitness band), along with the supporting software, can gather information like steps, sleep, heart rate, sun exposure, and calories. Most of these wearable tracking devices are attractive, easy to use, accessible, and comfortable. However, they may lack health and medical care grade data quality (DQ), which slows down their commercialisation for clinical use. Such shortcomings include accuracy, consistency, data security, and credibility [12]. Many wearable-device makers believe that user engagement is more important than accuracy [13]. Most of these devices are introduced and their applications explained in detail in other chapters of this book.

## 14.3 BAN Programming Framework

In a BAN platform, often a certain programming framework is selected which in the best and fastest way provides access to the BAN components (i.e. sensors, network, processing units, and finally visualisation blocks). These also include gateways and security issues.

Most of the BAN architectures are divided into three or four layers (also called tiers). In a three-tier architecture there are body sensors, mobile devices, and archival systems. In a four-tier BAN architecture, the forth tier represents the visualisation, processing, and diagnostic tools and bodies (such as doctors and nurses).

## 14.4 Commercial Sensor Node Hardware Platforms

A number of platforms for BANs have been on the market for the past decade and the number of suppliers is increasing rapidly. This clearly shows the rising demand and importance of such platforms for capturing and assessing various human biomarkers.

Among the characterising indicators of BANs, node lifetime is one of the key parameters for the selection of wireless sensor communication board called remote (or mote in brief). The technical specifications of the mote directly or indirectly affect the lifetime of a sensor node. This parameter becomes particularly important for 'power-hungry' nodes deployed in remote environment. Usage of controllers in the hardware module with less power consumption enhances the availability of residual energy in the node to a greater extent. Radio frequency (RF) modules with increased efficiency in transmitters and receivers are obviously considered a trade-off with the node lifetime. Less-power-consuming processors, high-quality power supplies, and suitable (low) sampling frequency with lower computational cost are the favourable characteristics of BANs [14].

For example, TelosB uses an MSP430 processor, which supports five power-down states. An auxiliary module within TelosB supports power-down and sleep modes. It also enables the cutting-off of the RF module connection and the processing element. This reduces the battery drain, thereby increasing node lifetime. A number of wireless motes are described and, depending on their application usage, technical specifications, etc., classified in the following sections. The older motes are described first and the recent ones later.

### 14.4.1 Mica2/MicaZ Motes

The boards for early motes Mica2 and MicaZ are shown in Figure 14.2. In some applications the motes are combined with or connected to the sensor boards. These motes belong to the second- and third-generation mote technologies from CrossBow Technology. Mica2 and MicaZ use an Atmega128L controller along with a CC1000/CC2420 RF module, respectively. Mica2/MicaZ are equipped with humidity, temperature, and light sensors. These sensors are connected to the mote via an interface. These motes are capable of measuring barometric pressure, acceleration/seismic activity, and some other variables. The motes are

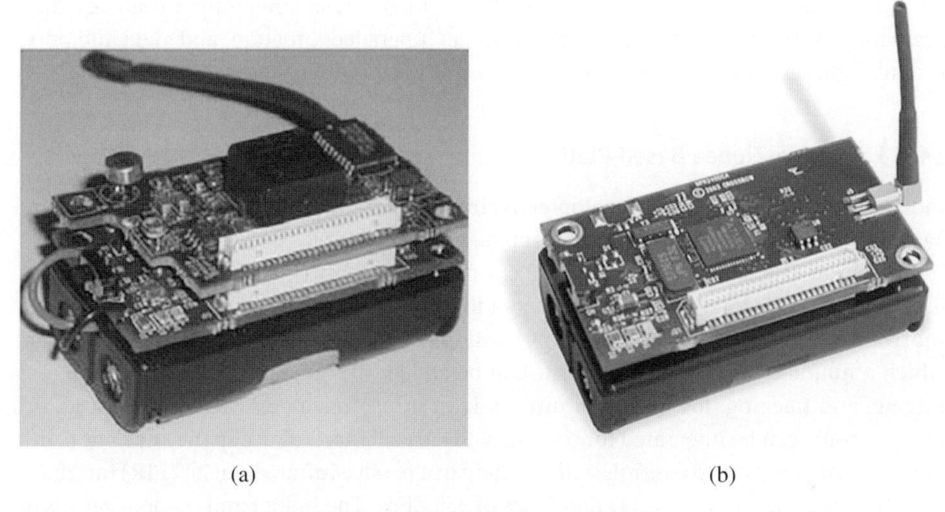


**Figure 14.2** (a) Mica2 mote with its sensor board; (b) MicaZ mote.

**Figure 14.3** TelosB mote.

powered by two external AA batteries within an operating range of 2.1–3.6 V DC. ZigBee is usually used as the short-range communication network for these motes.

### 14.4.2 TelosB Mote

TelosB mote is shown in Figure 14.3. It was initially developed by the University of California, Berkeley. This mote embeds an 802.15.4-compatible CC2420 radio chip from Texas Instruments. It provides onboard humidity and temperature IC type sensors (SHT2x from Sensirion). The accuracy of relative humidity and temperature readings are, respectively, around 3% and 0.4 °C. The motes are powered by two external AA batteries with an operating range of 2.1–3.6 V DC. Additionally, XM1000 wireless motes have been designed according to TelosB specifications but with an upgraded program and data memory. In-built light sensors have also been introduced in this product.

### 14.4.3 Indriya-Zigbee Based Platform

Indriya is a capable hardware development environment for building ambient intelligence based wireless sensor network (WSN) applications [15]. It features a low-power MSP430 core with an IEEE 802.15.4-based CC2520 from Texas Instruments. On-board sensors include an accelerometer and light sensors together with many other optional add-ons. Indriya has many potential applications, including (a) indoor air quality management for which a humidity and/or a $CO_2$ sensor can be added; (b) range measurement, direction finding, and tracking, for which an ultrasonic or a magnetometer can be interfaced; (c) image sensors can be integrated for security and surveillance; (d) occupancy detection and human occupancy-based controls with the help of a passive infrared sensor (PIR) interface. The RF module offers achievable data rates of 250 Kbps. The indoor and outdoor ranges of this mote vary, respectively, between 20–30 m and 75–100 m [16].

### 14.4.4 IRIS

An IRIS mote is shown in Figure 14.4. It is one of the available wireless node platforms which offer a higher communication range – close to 500 m in line-of-sight (LOS). It uses a 2.4 GHz IEEE 802.15.4 wireless module. The mote operates using a TinyOS operating system on an ATmega1281 based low-power micro-controller. The IRIS mote supports integrating sensor boards through a standard 51-pin expansion connector. It consumes 10–17 mA current in the transmitter and up to 16 mA in the receiver. The communication range varies above 300 and more than 50 m, respectively, for outdoor and indoor ranges (for LOS) [17]. Owing to its higher communication range, the motes can be deployed underground for agriculture and soil monitoring. Although soil provides higher interference to

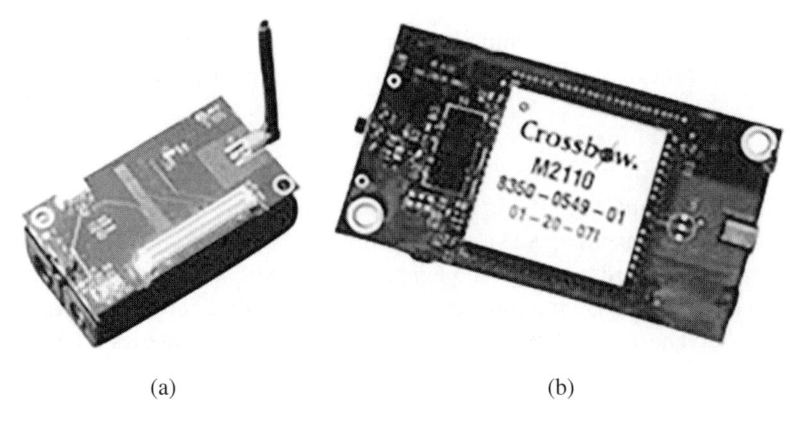


(a) (b)

**Figure 14.4** IRIS Mote: (a) top and (b) back views.

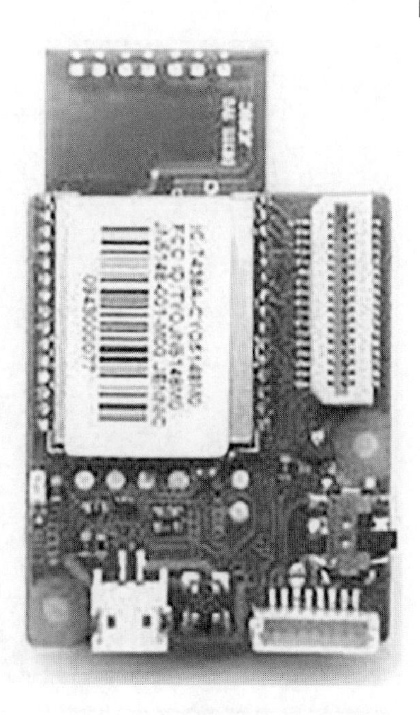

**Figure 14.5** iSense core.

RF communication, it is possible to implement a network of underground motes within a mote-mote communication range of approximately 30 m underground. Moreover, IRIS motes can be integrated with a MIB600 TCP/IP Ethernet network, which can perform as a base station.

### 14.4.5 iSense Core Wireless Module

An iSense core module (Figure 14.5) combines the controller with an RF transceiver in a single housing. The microcontroller runs on a powerful 32-bit RISC core with a shared 128 Kbyte program and data memory. It is possible to reach the data rates as high as 667 kbps. The RF module facilitates a receiver sensitivity of −95 dBm and a transmit power of +2.5 dBm. An extra power module can also be added to increase the receiver sensitivity to −98 dBm and a transmit power level of +10 dBm (www.coalesenses.com).

### 14.4.6 Preon32 Wireless Module

A Preon32 Sensor node, depicted in Figure 14.6, is a universally applicable sensor and actuator platform for realisation of sophisticated implementations within short-range wireless networks. With a high-performance Cortex-M3 controller, it has an IEEE 802.15.4-compliant RF module. As a major advantage of this mote, the developers benefit from object-oriented languages like Java to program the wireless module. It also facilitates connection to external interfaces like CAN (controller area network), USB (universal serial bus), and SPI (serial peripheral interface) (www.virtenio.com). There are many applications for Preon32, including home automation, agricultural, habitat, and road traffic monitoring.

**Figure 14.6** Preon32 wireless module.

### 14.4.7 Wasp Mote

Wireless sensing has become a demanding entity to attain a 'smart environment'. One of the prevalent examples for this is a 'smart water sensor' utilising a Wasp mote (Figure 14.7), introduced by Libelium. A smartwater wireless sensing platform has been deployed to simplify the monitoring of water quality. The module enables measuring certain water quality parameters like PH levels, conductivity, dissolved ion content, and oxygen level. The nodes can be connected to the Internet or a cloud network for real-time monitoring and control. This platform uses an ultra-low-power sensor node for use in rugged environments. The sensor network connection can be via long range 802.15.4/ZigBee (868/900 MHz). Most importantly, this smartwater sensor can accommodate energy harvesting using solar panels.

### 14.4.8 WiSense Mote

An interesting platform for WSN and Internet-of-things (IoT) implementation is the WiSense platform illustrated in Figure 14.8. Apart from providing hardware modules for developers, WiSense provides a framework through which researchers and developers can build their own networks through a graphic user interface. The software makes use of an easy-to-use Eclipse platform involving an IEEE 802.15.4 protocol stack implementation. This system provides an extended support to the developer through hardware and software interfaces. The hardware platform involves an MSP430 low-power controller from Texas Instruments. The mote uses an 8/16 MHz clock together with a CC1101 RF module. With the usage of CC1101, possible applications of this mote extend to home automation, automated food ordering system in restaurants, campus network, industrial automation, green cities, and a variety of other home, industrial, and clinical applications. The data and program memory are set to 4 and 56 kB, respectively (www.wisense.in).

**Figure 14.7** Wasp mote.

**Figure 14.8** WiSense mote.

### 14.4.9 panStamp NRG Mote

The panStamp NRG relies on a CC430F5137 processor and an in-built CC11xx RF module. The main advantage of this module is that it needs much less space compared to the other available motes. It offers a programable speed of 8–24 MHz with flash and RAM capability of 32 and 4 kB, respectively. With the inclusion of a three-axis accelerometer and support for AES security encryption, this mote outperforms the standard available motes in the market. There are also options provided for including a dual temperature/humidity sensor at the bottom layer. Since the bottom layer is used for linking to optional sensors, it makes the sensing more efficient by reducing the effect of dust and other environmental parameters on the sensed values. Most importantly, panStamp allows the developers to integrate the mote with Raspberry Pi as a shield, making it the most supportive mote available (www.panstamp.com). The panStamp NRG 3 mote, shown in Figure 14.9, is a very recent version of the panStamp mote.

Collaborative BSNs which can combine or enable interaction between two or more BANs become important in order to monitor, recognise, or model group activities with the same goals.

Recently, integration of BANs and WSNs have become of great interest to healthcare system developers. This is particularly important since the mutual effects of the body and the environment in the case of, for example, a patient walking in or through air polluted regions, needs to be assessed or the pollution problems mitigated.

In this regard, some platforms which properly integrate multiple systems as well, as archiving resources such as the cloud, have emerged. Such platforms are mostly software-based and use the available hardware motes and modules to enhance their applications. Below, one of these platforms is addressed.

### 14.4.10 Jennic JN5139

This is a popular microcontroller often used at the sensor node [18]. The JN5139-Z01-M00R1T is a low-cost 2.4 GHz ZigBee mote with integrated antenna. It is a member of the JN5139-xxx-Myy IEEE802.15.4/ZigBee module range and is a surface-mount module that

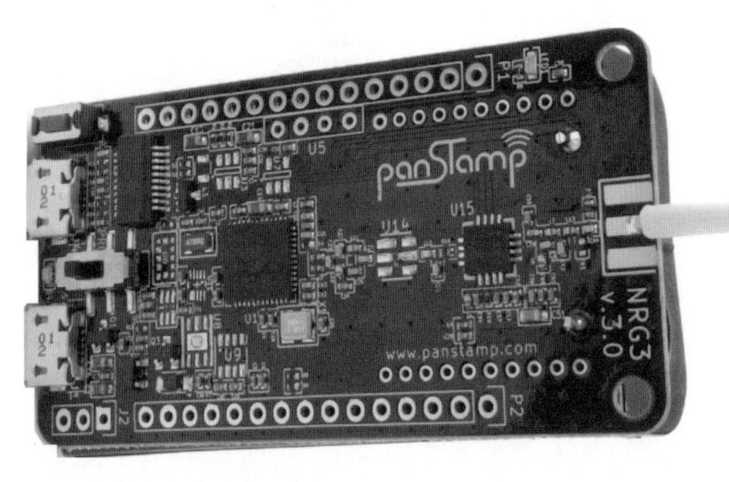


**Figure 14.9** panStamp NRG 3 mote.

**Figure 14.10** Jennic JN5139-Z01.

enables users to implement systems using a JenNet networking stack with minimum time to market. Using this mote, the lengthy development of custom RF board designs and test suites are avoided. JN5139 wireless microcontroller has been deployed to provide a comprehensive solution with high radio performance. Using this microcontroller, all that is required to develop and manufacture wireless control or sensing products is to connect a power supply and peripherals such as switches, actuators, and sensors which simplifies the product development. The JN5139 uses hardware MAC and highly secure advanced encryption standard ASE accelerators for low power and minimum processor overhead. A reasonably low power consumption is enabled using an integrated sleep oscillator and the necessary power saving facilities. The JN5139 module can be seen in Figure 14.10.

## 14.5 BAN Software Platforms

In order to power up the potentials and applications of the hardware and integrated modules powerful software including graphic user interfaces are needed. The approaches to achieve this goal can be either application- or platform-specific programming meant for particular applications, automatic code generation aiming at more generally applicable platforms, and middleware-based programming which allows for ease of and speedy access for the developers [19]. The main current software frameworks involved in BAN development are discussed below.

### 14.5.1 Titan

'Titan' stands for tiny task network [20] and is conceived to enable dynamic context recognition on the BAN. It provides the users/developers with a library of predefined tasks, each representing a particular operation such as a sensor reading, a processing function, or a classification algorithm. It is often represented as an interconnected graph representing a task execution over different blocks.

### 14.5.2 CodeBlue

This is a sensor network structure used as a decision support system [21] which supports medical assistive technology for both indoor and outdoor patient care applications. It uses TelosB and MicaZ as the base for BAN communication, processing, and security applications. CodeBlue is also based on a flexible publish/subscribe data delivery model providing a scalable and robust coordination.

### 14.5.3 RehabSPOT

It is another BAN platform [22] based on Sun SPOT sensor nodes developed for assisting physical therapists and patient limb rehabilitation treatment (www.sunspotdev.org). It allows for adaptive data collection, online processing, and display. It uses a centralised start-topology-based mesh to connect the sensors to a coordinator.

### 14.5.4 SPINE and SPINE2

SPINE is an open-source software for BAN [23, 24] and allows for programming over some popular hardware platforms such as TinyOS, Tmote Sky/TelosB, MicaZ, and Shimmer (www.shimmersensing.com). It also allows for ZigBee short-rang communication devices as well as Sun SPOT sensors.

Evolving from SPINE, SPINE2 is a task-oriented platform-independent programming framework [25, 26]. Following a layering approach, this allows an easier and faster adaptation of this system to the C-like sensor platforms.

### 14.5.5 C-SPINE

This has been designed to enable and support distributed or collaborative body sensor (or area) networks (CBANs) [27, 28]. The C-SPINE architecture includes the SPINE sensor-side and base station-side software components, with the addition of particular CBAN modules enabling wider applications including inter-CBAN communication, BAN proximity detection, BAN service discovery or selection, and application-specific protocols and services.

### 14.5.6 MAPS

It is a Java-based framework [29, 30] which allows for agent-oriented programming over sensor networks. MAPS facilitates the necessary bases for agents, including messaging, agent creation, cloning, and migration as well as timer handling and easy access to sensor node resources.

### 14.5.7 DexterNet

This BAN platform is open-source [31] and supports scalable and real-time monitoring for indoor and outdoor over heterogeneous wearable sensors. DexterNet includes three layers: body sensor, personal network, and global network layers.

## 14.6 Popular BAN Application Domains

Electronic health, games, entertainment, sport, emergency, social network, driving, and interacting with industrial systems are the most popular BAN applications. In Table 14.1 a summary of these BAN systems can be viewed.

**Table 14.1** Popular BAN systems [19].

| Design title | Application | Sensors | Hardware and accessories | Node platform | Communication protocol | Operating system/ programming language |
|---|---|---|---|---|---|---|
| Real-time arousal monitor | Emotion recognition | ECG, respiration, temperature, GSR | Chest-belt, skin electrodes, wearable monitor, USB dongle | Custom | Wired sensor network | N/A, C-like |
| LifeGuard | Medical monitoring in space and extreme environments | ECG, blood pressure, respiration, temp., accelerometer, $SpO_2$ | Custom microcontroller device, commercial biosensors | XPod signal conditioning unit | Bluetooth | N/A |
| Fitbit | Physical activity, sleep analysis, heart monitoring | Accelerometer, heart rate | Waist/wrist-worn device, PC USB dongle | Fitbit node | RF proprietary | N/A |
| VitalSense | Physical activity, heart monitoring, in- and on-body temperature | ECG, temperature, respiration, accelerometer | Wearable monitor, wireless sensors, skin electrodes, ingestible capsule | VitalSense monitor | RF proprietary | Windows mobile |
| LiveNet | Parkinson symptom, epilepsy seizure detection | Blood pressure, ECG, EMG, respiration, GSR, $SpO_2$ temperature | PDA, microcontroller board | Custom physiological sensing board | Wires, 2.4GHz radio, GPRS | Linux (on PDA) |

**Table 14.1** (Continued)

| Design title | Application | Sensors | Hardware and accessories | Node platform | Communication protocol | Operating system/ programming language |
|---|---|---|---|---|---|---|
| AMON | Cardiac and respiratory diseases | Blood pressure, ECG, accelerometer, $SpO_2$, temperature | Wrist-worn device | Custom wrist-worn device | Sensors connected through wires-GSM/UMTS | C-like/JAVA (on the server station) |
| MyHeart | Prevention and detection of cardiovascular diseases | Blood pressure, ECG, EMG, respiration, GSR, $SpO_2$ temperature | PDA, textile sensors, chest-belt | Proprietary monitoring station | Conductive yarns, Bluetooth, GSM | Windows on mobile/PDA |
| Human++ | General health monitoring | ECG, EEG, EMG | Low-power BAN nodes | ASIC | 2.4 GHz radio/UWB modulation | N/A |
| HealthGear | Sleep apnoea detection | Heart rate, $SpO_2$ | Custom sensing board, commercial sensors, cell phones | Custom wearable station (including XPod unit) | Bluetooth | Windows on mobile |
| TeleMuse | Medical care and research | ECG, EMG, GSR | ZigBee wireless motes | Proprietary | IEEE 802.15.4/ZigBee | C-like |
| Polar heart rate monitor | Fitness and exercise | Heart rate altimeter | Wireless chest-belt, watch monitor | Proprietary watch monitor | Polar OwnCode (5 kHz) – coded transmission | N/A |

GSR: galvanic skin response; $SpO_2$: blood oxygen saturation.
Source: Courtesy of Fortino, G., Gravina, R., and Galzarano, S.

There are certainly many more application platforms developed on daily basis. Their description, however, is beyond the scope of this chapter.

## 14.7 Conclusions

To facilitate fast and easy BAN setup, good platforms (software and hardware) are necessary. Various WBAN platforms include sensors, processing, and communication modules. The networking strategies are limited but can accommodate various network topologies. The BAN hardware sensor and processing modules, however, have been under growing development and are well adapted to the new technology in electronics, sensing, data mining, archiving, and communications. The software platforms seem to be easier to expand. Their growth, however, requires the help of data processing and machine learning experts. In addition, the current platforms are mostly centralised. Even collaborative BANs, which include few relays, are not yet fully distributive. In future, decentralised systems are expected to exploit the capability of smart sensors to grow and allow more flexible, versatile, and user-specific BAN–BAN and BAN–WSN connections.

## References


1 Latré, B., Braem, B., Moerman, I. et al. (2011). A survey on wireless body area networks. *Wireless Networks* 17 (1): 1–18.

2 IEEE Std 802.15.6-2012 (2012). *IEEE standard for local and metropolitan area networks – Part 15.6: Wireless body area networks.* New York: IEEE.

3 Otto, C., Milenkovic, A., Sanders, C., and Jovanov, E. (2006). System architecture of a wireless body area sensor network for ubiquitous health monitoring. *Journal of Mobile Multimedia* 1 (4): 307–326.

4 Jovanov, E., Milenkovic, A., Otto, C., and de Groen, P.C. (2005). A wireless body area network of intelligent motion sensors for computer assisted physical rehabilitation. *Journal of Neuroengineering and Rehabilitation* 2 (1): 16–23.

5 Gao, T., Greenspan, D., Welsh, M. et al. (2005). Vital signs monitoring and patient tracking over a wireless network. In: *27th IEEE International Conference of the in Engineering in Medicine and Biology Society*, 102–105. Shanghai: EMBS.

6 Lorincz, K., Malan, D.J., Fulford-Jones, T.R.F. et al. (2004). Sensor networks for emergency response: challenges and opportunities. *IEEE Pervasive Computing* 3 (4): 16–23.

7 Venkatasubramanian, K., Deng, G., Mukherjee, T. et al. (2005). Ayushman: a wireless sensor network based health monitoring infrastructure and testbed. In: *Distributed Computing in Sensor Systems, DCOSS, LNCS*, vol. 3560 (eds. V.K. Prasanna, S.S. Iyengar, P.G. Spirakis and M. Welsh), 406–407. Berlin: Springer.

8 Gyselinckx, B., Vullers, R., Hoof, C.V. et al. (2006). Human++: emerging technology for body area networks. In: *IFIP International Conference on Very Large Scale Integration, Nice*, 175–180.

9 van Halteren, A., Bults, R., Wac, K. et al. (2004). Mobile patient monitoring: the mobihealth system. *Journal on Information Technology in Healthcare* 2 (5): 365–373.

10 Falck, T., Espina, J., Ebert, J.P., and Dietterle, D. (2006). BASUMA: the sixth sense for chronically ill patients. In: *International Workshop on Wearable and Implantable Body Sensor Networks, BSN, Cambridge, MA, USA*, 57–60.
11 Farella, E., Pieracci, A., Benini, L. et al. (2008). Interfacing human and computer with wireless body area sensor networks: the WiMoCA solution. *Multimedia Tools and Applications* 38 (3): 337–363.
12 Metz, R. (2015). *The Struggle for Accurate Measurements on Your Wrist*. MIT Technology Review http://www.technologyreview.com/review/538416/the-struggle-for-accurate-measurements-on-your-wrist.
13 Comstock, J. (2015) Fitness device makers say engagement, not accuracy, is most important. *MobiHealth News* (18 February). http://http://mobihealthnews.com/40646/fitness-device-makers-say-engagement-not-accuracy-is-most-important (accessed 14 December 2019).
14 Narayanan, R.P., Sarath, T.V., and Vineeth, V.V. (2016). Survey on motes used in wireless sensor networks: performance & parametric analysis. *Wireless Sensor Network* 8 (4): 67–76.
15 Rault, T., Bouabdallah, A., and Challal, Y. (2014). Energy efficiency in wireless sensor networks: a top-down survey. *Computer Networks* 67: 104–122.
16 Indrion (2020) Indriya_DP_01A11. www.indrion.co.in/datasheet/Indriya_DP_01A11.pdf (accessed 6 January 2020).
17 Crossbow Technology (2020) IRIS: Wireless measurement system. http://www.nr2.ufpr.br/~adc/documentos/iris_datasheet.pdf (accessed 6 January 2020).
18 Fernandez-Lopez, H., Afonso, J.A., Correia, J.H., and Simoes, R. (2014). Remote patient monitoring based on ZigBee: lessons from a real-world deployment. *Telemedicine Journal and E-Health* 20 (1): 47–54.
19 Fortino, G., Gravina, R., and Galzarano, S. (2018). *Wearable Computing from Modeling to Implementation of Wearable Systems Based on Body Sensor Networks*. Wiley-IEEE Press.
20 Lombriser, C., Roggen, D., Stager, M., and Troster, G. (2007). Titan; a tiny task network for dynamically reconfigurable heterogeneous sensor networks. In: *Kommunikation in Verteilten (KiVS)*, 127–138. New York: Springer.
21 Malan, D., Fulford-Jones, T., Walsh, M., and Moulton, S. (2004) CodeBlue: an ad hoc sensor network infrastructure for emergency medical care. Proceedings of the International Workshop on Wearable and Implementable Body Sensor Networks. https://dash.harvard.edu/bitstream/handle/1/3191012/1242078272-bsn.pdf?sequence=2 (accessed 14 December 2019).
22 Zhang, M. and Sawchuk, A. (2009). A customizable framework of body area sensor network for rehabilitation. In: *Second International Symposium on Applied Science on Biomedical and Communication Technologies, (ISABEL)*. Bratislava, Slovak Republic (24–27 November 2009), 1–6. IEEE.
23 Fortino, G., Giannantonio, R., Gravina, R. et al. (2013). Enabling effective programming and flexible management of efficient body sensor network applications. *IEEE Transactions on Human-Machine Systems* 43 (1): 115–133.
24 Bellifemine, F., Fortino, G., Giannantonio, R. et al. (2011). SPINE: a domain-specific framework for rapid prototyping of WBSN applications. *Software Practice and Experience* 41 (3): 237–265.

**25** Hui, J.W. and Culler, D. (2004). The dynamic behaviour of a data dissemination protocol for network programming at scale. In: *Proceedings of 2nd International Conference on Embedded Networked Sensor Systems*, 81–94. ACM.

**26** Raveendranathan, N., Galzarano, S., Loseu, V. et al. (2012). From modeling to implementation of virtual sensors in body sensor networks. *IEEE Sensors Journal* 12 (3): 583–593.

**27** Fortino, G., Galzarano, S., Gravina, R., and Li, W. (2015). A framework for collaborative computing and multi-sensor data fusion in body sensor networks. *Information Fusion* 22: 50–70.

**28** Augimeri, A., Fortino, G., Galzarano, S., and Gravina, R. (2011). Collaborative body sensor networks. In: *Proceedings of the IEEE International Conference on Systems, Man, and Cybernetics (SMC)*, 3427–3432. IEEE.

**29** Aiello, F., Fortino, G., Gravina, R. et al. (2011). A Java-based agent platform for programming wireless sensor networks. *The Computer Journal* 54 (3): 439–454.

**30** Aiello, F., Bellifemine, F., Fortino, G. et al. (2011). An agent-based signal processing in-node environment for real-time human activity monitoring based on wireless body sensor networks. *Journal of Engineering Applications of Artificial Intelligence* 24 (7): 1147–1161.

**31** Kuryloski, P., Giani, A., Giannantonio, R. et al. (2009). DexterNet: An open platform for heterogeneous body sensor networks and its applications. In: *Sixth International Workshop on Wearable and Implantable Body Sensor Networks*, 92–97. IEEE.

# 15

# Conclusions and Suggestions for Future Research

## 15.1 Summary

This book has pursued four major objectives. First of all, the measurable and recordable biomarkers of human body are introduced. Physical, physiological, mental, and biological indicators have footsteps in many living human recordings. Hence, further to a brief introduction of how data look and how clinically important the underlying information for disease diagnosis are, human body measurable and recordable biomarkers are presented. Second, a wide spectrum of sensors, sensor technologies, and their available platforms are reviewed and their applications discussed. Power consumptions and wireless communication capability are the two important aspects of using these sensors in a body sensor network (BSN) domain. This implies solving the problem of energy harvesting, too. As the third objective, signal processing and machine learning techniques are presented to recover far more information than what could be deciphered by expert clinician naked eyes.

Finally, wireless communications between various points of a BSN, employing different short-range communication strategies, systems, protocols, and routing algorithms, are explored for different sensor arrangement scenarios. The importance of electronics and integrated circuit devices plus their major role in developing sensors (and the associated motes), communication systems, and networks must be emphasised with high service quality, safety, and security. The overall objective of this book is to familiarise readers with almost all BSN aspects and applications.

## 15.2 Future Directions in BSN Research

With no doubt, sensor networks (and BSNs in particular) have experienced rapid developmental progress and innovation recently. In addition to outstanding ongoing research topics presented throughout this book, we look into some of the most important BSN research directions and technologies adopted or those which could be embraced. These topics will be the future research and development agendas.

*Body Sensor Networking, Design and Algorithms,* First Edition. Saeid Sanei, Delaram Jarchi and Anthony G. Constantinides.

Companion Website: www.wiley.com/go/sanei/algorithm-design

### 15.2.1 Smart Sensors: Intelligent, Biocompatible, and Wearable

Sensing technology has had rapid growth in the past decade. For example, in ophthalmology, retinal implants, as multicolour light sensors, have made a drastic transformation in restoring vision (although, further research is required to form a benchmark) [1]. A flexible wireless electrocardiographic (ECG) sensor with a fully functional microcontroller has been developed by IMEC in Netherland. An e-textile system for remote, continuous monitoring of physiological and movement data are offered by Smartex, Italy. In this garment, the embedded sensors allow both ECG and electromyographic (EMG) data capturing. Additional sensors facilitate thoracic and abdominal signals data recording associated with respiration and movement. Smartphone-based ECG monitoring system has been commercialised by IMEC. The ProeTEX project by Smartex aims to develop smart garments for emergency responders. Researchers in the University of Houston, USA have reported the design of a multifunctional ultra-thin wearable electronic device, a mechanically imperceptible, and stretchable human–machine interface (HMI) device, worn on human skin to capture multiple physical data or worn on a robot to offer intelligent feedback, forming a closed-loop HMI. To alleviate the difficulties in wearing electroencephalography (EEG) sensor cap, some electrodes have been embedded in the glass handles in a recently introduced EEGlass [2].

Smart sensors receive their inputs from the physical environment and based on the data decide on what to do next. This can happen mainly due to the built-in computational resources which perform predefined functions. These sensors enable an automated and more accurate collection of environmental data with less error, more accuracy, high robustness, and perform self-diagnosis for human body and self-calibrating within the BSN. Such devices have onboard processing which can analyse the captured data and make decisions. Finally, as in many other sensors, they can communicate with other devices using wireless technology. Additionally, by incorporating artificial intelligence with the sensor technology, a new generation of sensors will be able to learn from the environment and the data. Moreover, smart sensors will promote a new generation of cooperative and decentralised (as opposed to centralised) BSN networks. These networks are central to future Internet-of-things (IoT) technology. Adaptive cooperative networks and cooperative learning are the two advanced techniques in signal processing and machine learning which focus on decentralised distributive systems.

There are enormous applications for smart sensors. For example, in wound care, it is crucial to keep the wound cover untouched unless its exposure is imperative. This reduces interruption to the wound healing process. To enable this, the sensors can be embedded within the bandage to constantly check the state of the wound and decide whether there is need for a change or for medication. Another example is the sensor used for food checking in supermarkets. Other smart sensors include smart needles to analyse and identify blood vessels or tissues in brain surgery which involves a great deal of signal and image processing; iTBra, for the detection of breast cancer by recognising the thermal abnormalities or tissue elasticity; the CADence System, for the detection of coronary artery blockages using cardiac sounds; UroSense, for urine monitoring with the help of core body temperature (CBT) and urine measures; Digital Pill, which senses electric signals in the stomach fluid to prescribe suitable medication [3]; and as another interesting device, a sociometric badge which is a wearable computing platform for measuring and analysing human behaviour, based on variability in a person's speech spectrum, in organisational settings [4]. Integration of these sensors within a BSN will certainly boost the effectiveness and inclusiveness of such networks for better human body monitoring.

Biocompatibility of sensors measuring internal body, biological, or metabolic activities is another issue in sensor technology. Some recent developments include biocompatible soft fluidic strain and force sensors for wearable devices [5]. This is a silicone-based strain and force sensor composed of potassium iodide and glycerol solution. Currently, a considerable amount of research on designing biocompatible sensors and devices has been undertaken. Recently, Harvard University researchers developed a soft nontoxic wearable sensor that unobtrusively attaches to the hand to measure the force of a grasp and the motion of the hand and fingers [6]. The sensing solution is made from potassium iodide, which is a common dietary supplement, and glycerol, which is a common food additive. Nontoxicity and highly conductive liquid solutions are two novel elements of this sensor. After a short mixing duration, glycerol breaks the crystal structure of potassium iodide, resulting in potassium cations and iodide ions. This makes the formed liquid conductive. Since glycerol has a lower evaporation rate compared to water, and the potassium iodide is highly soluble, the liquid is both highly conductive and stable across a range of temperatures and humidity levels. Similar to smart sensors, designing and using biocompatible devices can make a significant contribution to more effective BSN designs. This, to a large extent, depends on the replacement of traditional components with new ones made of polymers. Chinese researchers have developed a stretchable biocompatible metal-polymer conductor that has potential for wearable electronic circuits bridging electronics and biology [7]. Printable, highly stretchable, and biocompatible metal-polymer conductors have been introduced by casting and peeling off polymers from patterned liquid metal particles, forming surface-embedded metal in polymeric hosts. These products are the samples from the incredible world of biodegradable and biocompatible polymers for sensor designs [8].

The adoption of wearable sensors is one outstanding problem for BSNs. The sensor's measurement not only has a valuable impact on patient monitoring but also can be used by occupational safety and health professionals, athletes, and computer game players. Therefore, there are potential benefits of using such technologies in the workplace. Nevertheless, there are perceived barriers which prevent the widespread adoption of wearable sensors for most of these applications, particularly in industrial workplaces. Thus, many workplaces are hesitant to adopt these technologies.

In a survey of 952 valid responses from public, health, and safety professionals, over half of the respondents were in favour of using wearable sensors at their respective workplaces [9]. Nevertheless, concerns regarding employee privacy/confidentiality of collected information, employee compliance, sensor durability, the trade-off between cost and benefit of using wearables, and good manufacturing practice requirements were barriers and described as challenges hindering adoption. Based on this study, it was concluded that the broad adoption of wearable technologies appears to depend largely on the scientific community's ability to successfully address the identified barriers. For general applications of wearable sensor systems, several approaches may be followed. These include: (i) design of more miniaturised wearables, (ii) make less intrusive sensors, (iii) improve sensors appearances,

(iv) use lighter materials, (v) embed it within current multimedia gadgets such as wireless headphones, and (vi) more closely follow the needs of different users.

### 15.2.2 Big Data Problem

Identifying the health status of the human body is more accurate when there is access to many biometrics over a long duration. A combination of a large amount of data recorded using various modalities is valuable since there currently exist powerful mining, archiving, learning, and processing tools and algorithms for analysis of such data. The term 'big data' does not necessarily refer to the size of data; instead, it implies that the data cannot be processed in a single core computer or processing unit. Therefore, one aspect of future work on big data is to develop signal processing and machine learning techniques that can be run in a distributive manner over multicore computers or computing clusters [10].

### 15.2.3 Data Processing and Machine Learning

The underlying information derived from raw data is often very rich. Nevertheless, powerful signal processing algorithms have to be designed and implemented to uncover and extract this information. Current techniques in both diffusion (such as tensor factorisation or diffusion adaptive filters) and fusion (such as various data mining algorithms and transcription methods) can further be improved or regularised by incorporating subject (such as peripheral clinical and personal information), data (such as the nature of data, their dimension, colour, sparsity, smoothness, and variation boundaries), and environmental (such as temperature, system limitations, and noise) constraints into the algorithms.

The availability, multimodality, multidimensionality, and multiclass nature of data and computation capacity of the new machines make deep neural networks (DNNs) outperform other classifiers. While these classifiers still require very high computational power, the number of DNN architectures is also ascending rapidly. One new direction in DNN design is a deeper analysis of natural brain network functioning. One example of using this concept is by means of biological learning algorithms [11, 12]. These systems present a theoretical framework for understanding the regularity of the brain's perceptions, its reactions to sensory stimuli, and its control of movements. They offer an account of perception as the combination of prediction and observation: the brain develops internal models that describe what will/should happen next and then combines this prediction with reports from the sensory system to form a belief. The biological learning algorithms outperform the optimal scaling of the learning curve in a traditional perceptron. It also results in a considerable robustness to the disparity between weights of two networks with very similar outputs in biological supervised learning scenarios. The simulation results indicate the ability of neurobiological mechanisms and open opportunities for developing a superior class of deep learning algorithms. Since the new direction in machine learning system design is to sense and learn from the data then the more data available from a diverse range of sensors, the more precisely the state of human body can be described or identified.

### 15.2.4 Decentralised and Cooperative Networks

As soon as smart sensors become popular, there will be a shift from centralised systems (including a hub and a number of sensors directly connected to it) to a decentralised network, where there will be no central hub and the body sensors can process the data, cooperate with each other, make decisions, and also communicate with other sensors, including those outside the BSN. To facilitate this, new cooperative models through consensus and diffusion adaptation techniques should be developed which consider each sensor as an intelligent agent. Advances in graph theory and related signal processing techniques may be employed to best model the connectivity among the body and environmental sensors. Adaptive cooperative algorithms should be able to model the interactions between two or more BSNs and also between a BSN and an external wireless sensor network (WSN). As an example, consider a patient suffering from asthma, equipped with a wireless body area network (WBAN), walking in an area with a variable level of air pollution equipped with a WSN. A cooperative decision-making system can help this patient find their way through the less polluted regions.

### 15.2.5 Personalised Medicine Through Personalised Technology

In the context of human monitoring using wearable technology, 'personalised medicine' refers to linking an individual's internal (physiological or metabolic) and external (movement in a parametrised environment) bodily status to their monitoring system. Height, weight, diet, living style, amount of sport, a patient's daily activities, the environment they interact with, etc., can be used in a more comprehensive treatment plan. The smart sensors may then be regulated so that the wearable system best reveals the necessary diagnostic information through better and more accurate parameters setting. There is great advantage in such adoption for sport when, for example, different-sized athletes engage in completely different types of sport. Moving one step further, consider the advances and progresses in machine learning and signal processing from predictive models (autoregressive, singular spectrum analysis) to recurrent neural networks, and, very recently, long short-term modelling (LSTM) deep networks. Recurrent neural networks and LSTM exploit the history and background activities of the subject's data to assess their current and future status. Thus, an effective personalised medicine can benefit from these advancements in technology.

### 15.2.6 Fitting BSN to 4G and 5G Communication Systems

Current mobile systems adhere to long-term evolution (LTE) and 4G technology standards, while 5G is being offered by many communication system developers as a near-future mobile communication system. 5G systems are expected to provide bandwidth efficiency and higher accessibility by providing more intelligent communications and data retrievals. Although the body area network (BAN) short-range communication remains the same but for the data to be available over the World Wide Web as well as mobile systems, the BAN needs to be updated to meet the new generation of communication systems. Energy efficiency, interference mitigation, and wireless power transfer capability are probably the most important factors for making a WBAN suitable for integration within a 5G network [13]. More importantly, as mentioned in Section 15.2.4, in future decentralised systems the smart sensors will be able to communicate with devices outside the BAN local network individually. In such an inevitable scenario, the sensors have to be equipped with full Internet connection and comply with the new generation communication system standards and protocols.

### 15.2.7 Emerging Assistive Technology Applications

The most popular and effective BAN application is for assistive technology. Both rehabilitative assessments and patient rehabilitation should expand to cover a wider needy population. Vulnerable epileptic, stroke, dementia, depression, paralysis, and many other groups of patients should benefit from the outcome of this research direction now rather than tomorrow. Smart chair, assistive robot, fall detector, driver fatigue detector, hazard detector, and many other devices used by the military for life protection are only a few examples of assistive technology made available to the public.

Assistive technology covers a wide range of products and equipment furnished with assistive robots as the most advanced, effective, and intelligent ones. In the near future, these robots may be able to assist humans both physically and mentally, and offer patients a much better quality of life at home and at work.

### 15.2.8 Solving Problems with Energy Harvesting

Although new computerised systems are well equipped with low-power processors, in many applications where on-board processing is essential the sensors have serious demand for green energy harvesting. The conversion of thermal, mechanical, and chemical energy to electricity is always a prime objective when choosing the sensors. New systems should also allow energy to be delivered from sensor to sensor as well as the power in the sensor itself or those within its neighbourhood to be predicted. This raises many questions about how the energy can be shared between the sensors or transferred across a sensor network. We may think of a number of solar cell sensors with only a few of them exposed to sunshine for a certain period. Therefore, researchers should constantly look for new sources of energy to better operate the sensors at any required time.

### 15.2.9 Virtual World

Social affairs and interactions are now far more extensive and complex than before. This requires a more advanced technology to better cater for new demands and user requirements. In addition to a considerable amount of work, mostly in computer game design, using virtual reality (VR) [14], one of these technologies, called augmented reality (AR), which has applications in education, games, business, etc., has gained a prominent position among developers and researchers.

Exploiting AR, the users can see the 2D or 3D digital information (images, text, etc.) at the same time as seeing the rea -world by using a gadget (such as glasses or a cell phone). Through AR, humans interacts with sensor measurements (e.g. the scene captured by a video camera). Using this technology, a blend of augmented and virtual reality system using Holobody technology which, with the help of HoloLens glasses, is able to see the anatomy of the human body in real 3D space have come to practice [15].

Some applications of AR include, but are not limited to, reducing the schizophrenia stigma [16] to improve sufferers' social interactions, fishing training computer games [17], civil engineering education [18], and teaching chemistry for high school/university students [19]. In medicine, this technology is also used to treat disease. In [20] an approach based on AR for the treatment of pain created in the lost limb (phantom pain) is presented. Also, in [21], AR has been used to treat the people suffering from fear from small animals such as beetles and spiders. The results of this research have been quite useful and effective. In stomatology, AR has been suggested for dental treatment and implantology [22].

## 15.3 Conclusions

Research and development in sensors, wearable technology, and BSNs are fast progressing and expanding to cover more applications. The world of sensors is a fascinating one. New sensing methodologies are introduced and new sensors come to market on a monthly basis. The newly introduced quantum sensors may replace 3D imaging, providing much higher resolution and potential for more effective cancer treatment [23]. These systems will not be operational without the development and application of advanced machine learning and signal processing tools and algorithms. A small, wearable device kitted out with sensors for monitoring markers of stress – such as heart rate, sweat production, skin surface temperature, and arm movements – can be used to investigate and predict the onset of autism [24]. In the world of BSNs, human health monitoring and assistive technology are probably the most popular applications for the public and clinical departments. Therefore, for a healthier life and disease prevention, wearable sensor networks will soon become increasingly popular. Advances in sensor technology, data processing, machine learning, wireless communications, energy harvesting, and information security have no limit and are expected to cover more monitoring, diagnostic, and assistive applications. Although not discussed in this book, the importance of electronics in manufacturing sensors, data acquisition, and communication systems cannot be highlighted enough. After all, human body sensing and measurement have shifted from being invasive to being noninvasive (minimal contact). Therefore, it is time to focus on nonintrusive sensor design.

## References

**1** Lorach, H. (2019). The bionic eye. *TheScientist* http://www.the-scientist.com/?articles.view/articleNo/41052/title/The-Bionic-Eye, accessed 14 December 2019.

**2** Vourvopoulos, A., Niforatos, E., and Giannakos, K. (2019). EEGlass: An EEG-Eyeware Prototype for Ubiquitous Brain-Computer Interaction. In: *Proceedings of the 2019 ACM*

*International Joint Conference on Pervasive and Ubiquitous Computing (UbiComp).* ACM https://doi.org/10.1145/3341162.3348383.

3 Dutta Pramanik, P.K., Kumar Upadhyaya, B., Pal, S., and Pal, T. (2019). Internet of things, smart sensors, and pervasive systems: enabling connected and pervasive healthcare. In: *Healthcare Data Analytics and Management, Advances in Ubiquitous Sensing Applications for Healthcare*, 1–58. Academic Press.

4 Olguín Olguín, D., Waber, B.N., Kim, T. et al. (2009). Sensible organizations: technology and methodology for automatically measuring organizational behavior. *IEEE Transactions on Systems, Man, and Cybernetics: Part B: Cybernetics* 39 (1): 43–55.

5 Xu, S., Vogt, D.M., Hsu, W.-H. et al. (2019). Biocompatible sensors: biocompatible soft fluidic strain and force sensors for wearable devices. *Advanced Functional Materials* 29 (7) https://doi.org/10.1002/adfm.201970038.

6 Harvard John A. Paulson School of Engineering and Applied Sciences (2018). A safe, wearable soft sensor: Biocompatible sensor could be used in diagnostics, therapeutics, human-computer interfaces, and virtual reality. *ScienceDaily* https://www.sciencedaily.com/releases/2018/12/181221142511.htm, accessed 14 December 2019.

7 Tang, L., Cheng, S., Zhang, L. et al. (2018). Printable metal-polymer conductors for highly stretchable bio-devices. *iScience* 4: 302–311.

8 Cao, Y. and Uhrich, K.E. (2019). Biodegradable and biocompatible polymers for electronic applications: a review. *Journal of Bioactive and Compatible Polymers* 34 (1): 3–15.

9 Schall, M.C. Jr.,, Sesek, R.F., and Cavuoto, L.A. (2018). Barriers to the adoption of wearable sensors in the workplace: a survey of occupational safety and health professionals. *Human Factors* 60 (3): 351–362.

10 Andreu-Perez, J., Poon, C.C., Merrifield, R.D. et al. (2015). Big data for health. *IEEE Journal of Biomedical and Health Informatics* 19 (4): 1193–1208.

11 Shadmehr, R. and Mussa-Ivaldi, S. (2012). *Computational Neuroscience: Biological Learning and Control: How the Brain Builds Representations, Predicts Events, and Makes Decisions.* Cambridge, MA: MIT Press.

12 Uzan, H., Sardi, S., Goldental, A. et al. (2019). Biological learning curves outperform existing ones in artificial intelligence algorithms. *Scientific Reports* 9 https://doi.org/10.1038/s41598-019-48016-4.

13 Jones, R.W. and Katzis, K. (2018). 5G and wireless body area networks. In: *IEEE Wireless Communications and Networking Conference Workshops (WCNCW)*, 373–378. IEEE.

14 Rizzo, A., Requejo, P., Winstein, C.J. et al. (2011). Virtual reality applications for addressing the needs of those aging with disability. *Studies in Health Technology and Informatics* 163: 510–516.

15 Santoso, M. and Jacob, C. (2016). Holobody galleries: blending augmented and virtual reality displays of human anatomy. In: *Proceedings of the 2016 Virtual Reality International Conference*, 32. ACM.

16 Silva, R.D.d.C., Albuquerque, S.G.C., Muniz, A.d.V. et al. (2017). Reducing the schizophrenia stigma: a new approach based on augmented reality. *Computational Intelligence and Neuroscience* 2017: 1–10.

**17** Chen, C.H., Ho, C.-H., and Lin, J.-B. (2015). The development of an augmented reality game-based learning environment. *Procedia – Social and Behavioral Sciences* 174: 216–220.

**18** Dinis, F.M., Guimarães, A.S., Carvalho, B.R., and Martins, J.P.P. (2017). Virtual and augmented reality game-based applications to civil engineering education. In: *IEEE Global Engineering Education Conference (EDUCON)*, 1683–1688. IEEE.

**19** Crandall, P.G., Engler, R.K. III,, Beck, D.E. et al. (2015). Development of an augmented reality game to teach abstract concepts in food chemistry. *Journal of Food Science Education* 14 (1): 18–23.

**20** Ortiz-Catalan, M., Gumundsdttir, R.A., Kristoffersen, M.B. et al. (2016). Phantom motor execution facilitated by machine learning and augmented reality as treatment for phantom limb pain: a single group, clinical trial in patients with chronic intractable phantom limb pain. *The Lancet* 388 (10062): 2885–2894.

**21** Chicchi Giglioli, I.A., Pallavicini, F., Pedroli, E. et al. (2015). Augmented reality: a brand new challenge for the assessment and treatment of psychological disorders. *Computational and Mathematical Methods in Medicine* 2015: 1–12.

**22** Prochazka, A., Dostálová, T., Kašparová, M. et al. (2019). Augmented reality implementations in stomatology. *Applied Sciences* 9 (14) https://doi.org/10.3390/app9142929.

**23** MPO (2019). New quantum sensor could improve cancer treatment. https://www.mpo-mag.com/contents/view_breaking-news/2019-03-05/new-quantum-sensor-could-improve-cancer-treatment/65655#, accessed 14 December 2019.

**24** Lavars, N. (2019). Wrist-worn device detects autism outbursts 60 seconds ahead of time. https://newatlas.com/medical/wearable-detects-autism-outbursts-60-seconds, accessed 14 December 2019.

# Index



*Body Sensor Networking, Design and Algorithms,* First Edition. Saeid Sanei, Delaram Jarchi and Anthony G. Constantinides.

Companion Website: www.wiley.com/go/sanei/algorithm-design

***b***

***C***

### e

## f

### *h*

### *i*

### m

***n***

## O

## p

***S***

***t***

### *u*



### *v*